MORAINES & VARVES

PROCEEDINGS OF AN INQUA SYMPOSIUM ON GENESIS AND LITHOLOGY
OF QUATERNARY DEPOSITS / ZURICH / 10 - 20 SEPTEMBER 1978

Moraines and Varves

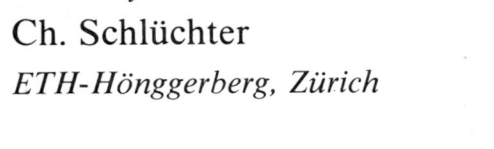

Origin / Genesis / Classification

Edited by
Ch. Schlüchter
ETH-Hönggerberg, Zürich

A.A.Balkema / Rotterdam / 1979

The texts of the various papers in this volume were set individually by typists under the supervision of each of the authors concerned.

ISBN 90 6191 039 0

Distributed in USA & Canada by MBS, 99 Main Street, Salem, NH 03079

Printed in the Netherlands

Preface

The Commission on Genesis and Lithology of Quaternary Deposits of the International Union for Quaternary Research (INQUA) has a well established tradition in organizing annual meetings to discuss the commission's aims: to contribute to the knowledge of glacial deposits, their distribution, their geology, genesis and classification and to encourage specific research. In 1978 the commission has left the lowland areas of Canada, USA, USSR, Poland, Sweden and Iceland where previous meetings have been held to convene from September 10-20 in the alpine environment of Switzerland. - At this meeting, mainly the members of two commission work groups were involved: Work Group (1): "Genetic classification of tills and criteria for their differentiation" and Work Group (4): "Glacio-lacustrine deposits, their genetic classification and methods of investigation". This twofold topic of the meeting has led to its naming "Moraines and Varves". - The symposium "Moraines and Varves" is the first reunion of an official INQUA-body ever to be held in Switzerland, since INQUA went into existence half-a-century ago.

The symposium has been organized jointly by staff members of the Federal Institute of Technology (ETH) and the University of Zurich with Prof. Dr. G. Furrer (President), Dr. M. Sturm and Dr. Ch. Schlüchter (Meeting Secretary). Officially, "Moraines and Varves" has been hosted at the Hönggerberg-Campus of ETH and the participants were given warm welcome-addresses by Prof. Dr. H. Jäckli, Prof. H. Grob (Vice-Chancellor of ETH) and by Prof. Dr. W.K. Nabholz (President of the "Swiss Geological Commission").

"Moraines and Varves" has attracted participants from all over the world: scientists from Argentina, Austria, Belgium, Canada, Denmark, England, Federal Republik of Germany, Finland, France, India, Ireland, Japan, The Netherlands, New Zealand, Norway, Poland, Spain, Switzerland, United States of America and Venezuela gathered at ETH. With this wide geographic coverage all the main mountainous and glaciated areas of the world, except Greenland, were represented. As broad as the geographic distribution were the fields of research: Pleistocene and Glacial Geology, Sedimentology, Applied Geology, Glaciology, Geomorphology, Hydrology, Pedology, Limnology and Palynology, thus creating an open-minded and comfortable atmosphere for the discussions.

During 9 sessions from Sept. 11-13, a total of 39 papers were delivered. The contributions to the problems on rhythmic sedimentation and processes in glacio-lacustrine basins have been staged as "Colloquium on Varves" on Sept. 12. It has been scheduled and organized by the President of Work Group (4), Dr. M. Sturm. - The organization of the sessions has greatly benefitted by the chairmanship of colleagues who have served as session-leaders. We extend our thanks for their help to Prof. Dr. G.M. Ashley, Dr. G.S. Boulton, Prof. Dr. A. Dreimanis, Prof. Dr. G. Furrer, Dr. F.-E. Grube, Dr. Ch. Schlüchter, Prof. Dr. R. Souchez, Dr. M. Sturm and Prof. Dr. K. Virkkala.

Without a considerable financial and great moral support from the President and Vice-Chancellor of the Federal Institute of Technology (ETH) and from Prof. H.J.

Lang, Director of the Institute of Foundation Engineering and Soil Mechanics, the symposium on "Moraines and Varves" could never have taken place. - Organizing the symposium and the field trip has been made an easy task by the limitless help of the technical staff: Th. Frei, U. Frei and D. Mark acted as secretaries; E. Durrer, H. Gehri, G. Kägi and C. Smith as drafting personnel; H. Schenkel as photographer and E. Hug and K. Wigert as technical aides. - Invaluable help has been given by the assistants M. Maisch, J. Suter and N. Zingg by keeping the field-party together in every situation. - We are very thankful to T. Jenal for carrying out the ladies' programme and to P. Honold for his patience during all stages of disturbance during the preparations. Many good advice in delicate situations has been given by R. Wullimann.

The field excursion could not have been outlined as such a varied program without the spontaneous support of many colleagues. Contributions and help by D. Aubert (Lausanne), Dr. B. Blavoux (Thonon, France), Prof. Dr. M. Burri (Lausanne), Prof. Dr. M.J. Clark (Southampton, England), Prof. Dr. R. Hantke (Zurich), Prof. Dr. P. Kasser (Zurich), Dr. Ph. Olive (Thonon, France), A. Parriaux (Lausanne), Dr. F. Roethlisberger (Birmensdorf), Dr. C. Schindler (Zurich), Prof. Dr. R.J. Small (Southampton, England) and Dr. M. Sturm (Zurich) are gratefully acknowledged. We are specially indebted to PD Dr. H. Roethlisberger (Zurich) for his help organizing the excursions to the glaciers around Zermatt and Arolla.

The publication of this volume in the present form as well as the organization of the whole symposium has been made possible due to substantial financial contributions by the following institutions:

- Eidgenössische Technische Hochschule (ETH), Schulleitung
- Eidgenössische Technische Hochschule (ETH), Zentenarfonds
- Institut für Grundbau und Bodenmechanik (ETH)
- Schweiz. Naturforschende Gesellschaft
- Kies-AG-Aaretal
- Schweiz. Fachverband für Sand und Kies
- Regierungsrat des Kt. Bern
- Regierungsrat des Kt. Zürich
- Université de Lausanne, Faculté des Sciences
- Fondation de la Murithienne
- Migros-Genossenschaftsbund
- Swissair AG
- Schweiz. Volksbank
- Zürcher Kantonalbank
- Schweiz. Bankverein

Donations, contributions and support at large have been made by:

- Grande Dixence SA
- Force Motrice de Mauvoisin SA
- Stiftung Amrein-Troller, Gletschergarten Luzern
- Schweiz. Verkehrszentrale
- Verkehrsverein Zermatt
- Lindt & Sprüngli
- Schweiz. Kreditanstalt
- Schweiz. Landestopographie
- Schweiz. Geologische Kommission
- AG Hunziker & Cie
- Mineralquellen Eptingen AG

This wide ranging support has enabled an organization with substantial flexibility in our search for extra-arrangements and especially to publish the papers in the richly illustrated form. On behalf of the participants from abroad and of the Organizing Committee we express our sincerest thanks to the donators.

During all stages of preparing the meeting, the President of the INQUA-Commission on Genesis and Lithology of Quaternary Deposits, Prof. Dr. A. Dreimanis, has helped with good advice and has enabled sensitive decisions. His help and the limitless patience of my family deserve many special thanks.

Christian Schlüchter

Zurich, 17th April 1979

Table of contents

VARVES AND GLACIOLACUSTRINE SEDIMENTATION

CONTRIBUTIONS RELATED TO THE FIELD EXCURSION

List of contributors

Acomb, L.J., University of Wisconsin, Department of Geology and Geophysics, Madison, Wisconsin 53706, U.S.A.

Ahmad, N., Aligarh Muslim University, Department of Geology, Aligarh, 202 001, India

Aliotta, Guida, Fundacion de Bariloche, C.C. 138, 8400 San Carlos de Bariloche, Prov. de Rio Negro, Argentina

Archer, J., University of Alberta, Department of Geography, Edmonton, Alberta, T6G 2H4, Canada

Ashley, Gail M., Rutgers - The State University of New Jersey, Department of Geological Sciences, New Brunswick, New Jersey 08903, U.S.A.

Bläsig, H., Universität München, Institut für Allgemeine und Angewandte Geologie, Luisenstrasse 37, D-8000 München 2, BRD

Boulton, G.S., University of East Anglia, School of Environmental Sciences, Norwich, NR4 7TJ, Great Britain

Boyko-Diakonow, Maria, 3688 Parkview Street, Penticton, British Columbia, V2A 6H1, Canada

Chinn, Trevor J.H., Ministry of Works and Development, Water and Soil Division, P.O. Box 1479, Christchurch, New Zealand

Cohen, Jonathan M., Trinity College Dublin, Department of Geography, Dublin 2, Ireland

Degens, Egon T., Universität Hamburg, Geologisch-Paläontologisches Institut, Bundesstrasse 55, D-2000 Hamburg 13, BRD

DiLabio, R.N.W., Geological Survey of Canada, 601 Booth Street, Ottawa, Ontario, KIA OE8, Canada

Doppler, G., Universität München, Institut für Allgemeine und Angewandte Geologie, Luisenstrasse 37, D-8000 München 2, BRD

Dreimanis, Aleksis, University of Western Ontario, Department of Geology, London, Ontario, N6A 5B7, Canada

Drozdowski, Eugeniusz, Zaklad Fizjografii lG PAN, ul. Kopernika 19, PL-87-100 Torun, Poland

Edil, T.B., University of Wisconsin, Departments of Civil and Environmental Engineering and Engineering Mechanics, Madison, Wisconsin 53706, U.S.A.

Eschman, Donald F., University of Michigan, Department of Geology, 1006C.C. Little Building, Ann Arbor, MI 48106, U.S.A.

Evenson, Edward B., Lehigh University, Department of Geological Sciences, Bethlehem, PA 18015, U.S.A.

Eyles, N., University of East Anglia, School of Environmental Sciences, Norwich, NR4 7TJ, Great Britain

Fakhrai, M., Universität München, Institut für Allgemeine und Angewandte Geologie, Luisenstrasse 37, D-8000 München 2, BRD

Garnes, Kari, Universitetet i Bergen, Geologisk Institutt, avd. B, Allégt. 41, N-5014 Bergen-Universitetet, Norge

German, Rüdiger, Bezirksstelle für Landschaftspflege und Landschaftsschutz,

Nauklerstrasse 56/58, D-7400 Tübingen, BRD

Goroncek, K., Universität München, Institut für Allgemeine und Angewandte Geologie, Luisenstrasse 37, D-8000 München 2, BRD

Grimm, W.-D., Universität München, Institut für Allgemeine und Angewandte Geologie, Luisenstrasse 37, D-8000 München 2, BRD

Gripp, Karl, Klosterstrasse 22, D-2400 Lübeck, BRD

Habbe, Karl Albert, Universität Erlangen-Nürnberg, Institut für Geographie, Kochstrasse 4, D-8520 Erlangen, BRD

Hambrey, M.J., University of Cambridge, Sedgwick Museum, Department of Geology, Cambridge, CB2 3EQ, Great Britain

Hantké, René, Geologisches Institut, ETH-Zentrum, CH-8092 Zürich, Schweiz

Harland, W.B., University of Cambridge, Sedgwick Museum, Department of Geology, Cambridge, CB2 3EQ, Great Britain

Haselton, George M., Clemson University, Department of Geology, 238 Brackett Hall, Clemson, South Carolina 29631, U.S.A.

Hintermaier, G., Universität München, Institut für Allgemeine und Angewandte Geologie, Luisenstrasse 37, D-8000 München 2, BRD

Horie, Shoji, Kyoto University, Institute of Paleolimnology and Paleoenvironment of Lake Biwa, Omi-Takashima, Shiga-Ken, 520-11, Japan

Hsü, Kenneth J., Geologischés Institut, ETH-Zentrum, CH-8092 Zürich, Schweiz

Jäckli, Heinrich, Geologisches Institut, ETH-Zentrum, CH-8092 Zürich, Schweiz

Jerz, Hermann, Bayerisches Geologisches Landesamt, Prinzregentenstrasse 28, D-8000 München 22, BRD

Just, J., Universität München, Institut für Allgemeine und Angewandte Geologie, Luisenstrasse 37, D-8000 München 2, BRD

Kempe, Stephan, Universität Hamburg, Geologisch-Paläontologisches Institut, Bundesstrasse 55, D-2000 Hamburg 13, BRD

Kiechle, W., Universität München, Institut für Allgemeine und Angewandte Geologie, Luisenstrasse 37, D-8000 München 2, BRD

Kilger, Bernd, Im Kleeacker 13, D-7400 Tübingen-Kressbach, BRD

Knecht, Urs, Universität Zürich, Geographisches Institut, Blümlisalpstrasse 10, CH-8006 Zürich, Schweiz

Lambert, André, Versuchsanstalt für Wasserbau, Hydrologie und Glaziologie (VAW), ETH-Zentrum, CH-8092 Zürich, Schweiz

Lindner, A., Warsaw University, Department of Geology, Al. Zwirki i Wigury 93, PL-02-089 Warsaw, Poland

Lobinger, W.H., Universität München, Institut für Allgemeine und Angewandte Geologie, Luisenstrasse 37, D-8000 München 2, BRD

Ludewig, H., Universität München, Institut für Allgemeine und Angewandte Geologie, Luisenstrasse 37, D-8000 München 2, BRD

Ludlam, Stuart D., University of Massachusetts, Department of Zoology, Amherst, Massachusetts 01003, U.S.A.

Mader, Matthias, Dreikönigskeller 10, D-7312 Kirchheim/Teck, BRD

Mickelson, D.M., University of Wisconsin, Department of Geology and Geophysics, Madison, Wisconsin 53706, U.S.A.

Muzavor, S., Universität München, Institut für Allgemeine und Angewandte Geologie, Luisenstrasse 37, D-8000 München 2, BRD

Pakzad, M., Universität München, Institut für Allgemeine und Angewandte Geologie, Luisenstrasse 37, D-8000 München 2, BRD

Parriaux, A., Ecole Polytechnique Fédérale de Lausanne, Laboratoire de Géologie, Route Bussigny 26, CH-1023 Crissier, Suisse

Pasquini, Thomas A., Exxon Oil Company, Kingsville, Texas, U.S.A.

Rabassa, Jorge, Fundacion de Bariloche, C.C. 138, 8400 San Carlos de Bariloche, Prov. de Rio Negro, Argentina

Roethlisberger, Friedrich, Eidgenössiche

Anstalt für das forstliche Versuchswesen, CH-8903 Birmensdorf, Schweiz

Rubulis, Sigfrido, Fundacion de Bariloche, C.C. 138, 8400 San Carlos de Bariloche, Prov. de Rio Negro, Argentina

Ruszczynska-Szenajch, Hanna, Warsaw University, Department of Geology, al. Zwirki i Wigury 93, PL-02-089 Warsaw, Poland

Schlüchter, Christian, Institut für Grundbau und Bodenmechanik, ETH-Hönggerberg, CH-8093 Zürich, Schweiz

Schneebeli, Walter, Universität Zürich, Geographisches Institut, Blümlisalpstrasse 10, CH-8006 Zürich, Schweiz

Schove, D.J., St. David's College, Beckenham, Kent, Great Britain

Schubert, Carlos, Centro de Ecologia, I.V.I.C., Apartado 1827, Caracas 101, Venezuela

Schwarz, U., Universität München, Institut für Allgemeine und Angewandte Geologie, Luisenstrasse 37, D-8000 München 2, BRD

Seret, Guy, Université de Louvain, Institut Géologique, Place L. Pasteur 3, B-1348 Louvain-la-Neuve, Belgique

Serrat, D., Universidad de Barcelona, Facultad de Ciencias, Departamento de Geomorfologia y Geotectonica, Avenida José Antonio 585, Barcelona -7, Espana

Shaw, John, University of Alberta, Department of Geography, Edmonton, Alberta, T6G 2H4, Canada

Shilts, W.W., Geological Survey of Canada, 601 Booth Street, Ottawa, Ontario, KlA OE8, Canada

Sidiropoulos, T., Universität München, Institut für Allgemeine und Angewandte Geologie, Luisenstrasse 37, D-8000 München 2, BRD

Stephens, George, George Washington University, Department of Geology, Washington, D.C., U.S.A.

Stewart, Robert A., University of Western Ontario, Department of Geology, London, Ontario, N6A 5B7, Canada

Sturm, Michael, EAWAG-ETH, Ueberlandstrasse 133, CH-8600 Dübendorf, Schweiz

Suarez, Jorge, Fundacion de Bariloche, C.C. 138, 8400 San Carlos de Bariloche, Prov. de Rio Negro, Argentina

Van der Meer, J.J.M., Universiteit van Amsterdam, Fysisch geografisch en bodemkundig Laboratorium, Dapperstraat 115, Amsterdam-Oost, The Netherlands

Van Husen, Dirk, Technische Universität Wien, Geologisches Institut, Karlsplatz 13, A-1040 Wien, Oesterreich

Warren, W.P., Geological Survey of Ireland, 14 Hume Street, Dublin 2, Ireland

Introduction

Ch. SCHLÜCHTER
ETH-Hönggerberg, Zurich, Switzerland

"We are still confused
- but on a higher level"

The articles published in this volume represent the majority of the contributions delivered during the sessions at the Zurich-Symposium. It is about half of the papers previously submitted. We are glad to refer to the Volume of Abstracts containing all originally submitted papers which is still available and can be obtained from the Meeting Secretary.

A few articles are published even if they have not been presented as papers at the symposium because, unfortunately, their authors could not be present: Maria Boyko-Diakonow, Prof. Dr. K. Gripp, Prof. Dr. St. Ludlam, Prof. Dr. Mickelson and Dr. D.J. Schove. - We have considered these articles to be most valid contributions and therefore, gladly accepted for publication. We are specially referring to the article by Prof. Dr. K. Gripp on the dynamics of "pressed scales in frontal and/or lateral position" as an example could be studied during the field excursion at the Glacier de Tsidjiore Nouve (see also Introduction to part 3 of this volume).

We have done our best in fulfilling the philosophy of accepting all manuscripts submitted for publication. This has caused us a considerable amount of extra work as quite a number of manuscripts had to be reorganized graphically or even retyped. - The substantial additional costs caused by this procedure have been covered generously by the Institute of Foundation Engineering and Soil Mechanics (ETH). The delay in publication is, therefore, set by these circumstances. - There are ideas and interpretations expressed in some papers which do not necessarily coïncide with the Editor's opinion and with the standards set by the previous commission publications. But they are considered as being a stimulus for further, more precise discussions, hopefully.

The quality of this publication has been greatly improved by those persons who have acted as critical and constructive reviewers: Prof. Dr. G.M. Ashley, D. G.S. Boulton, Prof. Dr. A. Dreimanis, Prof. Dr. R. Hantke, Dr. K. Kelts, Dr. G. Kukla, Dr. A. Lambert, Prof. Dr. A. Lerman, Prof. Dr. St. Ludlam, Doc. Dr. J. Lundqvist, Dr. F. Madsen, Prof. Dr. A. Matter, Prof. Dr. B. Messerli, PD Dr. F. Schweingruber, Dr. Ch. Schlüchter and Dr. M. Sturm.

Many of the readers of the following articles - and of the title perhaps - are confused by the use of the term "moraine". It is used as a translation of the German term "Moräne" (or French "moraine"). The confusion may be substantial as "Moräne" is used with a threefold meaning:

(1) "Moräne" as a sediment. - deposited by the direct activity of a glacier. - (genetic)

(2) "Moräne" as a sediment with certain characteristics: specific grain size distribution + characteristic shape and surface texture of the clasts + "erratic" petrography. - (descriptif)

(3) "Moräne" as a geomorphic feature (Landschaftsform, "élément constitutif du paysage"). - (geomorphology)

The term "Moräne" used to describe a sediment from the genetic and descriptif point of view is thus the equivalent of the English term "till".

Obviously, it is the "non-linear use" of the term "Moräne" which has prevented the establishment of a coherent classification of glacigenic deposits and landforms in the Alps, or in other words, in the German and French literature. The existing classifications are one-sided morphology oriented and dominated.

The contributions to this volume are organized in three main parts: (I) Papers on the Geology of Moraines, (II) Contributions on Varves and Glaciolacustrine deposition and (III) on the Field Excursion. - The contributions to the Geology and Genesis of Moraines comprises three chapters: (A) Valley Glacier Environment, (B) General Aspects and Lowland Environments and (C) Progress Reports.

Chapter (A) is introduced with a proposal for the classification of the sediments produced in a valley glacier system (G.S. Boulton). Most of the contributions then deal with the occurrence and description of glacial deposits in mountainous areas of South America (C. Schubert, S. Rabassa et al), India (N. Ahmad) and New Zealand (T.J.H. Chinn). Sedimentological aspects of predominantly valley glacier deposits are discussed by J. Rabassa & G. Aliotta, D. van Husen and R. German et al ..

Emphasis on the litho- and morphostratigraphic development ("Landschaftsentwicklung") is given by W.-D. Grimm et al. and by K.A. Habbe. The discussion on the formation of glacial deposits is enlarged by the paper on applied glacial geology (to mineral exploration, E.G. Evenson et al.) and by the discussion of rock glacier morainic deposits (D. Serrat).

The second chapter on "General Aspects and Lowland Environments" is introduced by a study on debris dispersal and characteristics in modern glaciers in the Canadian Arctic (R.N.W. DiLabio & W.W. Shilts). Of importance in explaining specific aspects of glacitectonics in Pleistocene deposits is the article on "pressed scales" (K. Gripp). The confusion in terminology exists not only for sediments of the valley glacier milieu: A. Dreimanis unveils an almost babylonian assemblage of terms for "Waterlain tills" and proposes a clarified approach. - Geotechnical parameters as criteria for differentiating tills are given with an example from Eastern Wisconsin (D.M. Mickelson et al.). Discussions of basic geomorphic processes are the contributions on the formation of Drumlins (G. Seret) and on the "deglaciation ridges" in Glacier Bay National Monument (G. Haselton). - Emphasis on the climatic development during a glacial cycle and the resulting shifts in ice-sheet behaviour and the melting of the ice-body reflected in glacial deposits ("till-stratigraphy") is given in three articles from central and peripheral parts of the Scandinavian ice-sheet (K. Garnes, E. Drozdowski, A. Lindner & H. Ruszczynska-Szenajch).

The relationship between ice-sheet development, local cirque glaciers and deposition in ice-dammed lakes and resulting Varve-deposition is demonstrated with an example from south-west Ireland (W.P. Warren).

The Progress Reports reflect continuing work on the filling of overdeepened glacial basins in southern Germany (H. Jerz), on "more waterlain tills" in Michigan (D.F. Eschman) and on till complexes in the Swiss Plain (J.J.M. van der Meer). A reference to the importance of contacts between INQUA and the IGCP-Project on "Pre-Pleistocene Glacigenic Rocks" concludes the articles on the geology of moraines.

If we have not arrived yet at (a.) a genetic, (b.) a descriptif and (c.) a geomorphological classification of glacial deposits in the alpine environment (in the German and French languages) which is as advanced and challenging as that introduced by Dr. G.S. Boulton - then the Symposium "Moraines and Varves" has acted as the stimulus to do so in the future. If we are still confused by the extreme variety of "moraines" - then it is on a much higher level!

The following two figures may serve as a visual introduction to the contributions at the symposium on "Moraines and Varves".

The Valley Glacier Environment: view from the Gornergrat near Zermatt to the south towards the Gornergletscher (foreground), Unterer Theodulgletscher (Th) and Triftjigletscher (T); the build-up of lateral moraines as geomorphologic features is clearly demonstrated. The medial moraine on Theodulgletscher can be traced above the ice-fall (arrow). On the Gornergletscher well developed longitudinal moraines ("Mittelmoränen", "debris septa") and the unique features called "entonnoirs"(*) can be observed.

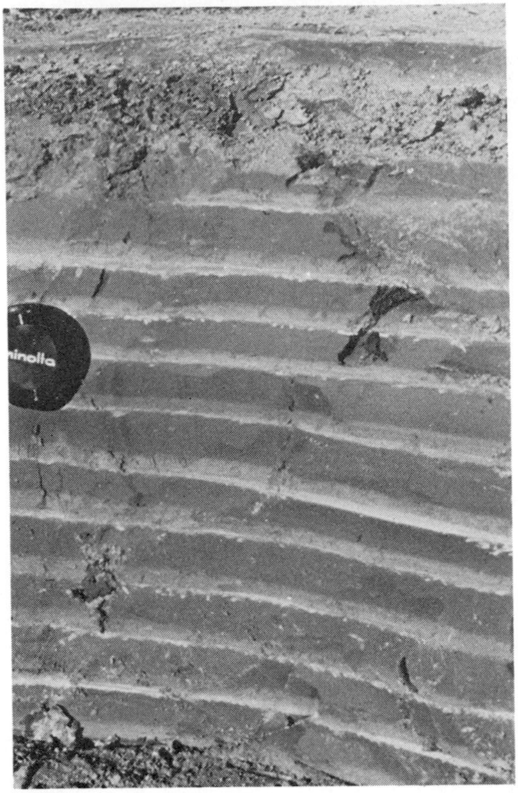

Glaciolacustrine Varves: this is an example considered to represent the classical setting of varves from Ontario, Canada (cf. paper by M. Sturm this volume), with alternating light/coarse and dark/fine laminae - each couplet representing one year of sediment accumulation in a "Late Glacial Great Lake". - Photograph by Steve Blasco, Geological Survey of Canada.

Opening address

HEINRICH JÄCKLI
Zurich, Switzerland

Herr Vorsitzender,
meine sehr verehrten Damen und Herren,

Ich freue mich, Sie alle hier in der Schweiz zum INQUA-Symposium "Genese und Lithologie von Moränenablagerungen" sehr herzlich begrüssen zu dürfen, ist es doch das erste Mal, dass sich Ihr INQUA-Symposium im speziellen dem Gebiet der alpinen Vergletscherung widmet.

Ich muss Ihnen gestehen, dass wir Schweizer ein ganz klein wenig stolz sind auf unsere Moränen. Nicht dass wir behaupten möchten, wir hätten diese selbst gemacht. Aber die Schweizer Alpen mit ihrer sehr grossen Zahl rezenter Gletscher hatten reichlich Gelegenheit geboten - und tun das übrigens auch heute noch - die Bildung der Moränen als glazigenes Sediment zu erforschen.

Unsere Vorgänger hatten dadurch schon in der ersten Hälfte des letzten Jahrhunderts die seltene Gelegenheit, ausgehend vom Studium der rezenten Moränen an unseren Alpengletschern, blockreiche wallförmige Ablagerungen im schweizerischen Mittelland richtigerweise ebenfalls als glazigene Bildungen einer grossen pleistozänen Vereisung zu erkennen. Darf ich unter anderem an Ignaz Venetz, Kantonsingenieur des Kantons Wallis, Jean de Charpentier, Minendirektor des Kantons Waadt, oder Louis Agassiz, Professor in Neuchâtel erinnern, die ganz konsequent die Regeln des Aktualismus befolgten und, als Kenner der rezenten Gletscher und Moränen der Alpen, die pleistozänen Moränen und erratischen Blöcke im Mittelland richtig deuteten.

Eine Generation später verfasste dann 1885 Albert Heim, Geologieprofessor hier in Zürich, immerhin als erster im deutschen Sprachbereich ein "Handbuch der Gletscherkunde"; welches für die damalige Zeit das Standardwerk darstellte.

In Gebieten wie in unserem schweizerischen Mittelland, die einst von einem pleistozänen Eisschild überdeckt waren, spielen Moränen auch in der Ingenieur-Geologie eine bedeutende Rolle. Sie stellen vorbelastete und deshalb gut konsolidierte Böden dar, die sich als Baugrund vorzüglich eignen und für Dämme ein gesuchtes Baumaterial darstellen.

Der Ingenieur wie auch der Architekt sind deshalb recht oft Konsumenten von Moränen. Der erstere, indem er Moränen wegbaggert und als Baumaterial verwendet, der letztere, indem er die schönsten Moränenwälle mit Bauten aller Art bedeckt. Beide zerstören bei dieser Gelegenheit einen typisch glazigenen geologischen Körper unseres Planeten, auf dem wir leben. Konsequenterweise müssten wir Geologen deshalb bestrebt sein, diese Moränen zu schützen, wie wir Seehunde oder Edelweiss schützen, denn intakte Moränenwälle gestatten uns eine Menge geologischer Probleme zu lösen, an denen wir alle beteiligt sind.

Einen ersten Schritt in der Richtung des geologischen Umweltschutzes haben unsere Vorgänger vor rund hundert Jahren getan, an ihrer Spitze Fritz Mühlberg von Aarau, als sie in unserem Lande begannen, die erratischen Blöcke von Gesetzes wegen zu schützen. Diese erratischen Blöcke, besonders soweit sie aus Granit und Gneis bestanden, waren bis damals als hochwillkommener Rohstoff für Randsteine und Pflastersteine in grossen Mengen abgebaut und über das ganze Land verkauft worden; selbst auch in der Stadt Zürich sind viele Randsteine, die im letzten Jahrhundert in unseren Strassen eingebaut wurden, aufgearbeitete erratische Granitblöcke aus dem Aargau. Auch naturfor-

schende Gesellschaften begannen vielerorts, besonders grosse Exemplare solcher Blöcke direkt zu kaufen, um sie auf diese Weise vor zukünftiger Zerstörung sicher zu schützen. Heute geniessen diese Blöcke in den meisten Kantonen auch einen gesetzlichen Schutz.

Aber das genügt nicht. In einem nächsten Schritt sollten wir nicht nur die einzelnen Blöcke, sondern auch die gesamten Moränenwälle als typische Elemente des glazigenen Formenschatzes schützen, und wir hoffen, dass das zukünftige schweizerische Planungsrecht uns das gestatten wird, vorausgesetzt, dass wir Naturwissenschafter es verlangen. Wir Geologen fühlen uns verantwortlich für die Natur, und Moränenlandschaften sind auch Teile dieser Natur, die wir nicht bloss nur deshalb schützen wollen, weil wir sie etwa für unsere speziellen geologischen Studien benötigen, sondern weil wir sie ganz einfach auch lieben.

*

Monsieur le président,
mesdames et messieurs,

Je suis bien heureux de pouvoir vous souhaiter la bienvenue à l'INQUA-symposium "genèse et lithologie des moraines", qui à lieu la première fois aux régions alpines.

Il faut que je vous confesse que nous Suisses sommes assez fiers de nos moraines. Nous n'osons pas prétendre que nous les ayons faites nous-mêmes. Mais les Alpes suisses avec leur grand nombre de glaciers ont offert l'occasion - et l'offrent encore aujourd'hui - d'étudier sur place la genèse des moraines comme sédiments glacigènes.

Nos prédécesseurs avaient la chance, partant de l'étude des moraines récentes des glaciers alpins, de reconnaître les cordons et les arcs de glaise graveleuse avec des blocs d'origine alpine au plateau suisse et le long du pied du Jura comme formations glacigènes, crées dans un passé où les glaciers alpins ont eu une très grande extension. Pensons à Jean de Charpentier, directeur des mines vaudoises, introduisant le mot savoyard "moraine" dans la terminologie scientifique, et Ignaz Venetz, ingénieur cantonal valaisan, et Louis Agassiz, professeur de Neuchâtel, étant pionniers tous les trois en leur temps, grâce à leur connaissance profonde des glaciers récents alpins et leur moraines.

Dans des régions couvertes auparavant d'une calotte glacière, comme c'est le cas pour de grandes parties du plateau suisse, les dépôts morainiques jouent aussi un grand rôle dans la géologie de l'ingénieur. Ils représentent des sous-sols préchargés et pour cela bien consolidés, se prêtant aussi comme terrain de fondation que comme matériaux de construction pour des digues, des barrages et des routes.

L'ingénieur civil ainsi que l'architecte consomment des moraines: le premier en dragant du matériel morainique pour pouvoir l'utiliser comme matériel de construction, le deuxième en les couvrant et les cachant avec des maisons et des autres constructions de toute sorte.

A cette occasion tous les deux détruisent un corps géologique typiquement glacigène de notre planète sur lequel nous vivons.

Par conséquent nous géologues devrions être obligés de protéger ces moraines, comme nous protégeons les phoques et les edelweiss, car des cordons et arcs morainiques nous permettent de résoudre une quantité de problèmes géologiques auxquelles nous nous sommes tous engagés.

Un premier pas dans la direction de la protection des objets géologiques fut exécutés par nos ancêtres il y a plus de cent ans, en protégeant les blocs erratiques géants.

Au siècle passé des quantités de ces blocs furent détruits en les taillant pour les utiliser comme bordure. Par exemple en Argovie les granites erratiques, démontrant l'extension du glacier de la Reuss, étaient vendus dans toutes les régions de la Suisse, surtout dans la ville de Zurich. Sous l'initiative de Fritz Mühlberg, géologue de Aarau, ces blocs furent protégés par la loi cantonale et leur taille défendue. Dans des autres cantons beaucoup de blocs géants furent achetés par des sociétés des sciences naturelles.

Mais aujourd'hui, 100 ans plus tard, ça ne suffit plus. A l'avenir il faudra que nous protégions non seulement les blocs, mais aussi les cordons morainiques entiers comme éléments morphologiques typiquement glacigènes. Nous espérons que la législation future nous le permettra, supposé que nous scientistes le demandions.

Nous nous sentons responsables envers la nature, et même les paysages de moraines représentent une partie de la nature. C'est pourquoi nous aimerions les protéger, car ils ne sont pas seulement l'objet de nos recherches scientifiques, mais aussi de notre amour.

Mr. Chairman,
Ladies and Gentlemen,

I should like to welcome you all to Switzerland to the INQUA-Symposium on the genesis and lithology of morainic deposits in an alpine environment.

I must confess that we Swiss geologists are rather proud of our moraines. Not that we would pretend to have made them ourselves. But Switzerland with its large number of alpine glaciers has offered - and still offers today - a rich variety of opportunities to study the recent formation of moraines as glacigenic sediments.

Following Charles Lyell's well known principal idea of actualism "The present is the key to the past", our predecessors persued their studies in the second quarter of the last century from the areas of the recent alpine moraines to those of the older ones, and they recognized in this way the boulder clay and the erratic boulders in the lowland regions as being evidence of pleistocene glaciation. May I name among others Louis Agassiz from Neuchâtel, from 1846 - 1873 a famous professor of natural history at Harvard University in Cambridge Mass., who in his younger days spent weeks on end with his geologist colleagues in primitive bivouacs on the Grindelwald glacier to study on site ice movement and moraine formation, and whose tombstone in Cambridge is an erratic boulder from Switzerland.

In regions once covered by a pleistocene ice-sheet, as was the case for the greatest part of Switzerland, moraines are also of importance from the point of view of engineering geology: They represent a usually preloaded and therefore well compacted ground for fondations, and on the other hand, as a boulder clay of suitable composition a useful building material for dams and embankments.

The engineer and the architect are therefore often enough "consumers" of moraine materials, the former by dragging them away, the latter by covering them with buildings and houses of all kinds; both destroy a typically glacigenic geological body of the planet on which we live.

Consequently we scientists should be eager to protect them - as we should protect seals or edelweiss - for moraines allow us to solve many problems of glacial morphology, of earlier climatic changes, of earth history.

In our country, we took a first step in this direction 100 years ago, when Fritz Mühlberg, a geologist in Aarau, campaigned to protect the giant erratic boulders, which until then had been used to make kerb stones and sold all over Switzerland. Scientific societies troughout the country then began to buy erratic boulders in order to protect them as a unique relic of the pleistocene ice age.

But that is not enough! In a second step, we should also protect not only the boulders, but also the entire feature as a typical glacigenic form, and we hope, that the new Swiss planing legislation will permit this, if we, geologists and others, request it.

We scientists feel responsible for nature - and even moraines are parts of nature - and as we not only need moraines for our research into the history of the earth, but as we also love them as a part of nature, we are obliged to protect them.

With this wish it is my pleasure to declare our INQUA-Symposium "Genesis and Lithology of Morainic Deposits in an Alpine Environment" open.

Geology and genesis of moraines

A. Valley glacier environment

Sedimentation by valley glaciers; a model and genetic classification

G.S. BOULTON & N.EYLES
The University of East Anglia, Norwich, England

ABSTRACT

A genetic classification of the sediments and landforms produced by valley glaciers is suggested, which is based primarily on the grain-size distinction between supraglacially-derived and subglacially-eroded debris, and secondly on the mode of deposition of that debris as till.

The initial grain size contrast is produced by the fact that supraglacially-derived debris, characteristic of valley glaciers, has not undergone a phase of tractional comminution at the glacier bed. This helps to define supraglacial morainic till.

This till is deposited in two principle facies. The first, Facies A, deposited during glacier retreat, produces sub-facies A_1 and A_2, depending on the discharge by the glacier of supraglacially-derived debris. Sub-facies A_3 comprises intimate mixtures of supraglacial morainic till and outwash sediment in till complexes.

Facies B is produced during glacier standstill or advance, and forms major dump moraines.

Intimately associated with supraglacial morainic till we find proximal outwash, a matrix-supported sediment which is often extremely difficult to distinguish from this till.

INTRODUCTION

It has been suggested that three principal sediment/landform associations can be used to describe most of the variation of sedimentary and geomorphic products in terrestrial glacial environments (Boulton 1976; Boulton and Paul 1976). Two of these, the supraglacial, and subglacial/proglacial sediment/landform associations are largely distinguished on the basis of the mode of deposition of debris eroded from the glacier bed. The third, the glaciated valley sediment/landform association is defined because of the character of the sediments and landforms produced in cirque and valley glaciers where much of the debris in transport is derived supraglacially from adjacent valley walls (Fig. 1). This third system may be superimposed on either one of the other two.

A reasonably good classification and description of sediments and landforms exists for the first two systems as a result of work over the last decade or so but the traditional classifications still in use for the glaciated valley land system and sediment association no longer do justice to the more refined picture of sediment and landform evolution that can now be built up as a result of recent work. Existing classifications (Böhm 1961; Sharp 1948, Charlesworth 1957) are not genetic and thus provide little information about the processes at work in the glacier debris transport system. In addition they do not distinguish between sediments and landforms deposited by valley glaciers and the very different sequences deposited by ice caps and ice sheets.

In this paper we propose a genetic classification of the glaciated valley land system and sediment association, based on a genetic model built up from observations of modern glaciers and we suggest criteria which allow these different sediments and landforms to be identified.

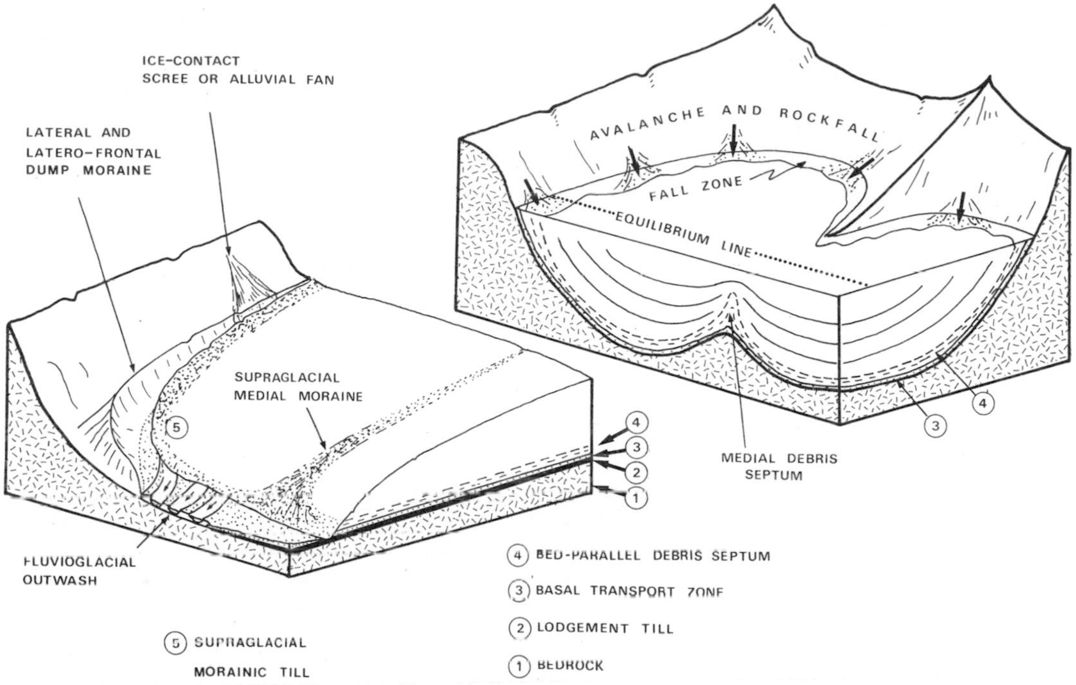

Figure 1. Transport paths of debris through temperate confluent valley glaciers with rock headwalls. Subglacially-eroded debris goes almost exclusively to form lodgement till via the basal transport zone, although some might find its way into the medial debris septum at a glacier confluence. Supraglacially-derived debris passes into the bed-parallel debris septum. From there it may (i) pass into the basal transport zone from where it is likely to be deposited as lodgement till, (ii) continue to be transported in the bed-parallel septum to be deposited frontally or laterally as supraglacial morainic till, (iii) go into the medial debris septum, finally to be deposited as supraglacial morainic till.

REASONS FOR ADOPTING A NEW CLASSIFICATION

Whereas we are aware how deeply engrained are several of the usages which we wish to overturn, we believe it is important to do so in the interest of clarity. For instance the almost ubiquitous use of the term moraine hides many different connotations. It is used as a description of segregated debris in ice as in medial moraine and lateral moraine, and is also used to describe supraglacial and deposited landforms. The term lateral moraine is also applied to ice-contact screes and fans and to deposited teral moraines. In addition the term moraine is used to describe till. Thus moraine is used to describe debris in transport, landforms of several types, and as a material description. There are also a wide variety of terms used indiscriminately to refer to till deposited on the surface of a glacier irrespective of glacier regime. Terms such as supraglacial, superglacial, superficial, ablation, ablational, perluvial, collapsed, disintegration, stagnation, high-level and upper, are robbed of much gentic signification when employed to describe deposition from the glacier surface both at the margins of cold glaciers where tills are derived from the bed and from valley glaciers where till is derived from the valley sides and has never experienced comminution at the glacier bed (Fig. 1). In view of this confusion new genetic terms are appropriate.

CLASSIFICATION OF COMPONENT SEDIMENTS

Firstly we wish to distinguish between debris and till. Glacial debris is a dispersion of particles in the glacier or on its surface, acquired by no matter what means. It may for example, be windblown dust or ash, or material derived from disaggregated englacial fluvial deposits, etc.,

but the two dominant sources are from erosion of the glacier bed or from material which falls onto the glacier surface from valley sides. This latter is a fundamental distinction, as the grain size distribution of these two types differs considerably.

Subglacially-derived debris has undergone a phase of traction at the glacier/bed interface which produces comminution and generates a relatively high proportion of silt and find sand, whereas supraglacially-derived debris is relatively impoverished in these fractions (Fig. 2). We believe this distinction to be so important that it should form a basic distinguishing element in the classification.

Till we define as "an aggregate whose components are brought together by the direct agency of glacier ice which though it may suffer deformation by flow, does not undergo subsequent disaggretion and redeposition" (Boulton 1972). On the basis of grain size distribution we believe that there are two fundamentally different populations of till produced in areas of recrystallised bedrock (Boulton 1979). These are tills prodominantly aggregated from subglacially-derived debris and which may be subdivided into several sub-types on the basis of other criteria (lodgement, melt-out, flow and lee-side tills) and those predominantly aggregated from supraglacially-derived debris for which the term supraglacial morainic till has been proposed (Boulton 1976).

EVOLUTION OF SUPRAGLACIAL MORAINIC TILL AND ASSOCIATED SUPRAGLACIAL LANDFORMS IN VALLEY GLACIERS

Debris falls onto the glacier surface either as a result of rockfalls or as a part of snow or ice avalanches and is normally dispersed over a relatively wide area, the fall zone (Fig. 1). In the accumulation area this debris is buried and subsequently transported englacially. Although strain within the glacier may considerably alter inter-particle distances, grain contacts are rare and little comminution occurs (Whalley and Krinsley 1974; Eyles and Rogerson 1978; Boulton 1979). The grain size distribution remains that of the parent rockfall. The debris is initially entrained within the glacier near the margin from which it is derived, and thus lies near to a basal flow-line, although some avalanches and rockfalls may move several hundred metres onto the glacier. This debris, derived supraglacially from flanking valley walls, is entrained along foliation planes parallel to the glacier bed and for some distance above it. The term lateral debris septum first attracted us for this debris horizon, as it is most frequently exposed along the lateral margins of valley glaciers. However the use of the word lateral may hide the real disposition of this debris in the ice, schematically shown in Figure 1. In outlet valley glaciers, this debris septum may primarily be lateral, but in a valley glacier derived from a rock headwall it forms an almost continuous stratum of dispersed supraglacially-derived debris some distance above the bed and parallel to it (Fig. 1). For this we propose the term bed prallel-debris septum. It tends to outcrop near to the lateral valley wall in the ablation area and may also outcrop frontally (Fig. 1).

Where two valley glaciers coalesce, there is a tendency for the bed-parallel debris septa from each glacier to merge to produce a medial debris septum which is normal to the glacier bed and which may develop an intrusive relationship to flanking glacier foliation (Fig. 1). If one of the confluent glaciers is dominant it may be inclined (Sharp 1948). Multiple confluences produce multiple debris septa.

Where debris falls onto the glacier surface in the ablation area, it is transported entirely supraglacially and does not form a debris septum.

When sufficient debris from an englacial debris septum has melted out onto the glacier to form a continuous aggregate, we employ the term supraglacial morainic till. This till is often no more than a few centimetres in average thickness, but where rates of supply of supraglacial material are high, thicknesses up to 6 m may be accumulated (Eyles 1978). The term supraglacial morainic till is also employed for aggregated, supraglacially-derived debris that accumulated on the glacier surface in the ablation area and has not undergone a phase of englacial transport.

The insulating effect of a stratum of supraglacial morainic till reduces the rate of ablation of underlying ice below that of adjacent clean ice, so that it tends to stand above the adjacent ice as a ridge. We suggest that the terms supraglacial medial moraine and supraglacial lateral moraine should be applied as geomorphic terms to these ridges of supraglacial morainic till derived from melting out of debris in medial and laterally outcropping bed-parallel septa (Fig. 1).

DISTINGUISHING CRITERIA FOR SUPRAGLACIAL MORAINIC TILL

The distinctive sedimentological and geotechnical properties of this till type, which separate it from tills derived from the basal transport zone of a glacier, are based on the fact that it has not been through a tractive phase at the glacier bed. It has a silt-clay content generally below 15 %, which increases downward in section in response to washing: predominance of angular, oxidised clasts that lack striations, glacial shaping and preferred orientation (Sharp 1949; Drake 1971; Flint 1971; Boulton 1979); a coarse mean size, enhanced sorting and fine-skewness relative to subglacially-aggregated tills; a high angle of friction, low density, frequent voids and low coefficient of compaction. The particle-size of supraglacial morainic till is independent of parent bedrock (Slatt 1971; Eyles and Rogerson 1978). "Terminal grades" of each component mineral indentified for basal tills by Dreimanis and Vagners (1971), cannot be recognised. Distinct lithologies remain unmixed and are usually far travelled from the extremities of the basin, weak lithologies survive transport in the absence of comminution (Young 1953; Portmann 1960; Dutt 1961). Organic material is commonly present as are frequent deformed lens and inclusion of sands, silts and clays (Okko 1955). Supraglacial morainic till is comparable to those sediments produced by laboratory experiments on the susceptibility of different rock types to freeze/thaw (Martini 1967; Potts 1970; Lautridou 1971; Brockie 1973) and those found on mountain top blockfields and felsenmeer (Dahl 1966; Strømquist 1973).

MODES OF DEPOSITION OF SUPRAGLACIAL MORAINIC TILL

We suggest that there are two principle depositional facies of supraglacial morainic till which are determined largely by the state of health of the glacier. Facies A accumulates during glacier retreat and tends to produce relatively muted landforms, whilst Facies B accumulates in dump moraines which are often extremely large formed when the glacier is stationary or advancing.

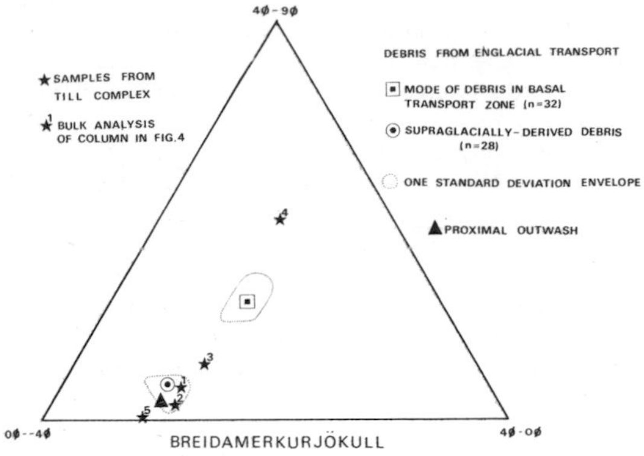

Figure 2. Grain size distribution of debris from the basal transport zone and from supraglacially-derived debris in high level englacial transport. Note the enrichment of the former in fines. A sample of proximal outwash is also plotted, and samples (1-5) from the till complex shown in figure 8.

Facies A - Deposited along the frontal margin during retreat

Three sub-facies can be distinguished:

1) Where the till cover on the glacier is thin it is deposited directly by being dumped from the ice surface (Figs. 4, 5). Deposited medial moraines along the valley floor may consist of closely-spaced dump ridges transverse to ice-retreat direction (Fig. 5) and cones which may show internal scree type bedding as a result of gravity sorting during deposition as ice-contact screes (Fig. 4). Hummocky, high relief-amplitude dump cones form when the rate of glacier recession is relatively slow and may impart an esker-like form to the medial moraine along the valley floor. On the other hand a coarse-textured carpet of till lacking any pronounced relief results from enhanced rates of glacier recession. Along many valley floors deposited crevasse fills, or the fillings of ice marginal supraglacial gullies, form linear bouldery ridges and record the orientation of crevasses or gullies on the depositing glacier, though these are often difficult to discern, especially in areas long deglaciated. This subfacies often comprises merely a sporadic collection of boulders dumped onto a progressively exposed subglacial bed. Winter readvances often concentrate this material in pushed ridges.

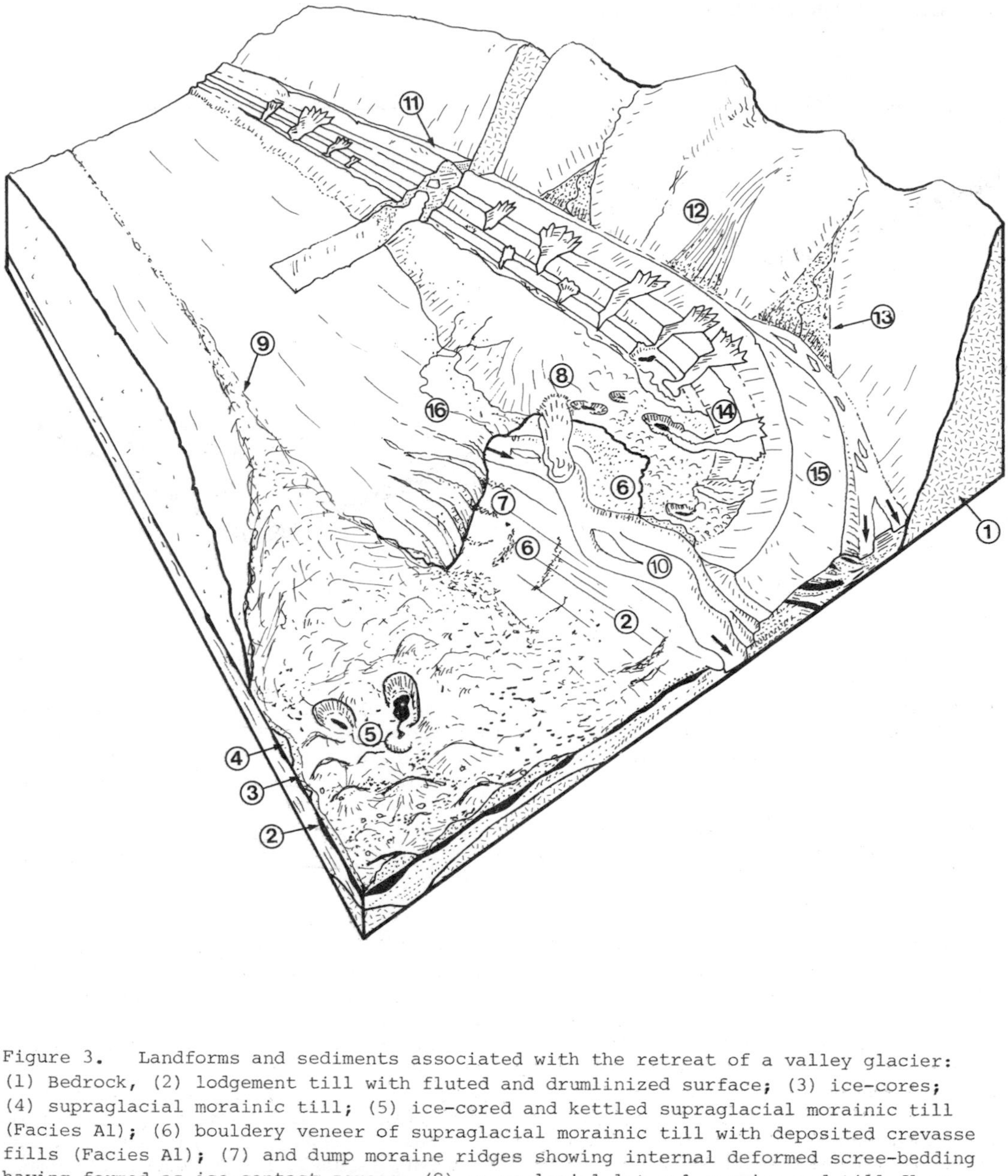

Figure 3. Landforms and sediments associated with the retreat of a valley glacier: (1) Bedrock, (2) lodgement till with fluted and drumlinized surface; (3) ice-cores; (4) supraglacial morainic till; (5) ice-cored and kettled supraglacial morainic till (Facies A1); (6) bouldery veneer of supraglacial morainic till with deposited crevasse fills (Facies A1); (7) and dump moraine ridges showing internal deformed scree-bedding having formed as ice-contact screes; (8) supraglacial lateral moraine and till flows into meltwater streams resultung in till complexes (Facies A3); (9) supraglacial medial moraine; (10) proximal meltwater streams; (11) kame terrace; truncated scree (12) and fan (13); (14) gullied lateral terrace; (15) latero-terminal dump moraine (Facies B); (16) fines washed from supraglacial moraine till into crevasses and moulins.

A. Summer.

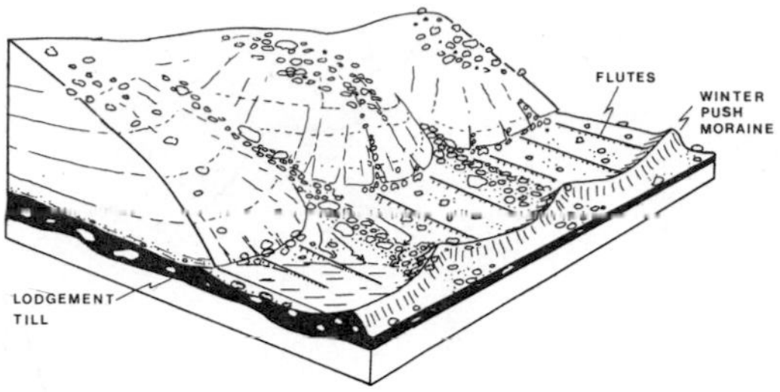

B. Winter.

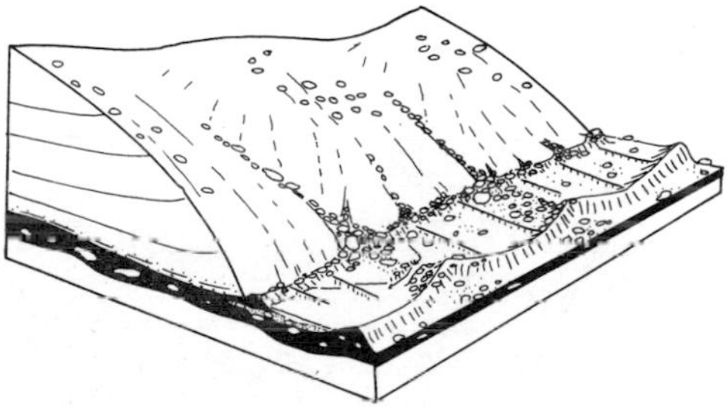

Figure 4a-b. Deposition of facies A1 of supraglacial morainic till during summer retreat (A) and winter readvance (B). Relatively thin masses of supraglacial morainic till slump off the glacier margin as it retreats, to leave a thin and sporadic veneer on the progressively exposed subglacial surface (till, bedrock, or outwash). Thicker longitudinal accumulations form where longitudinal gullies or crevasses in the ice margin cause the till to accumulate. Transverse accumulations may build up where glaciers of high activity undergo a winter readvance which sweeps up the boulders to incorporate them in a winter push moraine (B). Transverse accumulations may also form at the maximum of such a readvance due to dumping from the ice margin, or in glaciers of low activity where cessation of ablation in winter maintains the glacier front stationary allowing a dump ridge to build up without an accompanying push moraine.

2) Where the till cover on the glacier is thick enough to delay the melt of underlying ice, till is deposited in protracted fashion by being let down onto the subglacial surface, and results typically in hummocky stagnation topography indentified by thaw lakes and kettles (Fig. 6, 7). This sub-facies occurs either at the margin of an active valley glacier that is retreating up-valley by sequential stagnation of the margin, or at a glacier margin downwasting in situ over the whole glacier tongue area. Surging glaciers in their quiescent phase provide good examples of the latter. There are frequent reversals of relief as the ice core melts releasing melt-out till (Fig. 6) and the final relief amplitude of the till surface is controlled by the thickness of till accumulating as mudflows in hollows in the ice surface prior to final deposition.

3) A third type of till deposition occurs in association with proximal ice-contact outwash. Till deposition as mudflows may alternate with episodic meltwater incursions, commonly during flood events, such that within any one stratigraphic unit individual till and outwash horizons are discernible (Fig. 8). The term till complex, in this case a supraglacial morainic till complex, is used for such multiple sequences deposited in one episode of glacial sedimentation. With the melt of buried ice a pitted kame plain or outwash surface results, the relief amplitude of which is dependent upon the burial depth and spacing of ice-cores.

The grain size distribution within such a till complex is highly variable from sample to sample, because of the variability of the depositional processes. Figure 2 shows the variation in grain size distri-

Figure 5. Supraglacial morainic till of facies A1 being deposited as a small dump moraines at the margin of Breidamerkurjökull, Iceland (c.f. Fig. 4). Dump moraines parallel to ice flow form along the former lines of ice front gullies. Moraines transverse to flow form along the ice front.

bution of samples taken from within the till complex shown in Figure 8. However the bulk grain size distribution lies within a one standard deviation envelope for supraglacially-derived debris, suggesting that none of the sorting processes have been sufficiently powerful to preferentially export particular grain sizes from the system.

Relationship of supraglacial morainic till to proximal outwash

The braided stream model universally employed to predict the sediment sequences deposited by glacial meltwater streams (Church and Gilbert 1975) is of limited use in describing those sedimentary sequences accumulating where proximal meltstreams have access to a thick cover of supraglacial morainic till on the ice front. Braided streams form only a small part of the total hydraulic system associated with proximal meltwaters, in which discharges are much more variable, often flood prone, compared with downstream, distal locations. Further variety is added to the proximal outwash environment by the presence of ice-cores and the changing seasonal activity of the ice front (Fahnestock 1963). The frontal outwash aprons of valley glaciers that transport thick covers of supraglacial morainic till can be compared to semi-arid alluvial fans. Water flow is seasonal and flood-prone, and supraglacial morainic till is resedimented as mudflows and sub-aqueous flows.

The availability of large quantities of readily transported sedimentary particles and the rapid build up and decay to and from flood discharges has a strong effect on fluvial sediment character. Streams are often incompetent to transport the available till load, which is redistributed as planar, matrix-supported outwash beds with a particle size distribution that may be very little different from that of the parent till (Fig. 2). Massive unstratified proximal outwash is difficult to distinguish from supraglacial morainic till (Figs. 9 & 10) except by loss of oxidation skins on clasts, rapid destruction of weak lithologies and incipient clast rounding: During the flood phase all available particle sizes are transported and deposited simultaneously. Poor stratification at the top

Figure 6. Thick (1.75 m) supraglacial morainic till of facies A2 resting on dead ice at the margin of Fjallsjökull, south-east Iceland. A dark area to the right betrays the presence of an exposed ice core. Relief on the surface is 5-10 m.

Figure 7. Deposited supraglacial morainic till of facies A2 beyond the margin of Verkisjökull in south-east Iceland. The irregular till surface has a relief of 1-2 m. It is unlikely that much dead ice survives beneath this till blanket. The ridges on its surface reflect slow creep of the originally supraglacial blanket and some dumping at the ice margin.

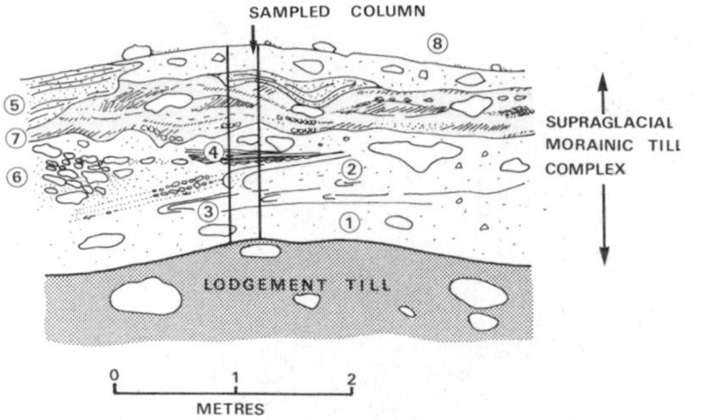

Figure 8. Supraglacial morainic till complex resting on lodgement till at the margin of Breidamerkurjökull in south-east Iceland. It comprises a basal unit of till (1) which accumulated on the glacier surface and was let down onto the underlying lodgement till; a mass of supraglacial morainic till which flowed in a relatively fluid state (2); mud-enriched horizons associated with the distal part (3) of the flow; a mud accumulation (4) washed out of the till; sands (5) deposited in ephemeral streams; a scree-like mass (6) dumped from the ice front; a bed of proximal outwash (7) and a massive till unit which flowed as a relatively viscous mass (8).

Figure 9. A section in proximal outwash at the margin of Fjallsjokull in south-east Iceland. This matrix supported sediment is difficult to distinguish from supraglacial morainic till. Indicators of origin are, incipient clast rounding and the loss of oxidation skins, and the poorly stratified uppermost bed produced during the waning flood stage.

of massive outwash beds consists of individual lamellae several clasts thick which develop as washed, armoured and often scoured horizons during waning flood stages, and reflect a very rapid decrease in discharge (Fig. 10). It is difficult to distinguish this fluvial sediment from supraglacial morainic till other than by the four criteria mentioned above. Saunderson (1977) has claimed that such matrix-supported sediments only form in enclosed pipe flow such as might occur in a subglacial or englacial esker. We suggest however that these sediments may be at least as important in sub-aerial proximal outwash where ephemeral streams of high discharge have access to an excess of readily eroded sediment.

SUB-SURFACE SEDIMENT SEQUENCES

The sub-surface relationship between the three sub-facies of supraglacial morainic till can be closely predicted as they are found in catenary sequence (Fig. 3). Thick accumulations of supraglacial morainic till which have been slowly superimposed onto an underlying lodgement till grade laterally into thinner, generally coarser and more disaggregated till horizons which have been dumped from the ice front. Through this bouldery drape may protrude streamlined landforms such as flutes and drumlins on the lodgement till surface (Fig. 4) or eskers that formed as subglacial channel fills. The melt of ice-cores within the supraglacial morainic till often results in a boulder pavement of boulders washed from the underlying lodgement till. Within the supraglacial morainic till, deformed mud lenses record the destruction of thaw lakes, during till deposition (Fig. 11).

Till complexes are found as deformed basinal inclusions only in the thicker sequences of supraglacial morainic till and record the resedimentation of supraglacial morainic till as mudflows into proximal

Figure 10. Kettled, ice-cored proximal outwash at Hrutajökull south-east Iceland. The section is 1.5 m thick. Note the island of ice-cored supraglacial morainic till (arrowed) partially buried by outwash.

meltwater streams (Fig. 3).

Frontal deposition results in this catenary sequence being exhibited within deposited medial moraines along the valley floor. The zone of thicker till cover is then seen as a 'controlled' longitudinal belt within which till complexes lie, grading into adjacent zones of thinner till cover. If the depth of till on the ice front is small then only a bouldery veneer is found along the valley floor.

'Uncontrolled' deposition of this catenary sequence occurs from glaciers that have a supraglacial morainic till cover over the entire ice margin (e.g. surging glaciers), or where the ice front is highly crevassed and the till cover is dispersed. In these cases sub-facies of supraglacial morainic till are found as an uncontrolled mosaic.

Deposition from lateral debris septa on the glacier surface results in the typical catenary facies sequence shown in Fig. 3 which passes up slope into valleyside lateral terraces.

Facies B - Deposition along lateral and latero-frontal margins during stationary or advance phases

Dramatic changes in the health of a valley glacier are reflected by major displacements of the glacier front, but against the steep lateral sides of the glacier channel large vertical movements of the glacier surface may only be accompanied by relatively small lateral changes in the plan position of the glacier margin. Because of the small displacements of lateral margins compared with frontal margins, thick localised sediment accumulations develop along lateral and latero-frontal margins.

Debris which accumulates around the periphery of a valley glacier in the accumulation area has a component of flow towards the centre-line of the glacier. In the terminal area of the glacier there is a component of flow away from the centre line. The net effect of these is to concentrate debris in the bed-parallel debris septum so that larger concentrations emerge from the ice in the latero-frontal zone than elsewhere. Thus a valley glacier with a roughly stationary ice margin will tend to produce latero-frontal dump moraines which are larger than frontal dump-moraines. This effect is often enhanced by the destruction of frontal dump moraines by powerful out-

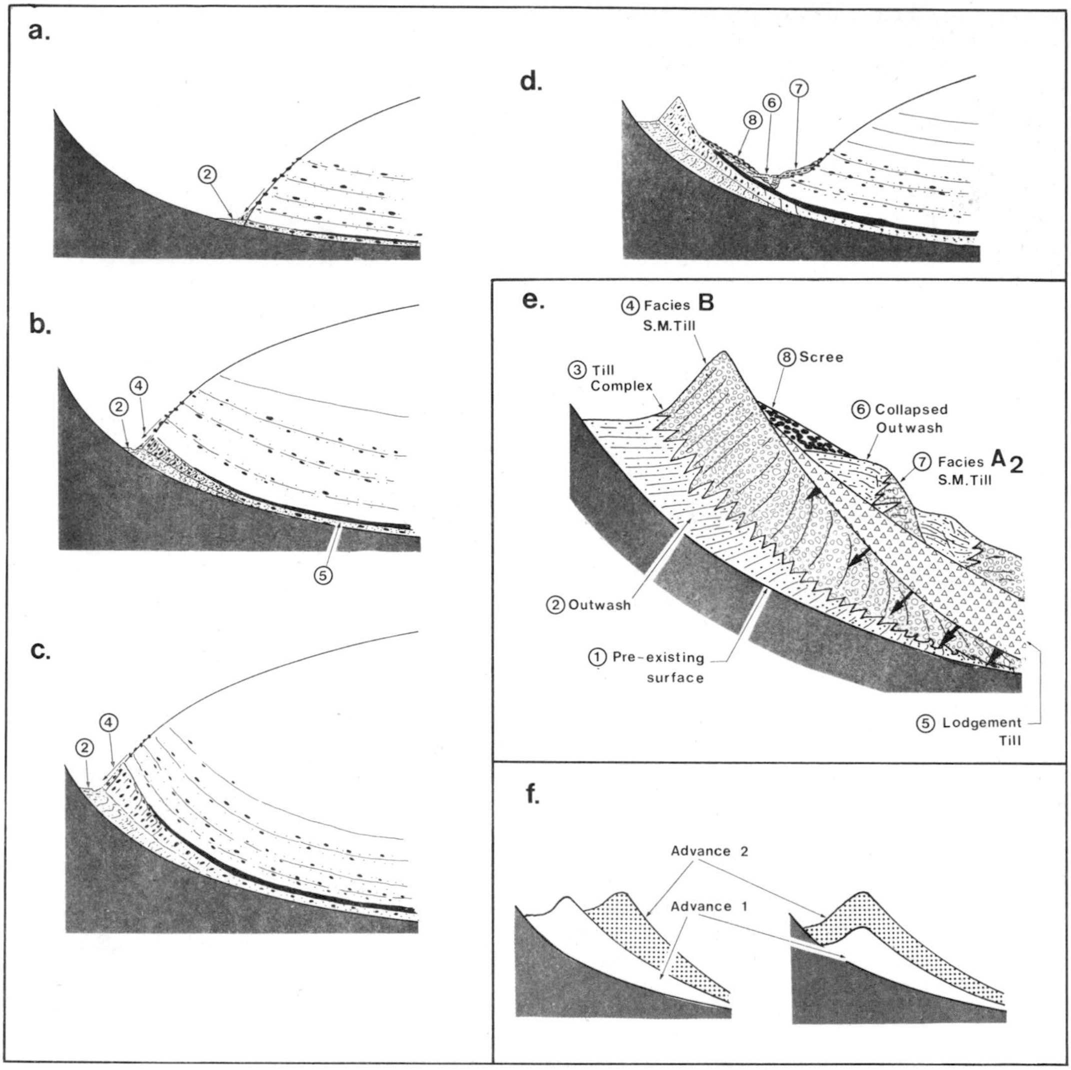

Figure 11a-f. The development of major lateral and latero-frontal dump moraines from supraglacially-derived debris a-d. The sequence of events during a single glacier advance-retreat stage. The key to numbers is contained in e.) e.) The distribution of sedimentary types typically associated with a major lateral dump moraine formed during a single phase of glacier advance and retreat. The arrows show a zone of glaci-tectonic deformation. f.) The relationship between two sediment suites associated with major dump moraines produced by two glacier advances. In the first, advance 1 is more extensive than advance 2; in the second, advance 1 is less extensive than advance 2.

washstreams flowing away from the front of the glacier.

Because debris almost exclusively moves away from the glacier margin, the supraglacial morainic till on dump moraines tend to have a well-defined course clast fabric, unlike the till of facies A, and crude internal bedding is ubiquitous (Fig. 12). Much of the material accumulates as scree from the steep glacier front, but mud flows and water washed sediments also accumulate in these dump moraines. The voids within these open-textured bouldery deposits are slowly filled by finer material which is washed in or flows in.

Several other types of sediment are frequently associated with the supraglacial tills of facies 2 which comprise dump moraines (Fig. 3).

During glacier recession, kamo terraces frequently form, reflecting lateral deposition of outwash between the dump-moraines and the glacier. All three sub-facies (a-c) of supraglacial morainic till facies A may also accumulate during recession on the proximal flanks of these moraines, though till complexes are particularly common.

Halts during the recession of a glacier down the proximal flank of a dump moraine or a valley wall frequently lead to the production of lateral till terraces (Fig. 3). Lateral terraces are commonly off-lapped on the inner walls of lateral moraines and are massively gullied when underlying ice cores melt. Winter readvances may also produce small dump moraine ridges at the limit of readvance on the proximal flanks of major dump moraines.

Scree cones and alluvial fans forming on glacial valley walls are blocked by a presence of the glacier and tend to accumulate at the glacier margin. If this is stationary or advancing, large lateral accumulations may form, which on glacier retreat form well-defined valley-side terraces.

REFERENCES

Boulton, G.S. (1976),'A genetic classification of tills and criteria for distinguishing tills of different origin' in Geografia, 12, pp. 65-80.

Boulton, G.S. (1979), Boulder shapes and grain-size distributions of debris as indicators of transport paths through a glacier and till genesis. Sedimentology, 25, pp. 773-799.

Boulton, G.S. and Paul, M.A. (1976), The influence of genetic processes on some geotechnical properties of glacial tills. Quarterly Journal of Engineering Geology, 9, pp. 159-194.

Brockie, W.J. (1973), 'Experimental frost-shattering' in Proceedings 7th Geographical Societies Conference Series No. 7, pp. 177-186.

Charlesworth, J.K. (1957), The Quaternary Era. Edward Arnold, London. 2 volumes, 1700 pp.

Church, M. and Gilbert, R. (1975), 'Proglacial fluvial and Lacustrine Environments' in Jopling A.V. and McDonald, B. S. (Eds) Glaciofluvial and Glaciolacustrine Sedimentation. Society of Economic Palaeontologists and Mineralogists Special Publication No. 23,320 pp. Tulsa, Oklahoma, USA, pp. 22-100.

Dahl, R. (1966), Blockfields, weathering pits and tor-like forms in the Narvik Mountains, Nordland, Norway. Geografiska Annaler, 48A, pp. 55-85.

Drake, L.D. (1971), 'Evidence for Ablation and Basal Till in East Central New Hampshire' in Goldthwait, R.P. (Ed) Till: A Symposium. Ohio State University Press, Columbus, Ohio, USA, pp. 237-250.

Dutt, G.N. (1961), 'The Bars Shigri Glacier, Kangra District, East Punjab India' in Journal of Glaciology, 3. pp. 1007-1015.

Eyles, N. (1978), Unpublished Ph.D. University of East Anglia.

Eyles, N. and Rogerson, R.J. (1977), 'Glacier movement, ice structures, and medial moraine form at a glacier confluence, Berendon Glacier, British Columbia, Canada, in Canadian Journal of Earth Sciences, 14, pp. 2807-2816.

Eyles, N. and Rogerson, R.J. (1978), 'Sedimentology of medial moraines on Berendon Glacier, British Columbia, Canada; implications for debris transport in a glacierized basin' in Geological Society of America Bulletin.

Fahnestock, R.K. (1973), 'Morphology and hydrology of a braided stream' in United States Geological Survey Professional Paper 422A. pp. 1-70.

Flint, R.F. (1971), Glacial and Quaternary Geology. John Wiley and Sons, Inc. New York, USA 892.

Lautridou, J.P. (1971), 'Recherches de gelifraction experimentale au centre de geomorphologie VI Schistes, gres et roches metamorphiques de Basse-Normandie' in Bulletin Centre de Geomorphologie, Caen, No. 10.

Martini, A. (1967), 'Preliminary experimental studies on frost weathering of certain rock types from the West Sudetes' in Biuletyn Peryglacjalny, 16. pp. 147-194.

Portman, J.P. (1960), 'Les inclusions rocheuses dans les glaciers' in Geographica Helvetiae 15, pp. 1-8.

Potts, A.S. (1970), 'Frost action in rocks: some experimental data' in Transactions, Institute of British Geographers, 49. pp. 109-124.

Saunderson, H.C. (1977), 'The sliding-bed facies in esker sands and gravels: a criterion for full-pipe (tunnel) flow?' in Sedimentology, 24. pp. 623-638.

Sharp, R.P. (1948), 'Constitution of valley glaciers' in Journal of Glaciology, 1. pp- 182-189.

Sharp, R.P. (1949), 'Studies of superglacial debris on valley glaciers' in American Journal of Science, 247, pp. 289-315.

Slatt, R.M. (1971), 'Texture of ice-cored deposits from ten Alaskan valley glaciers' in Journal of Sedimentary Petrology, 41, pp. 828-834.

Strømquist, L. (1973), Geomorfologiska Studier av Blockav och Blockfält i Norra Skandinavien. Uppsala Universitet Naturgeografiska Institutionen Report 22, 159 pp.

Young, R.A. (1953), 'Some notes on the formation of medial moraines' in Jökull, 3, pp. 32-33.

Systematic provenance investigations in areas of alpine glaciation: applications to glacial geology and mineral exploration

EDWARD B. EVENSON
Lehigh University, Bethlehem, Pa., USA

THOMAS A. PASQUINI
Exxon Oil Company, Kingsville, Tex., USA

ROBERT A. STEWART
University of Western Ontario, London, Ont., Canada

GEORGE STEPHENS
George Washington University, Washington, D.C., USA

1 INTRODUCTION

The purpose of this paper is twofold; first, to present a case study demonstrating the value of detailed provenance investigations in deglaciated mountainous regions and second, to present a rationale for the application of till provenance to mineral exploration in actively and previously glaciated alpine areas. Although the concepts are intimately interrelated, we will attempt to present them as separate entities for ease of presentation and understanding.

2 PROVENANCE INVESTIGATIONS IN THE ROCKY MOUNTAINS, IDAHO

2.1 Study area

The study area (Fig. 1) is located in the Pioneer Range of the Rocky Mountains of Central Idaho, U.S.A. Detailed investigations have been conducted along the headwaters of the Big Lost River over the past five years by students and staff of Lehigh University. To date, mapping has been completed in three areas known locally as: the Copper Basin (Wigley, 1976; Pasquini, 1976; Wigley, Pasquini and Evenson, 1978); Wildhorse Canyon (Stewart, 1977) and Summit Creek (Cotter, in prep.). See Figure 2 for locations.

2.2 Glacial geology

The age and distribution of glacial deposits and ice flow patterns in the study area are complex and have been reconstructed using detailed field mapping, pedologic and provenance investigations, and air photo interpretation. Glacial deposits of assumed Pre-Bull Lake (oldest), Bull Lake, Pinedale and Neoglacial (youngest) ages have been mapped and correlated using the criteria of Blackwelder (1915); Richmond (1948, 1965, and 1976); Mears (1974); Madole (1976), and numerous others. Because this paper is concerned primarily with the use of provenance as a technique for deciphering past glacial activity, no detailed review of the distribution, or criteria for relative age assignment, of glacial deposits will be attempted. In this paper, only those deposits of the Bull Lake (older) and Pinedale (younger) Glaciations will be considered when discussing the use of provenance as an interpretative tool. In addition, because the Pinedale and Bull Lake glaciers developed and spread from the same source areas, they differ only in morphologic character, volume and distance of ice stream movement. The provenance of Pinedale and Bull Lake glaciers emanating from the same canyons appears to be identical. In most areas, Pinedale deposits are completely enclosed by Bull Lake deposits indicating a maximum ice advance during Bull Lake time followed by Pinedale advances terminating at successively less extensive positions.

Moraines of the Bull Lake Glaciation are the oldest well-preserved glacial deposits in the study area. Generally, Bull Lake moraines extend 0.5 to 1.0 Km further down valley than those of Pinedale age. Bull Lake moraines are large, bulky, and well dissected by tributary streams. Moraine surfaces are irregular, but mature in appearance with filled or breached surface depressions (Fig. 3). Surface boulder frequencies are low to moderate, characterized by scattered, fractured, large boulders. Boulder surfaces are well pitted, removing all evidence of ice scour features.

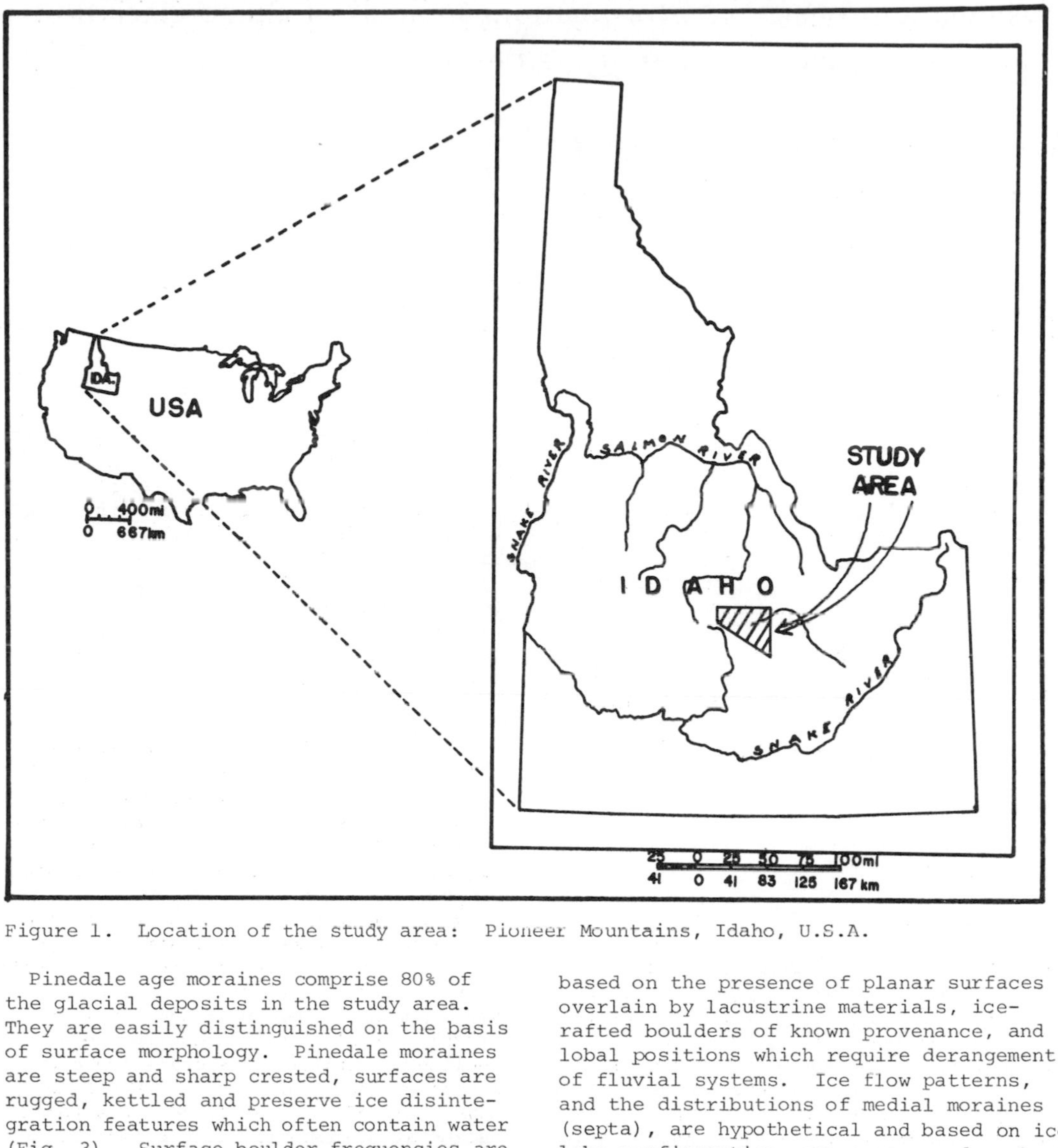

Figure 1. Location of the study area: Pioneer Mountains, Idaho, U.S.A.

Pinedale age moraines comprise 80% of the glacial deposits in the study area. They are easily distinguished on the basis of surface morphology. Pinedale moraines are steep and sharp crested, surfaces are rugged, kettled and preserve ice disintegration features which often contain water (Fig. 3). Surface boulder frequencies are very high, and boulders occasionally show ice scour features. Four sets of Pinedale moraines, and associated outwash terraces, can be mapped in the study area. The flow paths of Pinedale and Bull Lake glaciers were topographically controlled and differ only in extent and minor lobal complexities. Figure 4 is a generalized reconstruction of the study area at maximum Pinedale glaciation. Ice margin positions are reconstructed using the distribution of Pinedale morainal deposits, glaciofluvial deposits are based on the present distribution of elevated terrace remnants. The distribution of ice-dammed lakes is based on the presence of planar surfaces overlain by lacustrine materials, ice-rafted boulders of known provenance, and lobal positions which require derangement of fluvial systems. Ice flow patterns, and the distributions of medial moraines (septa), are hypothetical and based on ice lobe configuration, provenance and analogy with septa patterns in similar, actively glaciated, mountain areas.

2.3 Detailed provenance investigations

Although a generalized understanding of glacial flow patterns can generally be gained by traditional mapping techniques (lobal patterns, extent of glacial deposition and erosion, etc.) our provenance investigations have demonstrated that many aspects of past glacial flow, especially those that would go unnoticed without provenance investigations, can be indicated

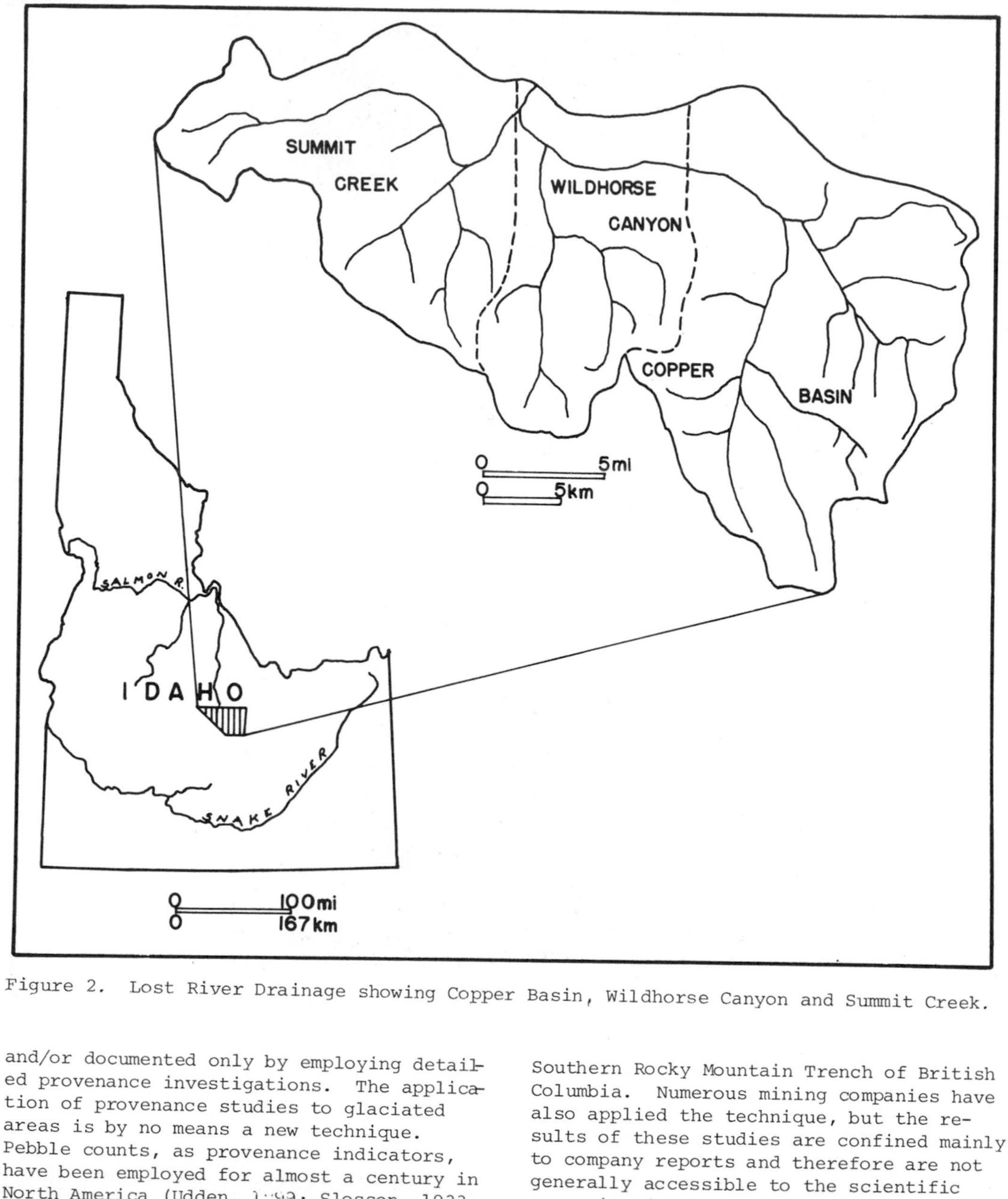

Figure 2. Lost River Drainage showing Copper Basin, Wildhorse Canyon and Summit Creek.

and/or documented only by employing detailed provenance investigations. The application of provenance studies to glaciated areas is by no means a new technique. Pebble counts, as provenance indicators, have been employed for almost a century in North America (Udden, 1899; Slosson, 1933; MacClintock, 1933; Dreimanis and Reavely, 1953; Anderson, 1955, 1957; Arneman and Wright, 1959) although the technique has most commonly been applied to decipher long distance transport in areas of continental glaciation. Pebble provenance investigations in areas of alpine glaciation are much less common although Clague (1975) successfully applied the technique to determine glacier flow patterns in the Southern Rocky Mountain Trench of British Columbia. Numerous mining companies have also applied the technique, but the results of these studies are confined mainly to company reports and therefore are not generally accessible to the scientific community (pers. comm. R. P. Goldthwait, 1977). Heavy mineral suites have also been utilized in conjunction with glacial investigations. Beginning with the classic work of Gravenor (1951) it was recognized that heavy mineral provenance provided a mechanism for deciphering glacial transport paths. In the Great Lakes region of the United States and Canada, numerous workers (Dreimanis and Reavely, 1953; Willman, Glass and Frye,

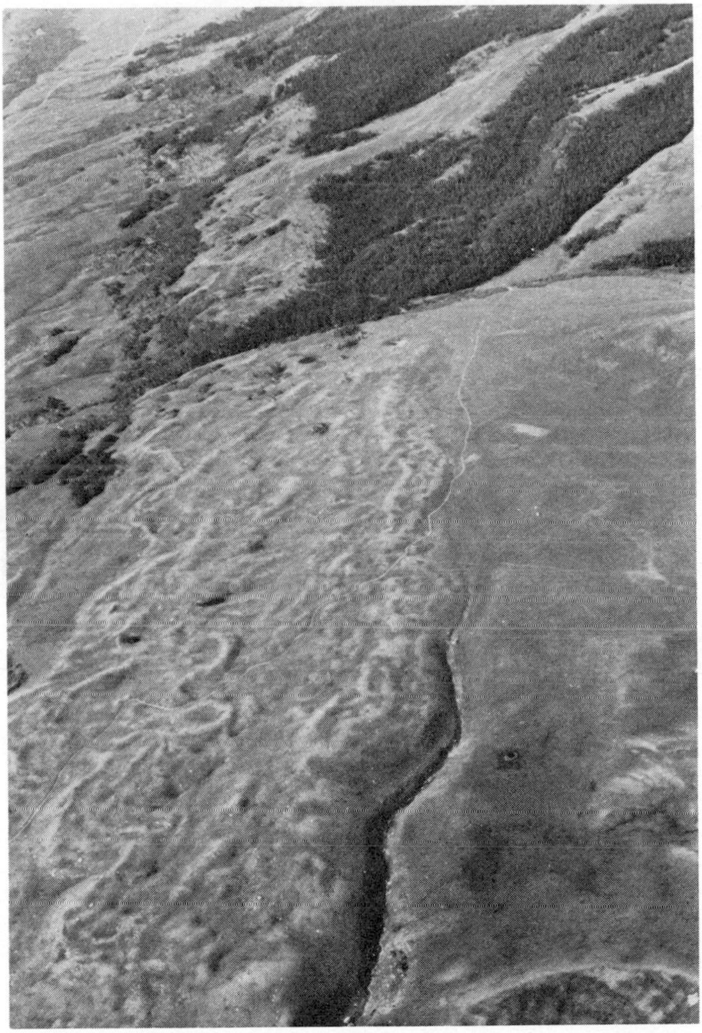

Figure 3. Photo of Copper Basin moraines. Pinedale age till (left) has a fresh, irregular surface. Bull Lake till (right) has a subdued topography.

1963; Frye, Willman, and Glass, 1964; Frye, Glass, Kempton and Willman, 1969) have successfully applied the technique to determine source areas of glacial deposits and to correlate deposits on the basis of the similarity of their heavy mineral suites.

Based on the success of the formentioned investigations, we decided to apply boulder, pebble and heavy mineral provenance investigations to the resolution of flow complexities in alpine areas of the Idaho Rocky Mountains. Much of the detailed work presented in this paper is excerpted from Pasquini (1976) who performed the first provenance investigations in the study area. Subsequently, several investigators (Stewart, 1977; Cotter, in prep.; Brugger, in prep,; and Repsher, in prep.) have used, or are using, provenance investigations to resolve ice distribution and flow patterns in the study area.

Pasquini (1976), working in the Copper Basin, performed a detailed provenance investigation utilizing boulders (>8 cm), pebbles (5-8 cm), and heavy mineral suites. At each of 161 sites (Fig. 5), 25 boulders within a 5 meter radius were selected and identified lithologically. In addition, at 84 selected suites, a pit 60 x 60 x 60 centimeters was excavated and 100 pebbles were collected and identified

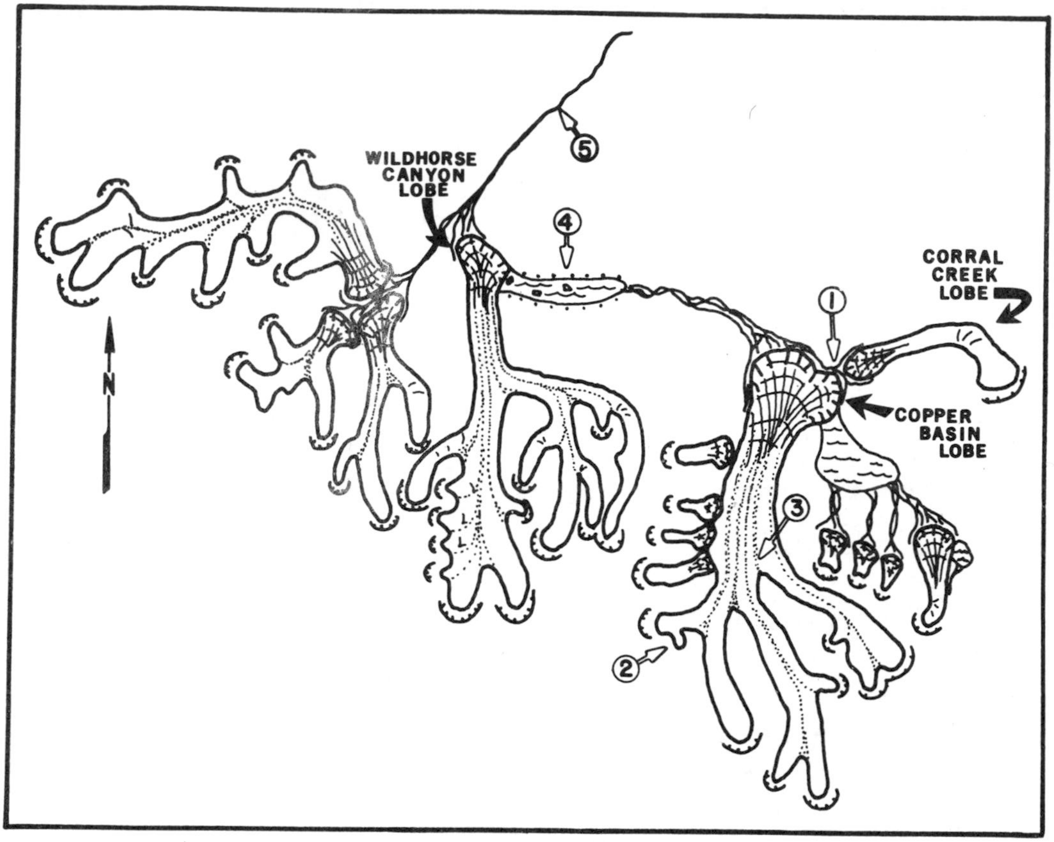

Figure 4. Reconstruction of ice margins, fluvial and lacustrine systems, at Pinedale glacial maximum. Numbers 1-5 indicate sites discussed in text.

and a sample of till matrix was collected (at 50 sites) from the base of the pit for heavy mineral analysis. The sample collected for heavy mineral identification was sieved to recover the 3 phi fraction. This fraction was separated using tetrabromoethane (S.G. 2.87) and 300 grains were identified using standard petrographic techniques. Visual inspection of the tabulated data revealed many trends and groupings. The data matrix was also subjected to varimax rotated R-mode factor analysis and Q-mode cluster analysis to further clarify various relationships and identify less obvious associations among the data. In addition to the detailed provenance study of Pasquini (1976), we have utilized boulder provenance investigations in several adjacent areas (Wildhorse Canyon and Summit Creek) to extend our understanding of past glacial activity in the region.

Provenance investigations attain their maximum usefulness when they are conducted in conjunction with field mapping. In such cases, provenance often indicates peculiarities of glacial flow (or meltwater discharge) that would otherwise go undetected. To date, we have detected and demonstrated five different "peculiarities" that serve to illustrate the value of provenance investigations. They are: 1) up-canyon glacier flow, 2) divide crossing, 3) readvance of individual ice lobes creating cross cutting relationships, 4) ice rafting into ice-dammed marginal lakes, and 5) jökulhlaup activity. Areas where provenance has proved valuable in deciphering unusual flow patterns are discussed individually in the succeeding sections of this paper.

2.4 Up canyon glacier flow

During the Pinedale and Bull Lake glaciations, ice filled the entire Copper Basin forming a composite piedmont glacier here

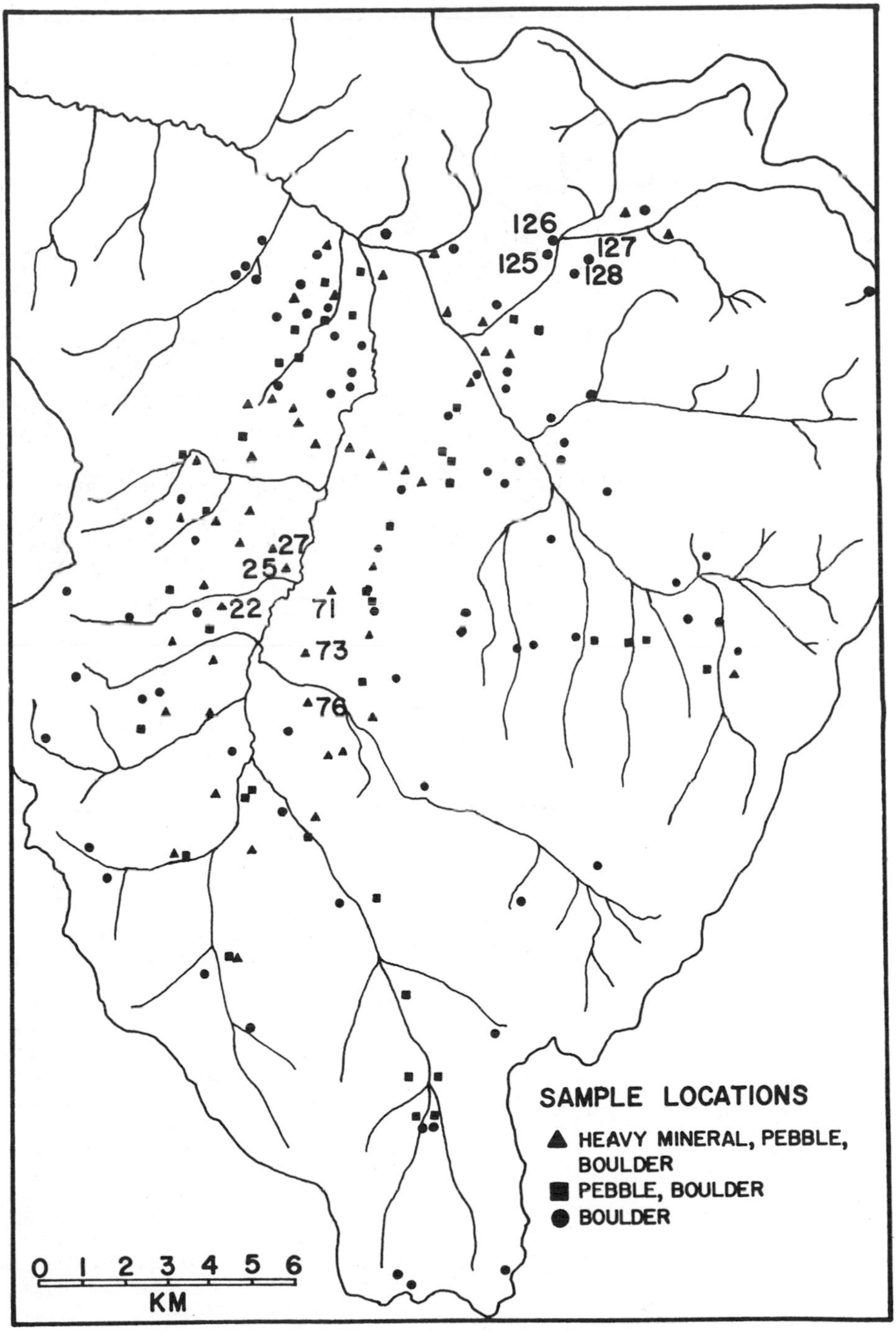

Figure 5. Location of sample sites for provenance investigation in the Copper Basin.

named the "Copper Basin Lobe" (Fig. 4). Ice flow was generally from south to north. At its maximum extent, this piedmont glacier traversed the entire basin floor and extended up valley into smaller tributary canyons (Corral Creek, Little Lake Creek, and Cabin Creek) which enter the basin from the north (Site 1, Fig. 4 and Fig. 6A). Cabin Creek and Little Lake Creek did not produce valley glaciers so the presence of till in the mouths of these canyons must be attributed to the Copper Basin Lobe. Provenance investigations conclusively support this interpretation. Corral Creek is of particular interest because, while ice of the Copper Basin Lobe flowed up canyon, a valley glacier from Corral Creek's favorably oriented cirques simultaneously flowed down valley and the two ice lobes coalesced or abutted. Our provenance investigation was able to differentiate deposits sourced from both the Copper Basin Lobe and the Corral Creek valley glacier and hence define the contact of two ice complexes. Because the till is continuous from the basin proper into Corral Creek, normal mapping techniques could not be utilized to define the intersection. However, provenance relationships provided a unique solution to the problem. Surface boulder counts within Corral Creek were particularly useful (Fig. 6A). Sample locations 126 and 127 have abundant boulders of Corral Creek provenance (granite) and no boulders sourced from the south. Sample locations 125 and 128 show the reverse relationship-- a southern provenance (quartz monzonite) and a complete lack of granites from Corral Creek. This provenance distribution demonstrates that ice from the northward flowing Copper Basin Lobe invaded Corral Creek and extended at least 2.5 Km up canyon into Corral Creek. Pebble lithologies and heavy mineral suites show the same relationship. Contour diagrams of heavy mineral percentages show very low percentages of diopside plus epidote (Corral Creek provenance suite) above the junction of the two lobes while, below the junction, diopside plus epidote percentages are high indicating a Copper Basin Lobe provenance.

2.5 Divide crossing

Divide crossing is not common in the study area, but one case can be documented in the Broad Canyon-Bear Canyon area of the Copper Basin (Site 2, Fig. 4). In this area the Broad Canyon ice stream breached a low col and spilled into Bear Canyon (Fig. 6B). This interpretation is supported by the presence of abundant granite boulders from the head of Broad Canyon in the Bear Canyon drainage. No granite outcrop exists in Bear Canyon. The ice flow, supported by geomorphic evidence (glacial modification of the col) and southwest trending striae was through a col in the divide separating the adjacent valleys. The ability of this sublobe to invade Bear Valley for a distance of almost one kilometer supports our interpretation that Bear Valley was ice free at the time of divide crossing.

2.6 Readvance and cross cutting of morainal deposits

During the Bull Lake and older Pinedale Glaciations, ice streams in the study area coalesced, and individual ice streams apparently flowed side-by-side as shown in Figure 4, and crowding precluded cross-cutting of one ice stream by those adjacent (tributary) to it. At these times, with large ice volumes, the piedmont glacier within any main valley probably advanced and retreated as one large lobe. By Late Pinedale time, the ice had thinned and retreated to form individual valley glaciers within the larger canyons. These smaller valley glaciers were more susceptible to climatic fluctuations and accumulation variations in their catchment area and no longer fluctuated sychronously. During Late Pinedale time, the Lake Creek Canyon valley glacier readvanced, crossed the basin flat and deposited till along the west wall of the basin (Site 3, Fig. 4). This readvance, and its maximum westward extent, can be documented by provenance investigations. Till deposited on the west side of the basin normally shows a high percentage of monazite (reaching 50% or higher) sourced from the Broad Canyon ice stream while till from Lake Creek Canyon is low in monazite (0-10%) and very high in diopside + epidote (21-50%). Cluster analysis of heavy mineral data groups samples 25, 73, 76 with deposits sourced from Lake Creek Canyon while samples 16, 22 and 27 group with samples sourced from the west side canyons (Fig. 6C). Pebble and boulder provenance, as well as moraine alignment also support this Late Pinedale cross-cutting relationship.

2.7 Ice rafting into ice dammed lakes

The extent of Bull Lake and Pinedale ice near the mouth of Wildhorse Canyon (Fig.

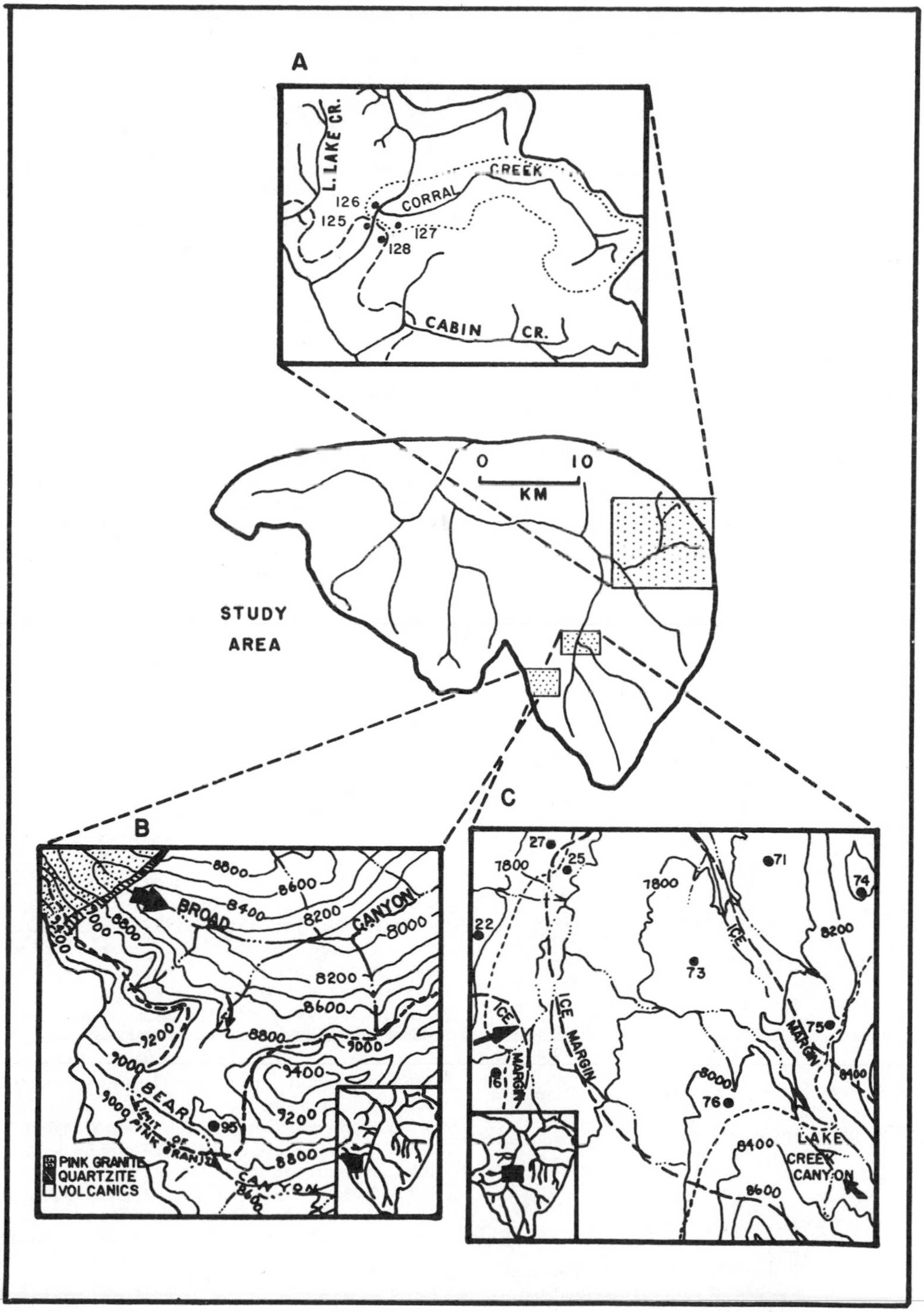

Figure 6. Examples of flow peculiarities detected by provenance investigations. (a) Up-canyon provenance flow of Copper Basin Lobe at the mouth of Corral Creek; (b) Divide crossing of Broad Canyon ice stream into Bear Canyon; (c) Limit of the Late Pinedale readvance from Lake Creek Canyon.

4) can be inferred from the distribution of moraines in the area. By reconstruction, it is clear that ice from Wildhorse Canyon repeatedly traversed and blocked drainage (along the East Fork River) from the Copper Basin. This blockage created an extensive ice-dammed lake along the East Fork River (Site 4, Fig. 4). Upon retreat of the ice from the area, the lake drained and through-drainage from the Copper Basin removed all evidence of lacustrine sedimentation and produced an extensive set of terraces. Ordinarily such erosion and sedimentation would obliterate all evidence of the extent and elevation of such an ice-dammed lake and its previous existence would go unnoticed. However, in this instance, boulder provenance investigations in the area of the ice-dammed lake have provided strong evidence attesting to the extent and depth of this lake. The headwaters of the Wildhorse Canyon drainage are underlain by Pre-Cambrian gneiss which provides abundant gneissic boulders to the till of this canyon. There is no Pre-Cambrian gneiss in the Copper Basin drainage. Boulder provenance studies in the area of the ice-dammed lake reveal gneissic erratics to an elevation of 7200 ft. (400 above the modern stream) and extending up valley along the East Fork system for a distance of 7 Km. Reconstruction of ice lobation patterns precludes the possibility that glacial ice from Wildhorse Canyon could have extended this far up valley to emplace the gneissic boulders as the lobe was free to flow, unimpeded, in a northward direction. Therefore, the only reasonable explanation for the boulders is that they were carried into the lake on icebergs from the calving edge of the Wildhorse glacier and were ice rafted to their present position where they were freed by melting. A concentration of boulders near the 7200 ft. level over a distance of 7 Km argues that the boulders were deposited by the melting of grounded icebergs along the margin of an ice-dammed lake. Our reconstruction of the location and extent of the lake shown at Site 4, Fig. 4, is based on this line of evidence.

2.8 Jökulhlaup activity

Evidence for Jökulhlaup activity in the study area is not unequivocal, but our demonstration of the existence of a large ice-dammed lake, coupled with the presence of large isolated boulders lying on terrace surfaces downstream from ice blockage of drainage, appears to argue for such a process.

Immediately downstream, and extending for a distance of 10 Km (Site 5, Fig. 4) from the ice-dammed lake discussed in the previous section (Site 4, Fig. 4) there are large boulders (up to 2 meters in diameter) lying on the surface of flat terrace remnants of Pinedale and Bull Lake age. The boulders are well beyond the limit of Pinedale and/or Bull Lake ice extent and appear to have been emplaced on top of terraces composed of pebble- to cobble-sized clasts. That they were transported to their present position by the same meltwater discharge that deposited the terrace material appears unlikely. Many lie in positions far removed from the valley walls precluding mass movement. In addition, many have a lithology that could not have been derived from the adjacent valley walls. The catastrophic drainage of the lake dammed by the Wildhorse lobe could easily have provided the discharge and velocity necessary to transport these boulders from the mouth of Wildhorse Canyon to their present position and if, as we assume, the lake discharged for the last time in Late Pinedale time, this would account for their location on terraces of Pinedale and greater age. Further work on the lithology of these boulders may allow us to relate them more directly to specific source areas and further enhance our understanding of the last deglaciation of the area.

In the preceeding sections, we have attempted to demonstrate the value of provenance as an interpretative tool in areas of alpine glaciation. Our experience indicates that provenance, when conducted in conjunction with standard mapping techniques, will initially indicate, and often document, phases of glacial activity that would go unnoticed if provenance techniques were not employed. The technique realizes its fullest potential when a detailed understanding of the small-scale complexities of glacial flow is required.

3 SUGGESTED APPLICATION OF PROVENANCE INVESTIGATIONS TO MINERAL EXPLORATION IN AREAS OF ALPINE GLACIATION

Ore minerals occurring as "float" in tills have been used as prospecting tools since the mid-eighteenth century, and excellent reviews of the history and application of the technique have been presented by Shilts (1976), Kujansuu (1976), Alley and Slatt (1976), and Dreimanis (1958), to name just a few. However, to date, almost all studies have been conducted in areas of

continental glaciation. We feel that provenance investigations also have application to mineral exploration in both actively glaciated and deglaciated alpine areas. In the following sections we will attempt to provide a rationale for the application of provenance investigation to mineral exploration in alpine areas. Initially, we will present results of our investigations in Idaho which, although not directly oriented to exploration, demonstrate that alpine tills provide an accurate, interpretable record of past ice flow paths of individual ice streams within a piedmont glacier complex, and subsequently, we will present a model for the application of provenance to mineral exploration in areas of active valley glaciation.

3.1 Exploration in deglaciated alpine areas

Valley glaciers are somewhat like rivers in that most have tributaries which contribute to the trunk system. This is readily apparent in most large compound valley glaciers which are commonly fed by tens of tributary ice streams. However, there is an important difference between river and glacier tributaries in that the individual ice streams do not intermix. Due to laminar, rather than turbulent flow, each ice stream within a compound glacier maintains its integrity and individuality as the compound glacier flows down its valley (Sharp, 1960; Flint, 1971; Levinson, 1974). An ice stream may become greatly thinned by compression and/or extension, but it does not mix with adjoining ice streams. The englacial debris transported by valley glaciers from high catchment areas is derived mainly from mass wasting of the valley walls and to a lesser extent from sub-glacial erosion. This material is transported in an unmixed manner by glacial flow from the high catchment basins to the ablation zone where it is deposited as lateral, end and medial moraines. Our investigations in the Copper Basin clearly document that till provenance can be utilized to determine the source areas of tills deposited up to 20 Km beyond the catchment basins in which the individual glaciers originated. For example, the distribution of diopside, epidote, monazite, topaz and zircon in till (Figs. 7a,b,c) clearly relates the heavy mineral suites to sources in individual tributary canyons. Figure 7a shows that diopside and epidote, occurring in abundance in a large lateral moraine on the east side of the basin, are derived from the headwaters of Lake and Anderson Canyons. Similarly, high percentages of monazite (Fig. 7b) occurring on the west side of the basin can be related to a large granitic pluton in the catchment area of the Broad Canyon ice stream, while the distribution of topaz plus zircon (Fig. 7c) is clearly related to a pluton in Starhope Canyon. The same relationships are demonstrated by boulder and pebble lithology (Pasquini, 1976).

Because all of the medial moraines in the study area have been destroyed or buried by subsequently fluvial activity, it is impossible to use their provenance suite to further refine the distribution of source canyons in the tributary valleys, however, in situations where medial moraines are preserved, they can provide a valuable source of information (see following section) and should be sampled. The fact that the lithology of alpine tills is clearly related to the bedrock underlying and eroded by the tributary glaciers is not surprising. What is surprising is the accuracy with which source areas can be identified and the infrequency with which alpine provenance has been systematically applied as an exploration technique. We feel that if provenance investigations were routinely conducted as a part of glacial geologic investigations that unanticipated mineralization would be detected in many cases and that our understanding of glacier flow patterns and ice activity would be greatly enhanced. In areas where the objective is not glacial geologic mapping, but rather assessment of mineralization potential, till provenance should be used in conjunction with standard stream sediment investigations (geochemical and heavy mineral) to further refine the accuracy of the assessment process.

3.2 Exploration in areas of active alpine glaciation (ICE S.T.R.E.A.M.)

The application of provenance investigations to mineral exploration may reach its fullest potential when applied to areas of active valley glaciation. In the following section we will present our rationale for application of the technique to actively glaciated areas and propose a model for data collection and analysis. To the best of our knowledge, there is no published report of a systematic investigation which has attempted to use the debris entrained in active valley glaciers to locate mineralized areas in the catchment

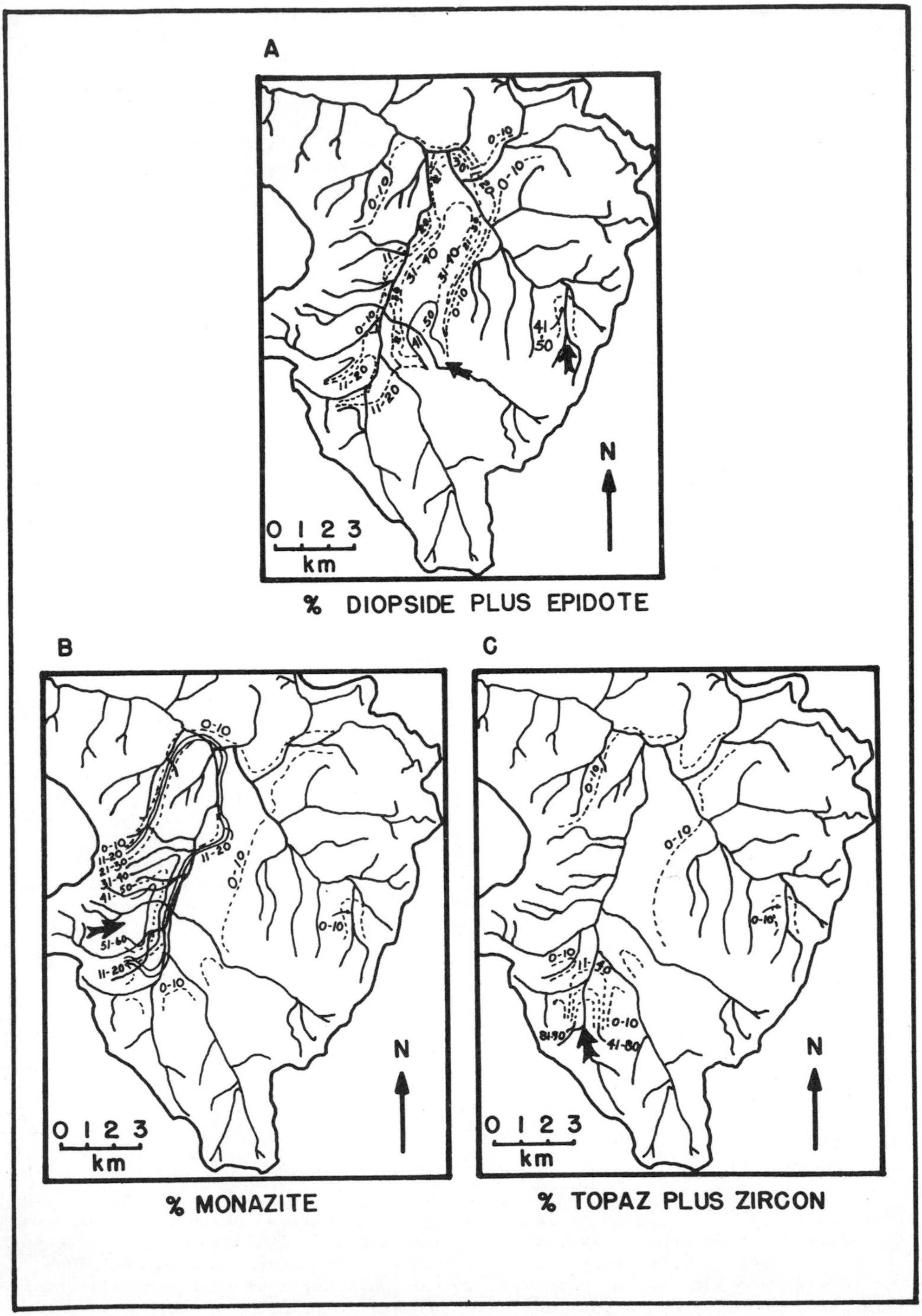

Figure 7. Heavy mineral distribution in tills of the Copper Basin. All contours are percent of the total heavy mineral suite. (a) Percent diopside plus epidote; (b) Percent monazite, and (c) Percent topaz plus zircon. For each diagram, source area and ice flow direction are indicated by heavy arrow.

region. We refer to the technique as *"ICE S.T.R.E.A.M."* (*ICE:* a *S*ystematic *T*echnique for *R*apid *E*xploration and *A*ssessment of *M*ineralization) and present it, for the first time, in this paper. It should be emphasized that we have not, as yet, verified the applicability of "ICE S.T.R.E.A.M." with a field application. We actively elicit suggestions and/or published references to similar applications.

In briefest terms, the proposed technique consists of sampling the englacial debris bands and ice streams in compound valley glaciers in an attempt to locate specific mineral concentrations in the high, rugged and often inaccessable source areas (catchment basins). The technique utilizes a combination of basic glaciological concepts and standard geochemical and geologic exploration techniques.

The debris lateral to any small tributary glacier is supplied by mass wasting of the valley walls and erosion of the substrate by glacial activity and therefore the mineralogy of this debris is directly related to that of the catchment basin. As tributaries coalesce to form larger ice streams, the material is moved to a medial position as shown in Figure 8. Once an ice stream and its associated debris load is shifted to a medial position there is little chance of it obtaining additional material from the valley walls--it is isolated and the ice stream and its debris load flow side-by-side acting like a series of conveyor belts transporting ice and debris in an orderly unmixed manner from the high catchment area toward the ablation zone. This flow is recorded by medial moraines (Fig. 9) which give the glacier its striped appearance. It is not uncommon for the contribution of 10-20 tributary glaciers from very large catchment systems (100's of km^2) to flow through narrow valley constrictions (1-3 Km) and yet maintain unmixed ice stream flow. For example, the major compound glaciers of the Western Chugach Range of Alaska drain the majority of the ice from a land area of nearly 10,000 km^2 (of which 4700 km^2 is ice covered; Field, 1975) through valleys with a total width of only 50-60 kilometers. Sampling the debris across these outlet glaciers yields an exploration efficiency of 200 km^2 evaluated per kilometer of traverse. Therefore, it is possible to evaluate the mineral potential of very large inaccessible areas very rapidly and efficiently and because of the individuality of ice stream flow, we can also accurately locate within the catchment basin, the source of ore minerals entrained within the ice and/or medial moraines. Because melting, folding and compressive flow-shearing, all complicate the geometry of the ice streams we suggest sampling a few kilometers up glacier of this disturbed zone in the zone of streamlined flow (i.e.; in the area shown in the foreground of Figure 9).

3.3 Field sampling

Using our technique, a field party would make a sampling traverse across the compound ice stream to collect boulder lithology data and to acquire samples for further analysis. At each debris band (medial moraine) or "dirty ice zone", a one kilogram sample of the finest grained debris available is collected. Field sieving (2 mm screen) is employed to ensure recovery of an adequate sample of the fine fraction. In addition, at each locality, the coarse debris is inspected for mineralized boulders and cobbles, and 100 pebbles are collected for later laboratory analysis. The characteristics of each lateral and medial moraine is described and accurately plotted on aerial photographs. In wider debris bands, samples are collected at regular intervals to determine variations within each debris band (as medial moraines are made of the combined lateral debris of two adjacent tributary ice streams). Within each ice stream a 1000 cm^3 sample of ice containing fine particulate sediments is collected and bagged. Detailed sampling on a 2-3 kilometer wide compound glacier such as the one shown in Figure 9 could yield 200-300 samples and require up to one week of sampling time.

3.4 Laboratory analysis

Samples collected from medial moraines and ice streams are analyzed, using standard laboratory techniques, as shown on the flow chart in Figure 10. The debris samples are split into five fractions (A, B-1 to B-4) and the ice samples into three fractions (Filtrate I, II and leached sediment, see Fig. 10). Fractionation of the samples into various grain sizes combined with analyses designed to detect different indications of mineralization will increase the likelihood that mineralization, if present, will be detected. This sampling and analysis procedure is designed to investigate all particle sizes from boulders to clay as well as dissolved ions within the ice itself.

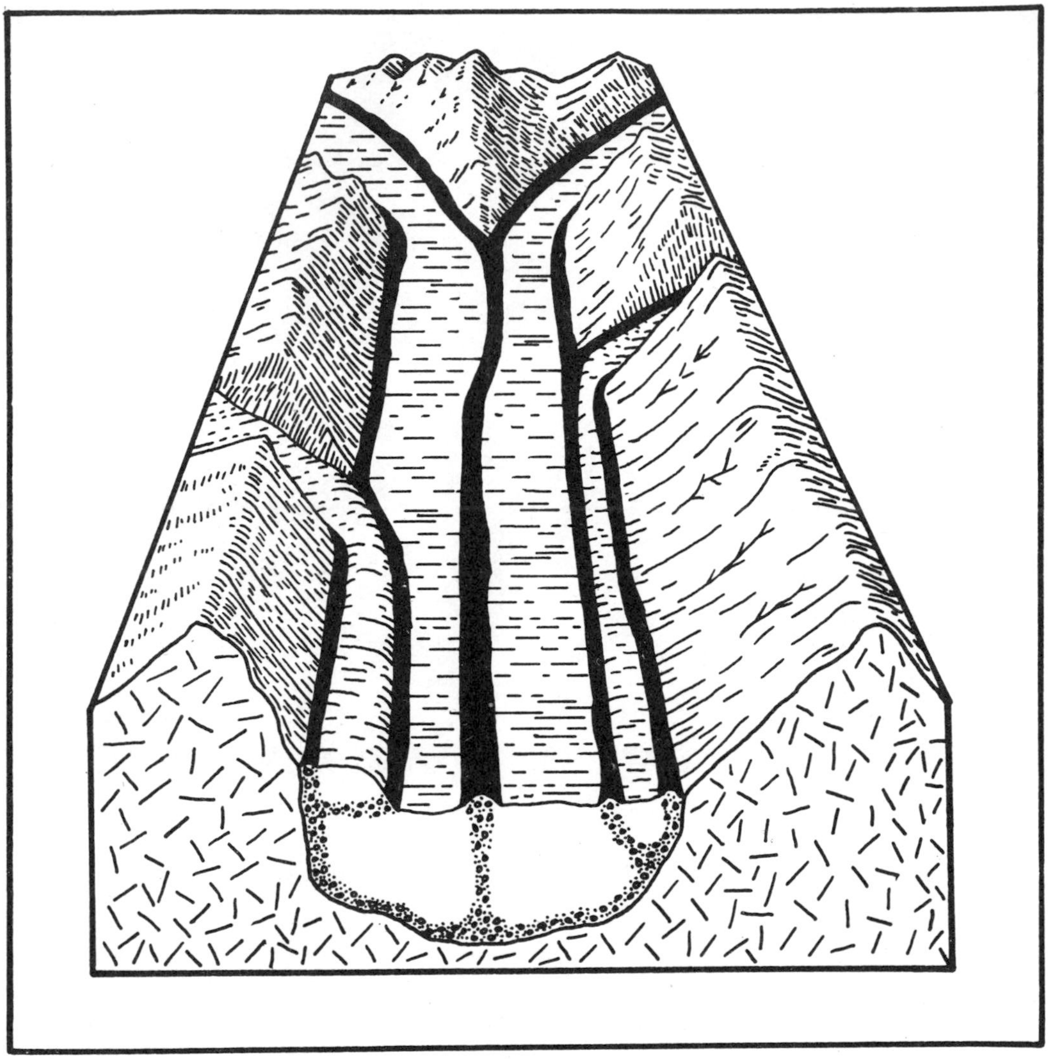

Figure 8. Diagramatic cross-section of a compound valley glacier showing distribution and entrainment of medial and lateral moraines. Modified from Sharpe, 1960, Fig. 2).

3.4.1 Analysis of coarse materials

Petrographic analysis of the coarser than 2 mm fraction (Fraction A, Fig. 10) and the 100 pebbles (4-8 mm) collected at each site takes three forms. Samples are split, and routine hand specimen, and binocular microscope identification and description of rock types and ore mineralogy are performed. For samples of more than routine interest (i.e.; extensively altered or mineralized pebbles), standard thin section and polished section samples are prepared. Thin section examination will yield a detailed evaluation of rock type and the degree and form of alterations present. Polished section examination will yield information on the nature and abundance of ore minerals present, the nature of mineral intergrowths, paragenetic relationships, etc. The presence of mineralization in the coarse fraction will be a direct indicator of mineralization in the catchment basin where the ice stream of medial moraine originates. The relative abundance of mineralized clasts provides an indication of the degree of mineralization. These examinations, combined with

Figure 9. Distribution of medial and lateral moraines on Yentna Glacier, Alaska. Note the parallel unmixed flow of individual ice streams. Tonal differences between individual medial moraines reflect the lithologies derived from individual catchment areas. (Photo by Austin Post, United States Geological Survey).

the geochemical results (see Section 3.4.2) should provide information on the genetic types of mineral deposits. In addition, sulfide and gangue assemblage data coupled with alteration type may provide indications of depth, and temperature range of formation, of the mineral deposits. The sand-sized portion (Fraction B-1, Fig. 10) is analyzed by standard heavy mineral techniques (Folk, 1968). The application of heavy mineral investigations to mineral exploration is reviewed by Levinson (1974), Huff (1971), Shilts (1976), and many others. Ores of many metals have a high specific gravity and many other heavy minerals are commonly associated with ore bodies. Levinson (1974) reports that all metals of interest will occur in one ore mineral phase (or another) in the heavy mineral fraction and for this reason they are invaluable as an ore exploration tool both as "direct" and "indirect" indicators of mineralization. The finer than 4Ø fraction (Fraction B-2, Fig. 10) is used for X-ray analysis. X-ray investigations are used to identify and quantify fine-grained material (mainly clays). X-ray analysis can be used to determine if the debris bands are undergoing weathering (and hence possible removal or transfer of metals) along their length. In areas of hydrothermal alteration, X-ray analysis provides information on clay mineral suites developed in alteration halos.

3.4.2 Atomic absorption analysis

Atomic Absorption Analysis is utilized to detect the presence of metallic and non-metallic ions of interest. The methods of sample digestion (or leaching) used in exploration geochemistry are extremely varied. To realize the full potential of exploration geochemistry, the type of extraction techniques employed must be de-

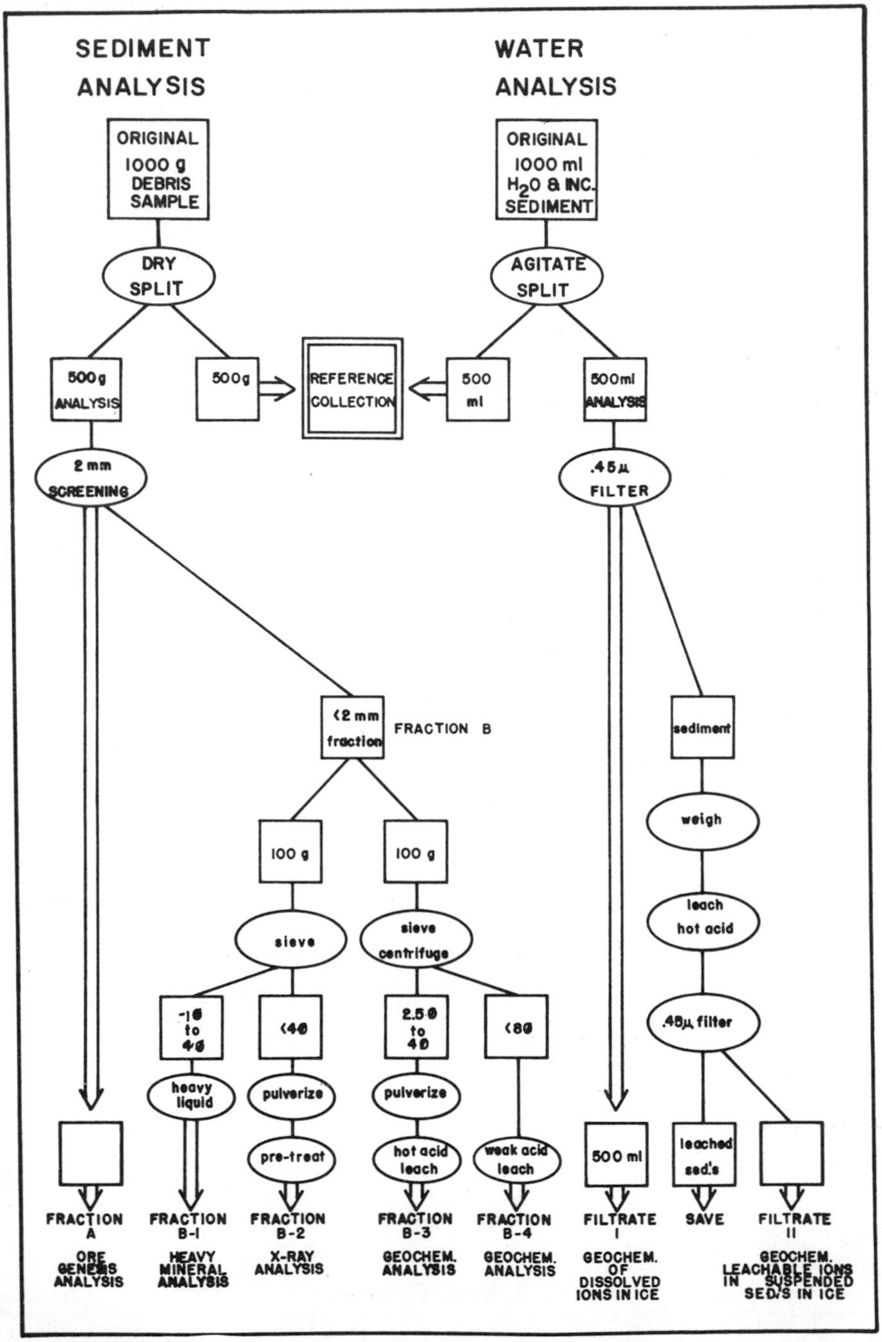

Figure 10. Laboratory flow chart.

signed to provide specific types of geochemical information. In a sediment, an element may occur in a number of forms, i.e.; as a rock-forming silicate, as a sulfide ore, or as loosely adsorbed onto clays or other weathered particles (Levinson, 1974). By employing different types of extraction techniques and treatments of various size fractions, the presence and amount of a given ion in each of these forms can be determined. The loosely adsorbed metallic ions on clay particles can be removed and determined by means of a dilute, cold acid leach (Fraction B-4, Fig. 10). The sulfide ions can be placed in solution and determined by means of a hot acid leach (Fraction B-3, Fig. 10). Weak acid leaches are designed to detect cations released by the weathering and oxidation of primary sulfides adsorbed by the cation-scavenging clays (Shilts, 1976; Levinson, 1974; Bradshaw et al., 1972) while strong hot acid leaches are designed to detect easily leached metals confined to the primary sulfide grains and perhaps some Fe- and/or Mn-coatings (Alley and Slatt, 1976). By establishing a lower size limit of 0.063 for analysis (Fraction B-3, Fig. 10), metals bound in clays, organic matter and secondary oxides are eliminated (Alley and Slatt, 1976; Canney and Wing, 1966; Govett, 1972; Shilts, 1976). Ice samples containing fine particulate matter are analyzed as shown in Figure 10 to detect easily leached sulfides in the suspended sediment within the ice and to detect dissolved ions present within the ice.

Although any element within the analytical capabilities of the Atomic Absorption technique can be determined, those that are actually determined will depend on the geologic setting of the area and the principal potential ores of the area. In a recent proposal, we suggested that the concentrations of 17 elements (copper, chromium, lead, zinc, cobalt, nickel, gold, arsenic, platinum, silver, bismuth, cadmium, mercury, molybdenum, tungsten, uranium, and tin) be determined for each sample. Arsenic is included as a pathfinder for gold and silver which often occur intimately associated with arsenide and sulf-arsenide minerals, but in such low concentrations as to make their direct detection difficult. Certainly other analytical techniques (X-ray fluorescence, optical emission spectrography, etc.) may be applicable to such geochemical investigations and each must be selected with reference to the analytical capabilities of the investigator and the geologic setting of the area.

4 SUMMARY

Recent investigations of the deglaciation history of parts of the Pioneer Mountains, Idaho, have demonstrated that provenance analysis of glacial deposits is an extremely effective tool for deciphering past glacial flow patterns. The research presented in this paper relies solely on previously-established techniques such as boulder and pebble provenance, heavy mineral analysis, and morphologic-stratigraphic relationships. It is emphasized, however, that these techniques are readily applicable to many presently glaciated and deglaciated alpine areas; and may greatly enhance our understanding of glacier flow and deposition for a particular region.

In addition to the commonly derived results of provenance analysis (e.g.; source area and ice flow paths) the examples discussed in this paper elucidate the value of provenance investigations for indicating and documenting more subtle phenomena (e.g.; up-canyon glacier flow, divide crossing, ice rafting, and Jökulhlaup activity) which commonly can be discerned, in alpine areas, only by systematic provenance investigations.

With an understanding of flow dynamics and deposition in actively glaciated areas, and our demonstration of provenance relationships in a formerly glaciated area (Pioneer Mountains, Idaho), we suggest that provenance analysis may be applied as a mineral exploration technique in areas of both former- and active-glaciation. Sampling is conducted across an active- or former-glacier terminus so as to provide representative material from the entire catchment area. An understanding of glacier flow permits the tracing of any resulting indicator of mineralization to the approximate source area. The analytical tools of such a study (i.e.; ICE S.T.R.E.A.M.) are based on principals established by many investigations, before and including this study. Analytical techniques and sampling patterns may be modified to suit the style of glaciation and the type of mineralization anticipated.

5 ACKNOWLEDGMENTS

This article is dedicated to Victor A. Johnson, his family, and the citizens of Mackay, Idaho, without whose assistance this investigation could never have been undertaken.

Susan Gawarecki, Alice Repsher, and Jack Ridge made the manuscript more readable and Lehigh University, through an initial

grant and summer teaching assistantships, provided financial support for the authors and many of the students who contributed so much to our understanding of the area.

Finally, we would like to thank the numerous scientists who, through informal discussion contributed to the concepts expressed herein. We owe a special debt of gratitude to A. Dreimanis, W. Shilts, R. Dilabio, T. Vorren, J. Parks, W. C. Wigley, J. Cotter and K. Brugger. Bobb Carson provided invaluable support with the analytical aspects. To those who provided input, and were not acknowledged, we offer our thanks and request their understanding.

6 REFERENCES

Alley, D.W. & R.M. Slatt 1976, Drift prospecting and glacial geology in the Sheffield Lake-Indian Pond area, north-central Newfoundland. In R.F. Legget (ed.), Glacial Till, p. 249-266. Roy. Soc. Can., Spec. Pub. 12, Ottawa.

Anderson, R. C. 1955, Pebble lithology of the Marseilles Till Sheet in northeastern Illinois, Jour. Geology 63: 228-243.

Anderson, R. C. 1957, Pebble and sand lithology of the major Wisconsin glacial lobes of the Central Lowland, Geol. Soc. Amer. Bull. 68: 1415-1450.

Arneman, H.F. & H.E. Wright 1959, Petrography of some Minnesota tills, Jour. Sed. Petr. 29: 540-554.

Blackwelder, E. 1915, Post-Cretaceous history of the mountains in central western Wyoming, Jour. Geology 23: 307-340.

Bradshaw, P.M.D., D.R. Clews & J.L. Walker 1972, Exploration Geochemistry. A series of seven articles reprinted from Mining in Canada and Canadian Mining Journal. Barringer Research Ltd., 304 Carlingview Dr., Rexdale, Ontario.

Brugger, K. in prep., Provenance investigation of glacial deposits in Wildhorse Canyon, Custer Co., Idaho, Unpublished M.S. Thesis, Lehigh University.

Canney, F.C. & L.A. Wing 1966, Cobalt: useful but neglected in geochemical prospecting. Econ. Geol. 59: 1361-1367.

Clague, J.J. 1975, Glacier flow patterns and the origin of late Wisconsinan till in the southern Rocky Mountain Trench, British Columbia, Geol. Soc. Amer. Bull. 86: 721-731.

Cotter, J.F. in prep., Glacial geology of the Summit Creek drainage, Custer Co., Idaho, Unpublished M.S. Thesis, Lehigh University.

Dreimanis, A. 1958, Tracing ore boulders as a prospecting method in Canada, Can. Inst. Mining and Metallurgical Bull., Feb., 73-80.

Dreimanis, A. & G.H. Reavely 1953, Differentiation of the lower and upper tills along the north shore of Lake Erie, Jour. Sed. Petr. 23: 238-259.

Flint, R.F. 1971, Glacial and Quaternary Geology. New York, John Wiley and Sons, Inc.

Folk, R.L. 1968, Petrology of sedimentary rocks. Austin, Hemphill's.

Frye, J.C., H.B. Willman & H.D. Glass 1964, Cretaceous deposits and the Illinoian Glacial Boundary in western Illinois, Illinois State Geol. Survey Circular 364, 28 p.

Frye, J.C., H.D. Glass, J.P. Kempton & W.B. Willman 1969, Glacial tills of northwestern Illinois, Illinois State Geol. Survey Circular 437, 45 p.

Govett, G.J.S. 1972, Interpretation of a rock geochemical exploration survey in Cyprus-statistical and graphical techniques. J. Geochem. Explor. 1: 77-102.

Gravenor, C.P. 1951, Bedrock source of tills in southwestern Ontario, Amer. Jour. Sci. 249: 66-71.

Huff, L.C. 1971, A comparison of alluvial exploration techniques for porphyry copper deposits. Geochemical Exploration. CIM Spec. Vol. 11, 190-194.

Kujansuu, R. 1976, Glaciogeological surveys for ore-prospecting purposes in northern Finland. In R.F. Legget (ed.), Glacial Till, p. 225-239. Roy. Soc. Can., Spec. Pub. 12, Ottawa.

Levinson, A.A. 1974, Introduction of Exploration Geochemistry. Calgary, Applied Publishing.

MacClintock, P. 1933, Correlation of the Pre-Illinoian drifts of Illinois, Jour. Geology 41: 710-722.

Madole, R.F. 1976, Glacial geology of the Front Range, Colorado. In W.C. Mahaney (ed.), Quaternary Stratigraphy of North America, p. 297-318, Dowden, Hutchinson & Ross Inc., Stroudsburg, Pennsylvania.

Mears, B., Jr. 1974, The evolution of the Rocky Mountain glacial model. In D. Coates (ed.), Glacial Geomorphology, p. 11-40, Publications in Geomorphology, Binghamton, New York.

Pasquini, T.A. 1976, Provenance investigation of the glacial geology of the Copper Basin, Custer Co., Idaho, Unpublished M.S. Thesis, Lehigh University, 136 p.

Repsher, A. in prep., Provenance investigation of glacial deposits in the Summit Creek drainage, Custer Co., Idaho, Unpublished M.S. Thesis, Lehigh University.

Richmond, G.M. 1948, Modification of Blackwelder's sequence of Pleistocene glaciation in the Wind River Mountains,

Wyoming, Geol. Soc. Amer. Bull. Abstracts, 59: 1400-1401.

Richmond, G.M. 1965, Glaciation of the Rocky Mountains. In H.E. Wright, Jr. & D.G. Frey (eds.), The Quaternary of the United States, p. 217-230. Princeton University Press, Princeton, New Jersey.

Richmond, G.M. 1976, Pleistocene stratigraphy and chronology in the mountains of western Wyoming. In W.C. Mahaney (ed.), Quaternary Stratigraphy of North America, p. 353-379, Dowden, Hutchinson & Ross Inc., Stroudsburg, Pennsylvania.

Sharp, R.P. 1960, Glaciers. Eugene, University of Oregon Books.

Shilts, W.W. 1976, Glacial till and mineral exploration. In R.F. Legget (ed.), Glacial Till, p. 205-224. Roy. Soc. Can., Spec. Pub. 12, Ottawa.

Slossen, C.B. 1933, The Jasper conglomeratc, an index of drift dispersion, Jour. Geology 41: 546-552.

Stewart, R.A. 1977, The glacial geology of Wildhorse Canyon, Custer Co., Idaho, Unpublished M.S. Thesis, Lehigh University, 101 p.

Udden, J.A. 1899, Some Cretaceous drift pebbles in northern Iowa, American Geologist 24: 389-390.

Wigley, W.C. 1976, Glacial geology of the Copper Basin, Custer Co., Idaho - a morphologic and pedogenic approach, Unpublished M.S. Thesis, Lehigh University, 126 p.

Wigley, W.C., T.A. Pasquini & E.B. Evenson 1978, Glacial history of the Copper Basin, Idaho: A morphologic, pedologic and provenance approach. In W.C. Mahaney (ed.), Quaternary Soils, p. 265-307, Geo. Abstracts Ltd., Norwich, England.

Willman, H.B., H.D. Glass & J.C. Frye 1963, Mineralogy of glacial tills and their weathering profiles in Illinois, Illinois State Geol. Survey Circular 347, 55 p.

Glacial sediments in the Venezuelan Andes

CARLOS SCHUBERT
Centro de Ecología, I.V.I.C., Caracas, Venezuela

1 INTRODUCTION

The Venezuelan Andes are an elongated mountain range in western Venezuela, located between 7° 30' and 10° 10' north latitude, 69° 20' and 72° 30' west longitude. It extends for over 400 km in a northeasterly direction, between the Venezuela-Colombia border and central-western Venezuela. Its highest elevation is just over 5000 m at Pico Bolívar, east of the city of Mérida (Fig. 1).

During the Late Pleistocene, the Venezuelan Andes were subjected to the Mérida Glaciation (Schubert, 1974) in areas above 3500 m elevation and to periglacial activity probably down to about 2000 m. The Late Pleistocene glaciated area covered approximately 600 km^2 and its periglacial zone covered approximately 2200 km^2. At present, glaciers exist only on the highest peaks of the Sierra Nevada de Mérida (Schubert, 1972) and occupy approximately 20 km^2 above 4700 m elevation. The present periglacial zone reaches down to about 3000 m and covers an area of approximately 1200 km^2. The glacial and periglacial morphology and geology have been described by Royo y Gómez (1959), Cárdenas (1962), Schubert (1970, 1976, 1977), Schubert and Valastro (1973), and Giegengack and Grauch (1975, 1976).

Glacial sediments have rarely been studied in detail in the Andes (see, for example, Dobrovolny, 1962; Flint and Fidalgo, 1964 and 1969; Marangunic and Thiele, 1971). These reports deal with certain aspects of the sedimentology of tills with the objective of determining provenance or age of the deposits. Recently, Spalletti and Gutiérrez (1976) published detailed grain-size analyses of till and fluvioglacial sediments from the Patagonian Cordillera of Argentina. Similar detailed sedimentological analyses of glacial deposits in tropical alpine environments of the Andes are inexistent, as far as I have been able to determine. It is the purpose of this report to partly fill this gap, so that comparisons with glacial deposits in alpine environments elsewhere can be made.

2 GLACIAL SEDIMENTS

2.1 Regional setting

Glacial sediments in the Venezuelan Andes consist almost exclusively of morainic till. The till is found in two main levels of moraines: a major morainic level beween 3000 and 3500 m elevation, and a possible older morainic level between about 2600 and 2800 m (Fig. 1). Schubert (1974) interpreted these two morainic levels as two possible stades within the Mérida Glaciation: the lower level is an Early Stade and the upper level is a Late Stade. No glacial sediments of clearly pre-Mérida Glaciation age have so far been found. This has been attributed to the effect of Quaternary uplift of the cordillera, during which sediments deposited by previous glaciations were obliterated by the following glaciations. This hypothesis, however, needs to be tested in view of the data of van der Hammen and others (1973), which indicate that the Cordillera Oriental of Colombia (whose northeastern extension are the Venezuelan Andes) had reached their present elevation by the Early Quaternary.

In view of the poor preservation of the Early Stade till, I shall concentrate in this report on the till deposited during the Late Stade. This till is contained within spectacular morainic complexes (Fig. 2, A and B), which are located at the mouth of glaciated valleys. The main complexes are located (Fig. 1) in the Sierra Nevada de Mérida, Sierra de Santo Do-

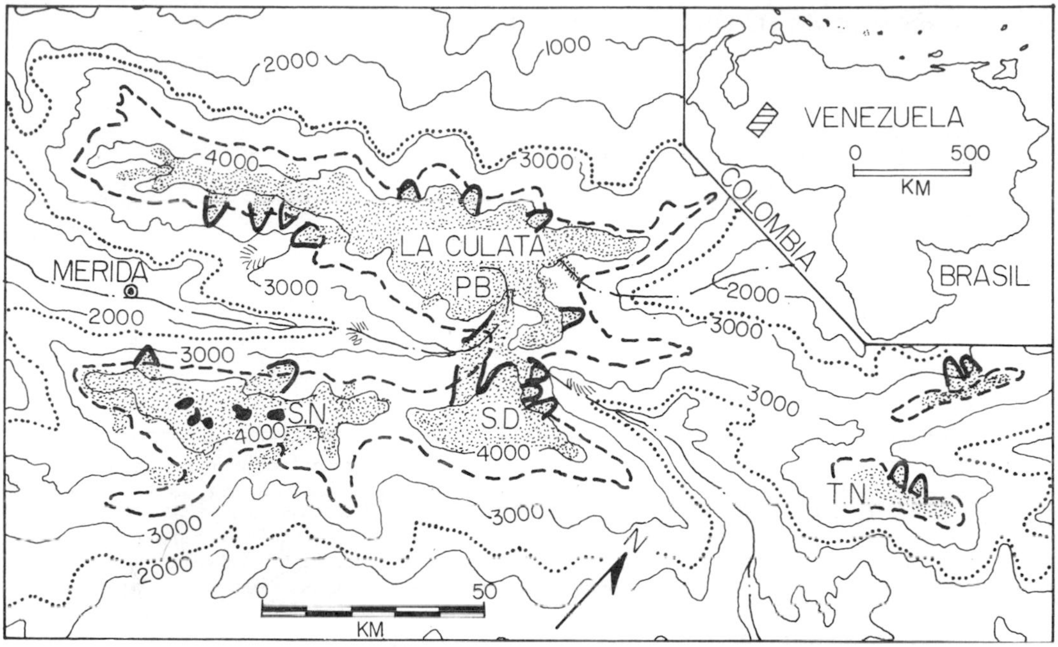

Figure 1. Preliminary glacial sketch map of the central Venezuelan Andes. S. N.: Sierra Nevada de Mérida; P. B.: Páramo de Piedras Blancas; S. D.: Sierra de Santo Domingo; T. N.: Teta de Niquitao (Sierra de Trujillo). Present glaciers are shown as small black areas in the Sierra Nevada de Mérida. Late Pleistocene glacier extent is shown as stippled areas. Major morainic complexes (3000 to 3500 m) are shown as heavy black lines. The dotted and broken lines represent the Late Pleistocene and present periglacial limits, respectively. Hatched lines indicate the location of probable lower morainic level (2600 to 2800 m).

mingo, Sierra de Trujillo (mainly in the Teta de Niquitao massif), Páramo de Piedras Blancas, and Páramo de La Culata. Of these, the best studied morainic complexes are those of Páramo de La Culata and the northern flank of the Sierra de Santo Domingo (Schubert, 1970; Schubert and Valastro, 1973; Giegengack and Grauch, 1976).

The criteria used in this report to classify the diamictos of the central Venezuelan Andes as till, are those listed by Flint (1971: 182), which include: wide range of grain-size; lack of sorting; relatively fresh, faceted, striated, and grooved clasts; presence of fabric; variable lithology; and morphology of the deposits. As will be shown below, most of these criteria are met by the tills of the Venezuelan Andes.

The glaciated areas of the central Venezuelan Andes are underlain principally by Precambrian (?) gneiss, schist, and amphibolite of the Iglesias Group; and to a lesser extent by metamorphosed Late Paleozoic sediments and granitic intrusive rocks, and unmetamorphosed Jurassic to Cretaceous sediments (Shagam, 1972).

In Páramo de La Culata, seven morainic complexes were identified (Schubert and Valastro, 1973) above 3000 m, named La Culata Moraines I to VII. Individual moraines, both lateral and terminal, reach heights of up to 150 to 200 m above the valley floor, suggesting similar thicknesses of till. The depth to bedrock is variable, but probably is shallow, as shown by several outcrops of metamorphic rocks in the river beds. Probable basal till was found at several localities below the morainic till. A moraine, possibly older than the seven complexes mentioned above, called Middle Mucujún Moraine, was found in the lower part of Páramo de La Culata, at 2600 m elevation. Its morainic morphology is subdued and it is densely covered by vegetation.

In the northern flank of the Sierra de Santo Domingo, five morainic complexes were identified above 3000 m (Schubert, 1970), two of which, Victoria and Zerpa Moraines, will be delt with in this report. Giegengack and Grauch (1976) subdivided the till in this area into two units: Mucubají and Santo Domingo Tills,

A

B

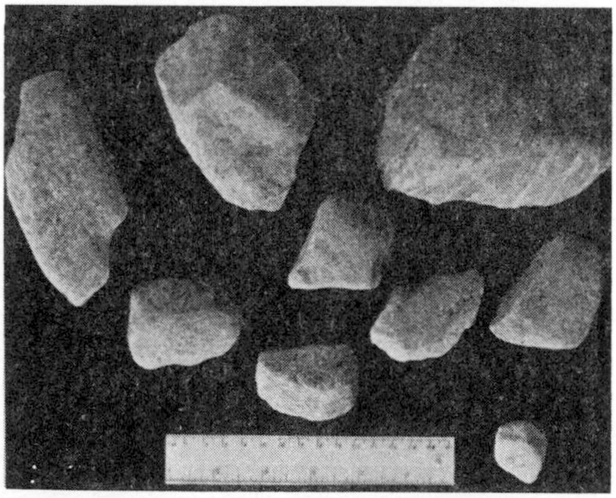

C

Figure 2. A: Typical moraine of the 3000 to 3500 m level in Páramo de Piedras Blancas. B: Typical till outcrop of the major morainic level. C: Typical faceted clasts of the till (ruler is 15 cm long).

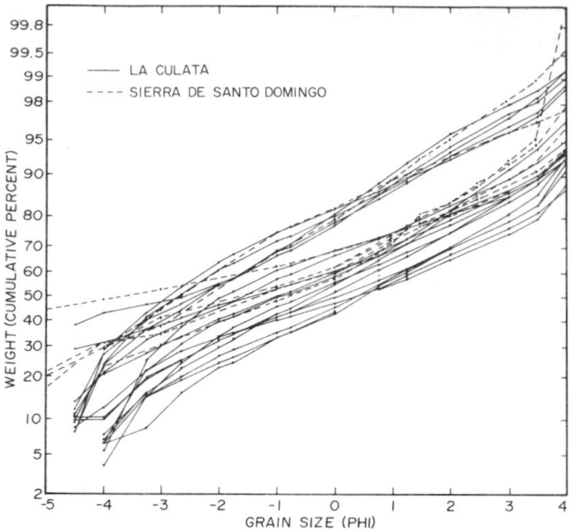

Figure 3. Grain-size distributions of morainic till from the Venezuelan Andes.

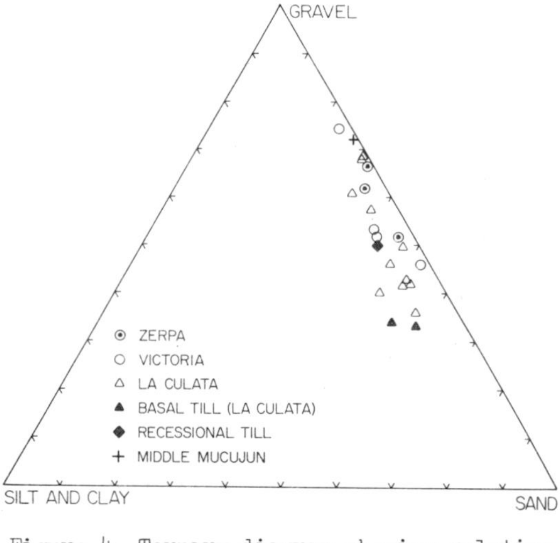

Figure 4. Ternary diagram showing relative size composition of till samples from the Venezuelan Andes (grain size after Wentworth, 1922).

located on the southern and northern flanks of the Santo Domingo river valley, respectively, and the divide which separates this valley from the central Chama river valley; these two tills have contrasting lithologies suggesting different source areas. An older diamicton, the Apartaderos Diamicton, was also identified by Giegengack and Grauch (1976). The tills of the Victoria and Zerpa Moraines are identical with the Mucubají Till.

TABLE 1. Distribution of shape, roundness, and lithology of morainic clasts of the Central Venezuelan Andes.

	Páramo de La Culata				Sierra de Santo Domingo			
	Middle Mucujún		La Culata I		Victoria		Zerpa	
	No, clasts	%	No, clasts	%	No clasts	%	No clasts	%
1. Shape (after Wentworth, 1936)								
Prismoidal	1	1.3	30	11.0	185	61.7	192	64.6
Trapezoidal	3	4.0	19	7.0	12	4.0	4	1.4
Pyramidal	1	1.3	9	3.3	17	5.7	11	3.7
Elongated prismoidal	54	72.0	106	38.8	13	4.3	26	8.8
Tabular	12	16.0	50	18.3	61	20.3	51	17.1
Triangular	-	-	12	4.4	8	2.7	13	4.4
Prismatic pointed	4	5.3	23	8.4	-	-	-	-
Irregular	-	-	23	8.4	3	1.0	-	-
Round	-	-	-	-	1	0.3	-	-
Total	75	99.9	273	99.6	300	100.0	297	100.0
2. Roundness (after Pettijohn, 1957, p. 59)								
Subangular	15	20.0	195	71.4	210	70.0	222	74.8
Subrounded	43	57.3	76	27.8	84	28.0	72	24.2
Rounded	17	22.7	2	0.7	6	2.0	3	1.0
Total	75	100.0	273	99.9	300	100.0	297	100.0
3. Lithology								
Gneiss	40	53.3	85	30.9	249	83.0	249	83.8
Amphibolitic gneiss	-	-	-	-	3	1.0	-	-
Shist	11	14.7	45	16.4	37	12.3	42	14.1
Schistose gneiss	-	-	-	-	9	3.0	-	-
Granitic rock	22	29.3	141	51.3	2	0.7	3	1.0
Pegmatite	2	2.7	2	0.7	-	-	-	-
Quartz	-	-	2	0.7	-	-	3	1.0
Total	75	100.0	275	100.0	300	100.0	297	99.9

In the following discussion, the moraines studied will be referred to simply by their name: Middle Mucujún, La Culata, Victoria, and Zerpa.

2.2 Sedimentological analysis of the till

Till samples were collected on cleared surfaces in road cuts or slide scars in the moraines, taking great care that no fine sediment was lost. At seven localities, the till fabric was analyzed, following the methods of Holmes (1941) and Andrews and Smith (1970). Only clasts of more than one centimeter in length, and with a definite elongation, were measured.

In the laboratory, grain-size distributions were determined by sieving in a RoTap machine, and plotted on probability graphing paper. The fabric data was plotted on equal-area stereographic projections (lower half of the sphere).

Fig. 3 and 4 represent the size distributions and composition, respectively, of 25 representative till samples. Fig. 5 shows the till fabrics. Tables 1 and 2 show the distribution of shape, roundness, lithology, and surface features of clasts within these samples and at the localities of fabric analysis.

The typical morainic till consists of a pale brown to light gray (5 YR 5/2 to N 7), unsorted mixture of clay- to boulder-size grains. No bedding is apparent in these deposits, except where small sand bodies are found among the till boul-

Table 2. Distribution of striated, grooved, faceted, and polished clasts in morainic till from the central Venezuelan Andes.

	Striated and grooved %	Faceted %	Polished %
Middle Mucujún	0	20.0	0
La Culata I	1.1	15.6	0
Victoria	2.7	8.7	2.3
Zerpa	2.4	10.1	0

ders. These small bodies are laminated and probably represent channels in which water flowed through the moraine, during ablation of the glaciers. In general, the boulders are prismatic and equidimensional with rounded edges and corners. Striation of the boulder faces is rare; crescentic gouges are common. The most common clast shape (Table 1) is elongated prismoidal; shape ranges from well shaped prisms to ellypsoidal. The most common lithologic types in the morainic clasts are gneiss, schist, and granitic rocks. These rock types are the most important in the source areas of the till: the Sierra de Santo Domingo and Páramo de La Culata. They belong to the Sierra Nevada Formation (Iglesias Group) of Precambrian (?) age, which forms the core of the Venezuelan Andes, and which underlies its highest parts (Schubert, 1969; Kovisars, 1971; Shagam, 1972). Facets are well developed in the clasts (Fig. 2C); their proportion ranges from 8.7 to 20.0% (Table 2). Striation and grooving is less common, up to 2.7% of all clasts examined; it usually is parallel or subparallel to the long axis of the clast. The relatively low percentage of surface markings on the clasts may be attributed to two causes: 1. hardness of the rocks; and 2. relatively short distance of travel by the clasts (3 to 10 km, at most).

Grain-size distributions of 25 till samples are shown in Fig. 4. From these curves. it is apparent that the tills contain very little material finer than sand-size (finer than 4 phi or 0.062 mm), usually less than 5%. Great care was taken during sampling in order not to loose any fine-grained sediment. This lack of fine-grained sediment in the till indicates probable drainage during or after glacier ablation and retreat. Such an explanation was also postulated for a similar type of till by Spalletti and Gutiérrez (1976). In fact, within the morainic loops and just below them downvalley, fluvio-glacial terraces (probably outwash deposits) contain sediments with a high proportion of fine-grained material. These may represent the fine-grained portion lost by the tills. Fig. 4 shows a ternary diagram which expresses the grain-size of the tills more graphically. The lack of silt- and clay-sized material is readily apparent.

Statistical grain-size parameters were calculated according to the formulas of Folk and Ward (1957). Mean size ranges between -2.25 phi (4.8 mm) and 0.55 phi (0.7 mm), a wide range of variation. Sorting (expressed as standard deviation) ranges between 2.30 and 3.80 which, according to the verbal scale of Folk and Ward (1957) mean that the till is very poorly sorted. Skewness ranges between -0.95 and 0.47; most samples are nearly symmetrical or positively skewed. Finally, kurtosis ranges between 0.83 and 1.02, which means that the till is mostly platykurtic to mesokurtic, that is, with a deficient to normal peakedness.

Two outcrops of probable basal till were found in Páramo de La Culata. In the northern flank of the Sierra de Santo Domingo, there are large areas above 3000 m, located among the morainic ridges, which are covered by possible basal till, as indicated by the presence of faceted and striated clasts. However, at this second locality, no obvious till (different from the morainic till) was found underlying the moraines of the valley within them. The composition of two samples of the basal till from La Culata is shown in Fig. 4. They clearly contain more fine-grained material than the morainic tills. The basal till is gray to mottled brown (N 6 to 5 YR 6/1) and very compact. Pebbles and cobbles within it frequently are fractured, and the matrix contains a significant amount of silt and clay. No fabric was observed in the basal till. One sample of recessional till (that is, till from recessional moraines above the main morainic level) was also analyzed (Fig. 4).

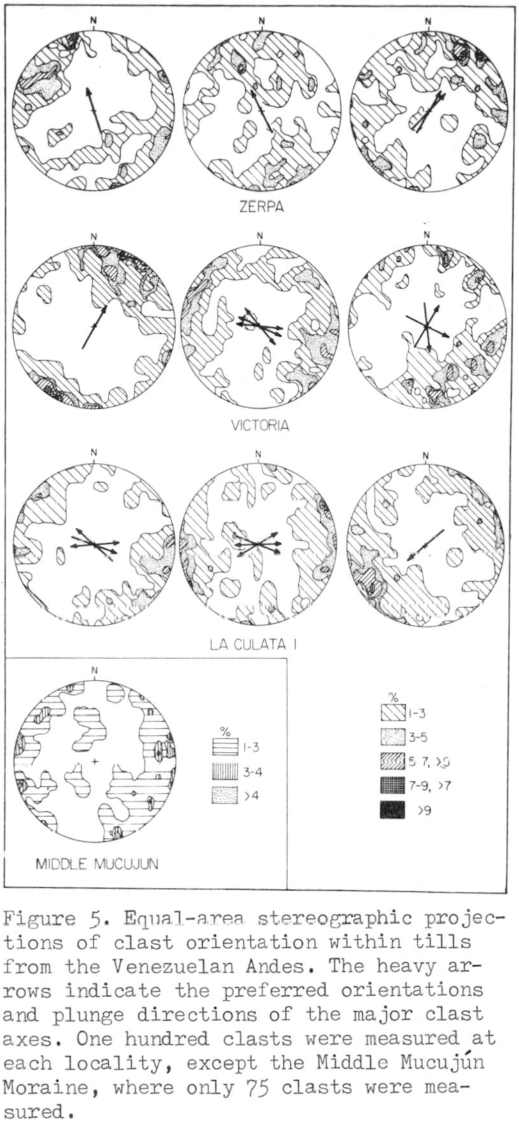

Figure 5. Equal-area stereographic projections of clast orientation within tills from the Venezuelan Andes. The heavy arrows indicate the preferred orientations and plunge directions of the major clast axes. One hundred clasts were measured at each locality, except the Middle Mucujún Moraine, where only 75 clasts were measured.

The fabric of the till in four moraines was studied: Zerpa and Victoria Moraines (Sierra de Santo Domingo), and La Culata I and Middle Mucujún Moraines (Páramo de La Culata). At each locality, three outcrops were analyzed, except for the Middle Mucujún Moraine, which is highly weathered and crops out very poorly. Fig. 5 shows the fabric diagrams. The arrows at the center of the diagrams indicate the principal orientation directions of the long axes of the clasts; each arrow points in a preferred direction of plunge . The fabric of La Culata I, Victoria, and Zerpa Moraines has a roughly monoclinic symmetry. The few maxima which are shown in the fabric diagrams and the arrows indicate that the clast orientation is, in general, parallel or subparallel to the inferred direction of glacier movement (downvalley), in a northeastern direction in the case of the Victoria and Zerpa Moraines, and in a southwestern direction in the case of the La Culata I Moraine.

Fig. 5 also shows the fabric data for till of the Middle Mucujún Moraine. The data is highly scattered and shows only minor maxima. This may be due to the smaller total number of clasts measured. A glacial origin of this sediment is suggested by the presence of striated clasts and the lack of bedding and arrangement of the particles.

The sampling problems encountered in fabric studies of till were discussed by Andrews and Smith (1970). I encountered similar problems, such as difficulty in clearing vertical faces of exposures due to hardness of the till, and the possible influence of exposure orientation on clast orientation and sample selection. The data presented here, however, support the general conclusion that till fabric in alpine regions reflects the direction of glacier movement.

3 CONCLUSIONS

Glacial sediments in tropical Andean alpine environments have not been studied in detail. Tills from Late Pleistocene moraines in the Venezuelan Andes are characterized by most of the features of tills in other regions and environments: 1. great variety in grain-size and poor sorting; 2. absence of bedding and arrangement of particles; 3. presence of faceted, striated, and grooved clasts; 4. presence of fabric which indicates the direction of movement of the glaciers; and 5. a typical alpine morainic morphology.

4 BIBLIOGRAPHY

Andrews, J. T. & D. I. Smith 1970, Statistical analysis of till fabric: methodology, local and regional variability, Quart. Jour. Geol. Soc. London 25: 503-542.

Cárdenas, A. L. 1962, El glaciarismo pleistoceno en las cabeceras del Chama, Rev. Geog. (Venezuela) 3: 173-194.

Dobrovolny, E. 1962, Geología del valle de La Paz, Min. Minas y Petról. (Bolivia) Bol. No. 3: 1-153.

Flint, R. F. 1971, Glacial and Quaternary geology. New York, John Wiley and Sons.

Flint, R. F. & F. Fidalgo 1964, Glacial geology of the east flank of the Argentine Andes between latitude 39° 10' S and latitude 41° 20' S, Geol. Soc. America Bull. 75: 335-352.

Flint, R. F. & F. Fidalgo 1969, Glacial drift in the eastern Argentine Andes between latitude 41° 10' S and latitude 43° 10' S, Geol. Soc. America Bull. 80: 1043-1052.

Folk, R. L. & W. C. Ward 1957, Brazos river bar: a study in the significance of grain size parameters, Jour. Sed. Petrol. 27: 3-26.

Giegengack, R. & R. I. Grauch 1975, Quaternary geology of the central Andes, Venezuela: a preliminary assessment, Bol. Geol. (Venezuela), Pub. Esp. 7, 1: 241-283.

Giegengack, R. & R. I. Grauch 1976, Late-Cenozoic climatic stratigraphy of the Venezuelan Andes, Bol. Geol. (Venezuela) Pub. Esp. 7, 2: 1187-1200.

Holmes, C. D. 1941, Till Fabric, Geol. Soc. America Bull. 52: 1299-1354.

Kovisars, L. 1971, Geology of a portion of the north-central Venezuelan Andes, Geol. Soc. America Bull. 82: 3111-3138.

Marangunic, C. & R. Thiele 1971, Procedencia y determinaciones gravimétricas de espesor de la Morena de la Laguna Negra, Provincia de Santiago, Univ. Chile Depto. Geol. Pub. No. 38: 1-25.

Royo y Gómez, J. 1959, El glaciarismo pleistoceno en Venezuela, Asoc. Ven. Geol. Min. y Petról. Bol. Inf. 2: 333-357.

Schubert, C. 1969, Geologic structure of a part of the Barinas mountain front, Venezuelan Andes, Geol. Soc. America Bull. 80: 443-458.

Schubert, C. 1970, Geología glacial del alto río Santo Domingo, Andes venezolanos, Asoc. Ven. Geol. Min. y Petról. Bol. Inf. 13: 232-261.

Schubert, C. 1972, Geomorphology and glacier retreat in the Pico Bolívar area, Sierra Nevada de Mérida, Venezuela, Zeit. f. Gletscherk. u. Glazialgeol. 8: 189-202.

Schubert, C. 1974, Late Pleistocene Mérida Glaciation, Venezuelan Andes, Boreas 3: 147-152.

Schubert, C. 1976, Glaciación y morfología periglacial en los Andes venezolanos noroccidentales, Soc. Ven. Cien. Nat. Bol. 22: 132/133: 149-178.

Schubert, C. 1977, Morfología glacial y periglácial de los Andes de Venezuela: informe de progreso, Mem. 5° Cong. Geol. Ven. 1: 149-166.

Schubert, C. & S. Valastro, Jr. 1973, Páramo de La Culata, Estado Mérida: glaciación del Pleistoceno Tardío, Asoc. Ven. Geol. Min. y Petról. Bol. Inf. 16: 108-142.

Shagam, R. 1972, Andean research project, Venezuela: principal data and tectonic implications, Geol. Soc. America Mem. 132: 449-463.

Spalletti, L. A. & R. Gutiérrez 1976, Estudio granulométrico de sedimentos glaciales, fluviales y lacustres de la región del Monte de San Lorenzo, Provincia de Santa Cruz, Asoc. Geol. Argent. Rev. 31: 95-117.

van der Hammen, T., J. H. Werner & H. van Dommelen 1973, Palynological record of the upheaval of the northern Andes: a study of the Pliocene and Lower Quaternary of the Colombian Eastern Cordillera and the early evolution of its high-Andean biota, Rev. Palaeobot. and Palynol. 16: 1-122.

Wentworth, C. K. 1922, A scale of grade and class terms for clastic sediments, Jour. Geol. 30: 377-392.

Moraine forms and their recognition on steep mountain slopes

T. J. H. CHINN
Ministry of Works and Development, Christchurch, New Zealand

ABSTRACT

In steep mountain country where moraines are either spilled over steep slopes or eroded by streams it is often impossible to infer the extents of past glaciers by conventional methods. Despite the tempo of activity in mountain environments, where numerous processes operate rapidly to modify glacial lineaments, traces of glacial activity do remain for considerable periods of time, albeit commonly in subtle and obscure forms. A study of these features, carried out in the Southern Alps, New Zealand, has shown that sufficient traces may be found to enable the position of the former ice margin to be established where no classical moranic forms have survived.

Modified and incised vestiges of glacial extents are described from the area studied, these include: the recognition of former ice extents by offset tributaries, double-ridges to the ends of spurs, outwash terraces, upper talus limits, break in valley side slopes, trim lines, spilled tills and the identification of till-like deposits of non-glacial origin. Large variations in the volumes of alpine moraines are caused by additions from landslides or by erosional losses in steep valleys.

A glacial chronology for the Holocene was constructed for the Waimakariri River headwaters based on rock weathering rind thicknesses and usual correlation methods. Some glacial events were apparently missing from the sequence in each valley and this led to the application of the above methods to determine whether or not the event had occurred in that particular valley. Normally evidence for the event was found, so that finally by counting the sequence within each valley, ages have been assigned with some degree of certainty to all glacial extents recognised in the area.

1 INTRODUCTION

Recognition of the number and extent of glacial events in a mountain valley is basic to making correlations and establishing a glacial chronology.

Extents of past glaciers are usually inferred from positions of lateral and terminal moraines marking limits of former ice margin stands. In steep mountain country however, many moraines are either spilled over steep slopes or are eroded contemporaneously with deposition, and have then undergone numerous subsequent processes that operate rapidly to modify the remaining ice-margin features. Despite the high tempo of geomorphic activity in mountain environments, traces of glacial activity frequently do remain, albeit commonly in subtle and obscure forms. Through direct and indirect evidence a careful study of these features can reveal indications of the position of the former ice margin with some degree of certainty.

This study was carried out mainly in the headwaters of the Waimakariri River, Southern Alps, New Zealand, where the rocks are almost entirely highly indurated Triassic sandstones and siltstones of the Torlesse Supergroup. These rocks have been extensively faulted and folded leaving them closely jointed and friable and thus subject to high erosion rates. The highest peaks rise to 2,200 m in the north-western sector, and the crests of the ranges descend to elevations of 1,600 m towards the north. Present glaciers are restricted to a few dozen small glaciers and permanent snow patches among the highest peaks, with the present glacial snowline at an elevation of 1,900 m.

The maritime temperate climate of the area is dominated by westerly winds which generate a strong föhn effect across the main divide, accompanied by a steep precipitation gradient from 5,000 to 10,000 mm a^{-1} in the west to 1,000 mm a^{-1} along the eastern margin of the mountains. The high annual precipitation values imply a high fraction of rainfall in the input into both present and past glaciers. Surviving glacial deposits are therefore characterised by evidence of ample fluvial action.

2 DESCRIPTION

The region has undergone extensive glaciation during the Pleistocene when major glacial features of the topography were sculptured and glaciers terminated some tens of kilometres downvalley. Cirque basins were formed at this time and these have been re-occupied a number of times during the Holocene. Immediately following the retreat of the Pleistocene ice, there was a dramatic change from glacial erosion to fluvial dissection.

The higher ranges towards the west display features associated with intensive glaciation, including "U"-shaped valleys with ice-scoured surfaces, asymmetrical cross-sections on bends and truncated and overridden spurs, etc. Few glacial features are visible on the main valley floors, which have been over-deepened and subsequently covered with valley train alluvium and large alluvial fans. On main valley floors, rôches moutonnées are rare, and glacial stairways non-existent. Exposed glaciated valley floors occur at higher elevations, towards the heads of tributary valleys, especially where the valleys head at a low pass. The streams in these higher valleys are usually deeply incised below their ice-trimmed valley floors.

Steep, actively eroding slopes and gullies occur between the alluvium-filled valley floors and higher-level ice-worn surfaces. The well-defined upper limit of fluvial dissection forms a boundary below which glacial deposits are unlikely to have survived. Extensive areas of ice-eroded landscape survive above this limit, the dominant forms being alp surfaces and cirque basins. Most Neoglacial deposits are found on these areas, which have as yet escaped destruction by fluvial erosion. Alp surfaces are widespread throughout the western part of the study area and frequently extend for considerable distances along the ranges between subsidiary ridges or cirques, indicating the existence of extensive areas of névé-sheathed slopes between tributary ice streams during Pleistocene times.

Cirque basins range in development from incipient hollows to mature cirques containing tarns. They are almost invariably discordant with the main valleys. Those at higher levels, that have been occupied by later glacial advances, are noticeably the more developed. Higher-level cirque basins developed at the heads of the large Pleistocene tributary ice streams by rotational ice-flow appear to have undergone little modification during the subsequent Holocene glacial advances. Evidence for the relatively minor amount of Holocene glacial erosion is seen in both the terminal positions and the volume of debris left by Holocene glaciers. The positions of moraines show that Holocene glaciers were generally discordant with the cirque morphology. Moraines either spill over the cirque lips, or are confined within the lower limits of the basins. Volumes of moraines are small compared with the volume of material removed to form the basin, suggesting that there has been limited modification of the basins since formation by Pleistocene ice erosion.

Near the Main Divide, where the highest peaks occur, the early Holocene moraines of the larger valleys are found at valley floor level. As summit elevations decrease along the main divide towards the north-east, the level of these moraines gradually rises until they rarely descend as low as the main valley floors. Later Holocene (or Neoglacial) moraine sequences are displayed up-valley from the early Holocene moraines in all of the higher ranges of the Main Divide area. In the north-east these sequences extend only into mid-Neoglacial times, while in the west they include recent moraines of existing glaciers.

3 RECOGNITION OF FORMER ICE LIMITS

The search for evidence marking past glacial limits is best carried out as a combination of fieldwork and stereoscopic study of aerial photographs. In addition to displaying a full coverage of the area, the photographs give the opportunity to detect certain broad features not readily discernible on the ground. Most of the features detected on the photographs should be field checked for a glacial origin, especially if there is forest cover. Where there is a forest cover the usefullness of aerial photographs is drastically impaired.

Problems in moraine recognition arise mainly from similarity with other widely distributed deposits of non-glacial origin, but which resemble glacial deposits in both form and composition. Landslide deposits in particular had to be distinguished from glacial deposits, as their moraine-like surface features occur commonly throughout the area. Where landslide deposits have been deeply dissected by streams, or have fallen on to moraines, differentiation can only be made on the content of the deposits. Other moraine-like features encountered in this work included bedding and other lithological contrasts, minor surface faults and slump scarps, and rare debris flows.

3.1 Older glacial deposits

Throughout the Upper Waimakariri River catchment preservation of glacial deposits is very fragmentary. No traces remain of many of the older deposits, while few of those on the most favourable sites have escaped at least some degree of modification. The degree of erosion and dissection of moraines was found generally to increase with increasing age and increasing gradient of the valley sides and floors.

3.2 Tills

In tills deposited by comparatively short alpine glaciers, the majority of the coarse material tends to be angular, with very few water-worn or striated clasts. To distinguish these tills from landslipe deposits and other diamictons, the inclusion of rare water-worn clasts will normally indicate a glacial origin except in some rare instances, eg, where a landslide has picked up alluvium when crossing a stream bed.

Although clasts in both till and landslide material are angular, boulders in landslides have undergone more intensive fracturing immediately prior to deposition. Consequently it is frequently a simple matter in the field to distinguish between tills containing angular boulders, and landslide or rockfall jagged sharp-cornered boulders.

Broad moraines of low relief frequently have swampy surfaces arising from the low permeability of tills. Such deposits in forested areas maybe suggested on aerial photographs as swampy clearings in the forest cover.

3.3 Offset tributaries

Streams tributary to mountain valleys generally flow parallel to one another, and approximately normal to the direction of the main valley. The lateral margin of a glacier will divert such streams to flow parallel to the glacier edge. Upon retreat of the ice, confining by the lateral moraine, and subsequent incision into the valley wall, can make the new stream direction permanent. This stream pattern may persist after all glacial deposits have been removed from the ridge separating the stream from the main valley. Such streams are very good indicators of past glacial margin positions, provided that the origin of the downvalley kink is not the result of structural control.

3.4 Double ridged spur ends

It is not uncommon for a glacier to terminate near the mouth of a valley where the end-moraine loop will be constructed on the spur ending the valley wall. Such a moraine then forms a second ridge to the end of the spur. This double-ridged spur-end is normally accentuated by further fluvial erosion and lies in a position of low erosion potential, favourable to preservation.

3.5 Outwash terraces

During a glacial advance the glacier melt-water stream-beds aggrade due to an increased sediment supply, leading to the construction of an outwash terrace. Large sub-angular boulders in such terraces indicate the close proximity of the ice front, and give a maximum extent for that particular glacial event where other evidence has not survived.

4 YOUNGER, STEEP GLACIAL FEATURES

Evidence of smaller glacial extents, in high precipitous mountains, rarely survives for long periods of time, and is infrequently well-defined, even at the time of deposition. There is, however, normally sufficient evidence to determine approximately the number and extent of events in most of these areas. When interpreting these past glacial margin positions, it is desirable to seek more than one piece of evidence before concluding an interpretation. Examination of sections for water-worn clasts was found a particularly useful secondary tool in this

Figure 1. Double and multiple ridged spur ends of Pleistocene glaciations.

Figure 2. Late Neoglacial upper scree limit, SL, and the associated lateral moraine, M, Ramsay Glacier.

respect. Evidence employed to assist demarkation of past glacial extents included:

4.1 Scree limit

Although the whole region studied has undergone extensive scree development by the usual scree building processes, towards the valley heads, some particular screes are interpreted to be of glacial origin.

These screes at the base of steep cliffs are frequently the result of degenerated moranic deposits rather than originating solely from the accumulation of rockfall and avalanche debris. During usual scree formation the debris accumulates in talus cones below gullies, frequently with a steep stream channel dissecting the cone. But between these gully bottom cones, less active screes may occur with their upper shoulders aligned along the former ice margin. These inter-gully scree shoulders can represent the position of former steep lateral moraines which have degenerated in form since the withdrawal of supporting ice. In many of the more recent cases these talus limits are seen to be associated with vegetation trim lines.

Figure 3. Offset streams, OS, and on upper scree limit, SL, in schist country. Mt Beaumont, Westland.

4.2 Break in slope

Glacial erosion of some valleys has led to over-steepening of the valley walls below the ice level, while, at the glacier margin, lateral moraines and kame terraces are deposited. Where no incised ice-marginal channel has developed, consequent streams will closely dissect the valley sides following withdrawal of the ice, but the break-in-slope marking the former glacier margin persists even after the bulk of the ice-margin sediment has been removed. This feature again may be associated with a vegetation trim line, and is frequently shown to best advantage on aerial photographs.

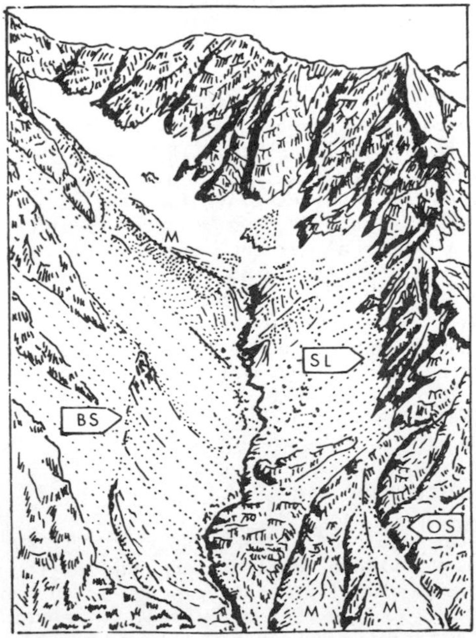

Figure 4. A Neoglacial ice extent outlined by break in slope, BS, upper scree limit SL, and offset stream, OS. Moranic ridges marked 'M'.

4.3 Trim lines

Vegetation trim lines form whenever glaciers advance below the vegetation limit in mountainous areas. Following ice withdrawal there will be an age difference between the original vegetation association and the younger group which colonises the newly bared ground. Identification of trim lines becomes more subtle the older they become, but in forest country in particular, they can remain visible for many hundreds of years.

4.4 "Spilled tills"

Debris from steep glaciers which descend in ice falls, is spilled over bluffs or snow cones to form conical shaped talus deposits often without an apparent source. These deposits have been termed "spilled tills" and may be identified by their position and composition. They occur either side of, instead of directly below the source gully; have frequently a discernible moranic crest running down the conical slope; and the material of these cones contains some water-worn clasts. The majority of spilled tills have been masked to some extent by the later addition of snow and rock avalanche debris.

5 VARIATIONS IN VOLUMES OF MORANIC DEPOSITS

With some field experience in a particular area, it is possible to predict approximately where a glacial terminus should have been in a particular valley. In many cases it is found that no deposits nor any trace of a past glacial margin can be found at a predicted location. In such instances the valley topography is invariably very steep with an incised stream flowing in a high gradient channel. A comparison of the topography of these areas, with that of existing steep narrow glaciers, shows that almost the entire discharge of moranic material should be transported to one position in the stream channel at the extreme tip of the glacier snout. From this site the high energy stream is competent to remove all debris supplied, and no moraine can be formed.

Late neoglacial moraines were found to vary considerably in volume and state of preservation. Many instances occur where, in adjacent basins with similar topography favourable to the preservation of moraines, one basin contains small moraines of relatively low volume, while in the adjacent basin, huge massive lateral moraine ridges have been constructed. Two reasons for this apparent discrepancy were found from a close study of the features. Small-volume moraines result from normal glacial erosion and deposition in the comparatively short period of less than 1000 years of late Neoglacial activity. Large-volume moraines result from either landslides or rock avalanches falling on to the glacier, or from the glacier over-riding and redepositing older moraine systems. Landslides of various ages are common throughout the area, and the majority of the massive moraines have resulted from these phenomena. Over-riding

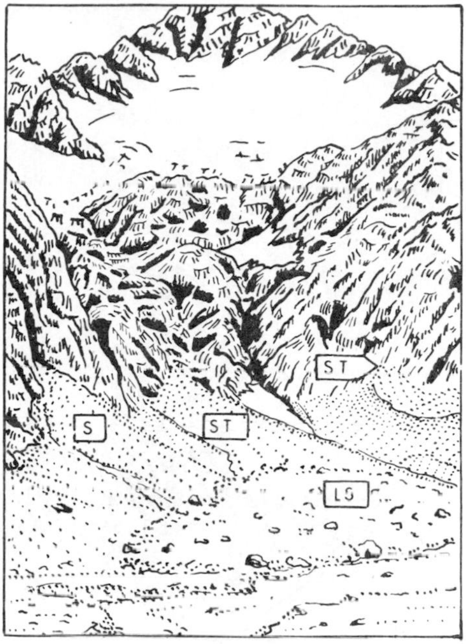

Figure 5. Spilled tills, ST, either side of a main gully below the Crow Glacier, Arthurs Pass area. S, normal gully - fed scree. LS, landslide deposit from opposite valley side.

and reconstitution of moraines from two separate periods of glacial activity into a single moraine system occurs when two glacial advances are similar in magnitude. Where two distinct glacial advances of similar magnitude have occurred in an area the usual evidence of these events are sets of closely concentric terminal moraines. In certain basins, however, small variations in the complex glacier dynamics can cause the later episode to extend further and over-ride the limits of the earlier episode to form a single massive moraine.

6 CONCLUSION

In any study of alpine glaciation it is rare to find a complete sequence for the extents of all glacial episodes in any one valley. It is therefore usual to resort to correlations between valleys by all methods available, and to count events up or down the sequence from known datums such as present glaciers or dated moraines. By using the methods described to map as

many glacial extents as it was possible to detect, a complete glacial chronology was constructed for the Waimakariri River headwaters area. By cross-correlation, and counting events, ages have been assigned with some degree of certainty to every glacial event recognised in the area.

7 ACKNOWLEDGEMENT

This paper is published with the permission of the Commissioner of Works, Ministry of Works and Development, New Zealand.

8 BIBLIOGRAPHY

Burrows, C.J. 1973, Studies on some glacial moraines in New Zealand, -2 Ages of moraines of the Mueller, Hooker and Tasman Glaciers. N.Z., Jour. Geol. and Geophys, Vol. 16, No. 4, p. 831-855.

Burrows, C.J. 1975, Late Pleistocene and Holocene moraines of the Cameron Valley, Arrowsmith Range, Canterbury, New Zealand. Arc. and Alp. Res., Vol. 7, p. 125-140.

Chinn, T.J. 1975, Late Quaternary snowlines and cirque moraines within the Waimakariri watershed. MSc thesis, University of Canterbury, Christchurch, New Zealand, unpublished.

Chinn, T.J. 1977, Holocene glacial history from the Waimakariri watershed. Abstracts, Tenth Congress. International Union for Quaternary Research. Birmingham, England.

Davis, J.L. 1969, Landforms of cold climates. Cambridge, M.I.T. Press, 323 p.

Embleton, C. and King, C.A.M. 1968, Glacial and periglacial geomorphology. New York, St. Martins Press, 608 p.

Gage, M. 1958, Late Pleistocene glaciation of the Waimakariri Valley, Canterbury, New Zealand. N.Z. Jour. Geol. and Geophys, Vol. 1, No. 1, p. 123-155.

Graf, W.L. 1970, The geomorphology of the glacial valley cross-section. Arc. and Alp. Res., Vol. 2, p. 303-312.

Luckman, B.H. 1971, The role of snow avalanches in the evolution of alpine talus slopes. In D. Brunsden (ed.), Slopes, Forms and Processes. Inst. Brit. Geog. Special Publication No. 3.

Madole, R.F. 1972, Neoglacial facies of the Colorade Front Range. Arc. and Alp. Res., Vol. 7, p. 33-37.

McGregor, V.R. 1967, Holocene moraines and rock glaciers in the central Ben Ohau Range, South Canterbury, New Zealand. Jour. Glaciol., Vol. 6, No. 47, p. 737-748.

Paterson, W.S.B. 1969, The physics of glaciers. Oxford, Pergamon Press, 241 p.

Penk, A. 1965, Glacial features in the surface of the Alps. Jour. Geol. Vol. 13, p. 1-17.

Porter, S.C. 1975, Equilibrium line altitudes of late Quaternary glaciers in the Southern Alps, New Zealand. Quat. Res., Vol. 5, p. 471-497.

Porter, S.C. 1975, Glaciation limit in the New Zealand Southern Alps. Arct. and Alp. Res., Vol. 7, p. 33-37.

Sharp, R.P. 1969, Semiquantitative differentiation of glacial moraines near Convict Lake, Sierra Nevada, California, Journ. Geol., Vol. 77, p. 68-91.

Speight, R. 1938, Moranic deposits of the Waimakariri Valley. Transactions of the Royal Society of New Zealand, Vol. 68, No. 2, p. 143-160.

Wardle, P. 1973, Variations of the glaciers of Westland National Park and the Hooker Range, New Zealand. N.Z. Jour. Bot. Vol. 11, p. 349-388.

Washburn, A.L. 1973, Periglacial Processes and Environment. London, Edward Arnold and Company, 320 p.

Morainic deposits in Kashmir Himalayas

N. AHMAD
Aligarh Muslim University, Aligarh, India

Himalayan glaciers are Alpine type. In this area glaciers never developed into ice caps, and their starting level have always been below 5500m. These glaciers owe their origin to the intensely folded mountain ranges that have risen above the regional snow-line, and they either originated from the zone of regenaration at the base of cols and aretes or from cirques along the valley wall. Their supply of snow and ice has also been limited to the basins and cirques of their origin.

All the living glaciers in Kashmir valley have now retreated high up in their valleys and they terminate between 3830-4290m. level, however during Pleistocene period, the terminal zone was much lower and at two levels. Glaciers coming down main or the trunk valleys terminated between 2760m-3060m and other glaciers coming down along steep valley walls terminated at 1680-2290m level. In Pleistocene period it also happened, that two glaciers moving in different part of the same valley terminated at two different levels. In west Liddar valley, Kolahoi glacier occupying the trunk valley, terminated at 2760m level and Aru glacier that joined the main valley below Kolahoi terminal end, terminated at 2600m level. In East Liddar and Sonamarg valleys, one glacier terminated at 2300m and 2760m levels respectively, and another that originated in the lower parts of the valley, terminated near 1840m. Some times these glaciers deposited their terminal moraines within short distance of each other. In such areas often deposites from two independent glaciers have been assigned to only one glacier or one of these deposits have been classified as outwash deposits. Dainelli (1922) reported roche moutonnee and sub-glacial moraines along one of the valley walls in Liddar valley near Kanjdori (1686m.), De Terra and Patterson (1939) did not accept these deposits as moraines and the structure as roche moutonne. Though, some distance up the valley and at a lower level they have reported terminal and sub-glacial moraine.

Holland (1907) classified Kashmir glaciers on the basis of their origin and the position of the terminal end, into Longitudinal and Transverse types. According to this classification, longitudinal glaciers occupy main or trunk valleys, running parallel to main mountain ranges. These glaciers carry large quantity of ice, they are the longest glaciers of the area, however moving at low gradient valley slopes these glaciers terminate at a higher level than the transverse glaciers. Transverse glaciers originate from small isolated cirques on the valley walls, these are generally small glaciers and they join the main valley from the sides. If these glaciers donot join a longitudinal glacier as a tributory, they terminate at a lower level than longitudinal glaciers.

In Kashmir valley width of the valley, gradient of the valley floor, and the changes in the direction of

valley are controlled by lithology and structure of the bed rock. East Liddar all along its course moves accross the strike of the bed rock and with the changes in the direction of the strike it has changed the direction of its course several times. Between the source and the terminal end E.Liddar valley has turned through 170°. Kolahoi Glacier starts at the nose of a plunging syncline, and lower down the valley it goes over volcanic lava flows (Permo-Carb.Panjal Traps) with low angle dip and hexagonal fractures. Between the source and the terminal end this glacier has turned round through more than 248°. Due to local structure some of the tributary glacier in Kolahoi valley have carved their hanging valleys facing upstream. In East Liddar and Sind valley some of transverse glaciers have either joined the main valley coming down at very steep slopes and at right angles, or are very acute angle coming down a valley running almost parallel to the trunk valley.

The lithology of the bed rock has always affected the width of the valley. When ever any of the glaciers have crossed, over a lava bed the width of the valley has decreased.

At the bands also these valleys become narrow, walls rise almost vertically, and the valley floor has riegels. All the riegels are along one valley wall, and between the riegel and the other valley wall the glacier went down a shallow U-shaped depression with very steep slopes. The lip of the riegel have roche moutonnee and whale backs, and begining from the depression on the lee side, meltwater streams have cut deep gorges in the riegels. In case of some high riegels at the foot of the drop, valley gets very narrow, glacier coming down along U-shaped depression turn at right angles to join the ice coming down the riegel, a tributary glacier joined the main valley at the end of narrows. In the longitudinal glacier the drop at the base of the riegels is rarely more than seven or eight meters, but in Transverse glaciers the drop at some places exceeds 612m. The gradient of the longitudinal glacier valleys nowhere exceeds 1 in 20, but that of transverse glacier valleys is always more than 1 in 9 and some of the tributary glacier come down, at slopes exceeding 1 in 3. Some of the Living glaciers that have receeded in their cirques,descend as ice falls.

During this study it was noted that the morainal deposits of the two types of glaciers also differed in many respect and possibly these differences were due to differences in the valley floor gradient,lithology of the bed rock, the quantity of the ice carried by the glacier, terminal level of the glacier and the steepness of the valley wall.

Some of the important differences between the two types of glaciers are as follows:

1. It was recorded that terminal moraines of the longitudinal glaciers were higher than those of transverse glaciers.

2. Matrix in the longitudinal glacier deposits was finer than that in transverse glacier deposits.

3. In longitudinal glacier moraines the percentage of facetted, rounded and striated clasts was higher than 20%, whereas in transverse glacier moraines it was less than .5%.

4. In transverse glacier moraines sphericity and roundness of more than 82% clasts was less than .3 whereas in longitudinal glacier deposits roundness and sphericity above .5 was observed in more than 31% clasts.

5. In longitudinal glacier moraines percentage of clasts were striated and it was noted that the clasts of the same lithology, in longitudinal glacier acquired striations and deep grooves with in three kilometres from the source where as in transverse glacier deposits for four and five kilometres these clasts were without any markings.

6. The terminal end of the transverse glaciers was at least 1000-1600m below the snow line and that of longitudinal glaciers it was never more than 600m.

Possibly in the transverse

glaciers, ice coming down at steep slopes, and at high speed may also have been at a higher temperature. This must be specially true for all the glaciers that originated in the lower parts of the valley, and those glaciers that terminated much below the snow line (3370m.). In these glaciers the ice being at a higher temperature may have been plastic and instead of flowing it may have spread out like mud. In this plastic ice clasts instead of being held firmly and rotated and rubbed against other clasts may have been transported as if floating in a dense medium. If the ice at the base was also not rigid enough to hold the clasts firmly chances of there being striated were reduced. Absence of striations on the clasts in transverse glaciers is all the more surprising when within two kilometres of the source of the glaciers, riegels, roche-moutonnees, whale backs and striated and polished surfaces are common in all these valleys.

These glaciers were thin and they descended farther below the snow line, therefore they remained active near the terminal end for only a short period, hence they deposited low terminal moraines.

Deep gorges at all levels of the valleys indicate that under these glaciers the rate of melting was high even in the upper reaches of the valleys. These large melt-water streams going down steep slopes must have washed and removed all the clay and silt from their morainal deposits. Moreover longitudinal glaciers having transported their load for longer distance from the source and having remained active for longer periods had greater oportunity to grind its sediments to finer size.

Greater hight of the terminal moraines in front of longitudinal glaciers may have been due to greater thickness of ice, and longer active period.

1. TERMINAL MORAINES

Terminal moraines of Sonamarg glacier at Gagangaur, and Kolahoi glacier at Aru are best examples of their type. The moraines are about 70m high, two kilometres wide and about four kilometres in lateral extent. Melt-water streams have cut through one end of these moraines, and now these are preserved only because they are resting above the river bed.

In these moraines, large clasts are embedded in clayey and silty matrix, and ice push and thrust structures are common.

Transverse glacier moraines were deposited under two conditions:
a). moraines deposited by individual glaciers coming down from isolated cirques.
b). moraines deposited by several glaciers after they had combined, to form a large glacier.

The first type is a small glacier which came down a high gradient slope and it was active for only short period, it has not deposited any true terminal moraine. At the lower end of its valley large angular blocks are scattered, which may be part of their terminal moraines. The second type has deposited terminal moraines at different levels. In East Liddar valley one such moraine was deposited jointly by all glacier that originated above Pahalgam, near the lower end of Pahalgam basin(2760m). The other moraine was deposited by another set of glaciers that originated from cirques below Pahalgam. Both these moraines were deposited in meltwater lakes and all the clay, silt and sand has been washed away. These moraines are marked by heaps of rounded and sub-rounded boulders lying accross the valley floor, and they are generally between 7 and 10 metres high. Wangat Nala moraine of Sind valley was deposited some distance above the lake levels and it has fine sandy matrix.

Some clasts in all these moraines may be larger than 10m in cross section but most of the clasts are less than 3 metres. Most of the large clasts are angular and sub-rounded, small clasts may be rounded and sub-rounded.

2. RECESSIONAL MORAINES

These moraines were deposited during halts in the retreat of the glaciers and minor advances. A metre or two high ridges of boulders mark the position of these moraines,

in the lower reaches of both the types of valleys.

In the upper reaches of the valleys, these moraines of longitudinal and transverse of glacier have differences.

In the upper reaches of longitudinal glacier valleys for several lengths of two or three kilometres, in both Kolahoi and Sonamara valleys, valley floor is covered with large angular blocks. These blocks have been deposited at base of hanging tributary valleys but they cover the entire width of the trunk valley. These blocks are lying loosely on the surfaces and they donot have any fine component mixed with them. Meltwater flows under these block. Near the snout of the glaciers recent recessional moraines form low ridges that are small, and discontinuous.

In the upper reaches of transverse glacier valleys these moraines have been eroded by meltwater streams and their position is marked by few isolated heaps of rock scattered at different places. Some times recessional moraines deposited behind the lateral moraine of a large glacier, are also preserved. These moraine form three or four metre high ridges, in which kettleholes, kame monds, eskers with small rounded pebbles and at some places current bedded sand lenses are also met with. In some cases these moraines with all these structures have developed with in three kilometres of the source.

In front of the living glaciers, at base of ice falls, piles of boulders have collected and from these also clay, silt and even fine sand have washed off, and the clasts are less that one metre in cross section.

3. GROUND MORAINES

Ground moraines were deposited in all the valleys but since the retreat of the glaciers these moraines have been eroded by the meltwater streams and now these moraines are exposed in isolated patches only. In these valleys Ground moraine is preserved only where the river has cut its valley below the glacial valley level or in small depressions where the till is overlain by lake deposits. Generally the thickness of these deposits does not exceed five or six metres, and in all the areas visited by me, till fabric was also not observed. At Liddarwat (Kolahoi valley) river has cut terraces in the ground moraine and it was observed that clasts of different sizes were embeded in clayey and silty matrix. One kilometre below this area all the fine clayey and sandy matrix of the ground moraine has been washed away and only large clasts of the moraine has been left.

In Pahalgam basin ground moraine is composed of large rounded clasts in fine and coarse sandy matrix. In East Liddar valley one kilometre up-stream from Pahalgam, ground moraine was possible deposited in a lake. This moraine has coarse sandy matrix, and poorly bedded layers of sand and boulders and pebbles. Overlying this moraine is another layer of morainal material with few boulder and pebbles and human artifacts in fine sandy matrix. This upper moraine was possibly deposited during a later advance, and is mostly composed of reworked earlier moraines deposited in the lake.

In transverse glacier valleys, these moraines were some times deposited in water and these are composed of poorly bedded badly sorted deposits resting on striated and polished floors. Glaciers coming down very steep slope have left behind a thick layer of loose angular boulders that cover the entire length of the valley as ground moraines.

4. LATERAL MORAINES

At the base of steep valley walls scree covers the lateral moraines. It is only at places wall is less steep that the lateral moraines are exposed for some distance along the slope. From clast composition it is difficult to distinguish between ground and lateral moraines.

Lateral moraines of some transverse glaciers encroche on the main valley floor and extend for some distance towards the middle of the valley. Such moraines were probably deposited after the retreat

of the main glacier and they are concave down stream which distinguishes them from terminal moraines.

In Zojipal basin, within five kilometres of the sources of Kohenhar glacier, valley takes a turn of about 30°. On the outward side of this bend, along the valley wall and for some distance up the slope lateral moraine has low drumlins. These drumlins were deposited behind the lateral moraine of a tributary glacier. In Sonamerg valley, during the last stage of the glaciation, the glacier was forced to turn at a sharp angle and move towards a gap cut by meltwater streams along one end of its terminal moraine. During this stage the terminal moraine was in the position of left lateral moraine, and on its slope low drumlins were deposited.

Small glaciers coming down from small cirques or from the cirques that developed near the lower limit of the snow line have deposited no lateral moraines. If these cirques were above the snow line and glacier have retreated within their cirques, they have deposited high lateral moraines on the two side of the cirques. These moraines are composed of only rock rubbles. In some large cirques these moraines form ridges that appear to have been pushed up in pulses.

The terminal moraine of Kolahoi glacier was also lateral moraine of Aru glacier, in the front side this moraine is also concave down stream.

5. MORAINES DEPOSITED IN LAKES

During maximum glacial advance terminal moraines were deposited at Mangam (7730m.) in Sind valley and at 2037m levels in Liddar valley and in Wangat Nala (a tributary of Sind). De Terra and Patterson (1939) have reported lake beds with invertebrate fossils at 2037m. in all the valleys. Possibly Mangam moraine was deposited under 100-300m of water and Liddar and Wangat moraines at the water level.

At Mangam the stratgraphic section of the glacial deposits is as follows:-

Down stream from the terminal moraines, for several kilometres extends a pebble bed with coarse sandy matrix. This bed passes upward into a boulder conglomerat bed with clayey and silty matrix. Boulder bed is overlain by yellowish brown fine sand and silt bands and poorly bedded coarse grained varves. A boulder bed with silty matrix overlie these varves. In this boulder bed, boulders are few and scattered, some times boulders and pebbles occur in bands and layers. The upper most bed is again a poorly bedded silt and fine sand varves.

The terminal moraine overlies lower boulder conglomerate bed and it is composed rounded and subrounded boulders in coarse matrix. Exposures of this terminal moraine are met with on the two valley walls and they laterally extend for half a kilometre.

Upstream from the moraine, the pebble bed is directly overlain by boulder bed with clayey and silty matrix.

Possibly during early stage of the glacial period the water level of the lake was low and the meltwater coming down the valley deposited the pebble conglomerate. With time, while the glacier advanced towards Mangam, water level of the lake may also have risen. During this period boulder bed with clayey matrix was deposited under partly floating ice shelf. After the deposition of this boulder conglomerate, may be the intensity of the glaciation increased and the rate of melting decreased, so that in front of the ice-shelf only poorly sorted and coarsely bedded varves were deposited, and boulders falling off the front of the shelf collected to form the Mangam moraine. In front of the moraine in deep waters clay and silt was being deposited, and on the surface ice bergs broke away from the ice front. These ice-bergs transported boulders and pebbles and as it melted these pieces were dropped in the clay bed. Some times overloaded ice-bergs, settled in water and melted. At the spot ice-berg was stranded, pebble and boulder lenses or layers were deposited.

On the upstream side of the

terminal moraines, boulder clay over-lying the pebble bed was possibly deposited under the floating ice shelf.

At the mouth of Wangat Nala in Sind valley and in Liddar valley probably the terminal moraines were laid at the water level. In Liddar valley the moraine overlies pebble conglomerate and it is overlying by yellow clay and silt varves, with invertebrate fossils. Wangat Nala beds are overlying bed rock and it is also overlain by silt and sand varves. Possibly by the time these moraines were deposited the water level had risen and poorly bedded varves were deposited on top of it.

In all the valleys small lakes had formed in depressions behind the riegels, and in these depressions clay and silt was deposited by meltwater streams, and now they form flat grassy plains in which braded and meandering streams are flowing.

During the maximum of the glacial period cirques had developed at the lowest snow line level. Since the snow line has retreated uphill these cirques are open and melt water has collected to form firly large lakes (Seshram Lake is 25 sq.kilometres in area). Meltwater streams bring clay and silt to these lakes and some of the lakes have been completely filled with this material and others are partly filled up.

During Pleistocene period meltwater from all the glaciers sorounding the Srinagar valley brought down fine silt and clay to the lake. In this lake about 700m thick varve beds were laid down. At places these varve beds have lenses of boulders and sand. These were possibly transported by meltwater when snow and ice suddenly melted along the slopes facing the lakes.

6. REFERENCES

Ahmad, N. & Hashmi, H. H. 1974, Glacial history of Kolahoi glacier, Kashmir, J.Glaciology. 13: 279-283.

Dainelli, G. 1922, Fillippi expedition to Himalayas, Karakoram and Chinese Turkistan (1913-14). Report Ser.2, 3: 659 (Italian Text).

Grilington, J. L. 1928, Former glaciation of East Liddar Valley, Kashmir. Geol.Surv.Ind.Mem.,49: Pt.2, 289-388.

Holland, T. H. 1907, A Priliminary survey of certain glaciers in N.W.Himalayas-Introduction. Rec. Geol. Surv. Ind., 35: Pt.3, 123-126.

Terra, H. De & Paterson, T.T. 1939, Studies on the ice age in India and associated human culture. Carnegie Inst.,Washington D.C., Publ. No.493.

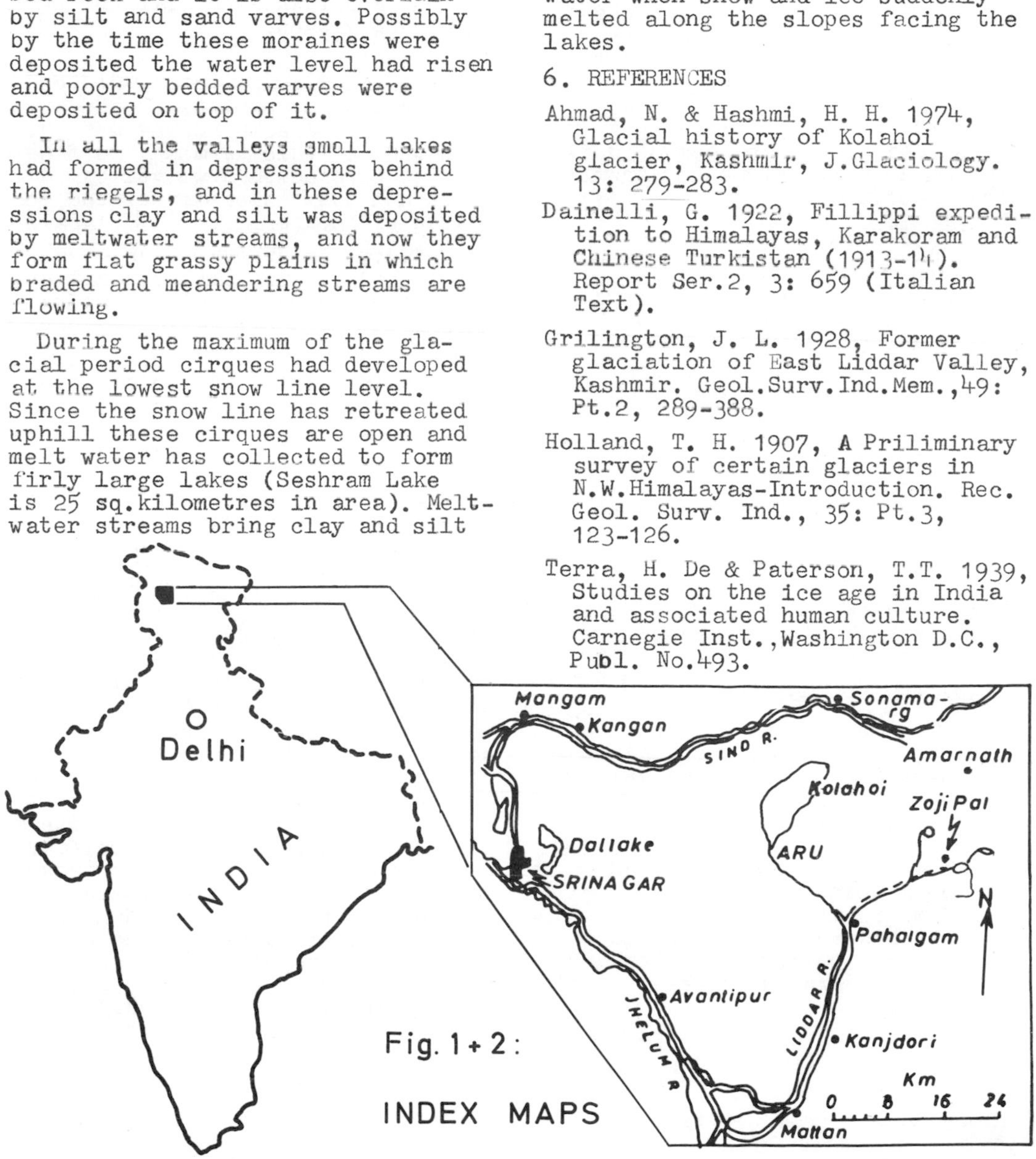

Fig. 1+2:

INDEX MAPS

Rate of formation and sedimentology of (1976-1978) push-moraines, Frias Glacier, Mount Tronador (41° 10'S; 71° 53'W), Argentina

JORGE RABASSA, SIGFRIDO RUBULIS & JORGE SUÁREZ
Fundación Bariloche, San Carlos de Bariloche, Argentina

ABSTRACT

Frías Glacier, a rapidly advancing (surging?) glacier of Mt. Tronador, reached in 1976 the levelled surface of its fluvioglacial plain, a former proglacial lacustrine plain, drained out some time between 1942 and 1953. The advancing glacier front has bulldozed the sediments forming the top of the fluvioglacial plain and has built up a push-moraine arc, discontinued only by the effect of the two major subglacial streams, coming out from beneath the glacier. These push-moraines have reached in 1977 an altitude of 2.5 m above the level of the fluvioglacial plain, after a one-year, 38-meter advance of the glacier front. In 1978, the process of moraine construction has evolved at a similar rate. The moraines are composed of rounded, clean gravels, which genetically correspond to a fluvioglacial environment. However, the marginal sectors of the morainic arc are also integrated by angular psephites both of colluvial-gravitational and nivation origin, which have been partially picked up by the glacier and have partially slid down to the glacier front from the glacier surface and/or the valley sides.

1. INTRODUCTION

Glacial activity comprises a varied complex set of physical processes having different characteristics, where ice action merges with melt-water reworking of surfaces and sediments together with nivation and mass-movement processes. This coexistence of glacial, ice-contact, and periglacial processes is particularly significant in an alpine-glaciated environment, where the valley floors are affected by the convergence of flow-directions of varying kinds. Therefore, the sediments formed under these conditions will reflect the complexity of the original environment in their sedimentological characteristics. The examination of glacial and associated processes at work at the border or beneath present-day glaciers has proved to be an effective way of unraveling the intricate relationships between depositional agents and their products in the study of glacial drift.

Our work on the glaciers of Mount Tronador and other areas of the Northern Patagonian Andes of Argentina was actually started in 1974 partly with this aim: to provide a reliable knowledge of glacial and associated processes that would enable us to understand the complexity of the sediments found in the Bariloche Moraine (Late Glacial; Flint & Fidalgo, 1964). This paper presents the results of our observations at Frías Glacier, an advancing glacier of Mount Tronador, and its push-moraines developed between March 1976 and March 1978 (end of the Southern Hemisphere Summer). The interpretation of the acting processes is based upon measurements and evidence collected in the field. A sedimentological characterization is attempted, with emphasis on the morphometry, granulometry and petrography of sands and gravels.

2. FRIAS GLACIER

Mount Tronador (71°53'W; 41°10'S; 3554 m.a.s.l.) is an ancient volcanic cone, superimposed on an erosional relief in the Northern Patagonian Andes. Its age is still uncertain, but is probably Upper Miocene to Lower Pleistocene. This mountain stands 1300-1400 m above the summit accordance of the Andes in this region. Its prominent height means that most of its upper portion is well above the regional snow-line and it

is completely covered by a thick ice-cap. Out of this ice-cap more than ten important ice-tongues descend, four of them into Argentine territory. From N to S: Frías Glacier, Alerce Glacier, Castaño Overo Glacier, and Río Manso Glacier. These last three drain into the Upper Río Manso Basin, to Lake Mascardi, and then, by the lower Río Manso through Chile to the Pacific Ocean. Frías Glacier drains into Lake Nahuel-Huapi and through the Río Limay, to the Atlantic Ocean (Figure 1).

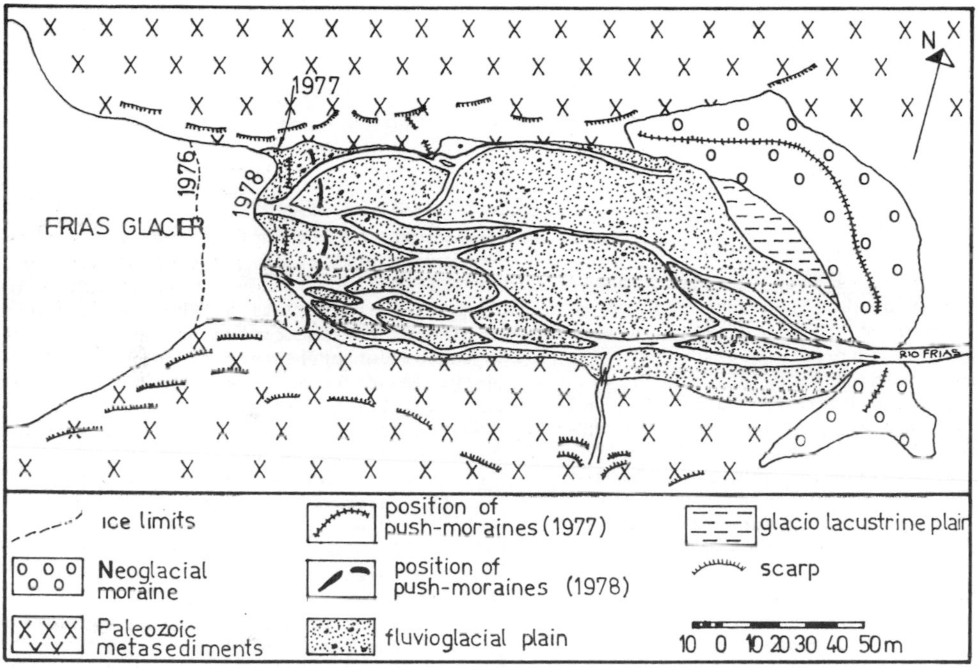

Figure 1: Map of Frías Glacier Snout

The glaciers of Mount Tronador have been studied by Lliboutry (1956), Mercer (1967), Colqui (1970), and Rabassa et al. (1975, 1977), among others. Frías Glacier is the northermost ice body of the Argentine sector of Mount Tronador. This glacier is also the only one of those mentioned here whose accumulation zone, ablation zone, and ice-movement direction are north-oriented. The characteristics and dimensions of Frías Glacier have been summarized in Table 1.

Table 1: Frías Glacier

Latitude	41° 10' S
Longitude	71° 53' W
Orientation accumulation zone	NE
Orientation ablation zone	NE
Area (total) (sq. km)	14.39
Length (km)	7.4
Minimum elevation (m.a.s.l.)	845
Maximum elevation (m.a.s.l.)	3200
Ice-falls	3
Ogives	8

Table 1. (cont.)

Supraglacial detritus	scarce
Intraglacial detritus	very scarce
Medial Moraine	no
Lateral Moraine	yes
Frontal Moraine (push-moraines)	yes
Neoglacial Morainic Arcs	3
Fluctuations of the ice-front (m) (*)	
1942-1953	- 352
1953-1970	- 166
1970-1976	+ 533
1976-1977	+ 32
1977-1978	- 7

(*) Source: 1942 - Aerial photographs, 1:40000 (Army Geographical Institute).
1953 - Aerial photographs, 1:50000 (Water and Energy Board).
1970 - Aerial photographs, blow-ups 1:6500 (National Parks Service).
1976 - 1978 - Our own plane-table map, 1:2000.

Figure 2: Front of Frías Glacier, seen from the western side of the valley. See the steep front of the glacier snout, the push-moraines, the false "trim-line" developed at the eastern side of the valley on metamorphic rocks. At the top of the slope, the Nothofagus pumilio (southern beech) forest. Río Frías is coming out from beneath the glacier (Photo by S.Rubulis, March 1978).

Frías Glacier has an important accumulation zone that starts from the Argentine Peak of Mount Tronador. It is a wide, gently sloping iceshed, descending from 3200 to 1600 m.a.s.l. At this altitude, the ice-mass moves over two rocky cliffs surrounding the mountain (i.e., the border of the volcanic plateau). Further down, the glacier comes down as an outlet valley glacier, deeply entrenched into a Late-Glacial, U-shaped valley, carved into Upper Paleozoic metasediments. There the ice surface is completely crevassed from side to side of the valley. Below this section, at 1100 m.a.s.l. approximately, a gently sloping portion shows a group of 8 ogives clearly separated, and a group of several more pressed one against the other. Then the ice-tongue moves over a second step, showing reactivated crevassing at the surface. The lower part of the tongue has reached the surface of an ancient proglacial lake, at the bottom of the valley. This lake, called "Lago Témpanos" (Lake of the Icebergs) in the old maps of the region (Bailey Willis, 1914), was a small depression between the ice-front and the Late Neoglacial Morainic Arc, fed by meltwaters coming from the glacier.

Lake Témpanos disappeared some time between 1942 and 1953 (according to the available vertical aerial photographs), in coincidence with a period of strong recession of the ice front. Probably the excess of meltwater raised the level of the lake and it poured heavily through the Neoglacial Morainic Arc, widening the spillway of the lake and provoking drainage of this basin. Afterwards, the lacustrine plain was mostly covered by fluvioglacial sediments, and a fluvioglacial plain was built on the ancient bed of the lake.

The front of the glacier is now more than 40 meters high at the snout, with a very steep ice-wall (40° to 75°) and with two major ice-caves where the Río Frías rises (Figure 2).

3. NEOGLACIAL MORAINES

Three Neoglacial Morainic Arcs have been identified at the bottom of the valley, at elevations varying from 836 m.a.s.l. (the youngest) to 760 m.a.s.l. (the oldest at Puerto Blest, Lake Nahuel-Huapi), but separated from one another by several kilometers. These morainic arcs have not yet been studied in detail, but it seems highly probable that they correlate with the three Neoglacial events described by Mercer (1968, 1970, 1976) in Chile and Southernmost Argentina.

The younger Neoglacial Moraine is a low, narrow arc, partially covered by small trees and shrubs, probably formed during the XVIII and XIX centuries (Rabassa et al. 1977). Meltwaters were dammed by this moraine and the Lake Témpanos originated. Some parts of the moraine lack drainage and glacial sediments outcrop in others. The moraine is composed of dark, sandy till, with large rounded boulders. The elevation of the Late Neoglacial Moraine is nowhere higher than 10 meters above the outwash-plain level.

4. 1976-1978 PUSH-MORAINES

During our visit to the snout of Frías Glacier in March 1976, we could observe that the ice front had reached a topographic position similar to the one shown in the 1942 aerial photographs, i.e., the bottom of the former proglacial lacustrine plain at the end of the rocky slope.

According to the 1970 aerial photographs, the ice front had advanced 533 meters since January 1970 but it is still not known which was the actual period of advance.

At all events, this advance of Frías Glacier has been very important and there are several characteristics that suggest a surge-type advance.

In March 1976, no moraines had been found in front of the glacier. This was possibly due to the very steep slope of the valley floor which prevented the concentration of debris in front of the glacier. The ice was also creeping up the southeastern side of the valley; some shrubs and small trees had been overwhelmed by the ice as the glacier was increasing in width.

At the northwestern side, the glacier showed a thin cover of supraglacial debris, coming from an eroded wall of volcanic rocks. In March 1977, the ice front had already advanced over the proximal part of the fluvioglacial plain (which had evolved on top of a former ice-contact lacustrine plain), bulldozing its upper sedimentary cover into a set of push-moraines but also contributing with some supraglacial detritus and rock-fragments falling on top of the glacier from the steep valley walls.

Three well-defined push-moraines were then observed: a western arc, 2.0 m high, 12 m long, 4 m wide; a central arc, 2.5 m high, 25 m long, 5.5 m wide; and an eastern arc, 2.0 m high, 10 m long, 3.0 m wide. The push-moraines were the result of an advance of the ice front of 32 m at the sides of the valley, and 45 m at the central part. In all cases, the ice was still in

Figure 3: Push-moraines in contact with the ice front (Photo by S. Rubulis, March '77).

Figure 4: Arcuate segments of the central morainic arc, in March 1978. Ice had already receeded from its maximum extent (Photo by S. Rubulis).

contact with the moraines at the end of the ablation season (Figure 3).

In March 1978, the moraines had been slightly modified by the continuity of the advance. The ice had advanced up to 16 m from its March 1977 position, probably during the Late Winter or Spring, but it finally receeded a maximum of 23 m towards the end of the Summer (i.e., a total retreat of - 7 m from the March 1977 position) (Figure 4). The year 1977 was exceptionally warm, according to our records (Rabassa and Rubulis, 1978). The western moraine has been transformed into a lateral moraine by the forward movement of the ice, a sharp ridge 22 m long, 2.8 m high and 3.5 m wide. The sediment forming this moraine is a mixture of the 1977 push-moraine, plus some very-angular debris slumped down from the "dirty" side of the glacier. Angular blocks up to 1.0 m in diameter, and smaller angular to sub-rounded clasts with abundant sandy matrix have been found. The central arc has split into two very arcuate segments: a western-, and an eastern portion, separated by a meltwater-originated gap. The western section is just a small till hummock, 2.2 m long, 1.0 m high and 1.0 m wide, sub-angular to rounded gravels with abundant sandy matrix, larger clasts up to 0.30 m in diameter. The eastern portion is composed of two segments: the first one is 7.5 m long, 1.5 m high and 3.5 m wide, and it is formed by rounded to sub-angular pebbles, with larger blocks up to 0.35 m in diameter. This segment of the push-moraine is accompanied by a group of 3-5 mini-moraines in the inner side of the arc, up to 0.50 m high and 1.00 m wide. These very small ridges may be related to minor readvances after the maximum extent, once the retreat had started. The second segment is very arcuate, 17 m long, 2.7 m high and 3.0 m wide. This push-moraine is composed of two different areas, on both sides of the inflexion point. To the west of it, the material is quite similar to the western portion and first segment of the eastern portion, rounded to sub-angular blocks and pebbles with very abundant sandy matrix, and the form is also lower, only 1.5 m high. The section to the east of the inflexion point is higher and wider than the other, and it is 12.5 m (out of the 17 m) long. The sediment is much more angular and matrix is coarser. Large blocks up to 1.00 m in diameter have been observed.

The eastern arc has also been pushed away by the glacier, but not as much as the other ones: the eastern border was still only 2.00 m from the ice front in March 1978. A significant contribution has been made to this moraine from snow and ice avalanches that had affected the eastern limit of the glacier. These avalanches have moved down the ice border, cleaning away the detritus and vegetation out of a belt surrounding the glacier. A false "trimline", 25-m wide, has evolved in this part of the glacier, and it should not be confused with actual glacier advances. The sediments forming this arc are very angular cobbles and pebbles without matrix. The sandy matrix has been washed away by meltwaters coming down by the avalanche track. Larger cobbles are more than 1.00 m in diameter, and the mean diameter of the pebbles is around 0.18 m. Between these arcs, meltwater streams coming from beneath the glacier have removed the sediment formerly integrating the moraines. Downslope from the morainic arcs, the outwash plain has kept its standard characteristics and there is no evidence of ice or water-pressure perturbations on the deposition of the gravels; a sharp break in the slope separates the push-moraines from the fluvioglacial plain.

5. SEDIMENTOLOGY OF THE PUSH-MORAINES

The push-moraines are formed by till, angular to rounded gravels with sandy matrix. The sedimentological analysis has been restricted to three samples, one from the western arc (M.A), one from the middle portion of the central arc (M.B), and one from the outer portion of the eastern section of the central arc (M.C). Because of the small dimensions of the morainic arcs, it is considered that the samples are representative.

Sampling was performed on the inner side of the arcs; after cleaning the upper 0.20 m of the surface with a shovel, a generous sample of the "matrix" was obtained. "Matrix" refers here to sediment particles smaller than 0.10 m in diameter. Observations were made on the distribution of larger clasts within the moraine. The samples obtained are the following:

Sample A: from central part of inner side of lateral, western moraine; total weight: 3890.7 g; larger, angular clasts of up to 0.30 m in diameter.

Sample B: from middle section of central moraine, inner side; total weight: 4376.6 g; sub-angular to rounded clasts up to 0.20 m in larger diameter and 0.07 m in shorter diameter.

Sample C: from outer portion of eastern central moraine, inner side; total weight: 2805.0 g; cobbles up to 0.12 m in larger diameter and 0.05 m in shorter diameter.

5.1. Granulometry

Samples were dried in an electric oven, weighed, and wet-sieved over an A.S.T.M. sieve series, at 0.5 phi grade intervals, from -4.0 phi to 4.5 phi. The psephitic fraction above -4.0 phi (i.e. above 16 mm in diameter) was separated for morphometric analysis. The granulometry of the pebble fraction shows a size distribution roughly similar between the three samples, although Sample C is slightly finer, no clasts in the -4.5 / -6.0 phi interval. The size distribution is shown in Table 2.

Table 2: Size distribution (% in weight)

	Phi Units	Sample A	Sample B	Sample C
Pebble	-6/-2	60.89	55.27	59.93
Granule	-2/-1	15.69	9.83	10.56
Sand	-1/4	21.43	33.16	27.41
Silt + Clay	< 4	1.99	1.69	2.11

The three samples are gravels, with a sandy matrix and very low fines content. A detailed representation of the distribution of size fractions after separation of the clasts larger than 16 mm is presented in the histograms and cumulative curves of Figures 5 and 6.

The observation of these figures shows that the three samples are polimodal, or perhaps it would be more correct to say they do not have any mode at all. However, they present an important concentration in the interval -3.5 / -3.0 phi (11.32 to 8 mm fine pebbles) and a curious deficiency at -3.0 / -2.5 phi (8 to 5.66 mm, very fine pebbles). It is not easy to find a proper explanation for this deficiency, but it seems more probably related to a characteristic of the source-rock (metamorphosed sedimentary rocks) than to the acting processes. Spalletti & Gutierrez (1976) studied active and recent moraines of San Lorenzo Glacier (Santa Cruz, Argentina; 47°30'S; 72°00'W). They present the results of four samples; three of them from recent moraines and the other from an active moraine, which is known to be in contact with the ice and considered by Spalletti (1975, 1976) as a recessional moraine although the reasons are not given in his papers. It is interesting to note that the sample obtained from the active moraine has a similar histogram and also shows a deficiency like those observed in our own samples, but in this case in the interval -3.0 / -4.0 phi (Spalletti & Gutierrez work on 1 phi unit interval). The other three samples have also polimodal distributions and distinguishable concentrations, but the histograms are substantially different.

The very small percentage of silts and clays in our samples suggests that a detailed investigation of these minor fractions would not be relevant. Thus, our effort has been concentrated on the gravel and sand fractions. The statistical parameters were calculated following Folk & Ward (1957), and the results are presented in Table 3.

The values obtained demonstrate that the matrix of these till is coarse, very poorly sorted, with slight positive asymmetry, and very flat curves. The sediments studied by Spalletti & Gutierrez, (1976) are finer than our tills (average of the mean size, 1.17 phi units; i.e., 2.3 mm) but the other statistical parameters are very similar to those obtained by us. C/M values (Passega, 1964) were plotted on the corresponding diagrams, and the samples fell within the field of the sediments formed by transport agents of great competency and viscosity.

5.2. Granulometry, morphometry and lithology of the gravel fraction

The psephitic fraction above -4 phi of these three samples has been investigated in terms of size, sphericity, roundness, shape, and lithology. The a, b, and c axes were determined with the aid of a calibre, following Krumbein (1941). Mean size (a+b+c/3), flatness (a+b/2c), b/a and c/b ratios, sphericity ($\sqrt[3]{b.c/a^2}$), and F-factor ($a.c/b^2$; Aschenbrenner, 1951) were calculated.

Roundness was estimated by visual methods, using a table proposed by Ruhkin (1961; in

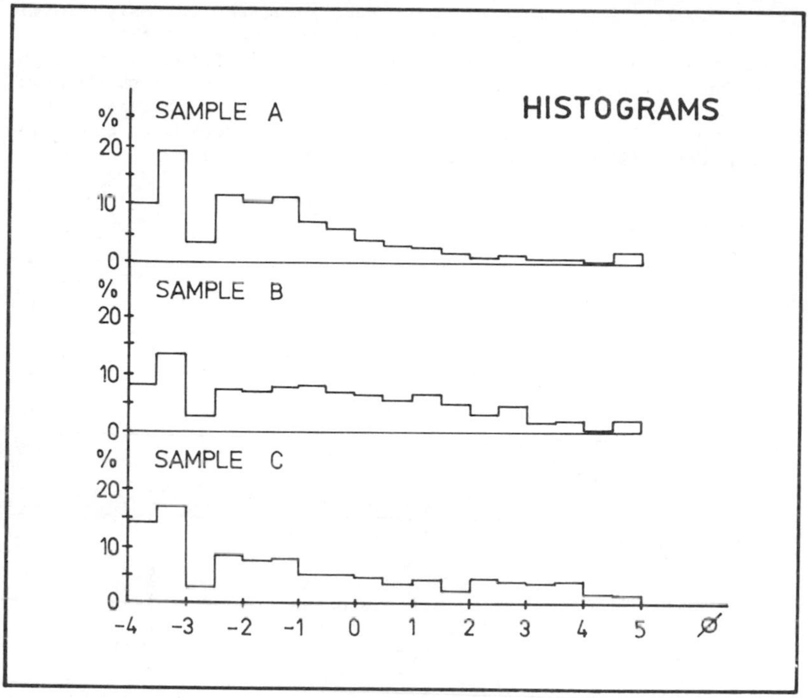

Figure 5: Granulometric analysis: histograms.

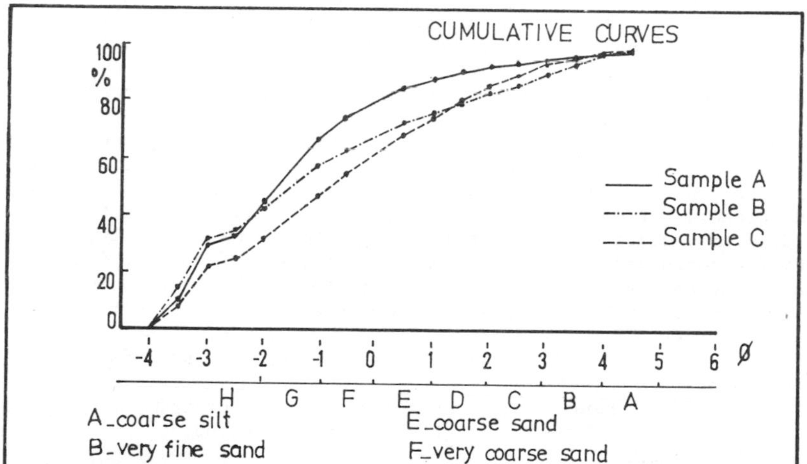

Figure 6: Granulometric analysis: cumulative curves.

Table 3: Statistical Parameters

	Sample A	Sample B	Sample C
Median	- 1.75	- 0.79	- 1.47
Mean	- 1.56	- 0.72	- 0.92
Standard Deviation	1.96	2.31	2.56
Skewness	0.27	0.11	0.33
Kurtosis	1.03	0.80	0.76

Marchese, et al., (1970). In cases where it was difficult to include a certain pebble within one of the five classes (0 to 4, the latter being "very well rounded"), it was considered as having an intermediate value between two of them, i.e., 0/1, 1/2, etc.

Shape was determined using the b/a and c/b ratios on Zingg's (1935) diagrams. A clast was considered "broken" when it showed fracture surface of a particular roundness value of 0 or 0/1 which was also lower than the roundness value of the rest of the clast surface.

Lithology of the pebbles was restricted to only two classes, "Metasedimentary rocks" and "Basaltic rocks", which represent the whole set of lithostratigraphic units present within Frías Glacier iceshed. The metasedimentary rocks (Upper Paleozoic) underlie the basalts (Pliocene); the contact is located approximately at 1700 m.a.s.l. and it is nearly horizontal.

The different parameters were computed for every clast and the average values for each sample are presented in Table 4, together with their standard deviations.

Table 4: Granulometric, Morphometric and Lithological Analysis of Pebbles

	Sample A	Sample B	Sample C
Number of clasts	39	50	37
c/b ratio (mean)	0.585	0.662	0.549
St.dev.of c/b	0.140	0.157	0.162
b/a ratio (mean)	0.802	0.772	0.729
St.dev.of b/a	0.137	0.115	0.174
Sphericity	0.713	0.726	0.653
St.dev.of sphericity	0.091	0.089	0.126
Flatness	2.066	1.876	2.453
St.dev.of flatness	0.526	0.562	0.882
F-factor	0.764	0.886	0.805
St.dev.of F-factor	0.278	0.282	0.350
Shape (%)			
Equidimensional	15.4	42.0	18.9
Oblate	64.1	40.0	51.3
Prolate	12.8	12.0	5.4
Laminar	7.7	6.0	24.3
Formula of shape distribution	$O_8E_2P_2L_1$	$E_7O_7P_2L_1$	$O_9L_4E_3P_1$
Broken clasts (%)	23.1	38.0	40.5
Roundness	0.86	1.57	1.08
St.dev.of roundness	0.72	0.98	0.97
Roundness distribution (%)			
Class 0	23.1	10.0	21.6
0/1	25.6	12.0	21.6
1	20.5	18.0	24.3
1/2	25.6	18.0	8.1
2	0	18.0	13.5
2/3	2.6	8.0	0
3	2.6	12.0	8.10
3/4	0	4.0	2.7
4	0	0	0

Table 4: (cont.)

	Sample A	Sample B	Sample C
Mean size (mm)	28.92	28.55	26.82
St.dev.of mean size	9.86	10.06	7.49
Mean size (phi units)	-4.63	-4.53	-4.45
Lithology (%)			
Metasedimentary rocks	79.5	86.0	86.5
Basaltic rocks	20.5	14.0	13.5

This Table presents an interesting set of similarities and differences between the studied samples. The c/b ratio is considerably lower than the b/a ratio for all samples. This, of course, is also reflected by the other parameters dealing with axes relationships. Sphericity and c/b ratio of Sample C are the lowest for this group, while flatness is the highest. F-factor is lowest for Sample A and highest for Sample B, but in all cases it is lower than the unity, showing that "oblate" clasts are predominant, i.e. the clasts plotted on the upper-left half of a Zingg's diagram, over the "prolate" clasts, i.e. those that fall within the lower-right half. Shape distribution exposes a pronounced difference between Sample B (very high content of equidimensional clasts) and the others. On the other hand, Sample C is rich in laminar clasts. The three samples have also a similar high percentage of oblate clasts. Shape distribution is further emphasized by Zingg's diagrams (Figure 7).

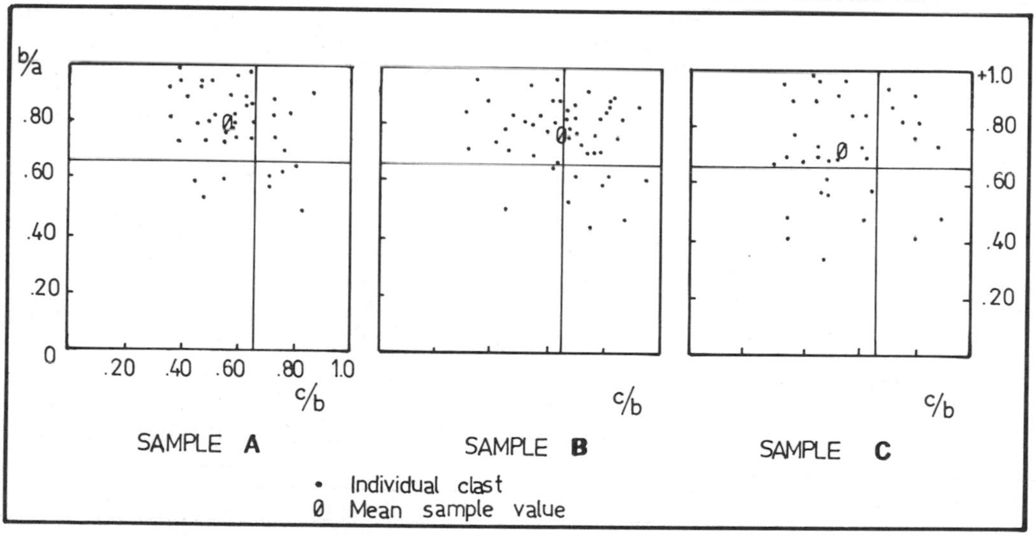

Figure 7: Pebble morphometry: Zingg's diagrams.

Curves that join points with the same percentage of clast concentration have been drawn on these diagrams, using a 10% by 10% network. A pattern of maximum concentration on two "poles", approximately located at b/a: 80/90 % - c/b: 60/70 %, and b/a: 60/80 % - c/b: 70/80 %, respectively, has been found for Sample A and B, whereas

Sample C shows only one"pole"at b/a: 60/70 % - c/b: 50/60 %. Two samples of Late Glacial till studied by the senior author (Rabassa & Aliotta, this volume) have a concentration pattern very similar to that of Samples A and B.

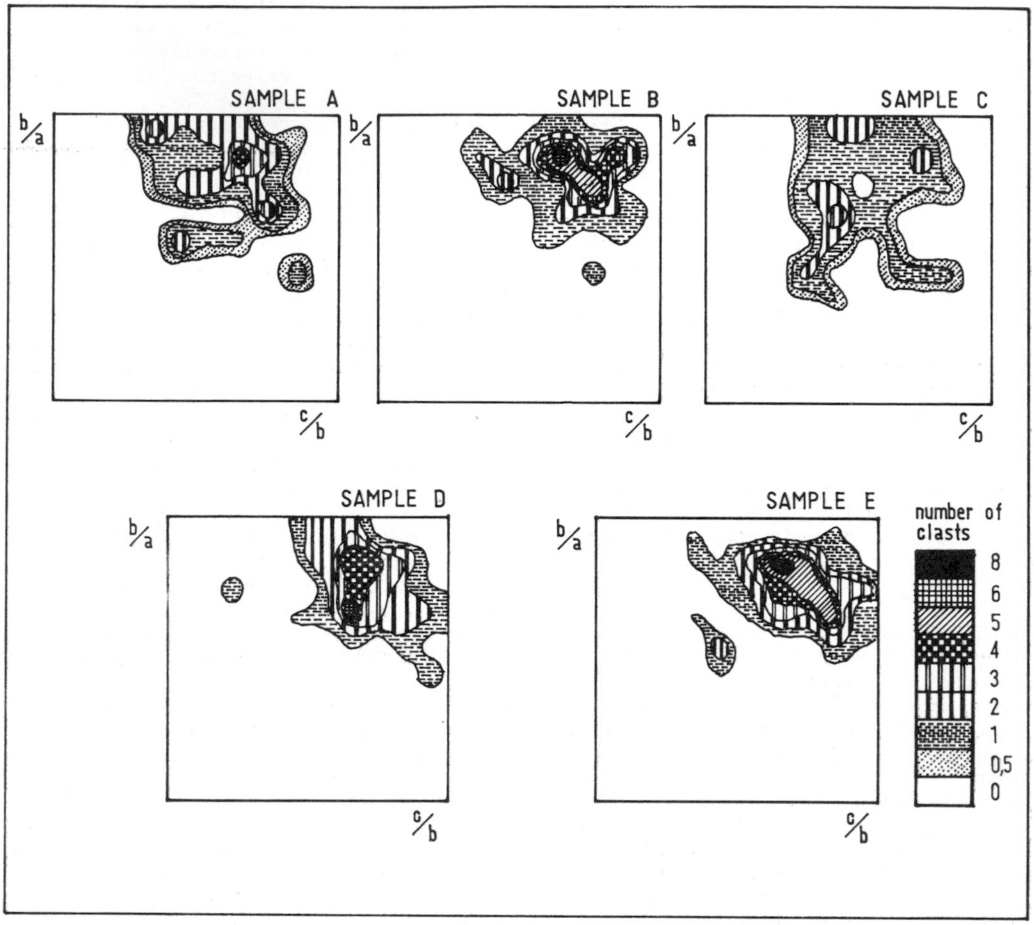

Figure 8: Clast concentration on Zingg's diagrams. A, B, C, Frías Glacier Push-Moraines; D, lodgement till, Bariloche Moraine; E, ablation till, Bariloche Moraine.

Roundness is also a clear discriminant parameter; it is very low for Sample A, low for Sample C, and moderate for Sample B. The distribution is quite significant: clasts of Sample A concentrate heavily on the most angular classes, whereas Sample B has a much wider distribution with important percentages of rounded-to-well-rounded particles.

Finally, the lithology of Sample A is quite different from the other two samples, with 50 % more of basaltic clasts.

5.3. Structure of the deposits

No sedimentary structures have been found in these sediments. They are massive, chaotic in appearance, and the excess of loose, coarse fractions does not favour the development and preservation of stratification and other sedimentary structures

either primary or secondary. It should be mentioned, however, that the morainic forms exhibit a concentration of boulders and cobbles at the toe of the front slopes. Others have been found to rest on top of the morainic ridges, but they do not stay on the slopes. The concentration of boulders at the toe of the slope is easily explained in terms of gravity movement of the larger clasts downslope over the steep, unstable sides of the newly-formed moraines. Those appearing on top of the ridges may have arrived there when the form was still in contact with the ice. We assume that these clasts will not remain long on top of the moraine and that degrading processes will take them to the bottom of the slopes in a short time.

Another structure that should be mentioned is the open-work character of the gravels forming the eastern morainic arc. It was not possible to get a sample of the matrix of the till in this section, for there are very few pebbles smaller than 120 mm (approximately - 7 phi) in diameter. The elimination of the finer fraction is a postdepositional process, under permanent washing-out by meltwaters from the glacier or from snow-accumulations.

6. INTERPRETATION

The process of formation of these moraines needs no interpretation, for the evolution of the ice-front has been followed for two complete years, under very favourable conditions, and the building-up of the push-moraines has been actually observed and photographed. Therefore, it is not necessary to look for alternative mechanisms for this kind of glacial feature, such as those suggested by Elson (1968) and Worsley (1974). Nevertheless there are several aspects that remain obscure to the observer and we think that geomorphic and sedimentological analysis may help in solving these questions.

The 1976-1978 push-moraines of Frías Glacier have been formed by an advance of the ice-front. The continuity of the morainic system is interrupted by depressions, most of them occupied by meltwater streams. The system is divided into three major ridges: the western arc, the central arc, and the eastern arc. The topographic and geomorphic position of these ridges seem to account for the differences in sedimentological and morphological characteristics. The western arc is formed by till, coarse, angular gravels with very scarce sandy matrix; pebbles of low sphericity and extremely low roundness; high content of basaltic clasts. These materials (Sample A) show the incorporation of supraglacial debris of basaltic origin (see Figure 2) coming from the western valley-wall. Part of this basaltic debris may be carried down to the moraine by nivation processes, on top of the glacier or along its western side. This interpretation explains the extremely low roundness values and the relative abundance of basaltic pebbles.

The central arc represents the most extended part of the push-moraines. Sample B comes from the middle segment of this arc. The till forming this part of the moraine is characterized by a higher content of sandy matrix, a clearly smaller mean size for the fraction lower than 16 mm, and most important, the highest values of sphericity, F-factor, c/b ratio, percentage of equidimensional pebbles and roundness, together with the smaller flatness value.

Percentage of broken clasts and lithology suggest an almost similar origin for Samples B and C, quite different from Sample A. Till of Samples B and C has been formed mainly by the pushing effect of the advancing glacier on the sediments of the proglacial plain, primarily of fluvioglacial origin, whereas till of Sample A is "contaminated" by materials brought into the moraine by nivation or gravitational processes. Sample C has lower values of roundness, sphericity, c/b ratio, F-factor and percentage of equidimensional pebbles than Sample B. It also has much higher values of median and flatness, the latter directly related to the high content of laminar clasts. The origin of these laminar clasts should be related to a significant supply of pebbles of metamorphic rocks, with very short or no transportation, to the eastern side of the fluvioglacial plain from the neighbouring rock walls of that composition. This material has mixed up with the sediments of true fluvioglacial or proglacial origin; this mixture explains the wide distribution of roundness classes in Sample C (cf. Table 4).

The till of these moraines has a common feature: the absence of finer materials as silt and clay in the matrix. This characteristic is related both to the pre-existing sediments and to the depositional and post-depositional processes that originated the till. Pre-existing sediments are fine, rounded gravels with sandy matrix of a fluvioglacial (proglacial) environment; these gravels grade laterally (i.e., towards the sides of the valley) into less-rounded gravels of proglacial origin mixed up with colluvial and/or nivation psephites, also with very scarce matrix. In this genetic type, fines have been eliminated by normal processes in a dynamic environment, controlled by

a fluid transportation agent. Depositional and postdepositional processes affecting the genesis of these push-tills are also heavily controlled by the abundance of meltwaters coming from the glacier or neighbouring snow-cover. The snout of the glacier is located at a very low altitude (845 m.a.s.l.), at the bottom of a protected valley where Summers are very temperate. Thus meltwater is most abundant and has washed the finer materials out of the till.

7. WAS THE ADVANCE OF FRIAS GLACIER A SURGE?

This question remains unsolved, unfortunately. We have not been able to obtain information on the intermediate years of the 1970-1976 period, when Frías Glacier advanced at a minimum average rate of 89 meters per year. However, it is possible that Frías Glacier had started its readvance during 1975, because the neighbouring Alerce Glacier was still receding in March 1975, when it reached the minimum extent of our records (Rabassa et al., 1977). Thus, the total advance of 533 meters would have been made in 1975-1976 only and continued during 1976-1977 and 1977-1978, although at a much reduced rate. This rate of movement is in accordance with the published data for glacier surges (Johnson, 1972; Meier and Post, 1969; Post, 1966, 1969, 1972; Embleton & King, 1975). A surge-type advance of Frías Glacier is also suggested by the following reasons:

(a) Frías Glacier has an extensive accumulation area (over 14,0 sq.km), but a very narrow outlet valley tongue of less than 0.5 sq.km (Rabassa et al., 1977). Thus, any change in the accumulation or storage rates in the upper zone would be reflected by the tongue, even if the changes were small.

(b) The strong slope of the upper portion of the outlet valley will favour any downward movement of the ice.

(c) The 1975 (?) -1976 advance has not been a gigantic slump of the ice tongue, because the continuity of the ice was not interrupted.

(d) Disruption of the ice is very intense in the upper portion of the tongue, with various crevasse sets in different directions, although this should be regarded with care because the glacier runs here over a very steep ground slope.

(e) The ice tongue is climbing up against the eastern wall of the valley (Figure 2) due to its rapid downslope movement which did not allow the glacier to take a smooth turn at the bottom of the slope.

(f) The snout of the glacier has constructed a set of push-moraines at the end of the advance and when topographical conditions and sediments were available in front of the glacier. These moraines are very arcuate, the convexity facing downslope the valley.

8. CONCLUSIONS

The study of the 1976-1978 push-moraines of Frías Glacier has shown the importance of coordinated field and sedimentological studies. Some features that have not been identified in the field were pointed out by granulometric, morphometric and lithological analysis of the pebble and sand fractions. Due to the restricted extent of these moraines, a few samples were enough to characterize the sediment type and to discriminate between different source depositional environments, such as the proglacial plain and the marginal plain mixed with colluvial and nivation materials.

Granulometry and morphometric properties were useful for the sediment characterization and amongst the latter, those related with sphericity and roundness. A quantitative formula of shape distribution such as the one used seems to be an adequate tool for rapidly describing a pebble population. Till forming these push-moraines was found to be substantially different from either ablation till or lodgement till of the Late Glacial Moraines in Bariloche (Rabassa and Aliotta, this volume), which are much richer in finer fractions and rounded to well-rounded pebbles. The proximity to the glacier, its recent nature and the different processes involved in its formation account for this difference. It is interesting to note that significant variations have been observed over very short distances within this same feature. Therefore, it should be kept in mind when studying ancient valley moraines, that the influence of the local relief is strong and that the superimposition of acting processes has given the sediments some special characteristics. As has been shown here, high values of roundness of till pebbles is not necessarily related to glacial transport, but may be inherited from other processes or environments.

The size and rate of formation of Frías Glacier push-moraines are similar to identical morainic systems in other parts of the world. The height of the moraine increased from 2.0 m in 1977 to 2.7 m in 1978. Height was found to be roughly proportional to maximum annual advance but it is primarily controlled by sediment availability. The value of 2.7 m is very similar to those

presented by Johnson (1972) for a push-moraine in front of Donjek Glacier, Alaska, which is much larger than Frías Glacier. This value is also almost identical to the height of some moraines described by Goldthwait (1973, p. 172), which were originally interpreted as annual (winter) push-moraines but later re-evaluated as crevasse filling deposits released during ice-recession.

In Argentina, very recent moraines in front of San Lorenzo Glacier (Spalletti, 1975) have quite similar dimensions, although they have been interpreted as "recessional" by this author.

There is some evidence supporting a possible relation of these moraines with a terminal phase at a surge of Frías Glacier. Further work should be performed to investigate the possibility of a similar origin to Late Neoglacial Moraines of Frías Glacier which are morphologically identical to these push-moraines, although larger. If the Late Neoglacial Moraines, which developed from 1700 A.D. to 1850 A.D. (Lawrence & Lawrence, 1959; Mercer, 1976; Rabassa et al., 1977) were also formed by a surge episode, then a possible (maximum) surge cycle of Frías Glacier would be of 120 years, approximately. There is neither geomorphological nor glaciological evidence of a similar event during the intermediate period, but it should be kept in mind that we lack information for the previous to 1914 and 1914-1942 periods. Studies should also be extended to other glaciers of the region to estimate the frequency of push-moraines in this mid-latitude, temperate climate. Push-moraines were identified in Esperanza Norte Glacier in March 1978, and probably a significant portion of the recent and Neoglacial morainic systems of this area may have a similar origin.

ACKNOWLEDGEMENTS

The authors are indebted to Ms. Margarita Bündel for the revision of the English version, and to Ms. Sonia Fremery for the careful typing of the paper.

BIBLIOGRAPHY

Elson, J. A. (1968). Washboard moraines and other minor moraine types. in: R. W. Fairbridge, ed., "Encyclopedia of Geomorphology", p. 1213-1219, Rheinhold Book Co., New York.

Embleton, C. & King, C.A.M. (1975). Glacial Geomorphology. John Wiley & Sons, New York.

Flint, R. F. & Fidalgo, F. (1964). Glacial geology of the east flank of the Argentine Andes between latitude 39°10' S and latitude 41°20' S. Geol.Soc.America Bull., v. 75, p. 335-352.

Folk, R. L. & Ward, W. C. (1957). Brazos River Bar: a study in the significance of grain size parameters. Journ, Sed. Petrol., v. 27, p. 3-26.

Goldthwait, R.P. (1974). Rates of formation of glacial features in Glacier Bay, Alaska. in: D. R. Coates, ed., "Glacial Geomorphology", p. 163-186, Publications in Geomorphology, State Univ. of New York, Binghamton, N. Y.

Johnson, P. G. (1972). The morphological effects of surges of the Donjek Glacier, St. Elias Mountains, Yukon Territory, Canada, J. Glaciol., v. 11, N°62, p. 227-234.

Krumbein, W. C. (1941). Measurement and geologic significance of shape and roundness of sedimentary particles. Jour.Sed. Petrol., v. 11, p. 64-72.

Lawrence, D. B. & Lawrence, E. G. (1959). Recent glacier variations in Southern South America. Office of Naval Res., Contract Nonr-641(04), Amer.Geograph.Soc., New York, 39 pp.

Lliboutry, L. (1956). Nieves y Glaciares de Chile. Univ.de Chile, Santiago, Chile, 470 pp.

Marchese, H. G.; Di Paola, E. C. & Spiegelmann, A. T. (1970). Métodos y técnicas para el estudio de muestras de perforación ("cuttings" y testigos). Argentina Asoc.Miner.Petrol.Sedim.Rev. (AMPS), v. 1, p. 93-116.

Meier, M. F. & Post, A. (1969). What are glacier surges? Canadian J. Earth Sci., v. 6, p. 807-817.

Mercer, J. H. (1967). Southern Hemisphere Glacier Atlas. U.S.Army, Natick Laboratories, Earth Sci. Lab., Tech. Report 67-76-ES, pp. 325.

Mercer, J. H. (1968). Variations of some Patagonian Glaciers since the Late Glacial. Am. J. Sci., v. 266, p. 91-109.

Mercer, J. H. (1970). Variations of some Patagonian Glaciers since the Late Glacial: II. Am. J. Sci., v. 269, p. 1-25.

Mercer, J. H. (1976). Glacial history of Southernmost South America. Quaternary Research, v. 6, p. 125-166.

Passega, R. (1964). Grain size representation by CM patterns as a geological tool. Jour.Sed.Petrol., v. 43, p. 168-187.

Post, A. (1966). The recent surge of Walsh Glacier, Yukon and Alaska. J.Glaciol., v. 6, N°45, p. 375-381.

Post, A. (1969). Distribution of surging glaciers in Western North America. J. Glaciol., v. 8, N°53, p. 229-239.

Post, A. (1972). Periodic surge origin of

folded medial moraines on Bering Piedmont Glacier, Alaska. J.Glaciol., v. 11, N°62, p. 219-226.

Rabassa, J. & Aliotta, G. (1978). Sedimentology of two superimposed tills in the Bariloche Moraine (Late Glacial), Río Negro, Argentina. This volume.

Rabassa, J. & Rubulis, S. (1978). An exceptional retreat of Mount Bonete South Cirque Glacier, Río Negro, Argentina. (Letter). J. Glaciol., in press.

Rabassa, J.; Rubulis, S.; Whewell, R. & Rodriguez-García, D. (1975). Castaño Overo and Alerce Glaciers, Mount Tronador, Río Negro, Argentina. Internat. Symposium on the Quaternary (abst.), Boletim Paran. Geociencias, v. 33, p. 38, Curitiba, Brasil.

Rabassa, J.; Rubulis, S. & Suárez, J.(1977). Los Glaciares del Monte Tronador, Parque Nacional Nahuel Huapi, Río Negro. Anales Parques Nacionales (arg.), in press.

Spalletti, L. (1975). Estudio del glaciar septentrional del Monte San Lorenzo, y del río del Oro (Provincia de Santa Cruz). I. Aspectos generales. Geomorfología. Arg.Geol.Asoc.Rev., v. 30, p. 17-43.

Spalletti, L. (1976). Sedimentología de gravas glaciales, fluviales y lacustres de la región del Cerro San Lorenzo (Provincia de Santa Cruz). Arg.Geol.Asoc.Rev., v. 31, p. 241-259.

Spalletti, L. & Gutiérrez, R. (1976). Estudio granulométrico de sedimentos glaciales, fluviales y lacustres de la región del Monte de San Lorenzo, Provincia de Santa Cruz. Arg.Geol.Asoc.Rev., v. 31, p. 95-117.

Willis, Bailey (1914). Northern Patagonia. Ministry of Public Works, Buenos Aires.

Worsley, P. (1974). Recent "annual" moraine ridges at Austre Okstindbreen, Okstindan, North Norway. J.Glaciol., v. 13, N°68, p. 265-277.

Zingg, T. (1935). Beitrag zur Schotteranalyse. Schweiz.Mineral.Petrog. Mitt., v. 15, p. 39-140.

Sedimentology of two superimposed tills in the Bariloche Moraine (Nahuel Huapi Drift, Late Glacial), Rio Negro, Argentina

JORGE RABASSA & GUIDA ALIOTTA
Fundación Bariloche, San Carlos de Bariloche, Argentina

ABSTRACT

The sedimentology of two superimposed tills of the Bariloche Moraine (Nahuel Huapi Drift, Late Glacial) has been studied at the type locality. The lower till is a lodgement till, with sandy beds of subglacial, meltwater origin. The upper till is a complex ablation till, composed of two till bodies and supraglacial fluvial sands. These units were deposited during a single glacial event, although there is the possibility of a short recession of the glacier followed by the deposition of marginal flow till.

1. INTRODUCTION

The region of San Carlos de Bariloche has been a classic zone for the study of deposits and landforms of glacial origin, in Southwestern Argentina. This is mainly due to its privileged location in an environment that has been repeatedly glaciated, and the distribution that the sedimentary bodies related to these processes show in the area.

This is why Feruglio (1941, 1949-50) noted the region for its glacial characteristics, and Flint & Fidalgo (1964) chose this area to start their glacial geology work in the Northern Patagonian Andes. Caldenius (1932), Auer (1950), Dessanti (1972), González Bonorino (1972, 1973), Greco (1974), and Paskoff (1976), among others, studied deposits or landforms of glacial origin around Lake Nahuel Huapi and the neighbouring area.

Feruglio (1941) describes two main morainic systems: an outer one, that extends up to the Pichileufu River, composed of glacial deposits corresponding to "one or more glacial episodes", and an inner system, that forms the eastern and southeastern shore of Lake Nahuel Huapi, and named by him the "Bariloche Moraine".

Flint & Fidalgo (1964) studied a group of glacial deposits, that correspond to three glacial episodes, probably three different glaciations, which they called Pichileufu Drift, El Cóndor Drift, and Nahuel Huapi Drift, the latter being the innermost and thus the most recent.

The age of these deposits has been considered by Flint & Fidalgo (1964) as Wisconsinan, except the Pichileufu Drift which could be older. They also recognize the existance of later glacial stages during ice recession, and Neoglacial readvances in some sections closer to the Andean Ranges.

Flint & Fidalgo (1964, p.312) mention that in no case two or more superimposed drift bodies have been observed; their stratigraphic and chronologic considerations have been based upon sedimentological and morphological criteria.

The aim of this paper is to present a sedimentological characterization of the first appearance of two superimposed tills at San Carlos de Bariloche. An extended, Spanish version of this paper has already been published in a local journal (Rabassa and Aliotta, 1976).

2. THE LOCALITY STUDIED

We have studied an artificial outcrop discovered by building procedures, located in the Avenida Costanera ("Lake-shore Avenue"), in front of the parking lot of the "New Port" of San Carlos de Bariloche (Figure 1).

3. SAMPLING METHODS AND SEDIMENTOLOGICAL PROCEDURES

Samples for sedimentological studies were obtained from Unit A and Unit B (see: "4. Description of the profile"). Psefitic frac-

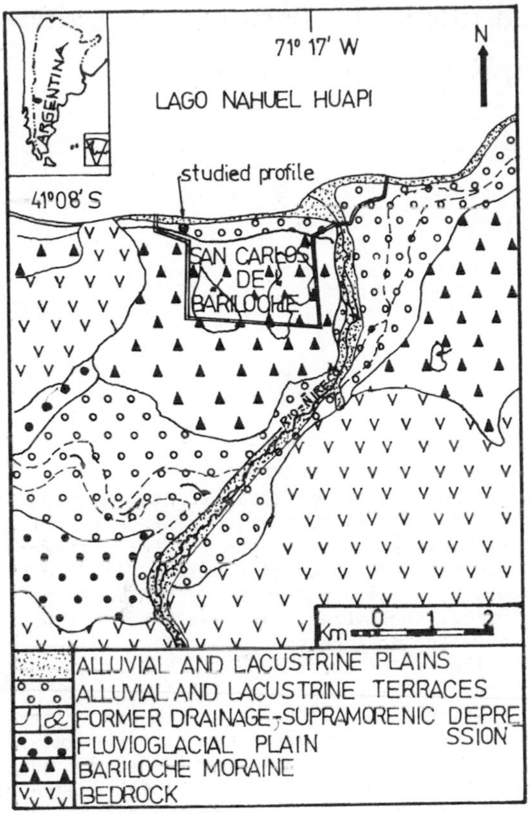

Figure 1: Location map and geomorphological sketch.

tion was sampled by taking all the clasts with diameter over 1 cm, that could be obtained in a 1 square meter of area on the vertical walls of the excavation, previously cleaned with a shovel. Once the major clasts had been retired, a sample of the matrix (1 kg approximately) was obtained.

Unfortunately, we visited the excavation during the erection of the building structure; thus, we were not able to get field data on dip and strike of the longer axis of the clasts. We do not have quantitative information referring to the fabric of these deposits.

Morphometry of the cobbles was studied with the help of a calibre, measuring the length of the principal orthogonal axes "a", "b" and "c", as indicated by Krumbein (1941). Sphericity values were obtained and plotted on Zingg's diagrams (Zingg, 1935). Flatness and other morphometric parameters were calculated for each clast, and other peculiar characteristics such as striae, breaking, facets, etc., were indicated.

Granulometry of the matrix was performed for sand, silt and clay fractions, by the usual methods of sieving and pipette.

Wet-sieving was done on A.S.T.M. sieves, 0.5 phi. The fraction under 44 μ was treated following the Robinson Pipette method (Milner, 1952, p.158; Strakhov, 1957, p323), using Na-hexametaphosphate (calgon) as a dispersant agent. Fractions were obtained by pipette at a constant room-temperature of 20°C.

The mineralogy of the sand and coarse-silt fractions was determined by optical methods, calculating percentages for each of the minerals or group of minerals present in the sample. X-ray diphractograms were performed on clay fractions, under 4 μ and under 2 μ. A Phillips PW-1300 diphractometer was used in the following conditions: 50 Kv, 25 mA, 4 x 10^2, TC 8, and 2° per minute.

4. DESCRIPTION OF THE PROFILE

The studied profile was observed at an excavation, 15 m by 20 m, with a maximum depth of 5 m. The locality is placed on the southern shore of Lake Nahuel Huapi, where terminal moraines of the Nahuel Huapi Drift (Flint & Fidalgo, 1964) lie on top of sedimentary and volcanic rocks of the Ñirihuau Formation (Upper Eocene-Lower Oligocene; González Bonorino, 1973), folded and faulted later in the Upper Oligocene or Lower Miocene. These terminal moraines have been eroded by wave action during moments of higher lake levels, and at least two terraces can be easily recognized. (Fig. 1).

The topographic surface at this locality has a slight inclination towards the lake (2°-3°). The ancient front of the site had been previously destroyed by the construction of the Costanera Avenue, which actually follows the top surface of the lowermost lake terrace.

The profile was observed on the western side of the excavation, and it is described as follows (Figure 2):

Base. Unit "A": 1.80-2.00 m of outcrop thickness; till, blueish grey, sandy to clayey, with cobbles up to 0.10 m in diameter, rounded to sub-rounded. Till lacks stratification or other structures, but it shows some layers of sand (coarse to very coarse) in its upper portion, grey to dark grey, very well stratified. Till base is not observed. Sample 1: till matrix; Sample 2: dark-grey coarse sand, intercalated at the top of the till; Sample 3: psefitic fraction of till.

Figure 2: Observed profile at the studied locality. See the lower till (dark grey) and the upper till (light grey).

Unit "B": up to 3,00 m of outcrop thickness; till, brown, sandy, with cobbles up to 0.15 m in diameter, rounded to subrounded. This unit shows some stratification, with preference in the upper section. Two sub-units or principal layers of till may be distinguished (B_1 and B_2), with an intercalated layer of brownish sand. This layer is very well stratified, well sorted, with very scarce small pebbles and some portions of highly compacted silts. Unit "B" overlies Unit "A" with a very clear uncomformity, which dips 20° to the North, that is, to Lake Nahuel Huapi. This peculiar disposition determines that the maximum outcrop thickness of Unit "B" is at the outermost portion of both lateral walls of the hole, while it progressively thins towards the back wall. Sample 4: matrix of the brownish till, lower portion of layer B_1; Sample 5: matrix of the brownish till, middle portion of layer B_2; Sample 6: brownish sand, well sorted, intercalated between till layers B_1 and B_2; Sample 7: psefitic fraction of till B_1.

Unit "C": 0.30-0.40 m of fine to medium gravels; matrix is very scarce and the sediment may be considered an "open-work gravel". Cobbles up to 0.10 m in diameter; a boulder of lithified tuff, 0.80 m in diameter, was observed in this unit, its longer axis lying in the stratification plane. This unit lies uncomformably over units "A" and "B", indistinctly. The surface of uncomformity dips gently (3-4°) towards the lake.

Unit "D": 1.00 m of very fine to fine, brownish volcanic ash, deeply weathered, with intercalations of colluvial debris and partly of loessoid nature. It overlies uncomformably on Unit "C", and it shows at the top development of soil profile, with variable thickness of the humic layer, up to 0.80 m in some parts of it. Sandy and psefitic materials of mass-movement origin have been incorporated into the humic layer. The soil profile does not exhibit clay-accumulation horizon, but lixiviation is so intense that material from this unit has been carried down to the underlying gravels, forming some sort of a "secondary matrix" as it has been deposited at the interstices among gravel clasts. The upper portions of the unit have been also affected by human activity.

5. RESULTS

5.1. Mechanical analysis

Granulometric analysis was performed for samples 1, 2, 4, 5, and 6. The histograms and cumulative curves are presented in Figures 2 and 3. Histograms of samples 1, 4, and 5, clearly show the polymodal character of these sediments. On the contrary, samples 2 and 6 are strongly bimodal. Obviously, cumulative curves reflect these characteristics. Samples 1, 4 and 5 (till matrix samples) present very flat curves, due to their very bad sorting.

On the other hand, curves of samples 2 and 6 clearly mark a better sorting, although their modes are completely different. Table 1 presents textural parameters of these samples, as calculated from cumulative curves following Friedman (1961).

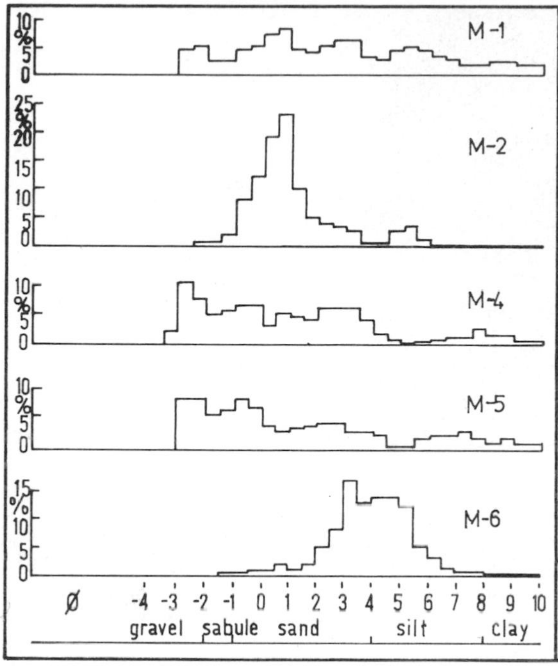

Figure 3:Granulometric analysis:histograms.

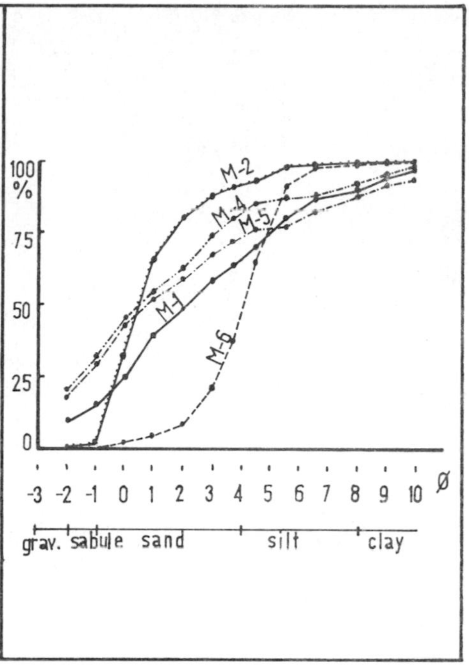

Figure 4:Granulometric analysis:cumulative curves.

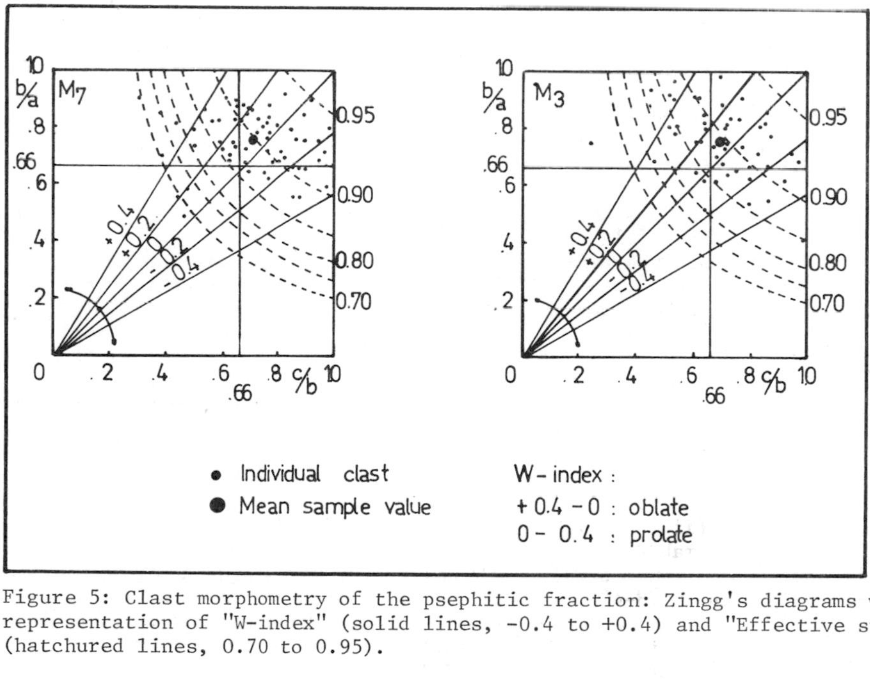

Figure 5: Clast morphometry of the psephitic fraction: Zingg's diagrams with the graphic representation of "W-index" (solid lines, -0.4 to +0.4) and "Effective sphericity" (hatchured lines, 0.70 to 0.95).

Table 1. Textural Parameters

	Q_3 mm	Q_1 mm	Md mm	Modality	Mean $\bar{x}\phi$	Sorting $\sigma\phi$	Skewness $\alpha 3\phi$	Kurtosis $\alpha 4\phi$
M 1	0.95	0.028	0.20	polymodal	+ 2.4	3.22	+ 0.4	2.37
M 2	0.95	0.35	0.625	bimodal	+ 1.05	1.8	+ 1.68	6.52
M 4	2.90	0.12	0.65	polymodal	+ 0.9	3.04	+ 0.92	3.39
M 5	2.5	0.04	0.55	polymodal	+ 1.3	3.37	+ 0.84	2.67
M 6	0.115	0.035	0.064	bimodal	+ 3.92	1.52	- 0.29	4.54

It can be seen in Table 1 that the polymodal character of samples 1, 4 and 5, exposed by the corresponding histograms, is coincident with very high values of standard deviation. It is also important to note the similarity between samples 4 and 5 (tills B_1 and B_2) and their discrepancy with sample 1 (till A). In relation with the textural parameters, $\bar{x}\ \phi$ shows that the matrix of the lower till is finer than the matrix of the upper till. The contrary happens with the sandy layers intercalated in them. Standard deviation ($\sigma\phi$) indicates that sorting is very poor in both tills and notably higher in the sandy layers. Skewness ($\alpha 3$) is positive in both tills, but it is completely opposite in the sandy beds, because that of Unit "A" has a very high positive asymmetry, a marked "tail" due to the distribution of fines, whereas that of Unit"B" exhibits negative asymmetry.

5.2. Morphometry of the psefitic fraction

The pebbles obtained for samples 3 and 7 were analyzed in detail, in consideration of their shape, roundness, and other particular characteristics such as breaking, faceting, striae, etc.

5.2.1. Shape

Pebble shape was determined in relation to their three main axes, a, b, and c, following Krumbein (1941). We have calculated the usual ratios (b/a ; c/b), and the obtained values have been summarized in Table 2.

From Table 2, we know that the adimensional coefficients are very similar in both samples, although mean values are higher in sample 3.

Table 2. Mean values for axes a, b, c, and ratios b/a and c/b

Sample	Number of clasts	$\bar{a}$	$\bar{b}$	$\bar{c}$	b/a	c/b
M - 3	53	40.5	30.6	20.8	0.751	0.693
M - 7	65	34.6	25.8	18.3	0.746	0.713

In Figure 4, we have represented Zingg's (1935) diagrams for both samples. It can be seen that more than half of the clasts fall in the area corresponding to the shape "equidimensional", i.e. b/a and c/b ratios both larger than 0.66. The corresponding mean values, as shown by Table 2, also fall within this sector. Similar percentages are distributed in the remaining zones, for both samples. More than 75% of the clasts of both samples have ratios that fall within the zones of "equidimensional" and "discoidal" (b/a higher than 0.66; c/b between 0 and 0.66). This is coincident

with Drake's (1970) observations, in the sense that these forms are favoured by glacial processes.

The problem of shape and sphericity determination in pebbles has been studied by many authors since Zingg (1935), and several methods have been proposed to estimate these parameters. Among others, Aschenbrenner (1956) developed the concepts of "effective sphericity" and "F-factor", which are functions of the ratios c/b and b/a and so they may be represented in Zingg's diagrams. F-factor ($a.c/b^2$) mean values are 0.9635 for Sample 3, and 0.9812 for Sample 7; both samples have very close values for this parameter, corresponding to the "prolate" type, but very close to the unity which is the limiting value for both types, "prolate" and "oblate". Williams (1965) proposed a new parameter, the "W-index", actually a modification of Aschenbrenner's F-factor. W-index has the advantage over F-factor that the range of values is from - 1 to + 1, whereas the latter ranges from 0 to infinity. We have represented in Figure 4 the areas limited by different values of W-index and effective sphericity. It can be observed that more than 90% of the clasts have an effective sphericity higher than 0.80, and also more than 90% of them fall within the range of - 0.4 to + 0.4 in Williams' scale for W-index. Sphericity of the clasts of these samples is thus very high and most of the pebbles are close to the limit of the "oblate" and "prolate" types. Flatness (Wentworth, 1922) values for these samples are: Sample 3, 1.811; Sample 7, 1.905. These are mean values with extreme marks of 1.150 and 2.840. If flatness is calculated without counting the broken or faceted clasts, the corresponding values drop to 1.719 and 1.610, respectively.

5.2.2. Roundness

Roundness was determined following Ruhkin's table for visual appreciation (1961; in: Marchese, Di Paola & Spiegelman, 1970). This table presents 5 possible degrees of roundness, from 0 --very angular-- to 4 --very well rounded--. The results are shown on Table 3.

Table 3. Roundness

Degrees of roundness	M - 3 n	M - 3 %	M - 7 n	M - 7 %
0	5	9.4	12	18.5
1	15	28.3	14	21.5
2	18	34.0	29	44.6
3	14	26.4	10	15.4
4	1	1.9	0	0

It is now possible to note the similarity of both samples and the absence of "very well-rounded" clasts. The mean values are 1.83 for Sample 3 and 1.57 for Sample 7, and the medians are 1.45 and 1.25, respectively. Note also that Sample 3 (Lower till) has the higher values for roundness.

5.2.3. Broken and faceted clasts

We have considered as "broken clasts" those showing fresh breaking surfaces whose own roundness values were 0 or 1. "Faceted clasts" are those which present abrasion surfaces that are not coincident with the general roundness of the clast; "pentagonal faceted clasts" are those that exhibit a basal, wide, abrasion surface, of pentagonal or sub-pentagonal shape.

Both samples have very similar percentages for these types; the values are also very high in relation to the total psefitic population.

5.3. Petrographical composition of the clasts of the different fractions

Table 4. Broken and faceted clasts

	M - 3		M - 7	
	n	%	n	%
Broken	15	28.3	18	27.7
Faceted	2	3.8	3	4.6
Pentagonal	1	1.9	3	4.6
Total	18	34.0	24	36.9

5.3.1. Lithology of the psefitic fraction

According to the lithologic types recognized in these samples, we have divided them into 6 groups:

1. Metamorphic rocks, schists, migmatites, etc.
2. Plutonic acid and mesosilicic rocks: granites, granodiorites, diorites.
3. Epiclastic and pyroclastic sedimentary rocks: sandstones, siltstones, tuffs.
4. Volcanic acid rocks: rhyolites, rhyodacites.
5. Volcanic intermediate rocks: andesites, lacites, etc.
6. Volcanic basic rocks: basalts, basandesites, etc.

Results are shown in Table 5.

Table 5. Lithology of the psefitic fraction

Lithologic type	M - 3		M - 7	
	n	%	n	%
1	2	3.8	1	1.5
2	4	7.5	5	7.7
3	23	43.4	29	44.6
4	5	9.4	10	15.4
5	13	24.5	18	27.7
6	6	11.3	2	3.1

Table 6 shows that there is a clear similarity between the lithologic content of both samples. This similarity suggests that the source area of the sedimentary material of both tills is essentially the same. It is also worth noting that the 6 recognized groups represent the totality of the studied rock sequences for this part of the Northern Patagonian Andes (see, for example, Gonzãlez-Bonorino, 1972, 1973; Greco, 1974; Dessanti, 1972; Cazau, 1972; Turner & Cazau, 1978; Uliana, 1978).

5.3.2. Mineralogy of the sand fractions

We have analyzed the mineralogy of the sandy fractions of Samples 1, 2, 4, 5, and 6 (Table 6).

The results exposed here show the similarity of the mineralogical content of Samples 4 and 5 (till B_1 and till B_2, respectively), and also the close percentage of heavy minerals of Samples 1 and 2 (till A and sandy bed of till A, respectively clearly different from the others. The observed differences in quartz and lithoclasts content between the sandy layers (Sample 2 and 6) and the related tills (Samples 1 and 4-5, respectively) may be due to the fact that these values

Table 6. Mineralogy of the sand fractions

Minerals or group of minerals	M-1 %	M-2 %	M-4 %	M-5 %	M-6 %
Quartz	20	5	13	6	25
Plagioclases	28	10	15	10	12
K-feldspar	8	9	6	11.5	8
Lithoclasts + glass shards	20	48	54	60	43
Opaques + maphics	24	28	12	12.5	12

are an average for all the sand fractions and in both cases the granometric mean of the sand layers is highly divergent from those of the associated tills. Thus, for example, a sample lacking fine-sand fraction will be relatively rich in lithoclasts, which appear in the coarser fractions.

5.3.3. Mineralogy of the clay fraction

We have prepared diphractograms of the clay fractions, under 4 μ and under 2 μ. The study of these diphractograms showed that Samples 1, 2, 4, 5 and 6 exhibit a very wide peak corresponding to montmorillonite. Samples 1, 4 and 5 also depict a series of minor peaks corresponding to quartz and feldspars. Curiously, Sample 2 and 6 (the intercalated sand layers) do not show the existence of quartz or feldspar in clay fractions. We interpret this difference as a result of the different depositional processes that acted upon these sediments. The existance of quartz and feldspar in significant quantities in the clay fractions, even in the one under 2 μ, represents the "rock flour", characteristic of till. These materials seem to have been washed out by water action in the sand layers, in relation to the sediments previously originated by glacial processes.

5.4. Structure of the deposits

Our considerations on the structure of the deposits have been based on field observations in the studied locality and other neighbouring outcrops.

Till A appears as a massive, highly compacted deposit, and sandy and finer intercalations are clearly noticeable in its outcrops. In some localities, deposits of similar nature that have been exposed to atmospheric agents exhibit a sort of subhorizontal jointing. Weathering profiles in this unit (or similar units) have developed some yellowish pigmentation. We have observed that, although the pebble content may be locally abundant, boulders are scarce, specially those of large dimensions.

Till B, on the contrary, frequently shows signs of stratification and intercalation of lenses of varied sedimentary types. These lenses may be of well-sorted sands, lacustrine pelites, banded till, etc. In several outcrops visited in the area, similar deposits exhibit weathering profiles of red-brownish colour, and large boulders of varied lithology.

The sedimentary structures observed in the associated sand beds of both units, such as cross-bedding, ondulites, lamination, etc., are indicators of running-water environments, with rapidly changing energy conditions.

6. OTHER FIELD OBSERVATIONS IN THE BARILOCHE MORAINE

Our sedimentological characterization is restricted to only one profile. However, a great number of observations have been made in other localities. Recently, two important excavations at San Carlos de Bariloche have been studied by one of us (Rabassa, 1978a; 1978b) for other reasons, but they have not yet been sedimentologically investigated.

In one of these, a hole excavated for the building of a large downtown hotel, a more complete sequence of 7 m of blueish till and lacustrine sediments (Unit "A") is overlain by 12 m of brownish till and an intricate interdigitation of sandy beds, water-lain gravels, and flow till (Unit"B"). The contact between both units is sharp but

no evidence of pushing or bulldozing has been observed on top of the lower till. The base of Unit "A" is resting on sedimentary rocks, glacially eroded, of the Lower Tertiary. From a geomorphological point of view, this profile is located at a more marginal position of the glacial sedimentary basin. Close to this profile, other outcrops have been identified as kame and kame-moraine deposits, representing small pockets of dominant melt-water environments between the ice front (at a semi-lateral position of the glacier snout) and the bedrock walls of the valley. It is worth mentioning here that Late-Glacial Nahuel-Huapi Glacier was more than 8 Km wide at the snout, around 80 Km long, and at least 750 meters thick in front of Bariloche.

The other excavation, also for a luxury hotel and discotheque, is located 300 meters to the East of the sedimentologically studied profile, at a similar elevation and geomorphological position. The sequence here is based on top of Tertiary sandstone, and it is formed by 4.5 meters of blueish till (Unit "A") overlain by 6.0 meters of the brownish upper complex (Unit "B"). In this case, it was observed that pebble distribution in till walls of Unit "A" was much more homogeneous than in till "B", where sub-horizontal layers of pebble concentrates were well defined. Horizontal jointing of Unit "A" was clearly noted, with a separation of 5 to 10 cm between planes.

At the northern foot of Cerro Otto, a relatively low peak 4 Km West of Bariloche, the till sequence was studied by Dominguez, Rabassa and Cabral (1978). Five meters of blueish, consolidated till of Unit "A" was found here lying on top of Tertiary sedimentary rocks. Unit "B" is formed by a thick (12 m) varve sequence, with abundant drop-stones, and 3 m of brown till, sands, and a very coarse conglomerate with a fine matrix, possibly flow till. Older (pre-Wisconsin?) till and stratified drift were observed at a higher elevation (+ 250 m) on the same slope of Cerro Otto.

At an outer position in the Bariloche Moraine, near the old bridge of Route 237 over Río Ñirihuau, the sequence is composed of 4.5 meters of outcrop thickness of blueish till, very compacted and endurated (Unit "A"), overlain by more than 10 meters of varves, sands and gravels (Unit "B"). Till B has been replaced here by outwash deposits.

At many other localities, the sedimentary sequence is observed only in a fragmentary way, i.e. only one of the units is present, but it has not been too difficult to correlate these outcrops within the Bariloche Moraine with the sedimentologically studied profile, according to the diagnostic characteristics found in the latter.

7. INTERPRETATION AND CONCLUSIONS

This study has presented a group of field observations and sedimentological parameters, with the aim of elucidating a two superimposed tills sequence. Although we have not intensified the study of Unit "C", overlying the till sequence, we think that it is an "open-work gravel", formed by fines elimination in a post-glacial lacustrine beach.

The problem of two superimposed tills, whose sedimentary characteristics appear clearly differentiated, has been extensively considered by Drake (1971). This author studied a great number of localities in New Hampshire, U.S.A., and he produced three alternative hypotheses to explain the sequence:

1) it is a single till unit, whose upper portion has been oxidazed and affected by pedogenetic processes during Postglacial times;

2) the upper till corresponds to a later glacial event;

3) the lower till is a lodgement till (basal till), whereas the upper till is an ablation till, genetically associated with stagnant ice.

After a very detailed investigation, Drake (1971, p.88) concludes that the obtained evidence strongly supports hypothesis (3) of ablation till.

Although we have studied the sedimentology of only one profile, we think that the information obtained from this study and other field observations in the surrounding area, strongly favours an interpretation of the sequence in the same way as was done by Drake (1971). This evidence is summarized as follows:

1) The lower till is compact, endurated, formed in a reducing environment, and, generally, it does not present stratification. The upper till is not so compacted, somewhat oxidazed, and it presents internal stratification, although poorly defined. The sand layers are intercalated in both units as discontinuous lenses, internally very well stratified.

2) The lower till has a psefitic fraction of shorter mean diameters, equidimensional in shape, and with mean roundness values a bit higher than the upper till. The granometric distribution is wider, displaying the conservancy of clay fractions.

3) The intercalated sand layers were deposited by running melt-waters, as is shown

by their high values of sorting in comparison to those of the tills.

4) In both tills a high percentage of broken and faceted pebbles can be seen that evidences the intensity of glacial attrition processes.

5) The lithology represented in the psefitic fraction of both units is essentially the same, and the mineralogical content of the sand fractions is also very similar, except for the percentage of lithoclasts + glass shards (higher in Till "B") and of heavy minerals (higher in Till "A"). Thus, a similar source area is estimated for both units. The difference in the content of glass shards may be related to the continuity of volcanic eruptions at the end of the Late-Glacial episode, when the sediments that later formed Till "A" were already isolated from the outside below the ice-mass, and those that formed Till "B" were still on top of the glacier or near it. The higher content of heavy minerals in Till "A" is not easy to explain, because ice viscosity is not likely to clearly favour a sort of gravity sorting, as would be expected in a more fluid environment. This aspect requires further studies to confirm and explain it.

There is a clear domination of materials of volcanic or volcano-sedimentary origin in comparison to those of "granitic" or "metamorphic" origin. This suggests that the maximum glacial erosion during Nahuel-Huapi Claciation was along the structural depression occupied by Lower Tertiary formations (González Bonorino, 1972).

6) Clay fractions in both tills exhibit a similar mineralogy. The high content of quartz and feldspars in these fractions is due to trituration processes of sub-glacial or en-glacial nature.

We understand that it is not possible to extend too much the interpretations based on only one profile, although field evidence from other parts of the Bariloche Moraine coincides with it. However, we think that the following conclusions are permissible:

1) The sedimentological analysis performed on the samples of this profile permitted an adequate characterization of the units, and constitutes a valuable support for field observations and correlation. It is our aim to extend this type of analysis to a larger number of sites to prepare a sedimentation model for Nahuel-Huapi Drift at the type locality. The methodology used enabled the differentiation of waterlain layers from those formed essentially by glaciogenic action. The distinction of genetic types of till had to be done based on field observations.

2) The studied profile is composed of a lodgement till at the base (Unit "A") and an overlying ablation till, partly a flow till (Boulton, 1971). Both units present signs of meltwater action, of sub-, en- and supraglacial origin.

3) Both units were deposited as part of a single glacial event. Lodgement till is subglacially derived and was simultaneously deposited with the glacial advance. Ablation till is mainly supraglacially derived and it was formed during the stagnant-ice period that followed to the maximum extent of the glaciation. It is probable that during this period, the general environmental conditions of Nahuel-Huapi Glacier may have been of the "pseudokarst" type, with ice hills protected by the debris layer, with "sink-holes" and subglacial streams. See, for example, the morainic depressions represented in the map (Figure 1), possibly originated in this way. Very similar conditions may be observed nowadays on the Río Manso Glacier, Mt. Tronador (Rabassa et al., 1978). These conditions are favoured by ice masses of large dimensions, either stagnant or with very slow movement, highly loaded with rock debris, and with very high ablation rates. Downmelting is predominant and is controlled by the distribution of debris within the glacier (Embleton & King, 1975).

4) The Bariloche Moraine, in its inner margin, has been shown to be composed of at least two till bodies, formed by different glacial processes but corresponding to a single glacial episode. The possible existance of more than one glacial pulsation for the Nahuel-Huapi Glaciation, as suggested by Flint & Fidalgo (1964, p.29), should not be confused with two superimposed tills sequences, such as we have described here.

ACKNOWLEDGEMENTS

The authors are greatly indebted to Professor Félix González Bonorino for his suggestions and criticism on preliminary drafts of the manuscript, to Ms. Margarita Böndel for the revision of the English version and to Ms. Sonia Fremery for the careful typing of the paper.

BIBLIOGRAPHY

Aschenbrenner, B. C. (1956). A new method of expressing particle sphericity. Jour. Sed.Petrol., v. 26, N°1, p. 15-31.
Auer, V. (1956). The Pleistocene of Fuego-

Patagonia.Part I:The Ice and Interglacial Ages. Annales Acad.Sci.Fennicae, Ser.A, III. Geol.Geogr., N°45, 226 pp.,Helsinki.

Boulton, G. (1971). Till genesis and fabric in Svalbard, Spitsbergen. in: R. P. Goldthwait, ed., "Till, a Symposium", Ohio State Univ. Press, p. 41-72,Columbus.

Caldenius, C. (1932). Las glaciaciones cuaternarias en la Patagonia y Tierra del Fuego. Geografiska Annaler, v. 14, p. 1-64.

Cazau, L. (1972). Cuenca de Ñirihuau-Ñorquincó-Cushamen. in: A.F.Leanza, ed., "Geología Regional Argentina", Acad.Nac. Cienc.Córdoba, p. 727-740, Córdoba, Argentina.

Dessanti, R. (1972). Andes Patagónicos Septentrionales. in: A.F.Leanza, ed., "Geología Regional Argentina", Acad.Nac. Cienc.Córdoba, p. 689-706, Córdoba, Argentina.

Domínguez, E.; Rabassa, J. & Cabral, R. (1978). Estudio de una cuenca aluvional en el Cerro Otto, Barrio Melipal, San Carlos de Bariloche. Secretary of Public Works, Bariloche County, Unpublished Tech.Report, 54 pp. To be presented at the IX Argentine Water Resources Congress, San Luis, May 1979.

Drake, L. D. (1970). Rock texture: an important factor for clast shape studies. Jour.Sed.Petrol., v.40, N°4, p.1356-1361.

Drake, L. D. (1971). Evidence for ablation and basal till in East-Central New Hampshire. in: R.P.Goldthwait, ed., "Till, a Symposium", Ohio State Univ.Press, p. 73-91, Columbus.

Embleton, C. & King, C. A. M. (1975). Glacial Geomorphology. John Wiley & Sons, New York, 573 pp.

Feruglio, E. (1941). Nota preliminar sobre la Hoja Geológica 40 b, "San Carlos de Bariloche", Y.P.F., Boletín Informaciones Petroleras, v. 18, N°200, Bs. Aires.

Feruglio, E. (1949-50). Geología de la Patagonia. Y.P.F., vols. I, II & III, Buenos Aires.

Flint, R. F. & Fidalgo, F. (1964). Glacial geology of the East flank of the Argentine Andes Between lat. 39°10'S and lat. 41°20'S. Geol.Soc.Amer.Bull., v. 75, N°4, p. 335-352.

Friedman, G. (1961). Distinction between dune, beach, and river sands from their textural characteristics. Jour.Sed. Petrol., v. 31, N°4, p. 514-529.

González-Bonorino, F. (1972). Geología de la región de San Carlos de Bariloche, provincias de Río Negro y Neuquén. Fundación Bariloche, unpublished, San Carlos de Bariloche.

González-Bonorino, F. (1973). Geología del área entre San Carlos de Bariloche y Llao-Llao, provincia de Río Negro. Fundación Bariloche, Publicaciones Departamento Recursos Naturales y Energía, N°16, 53 pp., San Carlos de Bariloche.

Greco, R. (1974). Descripción geológica de la Hoja 40 a, "Monte Tronador", Provincias de Río Negro y Neuquén. Servicio Nacional Geológico, unpublished, Buenos Aires.

Krumbein, W. C. (1941). Measurement and geological significance of shape and roundness of sedimentary particles. Jour. Sed.Petrol., v. 11, N°1, p. 64-72.

Marchese, H. G., Di Paola, E. C.& Spiegelmann A. T. (1970). Métodos y técnicas para el estudio de muestras de perforación ("cuttings" y testigos). Arg.Asoc.Mineral. Petrol.Sedim.Rev. (AMPS), v. 1, N°3/4, p. 93-116, Buenos Aires.

Milner, H. E. (1962). Sedimentary petrography. Volume I: Methods in Sedimentary Petrography. George Allen & Unwin Ltd.,London, 643 pp.

Paskoff, R. (1976). Les glaciations quaternaires dans les Cascades et dans les Andes méridionales aux latitudes temperées: bilen des recherches et comparisons. Revue Géographie Alpine, v. 2, p. 125-154, Grenoble.

Rabassa, J. (1978a). Informe pericial sobre los sucesos del 2-3 de Junio de 1977 en la excavación del Hotel Mónaco, San Carlos de Bariloche. Report to the Courts of the State of Río Negro, unpublished, 34 pp. San Carlos de Bariloche.

Rabassa, J. (1978b). Estudio de suelos en los lotes 34 y 35, Avenida Juan Manuel de Rosas, San Carlos de Bariloche. Secretary of Public Works, Bariloche County, Unpubl. Tech.Report, 24 pp., San Carlos de Bariloche.

Rabassa, J. & Aliotta, G. (1976). Sedimentología de dos tills superpuestos en San Carlos de Bariloche, Río Negro. Arg.Asoc. Mineral.Petrol.Sedim.Rev. (AMPS), v. 7, N°3/4, p. 47-62, Buenos Aires.

Rabassa, J.; Rubulis, S. & Suárez, J.(1978). Los glaciares del Monte Tronador, Parque Nacional Nahuel-Huapi, Río Negro. Anales de Parques Nacionales (Argentina), in press, Buenos Aires.

Strahkov, N. M. (1956). Méthodes d'étude des roches sedimentaires. Tome I. Bureau de Recherches Géologiques, Géophysiques et Miniéres (B.R.G.M.), Paris, 524 pp.

Turner, J. C. M. & Cazau, L. (1978). Estratigrafía del pre-Jurásico de Neuquén. VII Argentina Geol.Congress, Relatorio, Neuquén, p. 25-36.

Uliana, M. (1978). Estratigrafía del Terciario de Neuquén. VII Argentina Geol.Congress, Relatorio, Neuquén, p. 67-84.

Wentworth, C. K. (1922). A field study of

the shapes of river pebbles. U.S.Geol. Survey Bull., 730 C, p. 103-114.

Williams, E. M. (1965). A method of indicating pebble shape with one parameter. Jour.Sed. Petrol., v. 35, N°4, p.993-996.

Zingg, T. (1935). Beitrag zur Schotteranalyse. Schweiz.Mineral.Petrog.Mitt., v. 15, p. 39-140.

Rock glacier morainic deposits in the eastern Pyrenees

DAVID SERRAT
Universidad de Barcelona, Barcelona, Spain

1 INTRODUCTION

The part of the Pyrenean range we have studied is bounded on the East by a deep pass (Coll d'Ares 1400 m), and on the West by the tectonic depression of Cerdanya (approximately 1100 m). It includes the massifs of : Puigmal (2912 m), consisting mainly of the little-metamorphic schists of the Series of Canavelles and Jujols (P. Cavet 1957), belonging to the Cambro-Ordovician ; Carançà, with the maximum height in Puig de Bastiments (2875 m), consisting essentially of Carançà Gneisses (G. Guitard 1960) and the orthogneisses of Canigó and Freser which constitute the pre-Cambrian socle ; and the massifs of Costabona (2464 m), which in addition to the foregoing includes the presence of late-hercinian granites (Granites of Costabona).

This part of the Axial Zone of the Pyrenees was affected by the hercinian orogenesis in the form of various phases of deformation, acting subsequently as a socle in the Alpine orogenesis (fractures and deformations of a large radius of curvature).

The most relevant morphological characteristics consist in the presence of ample remains of tertiary erosion surfaces which form the watershed between the Spanish slope (South) and the French slope (North), and situated at altitudes varying between the 2200 metres of Pla de Salines, the 2400-2500 metres of Pla de Campmagre and Pla del Gorro Blanc, and

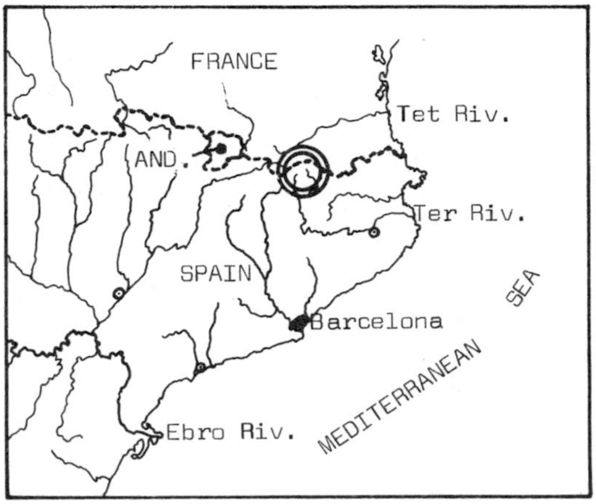

Fig. 1. Geographical situation of the massifs studied

the 2900 metres of Puigmal itself ; and the narrow valleys, drained by the river Ter towards the South and by the rivers Segre, Tec and Tet towards de North.

2 GLACIAL HISTORY

The geomorphological study which we have completed under the direction of Professors Jean Tricart of Strasbourg University and Lluis Solé Sabarís of the University of Barcelona, has demonstrated that the Quaternary glaciations affected this region of the Pyrenees in the following manner :

2.1 Pre-Würm

Although in this particular part of the Pyrenees we lack deposits and forms clearly associated with glaciations anterior to the Würm, in the Cerdanya and proceeding from the Carol Valley (tributary of the Segre), morainic materials were deposited which certainly belong to the Mindel and/or Riss. It is therefore obvious that these glaciations affected the massifs of Puigmal, Carançà and Costabona to a greater or lesser degree, though logically the sediments must have been reworked by later fluvio-glacial and periglacial dynamics, since the valleys are very narrow.

The existence of the fluvial terraces of the river Ter, in the Plana de Vic, at heights of 60 m, 80 m and more than 100 metres above the present course, appear to confirm this from a different point of view.

2.2 Würm

All the sediments, formations and forms of glacial origin still preserved in their original position appear to belong to this glaciation. The maximum of the Würm glaciation is characterized by the presence of valley glaciers with tongues whose lengths vary between 6 and 9 kilometres on both slopes, and a glaciation limit situated at approximately 2100 metres.

The surprising equality in the length of the Würm glaciers on both slopes indicates that were due to a great quantity of precipitation, apart from the proved existence of nival supercharge by the winds on the Southern slope by the action of the North winds on the topmost erosion surfaces already cited. This equality, however, has not noticeably influenced the resultant modelling, since the Southern valleys have hardly any obvious glacial remains, but retain their pre-glacial relief (typically fluvial gorges of the rivers Ter and Freser - Nuria).

The possible attribution of this maximum glaciation to Würm II or Würm III is not clear, though the discontinuity existing in the central part of some outcrops, together with the presence of peat, appears to indicate the existence of both pulsations.

A notable reduction in the precipitations at the close of the Würm, in addition to a new cold pulsation, mark the Tardiglacial episode of the Würm, which is characterized by the presence of numerous rock glaciers which form the subject of the present paper, together with a few cirque glaciers.

2.3 Present glacial dynamics

In spite of the altitude of some of the massifs studied (2912 metres), no type of glacial activity exists at present, except for a small neve (Congesta de la Llossa) at a height of 2480 metres and clearly oriented towards the South !, situated to the leeward of the Tramontana (North wind), which sweeps away the snow mantle on the mentioned erosion-surface of the Campmagre (2500 metres).

3 THE ROCK GLACIERS

Whe shall understand by rock glacier moraines those accumulations of blocks, with a greater or lesser quantity of matrix (normally scarce), situated at the foot of steep slopes (cirques and U-shaped valleys) which are evidently the result of a plastic displacement of the rubble accumulated at the foot of these walls, presenting a typical external morphology of ridges and furrows, very characteristic in vertical aerial photography, which mark the direction of their displacement. Their external sides show very steep slopes.

If an interpretation of the genesis of the present rock glaciers poses problems even to not knowing exactly what we can apply the name of rock glacier to (Østrem 1971, Barsch 1971), it is logical to think that such an interpretation according to deposited morainic remains could

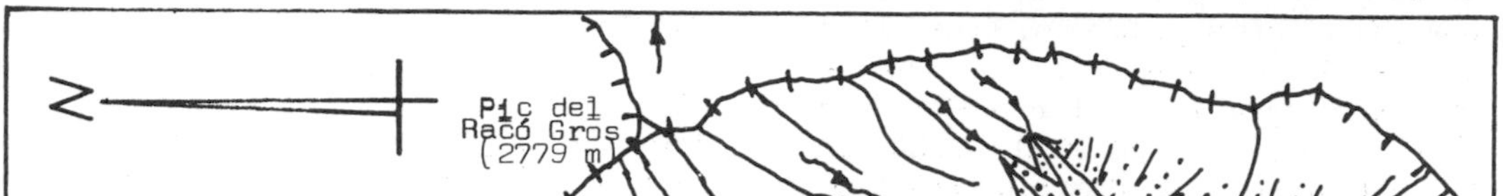

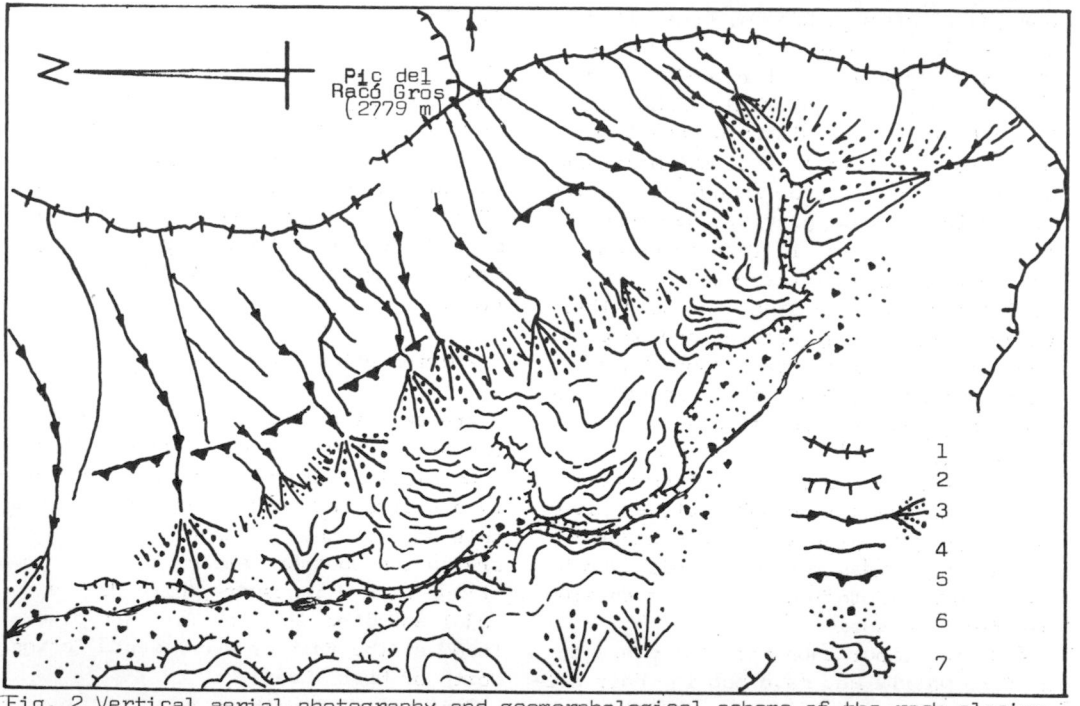

Fig. 2 Vertical aerial photography and geomorphological scheme of the rock glacier morainic deposits of the Coma Bailleta. Their light shades with the absence of vegetation contrasts with the dark shades of the würmian ground moraines over which are installed. 1.-crest; 2.-cirque; 3.-channel and fan of avalanches; 4.-spur; 5.-U shaped valley; 6.-Würmian ground moraines; 7.- rock glacier morainic deposits.

be chimerical. Obviously we cannot enter into the problem of whether it is a question of a nucleus or lenses of ice covered over or of intersticial ice which caused their movement. However whe shall study their lithology, orientation, form and dimensions in order to arrive at conclusions in accordance with the available data.

3.1 Lithology

In the region studied, two types of rock glaciers can be distinguished by their lithology. The rock glaciers developed in schists (Coma d'Eugassers 2650 m - fig.3 -, Fonts del Segre 2600 m and Conque de Planés 2550 motres) and those developed in gneiss and which form the majority (Cambres d'Ase 2300 - 2400 m, Coma Amagada 2380 m, Coma de Planés 2400 m, Coma dels Racons (Carançà) 2400 m, Coma de la Railleta 2200 - 2400 m - fig.2 -, Coma de l'Infern - 2550 m, Coma Mitjana 2250 - 2400 m, Coma de la Dona 2200 - 2500 m, Circ d'Ull de Ter 2450 m, and Circ del Gra de Fajol 2150 - 2300 m).

The feature that distinguishes them, apart from the lithology of the morainic blocks, is the greater abundance of fine matrix in the rock glaciers developed from schists, which in some cases has caused them to be interpreted as tongues or lobes of solifluction, but both their external morphology (aerial photography, and outer edge with steep slope), and their dimensions, make their interpretation more logical, we believe, as rock glaciers (Fig. 3).

This greater abundance of fine matrix has also brought it about that, at present, the rock glaciers developed from schists have been much more colonized by herbaceous vegetation than those developed from gneiss.

Thus the importance that the presence of fine matrix may have had in their origin is very problematic, though if we admit their displacement by ice in solid state (intersticial or in lenses), its importance must logically be slight.

The scarcity of rock glaciers developed from schists is associated with the scarcity of escarpments of glacial origin prior to the Tardiglacial pulsation of the Würm, over schist rocks in this part of the Pyrenees. We can therefore speak of a differential behaviour inhereited from the maximum of the Würmian glaciation.

Fig. 3 Vertical aerial photography of the Coma d'Eugassers (Puigmal). Here the contrast of shades between the Tardiglacial materials and those Würmian ground moraines of the maximum glacial is not clear because of the abundance of fine matrix in the rock glacier moraine has permitted the rapid installation of herbaceous plants.

3.2 Orientation

There can be no doubt the influence of the orientation on the genesis of the rock glaciers studied, since the majority are situated on the Northern slope of this part of the Pyrenees, and those that are on the Southern slope, such as those of the Coma d'Eugassers, Circ de Gra de Fajol and Circ d'Ull de Ter, are always found at the foot of escarpments facing North or North-East.

Contempory with these rock glaciers oriented mainly towards the North, some small cirque glaciers developed with orientation towards the South (without parallel on the Northern slope), which have

left us moraine remains (protalus ramparts) on the threshold of the cirques (Coma de Nou Fonts 2620 m, Coma de Vaca 2540 metres), but always situated at higher altitudes, since as we have already seen, the rock glaciers are found up to heights of 2100 metres.

These morpho-climatic contrasts between different orientations confirm the scarcity of precipitations which characterizes the final stage of the Würm, contrasting with the apparent equality in the maximum of the Würm which we mentioned when speaking of the glacial history.

3.3 Form and dimensions

The general characteristics that we gave to identify a rock glacier were : an abrupt face with a slope near to the slope of equilibrium (which indicates that it is an advance form) with a convex transverse profile, and a longitudinal profile characterized by the typical ridges and furrows clearly reflecting its dynamic movement. But we can group them into several different types according to their superficial disposition and dimensions :

A. Many have a simple lobe form and are generally small (maximum 100 - 150 metres in length), although in some cases they even have longitudinal ridges and furrows. In this group we can include, as examples of those with larger dimensions, the cases of la Coma d'Eugassers (Fig. 3), Coma de l'Infern and la Coma de Morens (Fig. 5).

B. We can only classify as a rock glacier in tongue form that which descends from the Coma dels Racons to the bottom of the valley of Carançà (Fig. 4), with an approximate length of about 750 metres, taking the form of a final small spatulate expansion over the ground and ablation morainic materials of the Würmian maximum, as can be seen in the morphological schemes (Fig. 2 and 5).

C. Others with clearer spatulate form are those found in la Coma de la Bailleta, tributary of the Valle de la Riberola, (Fig.2), although they achieved this

Fig. 4 Vertical aerial photography of the rock glaciers morainic deposits of the Coma dels Racons and a part of the Valle de Carançà (The North is on the right).

form after various single rock glaciers coalesced among themselves. In this way they installed themselves over previous morainic materials carpeting the floor of the valley.

The area occupied by these rock glaciers varies between less than 1 Ha. and more than 14 Ha, as in the last case we mentioned, la Coma de la Bailleta (Fig. 2), their dimensions depending directly on the altitudes and characteristics of the escarpments at the feet of which they have developed.

4 CONCLUSIONS

Now that we have analyzed and described these characteristics of the rock glaciers of the Eastern Pyrenees, I believe we are in a position to make a series of observations of a deductive nature.

The "mise en place" of these glaciers

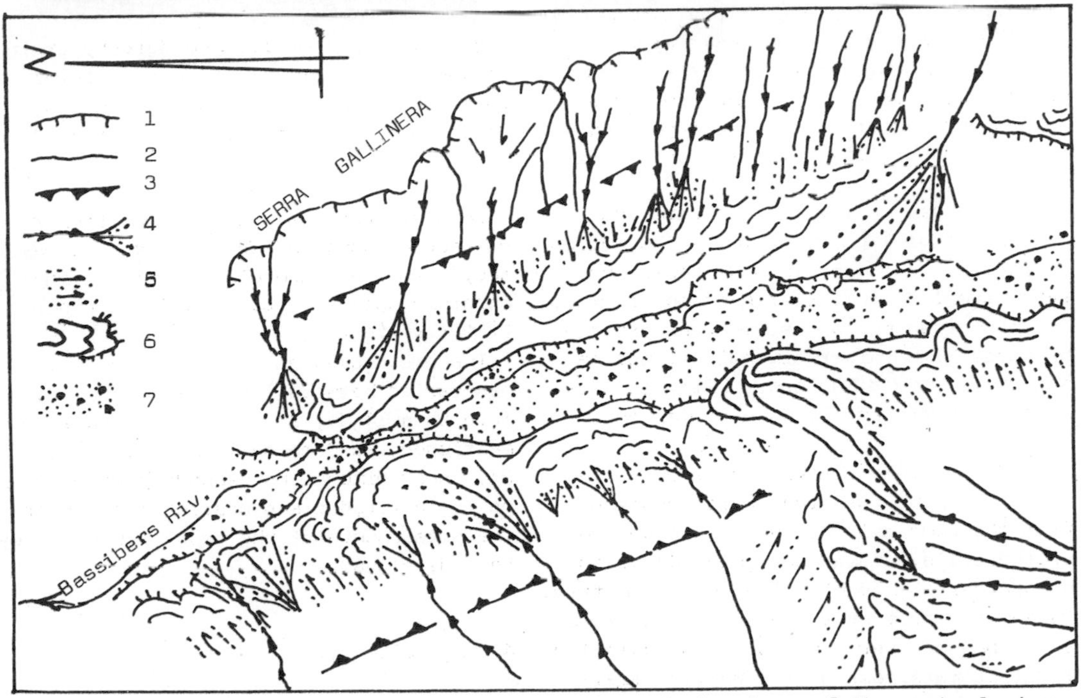

Fig. 5 Vertical aerial photography and geomorphological scheme of the rock glacier morainic deposits of the Coma de Morens and their relation with Würmian ground morainic deposits (dark). 1.- escarpment; 2.- spur; 3.- U shaped valley; 4.- avalanching channel and fan; 5.- rubbles; 6.- rock glacier morainic deposits; 7.- Würmian ground morainic deposits.

corresponds to a cooling of the climate after the disappearance of the ice of the Würmian maximum. There may remain some doubt as to whether the ice of the pre-previous glaciers remained buried under the rubble and that it was these that were the origin of the rock glaciers, but in the majority of cases the disposition of the rock glacier over the ground and ablation moraines seems to indicate that they disappeared completely.

The presence of the rock glacier moraines serves at the same time to confirm the nature of the dry cold that characterized the Tardiglacial pulsation of the Würm in this zone of the Pyrenees, as has been demonstrated by other authors and methods (Jalut 1974, Miskovsky 1974, Soutadé 1973, ...). The marked climatic inequality of the two slopes and a theoretical glaciation limit, with a difference of more than 400 metres between the North and South orientations is contrary to what occured in the pulsation of the Würmian maximum, with abundant precipitations and equality in the glaciation limit though not in the resultant modelling.

The presence of these rock glaciers was conditioned by the existence of escarpments previous to this cold pulsation, which were in turn conditioned by the modelling of the glaciers of the Würmian maximum and by the distinct behaviour of the different types of rock, as was the formation of U-shaped valleys and very steep cirques.

ACKNOWLEDGEMENTS

This study was carried out in the Department of Geomorphology and Tectonic of the Facultad de Ciencias Geológicas of the Barcelona University, under the patronage of the JUAN MARCH Foundation.

The author would therefore like to thank Professors Lluis Solé Sabarís, director of the Geomorphology Department, Jean Tricart, director of the Centre de Géographie Appliquée of Strasbourg, and José Fdez. de Villalta, director of the Sección de Ecología del Cuaternario of the Instituto Jaime Almera of the CSIC, for the facilities provided, and for their kind and interesting suggestions during the whole course of the study.

5 REFERENCES

Barsch, D. 1969, Studien und Messungen an Blockgletschern in Macun, Unterengadin, Zeitschrift für Geomorphologie, Supplementband 8:11-30.

Barsch, D. 1971, Rock glaciers and ice-cored moraines, Geogr. Annaler 53A:203-206.

Cavet, P. 1957, Le paléozoique de la zone axiale des Pyrénées orientales françaises entre le Roussillon et l'Andorre, Bull.serv.Carte Géologique France,55/25A.

Embleton, C., C.King, 1975, Glacial geomorphology, London, Ed.Arnold.

Flint, R.F. 1971, Glacial and Quaternary geology. New York, John Wiley.

Fontboté,J.M., L.Solé and H.Alimen 1957, Livret guide de l'excursion N1-Pyrénées. Madrid, V Con. INQUA.

Guitard, G. 1960, Lineations, schistosité et phases de plissement durant l'orogenèse hercynienne dans les terrains anciens des Pyrénées Orientales, Bull.Soc. Géol. France 7/2:862-887.

Jalut, G. 1974, Evolution de la végetation de l'extremité orientale des Pyrénées et variations climatiques durant les quinze derniers millenaires. Toulouse, Univ. Paul Sabatier.

Johnson, J.P. 1974, Some problems in the study of rock glaciers. In B.D.Fahey R.D.Thompson (ed.), Research in Polar and Alpine Geomorphology, p.84-94.

Marnezy, A.1977. Glaciers rocheux et phénomènes périglaciaires dans le Vallon de la Rocheure (Massif de la Vanoise), Rev.Géog.Alpine, 65/2: 147-167

Messerli, B. 1967, Die eiszeitliche und die gegenwärtige Vergletscherung im Mittelmeerraum, Geog.Helv. 3:103-228.

Miskovsky, J.C. 1974, Le Quaternaire du Midi Méditerranéen. Stratigraphie et paléoclimatologie, Etud.Quater. 3.

Østrem, G. 1971, Rock glaciers and ice-cored moraines, a reply to D.Barsch, Geogr. Annaler, 53/3-4: 207-213.

Santanach, P.F. 1974, Estudi tectònic del Paleozoic inferior del Pirineu entre la Cerdanya i el riu Ter. Barcelona, R.Dalmau ed.

Serrat, D. 1974, Nuevos datos sobre glaciarismo en el Pirineo Oriental, Trab. Neógeno-Cuaternario 2:175-180

Serrat, D. 1978, Quelques aspects des éboulis stratifiés fossiles des Pyrénées orientales espagnoles. In Coll. sur le Périglaciaire d'altitude du domaine mediterranéen et abords, Assoc. Géogr. d' Alsace, p. 147-156.

Soutadé, G. 1973, Aspects du modelé périglaciaire supraforestier des Pyrénées Orientales, Bull. Assoc.Fr.Etd. Quater. 10:239-254

Soutade, G. and A.Baudiere, 1974, Mutations phytogéographiques et variations climatiques durant l'holocène dans les Pyrénées Méditerranéennes françaises, Le Quaternaire (9 Con.INQUA), p.89-93.

Taillefer, F. 1969, Les glaciations des Pyrénées, Ft. Fran.Quater. (8 Con.INQUA) p. 19-32.

Tricart, J. 1971, Normes pour l'établissement de la carte géomorphologique détaillée de la France, Mém. et Doc. Nouv.Série 12:37-106

Tricart, J. and Cailleux, 1962, Le modelé glaciaire et nivale. Paris, SEDES.

Valadas, B. and Y.Veyret, 1974, Quelques aspects des modelés d'origin glaciaire, périglaciaire et nivale sur les confins méridionaux de la Margeride, Rev. Géom. Dyn. 23/4:163-177

Viers, G. 1971, Modélé glaciaire de type méditerranéen dans le massif granitique de la Carançà (Pyrénées Orientales), Photo Interpretation, 10/3:27-32

Quartärgeologische Untersuchungen im Nordwestteil des Salzach-Vorlandgletschers (Oberbayern)

W.-D.GRIMM und
H.BLÄSIG, G.DOPPLER, M.FAKHRAI, K.GORONCEK, G.HINTERMAIER, J.JUST, W.KIECHLE
W.H.LOBINGER, H.LUDEWIG, S.MUZAVOR, M.PAKZAD, U.SCHWARZ, T.SIDIROPOULOS
Universität München, München, Bundesrepublik Deutschland

ZUSAMMENFASSUNG

Geologische Kartierungen, sedimentpetrographische Untersuchungen und hydrogeologische Aufnahmen im Nordwestteil des Salzach-Vorlandgletschers und im Grenzbereich gegen den Chiemsee-Vorlandgletscher erbrachten neue, detaillierte Vorstellungen zur dortigen quartären Stratigraphie und Paläogeographie, insbesondere zum Vergletscherungsgeschehen und zu den Abschmelzvorgängen und -bahnen. Wichtige Befunde sind vor allem

- die etwa gleiche maximale Reichweite des günz-, mindel- und risszeitlichen Gletschers nach Norden; nur die Würmvereisung blieb hinter diesen Gletschervorstößen zurück;
- die weite Verbreitung der günzzeitlichen Vergletscherung mit Zweiteilung der Moränensedimente durch Seeablagerungen;
- der Aufbau der mindelzeitlichen Sedimentserie aus mächtigen Vorstoßschottern, darüber einem einzigen Endmoränenwall, der dem maximalen Gletscherstand entspricht, und einer nur geringmächtigen Moränenhaut im Rückland;
- die Anlage des Chiemseegletschers im Arbeitsgebiet erst zu Beginn der Risseiszeit;
- die Gliederung der Risseiszeit im Salzach- und Chiemsee-Gletschergebiet in 5 Phasen;
- die Möglichkeit einer zwischen Riss 3 und 4 eingeschalteten Gletscher-Oszillation (Kirchweidach-Tyrlachinger Oszillation);
- das Fehlen einer Frühwürm-Vergletscherung;
- die Einschaltung einer Gletscher-Oszillation zwischen Würm 2 und 3 (Radegund-Lanzinger Oszillation);
- die Gliederung des Spät- und Postglazials im Salzach-, Alz- und Inntal durch 4 bis 6 Terrassenstufen.

1 EINLEITUNG

Seit 1970 bearbeiten die Verfasser mit quartärgeologischen, sedimentpetrographischen und angewandt-geologischen Methoden den Nordwestteil des pleistozänen Salzachgletschers im südostbayerischen Alpenvorland. Die wichtigsten Ergebnisse seien hier kurz mitgeteilt; detaillierte Aufzeichnungen sind in 13 unveröffentlichten Diplomarbeiten (s.Kapitel 4.2) enthalten, die von 1972 bis 1978 unter der Leitung von W.-D.Grimm fertiggestellt wurden und werden (Diplomkartierungsgebiete s.Abb.). Die Namen einzelner Mitarbeiter werden im folgenden nur ausnahmsweise in Klammer hinter ihren wichtigsten Ergebnissen notiert. Den vorliegenden Text verfaßte W.-D.Grimm.

Eine vereinfachte, kleinmaßstäbliche Fassung der von uns erstellten geologischen Karte 1 : 25.000 der Gradabteilungsblätter 7741 Mühldorf/Südteil, 7742 Neuötting/Südteil, 7841 Garching, 7842/43 Burghausen, 7941 Trostberg und 7942/43 Tittmoning (insgesamt ca. 600 km^2) ist beigefügt.

Grundlegende Vorarbeiten für unsere Untersuchungen waren die Arbeiten von Brückner (1886), Penck & Brückner (1909), Troll (1924) sowie vor allem Ebers, Weinberger & Del Negro (1966).

Während unseren Untersuchungen erschien die Publikation von Eichler & Sinn (1974). Beide Autoren arbeiteten gleichzeitig mit uns, aber ohne gegenseitiges Wissen, im Raum Burghausen - Margarethenberg - Garching. Ihre Ergebnisse stimmen überein mit den Befunden unserer etwas früheren, aber unveröffentlichten Diplomarbeiten im selben Raum (Bläsig 1973, Fakhrai 1972, Just 1972, Lobinger 1973, Ludewig 1972, Muzavor 1973, Pakzad 1973). Wir sehen darin eine gute gegenseitige Bestätigung.

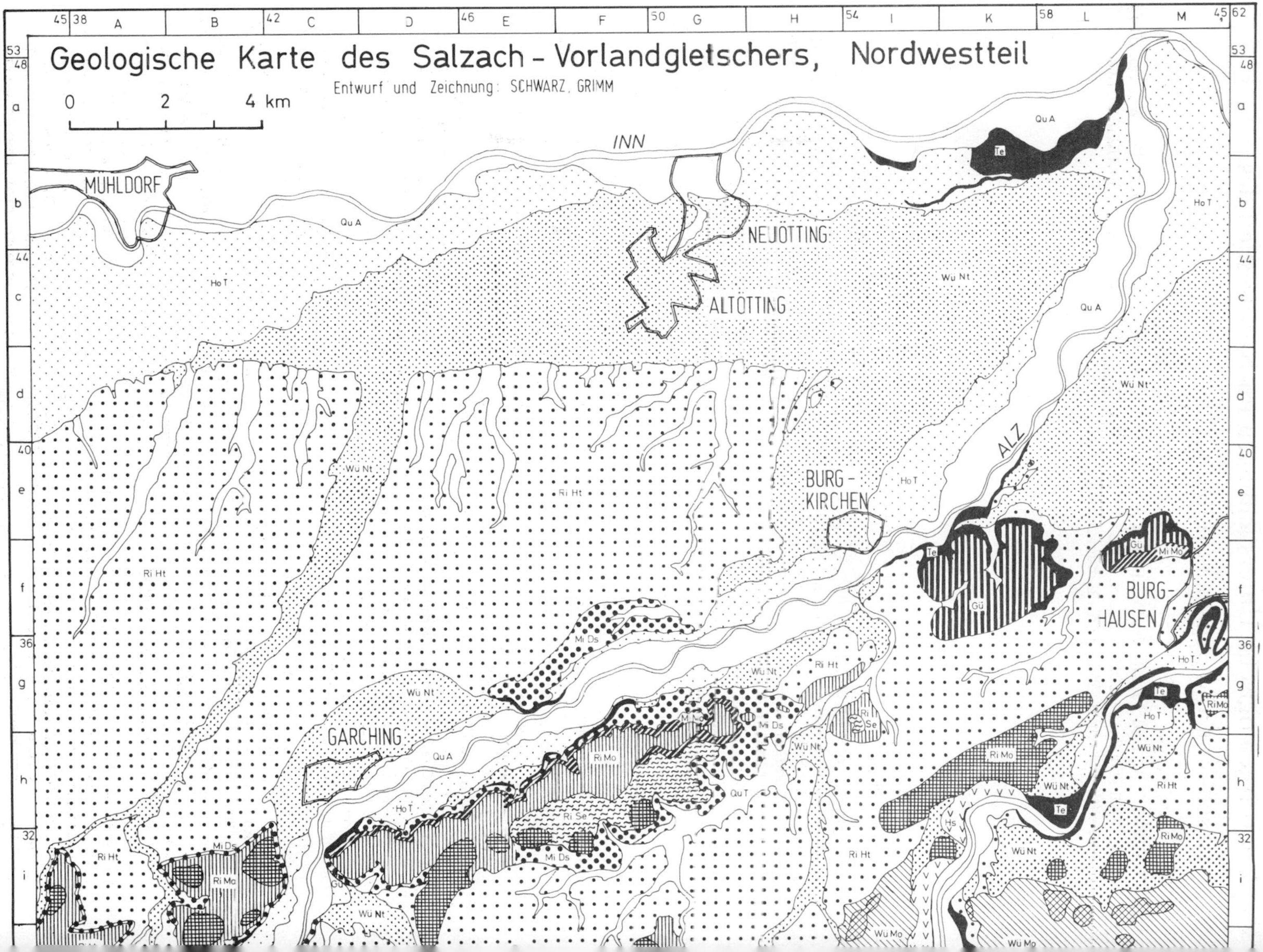
Geologische Karte des Salzach-Vorlandgletschers, Nordwestteil
Entwurf und Zeichnung: SCHWARZ, GRIMM
0 2 4 km
INN
MÜHLDORF
NEUÖTTING
ALTÖTTING
BURG-KIRCHEN
ALZ
BURG-HAUSEN
GARCHING
Qu A
Ho T
Wü Nt
Ri Ht
Ri Mo
Mi Ds
Ri Se
Qu T
Gü
Te
Mi Mo
Wü Mo
Hs

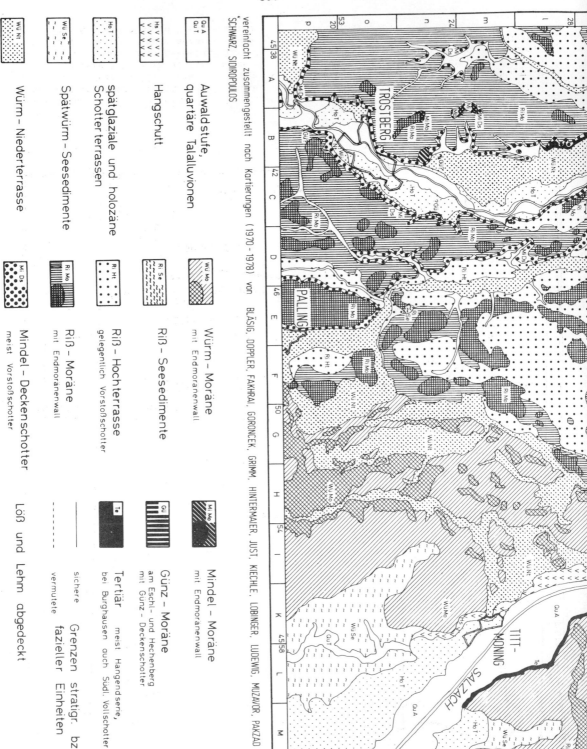

vereinfacht zusammengestellt nach Kartierungen (1970 - 1978) von BLÄSIG, DOPPLER, FAKHRAI, GORONCEK, GRIMM, HINTERMAIER, JUST, KIECHLE, LOBINGER, LUDEWIG, MUZAVOR, PAKZAD SCHWARZ, SIDIROPOULOS

Qu A, Qu T	Auwaldstufe, quartäre Talalluvionen	Wü Mo	Würm - Moräne mit Endmoränenwall	Mi Mo	Mindel - Moräne mit Endmoränenwall
Hs	Hangschutt	Ri Se	Riß - Seesedimente	Gü	Günz - Moräne am Eschl- und Hechenberg mit Günz - Deckenschotter
Ho T	spätglaziale und holozäne Schotterterrassen	Ri Ht	Riß - Hochterrasse gelegentlich Vorstoßschotter	Te	Tertiär meist Hangendserie, bei Burghausen auch Südl. Vollschotter
Wü Se	Spätwürm - Seesedimente	Ri Mo	Riß - Moräne mit Endmoränenwall	——— / - - - -	sichere / vermutete Grenzen stratigr. bzw. fazieller Einheiten
Wü Nt	Würm - Niederterrasse	Mi Ds	Mindel - Deckenschotter meist Vorstoßschotter		Löß und Lehm abgedeckt

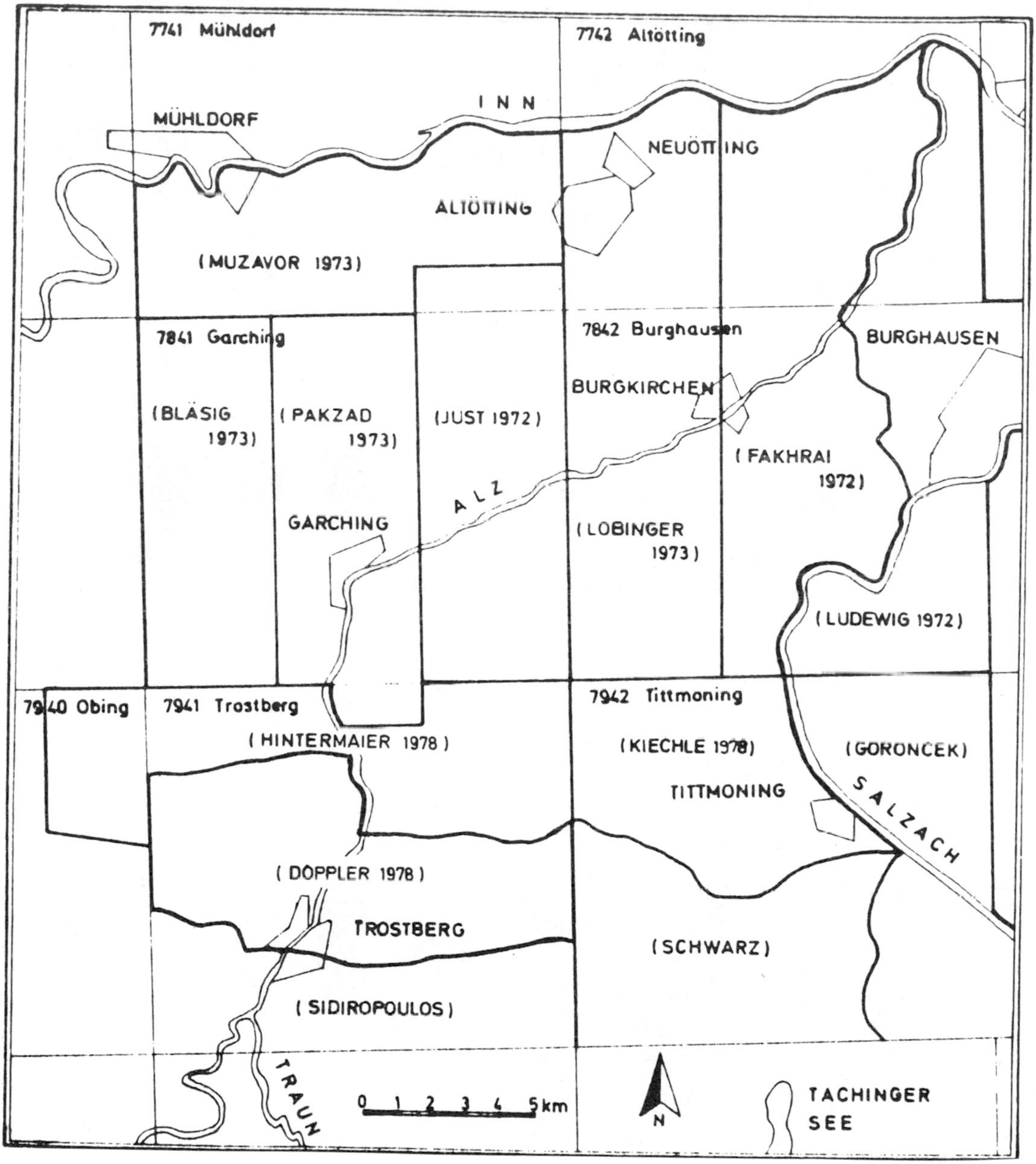

Kartiergebiete im Nordwestteil des Salzach-Vorlandgletschers

Wertvoll waren für uns ferner die gründlichen Beobachtungen von Traub (1953) und von Traub & Jerz (1975) sowie die Ergebnisse Zieglers (1977) im südlich angrenzenden Gebiet des Salzach-Vorlandgletschers. Wir danken diesen Autoren sowie Frau Dr.E.Ebers und den Herren Rektor J. Dirscherl / Garching, Prof.Dr.J.Fink / Wien, Prof.Dr.B.Frenzel / Stuttgart-Hohenheim, Prof.Dr.H.Heuberger / München, Dr.H. Kohl / Linz sowie allen Personen, Ämtern und Institutionen, von denen wir Hilfe, Anregungen und Kritik erhielten und die uns Unterlagen überließen.

2 ARBEITSMETHODEN

Die Untersuchungen wurden weniger extensiv als intensiv ausgeführt. Durch die Verwendung vielfältiger Gelände- und Labormethoden konnten die Ergebnisse gegenseitig kontrolliert und mehrfach abgesichert werden.

Die geologische Kartierung erfolgte mit den üblichen Geländeverfahren, unterstützt durch morphologische, pedologische und luftbildgeologische Methoden und bereichsweise präzisiert durch Handbohrungen und Sondierungen in engem Netz. Als Ergebnis der geologischen Kartierung liegen für das gesamte Arbeitsgebiet vor
- geologische Karten 1 : 25.000;
- Profilserien 1 : 25.000, 12,5 fach überhöht, Nord-Süd gerichtet, mit Profilabständen von jeweils 2 km; bereichsweise kommen Ost-West gerichtete Querprofilserien hinzu;
- Karten 1 : 25.000 der geologischen Aufschlüsse zur Bewertung der Aussagegenauigkeit der Karten und Profile.

Die Geländebefunde wurden überprüft und ergänzt durch sedimentpetrographische Gelände- und Laboruntersuchungen zur Korngrößenverteilung, zur Geröll- und Geschiebepetrographie, zum Schwer- und Leichtmineralgehalt, zum Ton- und Karbonatmineralbestand, zur Eisen- und Manganverteilung, zur Morphometrie und Morphoskopie von Geröllen und Sandkörnern, zur Gefügeregelung von Einzelkomponenten und Schichtverbänden sowie zum Verfestigungs- und Entfestigungsgrad. Die Daten wurden in Listen, Tabellen und Kartogrammen notiert und in Karten, Diagrammen und Textpassagen ausgewertet. Über das gesamte Arbeitsgebiet wurden erstellt
- Kartogramme 1 : 25.000 der maximalen Geröllgröße nach Wentworth (K_{max}- und K_m-Werte);
- Kartogramme 1 : 25.000 des qualitativen Geröllspektrums (4 Gesteinsgruppen).

Neben der geologischen und sedimentpetrographischen Bearbeitung erfolgte eine hydrogeologische Kartierung mit den üblichen Geländeverfahren und Brunnenmessungen und unterstützt durch Laboruntersuchungen, vor allem Analysen zum Grundwasserchemismus sowie zum Porenraum und zur Permeabilität der Sedimente. Als Ergebnis wurden für das gesamte Arbeitsgebiet gezeichnet
- Karten 1 : 25.000 der Grundwasseroberflächen (Isohypsen);
- Kartogramme 1 : 25.000 zur Grundwasserbeschaffenheit (Darstellung der Grundwassereigenschaften und -inhaltsstoffe);
- Karten 1 : 25.000 der hydrogeologischen Aufschlüsse (Brunnen, Pegel, Quellen, Vernässungshorizonte usw.) zur Kontrolle der Verläßlichkeit der hydrogeologischen Aussagen.

Ergänzend wurden die einschlägigen Gelände- und Labormethoden der Ingenieurgeologie angewandt, um die baugrundgeologisch relevanten Eigenschaften der Sedimente zu kennzeichnen.

Bei den sedimentpetrographischen Verfahren erwiesen sich die Kornverteilung und die daraus gewonnenen Parameter, die maximale Geröllgröße, das qualitative Geröllspektrum, die Schwermineralanalyse und das Schichten- und Einzelkorngefüge als besonders aussagekräftig zur Rekonstruktion der Herkunft, der Schüttungsrichtung, des Transportmediums, der Transportweite und des Ablagerungsmechanismus der Sedimente. Wichtig für die Klärung des stratigraphisch und faziell komplizierten Schichtenbaus der Quartärfolge war die Erfassung aller vorliegenden Bohrungen und Sondierungen; anders wären die Konstruktion enger und tiefreichender Profilserien und die Klärung des hydrogeologischen Stockwerkbaus nicht möglich gewesen. Die Grundwasserhydraulik trug zur Beantwortung stratigraphischer Fragen und zur Schichtenkorrelation entscheidend bei. Die baugrundgeologischen Untersuchungen erlaubten Aussagen zur Lagerungsdichte der Sedimente und gaben damit gelegentlich Hinweise, ob ein Sediment unter Eislast lag oder unvorbelastet ist.

3 STRATIGRAPHISCHE UND PALÄOGEOGRAPHISCHE BEFUNDE

3.1 Tertiärer Untergrund

Den unmittelbaren Untergrund der Quartärsedimente im Arbeitsgebiet bilden die Ablagerungen der miozänen Molasse. Sie umfassen die Obere Meeresmolasse, die Süßbrackwassermolasse zweier getrennter Verlandungsbecken und die Obere Süßwassermolasse, die örtlich bis mindestens an die Wende Miozän / Pliozän reicht. Die regionale Schichtenverteilung belegt insgesamt eine weitgespannte tektonische Mulde zwischen Alpenrand und unterem Inn.

Wichtig für den Stoffbestand der quartären Schichten ist eine altangelegte, bis heute wirksame tektonische Störungszone, die von Oberösterreich über Burghausen und Neuötting bis über Landshut hinaus durchzieht (Landshut-Neuöttinger Abbruch). Durch die Heraushebung des Bereichs nordöstlich der Störung gelangte

in Oberösterreich und bei Burghausen der grobe, quarzreiche Südliche Vollschotter (oberstes Miozän) bis in den Erosionsbereich und wurde in die quartären Schüttungen umgelagert. Südwestlich des Abbruchs dagegen - z.B. bei Burgkirchen und Garching - bildet die jüngere, feinerkörnige Hangendserie (Wende Miozän / Pliozän) mit überwiegenden Mergeln und Sanden die Sohle des Quartärs; dadurch fehlen dort die groben umgelagerten Quarzkiese aus dem unmittelbaren Untergrund in den eiszeitlichen Sedimenten.

Seit der Ablagerung der Oberen Meeresmolasse unter NN. erfolgte eine wohl kontinuierliche Hebung bis in mehr als 500 m ü.NN. Sie führte paläogeographisch zunächst zur Verbrackung (Süßbrackwassermolasse) und Verlandung (Obere Süßwassermolasse) des Molassebeckens, sodann im Pliozän zur Umgestaltung des Ablagerungsraumes in einen Abtragungsraum. Das bis zum Beginn des Quartärs durch fluviatile Ausräumung entstandene Tertiärhügelland legte die Bahnen des ersten Gletschervorstoßes fest. Dies präjudiziert jedoch keine Persistenz der Gletscherbahnen bis in die Würmeiszeit, da in der Mindeleiszeit eine Plombierung des Reliefs durch Vorstoßschotter zwischengeschaltet wurde (s.Kapitel 3.4).

Die fortgesetzte Hebungstendenz bedingt auch den Ablagerungsmechanismus während des Quartärs: Die Täler wurden sukzessive eingeschnitten; dadurch liegen die Ablagerungen aufeinander folgender Eiszeiten, vor allem die fluvioglazialen und fluviatilen Schüttungen am Außenrand des Vorlandgletschers, allgemein immer tieferen Basisflächen auf.

3.2 Generelles zum Pleistozän

Für das Pleistozän wurde die Gliederung " Mindel - Riss - Würm - Spät- und Postglazial " entsprechend den Arbeiten von Penck & Brückner (1909), Troll (1924) und Ebers et al. (1966) übernommen, ohne daß die zeitlichen Zuordnungen durch absolute Altersbestimmungen oder Leitfossilien oder durch Parallelisierungen mit benachbarten Gletschergebieten gesichert wären.

Neu entdeckte Moränensedimente und zugehörige fluviatile Schotter, die eindeutig älter sind als die Mindelablagerungen, wurden als "Günz" bezeichnet. Wir folgen damit der Gliederung, die Weinberger (seit 1950) durch seine Zuordnung der Altmoränen und -schotter am Nordostrand des Salzach-Vorlandgletschers, vor allem am Siedelberg, verwendete und entsprechen auch der Nomenklatur, die Eichler & Sinn (1974) gleichzeitig mit uns wählten. In Bohrungen im Raum Trostberg sind diese ältesten Pleistozänablagerungen als sehr mächtige Moränensedimente ausgebildet und durch Seeablagerungen zweigeteilt (s.Kapitel 3.3); ob es sich dabei um Günz I und II mit zwischengeschalteter Oszillation im Sinne eines Interstadials oder Intervalls handelt oder um eine selbständige ältere Eiszeit unter Günz, kann nicht entschieden werden; wir nehmen vorläufig erstere Deutung an.

Die jüngeren Eiszeiten - vor allem Riss und Würm - sind ebenfalls durch Rückzugsphasen und Eisrandschwankungen wechselnder Amplitude differenziert. Auch hier war mangels pedologischer Fakten und Vegetationsentwicklungen nicht zu klären, ob die ausgeprägten Oszillationen, die wir zwischen Riss 3 und 4 (Kirchweidach-Tyrlachinger Oszillation, s.Kapitel 3.5) vermuten und die wir für das Würm (Radegund-Lanzinger Oszillation, s.Kapitel 3.6) nachgewiesen haben, nur einem lokalen und kurzfristigen oder aber einem weiten und bedeutsamen Eisrückzug entsprechen.

Die Gesteinsveränderungen durch Verwitterung und Diagenese - vor allem die Bodenbildungen und der diagenetische Verfestigungs- und Entfestigungsgrad der verschiedenen Schichten - waren wichtige Hinweise auf das Alter der Sedimente. Beweisend sind sie aber keinesfalls, da diese Veränderungen auch unregelmäßig und unabhängig vom Alter der Sedimente sein können. Auffällig war vor allem, daß nichtbindige, gut durchlässige Ablagerungen (Aquifer), halbbindige, schwer durchlässige Ablagerungen (Aquitard) und bindige, stauende Ablagerungen (Aquiclud) sehr unterschiedlich auf die Einflüsse von Verwitterung und Diagenese reagieren. Das läßt darauf schließen, daß nicht nur der Stoffbestand und die Exposition der Sedimente sowie die Art und Dauer des chemischen und physikalischen Angriffs entscheidend sind, sondern auch die Wasserzirkulation und der Lösungstransport im Porenraum.

Im folgenden werden die Begriffe "Moräne" morphologisch, dagegen "Moränensediment" stofflich im Sinne von "Till" verwendet. Entsprechend wird als "Terrasse" die äußere Form, als "Terrassenmaterial" der Stoffbestand angesprochen.

3.3 Günz

Bis zu unseren Arbeiten waren im Bereich des Salzach-Vorlandgletschers nur die alten Moränen und Deckenschotter nördlich von Mattighofen / Oö., vor allem am Siedelberg, als günzzeitlich bekannt (Weinberger 1950).

Durch die Diplomarbeiten von Fakhrai

(1972) und Ludewig (1972) und durch die Publikation von Eichler & Sinn (1974) wurden auch der Hechenberg und Eschlberg nordwestlich von Burghausen als Günz-Deckenschotter in Wechsellagerung und Verzahnung mit Günzmoräne ausgewiesen. Dafür sprechen
- der hohe Tertiärsockel (Oberkante bei ca. 450 - 465 m ü.NN);
- die tief, bis unter 470 m ü.NN., herabreichenden Moränenablagerungen; (dagegen liegen die Mindelmoränen am Gletschernordrand im allgemeinen erst bei 495 - 500 m dem Mindel-Vorstoßschotter auf);
- die petrographische Zusammensetzung der Schotter;
- die mächtige Verwitterungsschicht, die in ausgedehnten Taschen mehr als 15 m tief ins Substrat hinabreichen kann.

Ca. 12 km westlich des Eschlberges - im Einschnitt der Tauernbahn am Brunnthal-Ausgang südwestlich von Wald - wurde ein weiteres eindeutiges Günzvorkommen entdeckt (Pakzad): Dort liegt grobblockiges, dicht konglomeriertes, ungeschichtetes Günz-Moränensediment mit ebener Grenzfläche unter kleinkörnigem, gut geschichtetem, z.T. schräg gelagertem Mindelvorstoßschotter. In Korrelation zu diesem Aufschluß ist auch eine grobblockige Moränenschicht, die in der ca. 2 km südlich gelegenen Brunnenbohrung bei Amsham zwischen tertiärer Hangendserie und Mindelschotter eingeschaltet ist (Traub 1953), eindeutig dem Günz zuzuordnen (Pakzad).

Weiter südlich im Alzbereich erschließen mehrere Bohrprofile tiefgelegene Moränensedimente und Seetone (Doppler). Sie dehnen sich in großer Mächtigkeit (bis 100 m) im weiteren Umkreis von Trostberg beiderseits des Alztales aus und sind, da sie teilweise unter Mindel-Vorstoßschottern liegen, ins Günz einzustufen. Zu diesem ausgedehnten Günzvorkommen gehören auch die am Alztalgrund zwischen Baumburg und Lengloh neu gefundenen Oberflächenaufschlüsse mit Grundmoränenmaterial und Seesedimenten (Alzufer unterhalb von Baumburg; Talhang südlich von SKW Trostberg; Straßenanschnitte in Eglsee westlich unterhalb von Heiligkreuz; Bergham) sowie die ausgespülten Findlingsblöcke an der Sohle der Niederterrassen-Kiesgrube Piedersdorfer südsüdöstlich von Tinning (Doppler; Sidiropoulos). Sie lassen zum Teil eine intensive und tief reichende Zersetzung sowie dichte Lagerung erkennen. Insgesamt weisen die oberirdisch und die in Bohrungen erschlossenen tiefgelegenen Moränen- und Seesedimente im Raum Trostberg auf 2 frühe Gletschervorstöße hin, die - mit den oben (Kapitel 3.2) dargelegten Unsicherheiten - vorerst zwei Günzstadien zugeordnet werden.

Ein weiterer, trotz seiner Auffälligkeit noch nicht beschriebener Aufschluß, der mit hoher Wahrscheinlichkeit Günz erschließt, liegt bei Leimhof / Oö. am Nordabfall des Adenberges in ca. 450 m ü.NN. (Grimm). Dort zeigt eine imposante Kiesgrube tertiären Quarzkies an der Basis, darüber ca. 10 m Moränenmaterial oder moränennahen Schotter mit großen Findlingsblöcken, darüber - über einem extrem dicht verfestigten Grenzband - Deckenschotter mindelzeitlichen Alters. Die tiefgelegene Moränenschicht ist in bizarren, riesigen Zapfen, die vom Grenzband aus stalaktitenartig in die Tiefe wachsen, verhärtet; daneben liegende, unverfestigte Partien sind zu Höhlen ausgewaschen. Die Lagerungsverhältnisse und der anormale Verfestigungs- und Verwitterungsgrad lassen auf Günz schließen.

Wenn alle genannten Vorkommen günzzeitlich sind, so verändert sich das frühere regionalgeologische und paläogeographische Konzept für die Günzeiszeit im Salzachgletschergebiet grundlegend: Zeugen für die günzzeitliche Vereisung sind dann nicht länger mehr auf den Nordostrand des Vorlandgletschers beschränkt, sondern reichen nach Westen kontinuierlich über Adenberg, Hechenberg und Eschlberg bis zum Alzknie bei Garching, von dort am Westrand des Gletschergebietes bis über Trostberg und Baumburg hinaus. Die Mächtigkeiten in den zentralen Bereichen können, wie die Bohrungen im Raum Trostberg erkennen lassen, beträchtlich werden. Damit wäre der Günzgletscher mindestens ebenso weitreichend, flächenhaft verbreitet und ablagerungsintensiv gewesen wie die nachfolgenden Mindel- und Rissgletscher.

3.4 Mindel

Die Mindelablagerungen im Arbeitsgebiet sind gut aufgeschlossen an den Steilhängen beiderseits des Alztales, am Rand des Brunnthales und in den autochthonen Nebentälern zur Alz und zum Brunnthal. Mögliche Reste mindeleiszeitlicher Moränen krönen den Hechenberg als Kappe über dem Günzsockel (Ludewig 1972; Eichler & Sinn 1974). Geologische Orgeln und Frostkeile sind fast gänzlich auf die Deckenschotter vor der Gletscherstirn beschränkt, wo die deszendente Verwitterung direkt am durchlässigen Schotter angreifen konnte.

Auffälligstes Ergebnis der Mindeleiszeit ist der weithin und gleichmäßig verbreitete Vorstoßschotter, der mehr als 50 m Mächtigkeit erreichen kann. Die Unterflä-

che und die primäre Oberfläche des Vorstoßschotters sind im Arbeitsgebiet meist flach oder sanft gewellt; nur gelegentlich ragt der Tertiärsockel auf oder tiefen sich Rinnenstrukturen in die Basisfläche ein. Der Vorstoßschotter-Charakter erweist sich vor allem am Nordrand des Verbreitungsgebietes in einer kontinuierlichen Korngroßenzunahme gegen das Hangende, im ganzen Gebiet in einem allmählichen Übergang des Schotters in moränennahes Material und dann Moränensediment. Die Moränendecke ist zumeist schmächtig und sehr lückenhaft; auf der beigefügten kleinmaßstäblichen Karte konnten die eng begrenzten und dünnen Moränenvorkommen nur ausnahmsweise eingetragen werden.

Der einzige deutliche Endmoränenwall über den Vorstoßschottern entspricht dem maximalen Gletscherstand. Dazu gehören links der Alz der Trostberg-Oberbrunnhamer Rücken mit dem Nunbichl (Doppler, Hintermaier), rechts der Alz der Höhenzug Wald - Margarethenberg (Just, Lobinger, Pakzad). Beide Höhenzüge bildeten ehedem einen geschlossenen Wall am West- und Nordrand des Gletschergebietes; heute ist er weitgehend von Rissmoränenmaterial bedeckt und durch das junge Alztal unterbrochen.

Im Trostberg-Oberbrunnhamer Rücken betragen die Mächtigkeiten der Mindel-Endmoränen bis 30 m. Im Vergleich hierzu nehmen wir - entgegen früheren Meinungen - an, daß auch im Kern des Höhenzuges Wald - Margarethenberg die Mindelmoränen unter der Riss-Moränendecke bis über 2 Zehnermeter mächtig werden können.

Die gleichmäßige Lagerung und Mächtigkeit der Mindelsedimente und die Übermacht der Vorstoßschotter werfen Fragen zum Schüttungsmechanismus auf. Zunächst müssen (Schmelz-)Wässer in großer Menge und über lange Zeiten hinweg breitfächernd und strömungsintensiv verfügbar gewesen sein, um die am Ende der Günzeiszeit bestehenden Reliefunterschiede erosiv zu glätten und dann gleichmäßig und mächtig mit Vorstoßschottern einzudecken. Sodann muß der Salzachgletscher sachte und auf breiter Front über seine eigenen Vorstoßschotter hinweggewandert sein. Der Maximalvorstoß war mit einem längeren Gletscherhalt verbunden und führte zur Akkumulation des Endmoränenwalles von Trostberg - Oberbrunnham - Wald - Margarethenberg (- Hechenberg). Das Abschmelzen dürfte einheitlich und sehr rasch erfolgt sein, so daß nur eine schmächtige und lükkenhafte Moränenhaut über dem Vorstoßschotter zurückblieb.

Die vorgeschobene Lage des begrenzenden äußeren Endmoränenwalles und die einheitliche Verbreitung der rückwärtigen Mindelgrundmoräne bis ins Gebiet westlich der Alz beweisen, daß der Salzachgletscher damals viel weiter nach Westen reichte als in den nachfolgenden Eiszeiten. Der Chiemseegletscher war noch nicht in Aktion! Der eisfreie Zwickel zwischen Salzach- und Inngletscher reichte etwa bis Trostberg zurück; von dort divergierten die äußeren Endmoränenwälle: der Wall des Salzachgletschers nach Norden, der des Inngletschers nach Westen.

3.5. Riss

Die Rissablagerungen bilden am Nordsaum des Salzachgletschers, außerhalb der Würm-Endmoränen, über weite Bereiche die Geländeoberfläche. Sie kommen in den Taleinschnitten häufig in natürlichen und künstlichen Aufschlüssen vor, finden sich gelegentlich aber auch auf den Hochflächen freigelegt.

Der Geröllbestand der Rissablagerungen mit relativ hohen Kristallin- und Quarzgehalten erlaubt zumeist eine signifikante Abtrennung von den karbonatreicheren Mindelsedimenten, dagegen nur eine undeutliche Unterscheidung von den Würmablagerungen. Salzach-, Chiemsee- und Inngletscher können in der Regel aufgrund ihrer Geröllspektren, vor allem der spezifischen Leitgesteine, charakterisiert werden. Auch innerhalb eines Gletschergebietes variieren die Geröllgesellschaften gesetzmäßig von Osten nach Westen und vom Rand gegen das Zentrum.

Die Lagerungsverhältnisse der Rissmoränen und -schotter und das Relief ihrer Basis- und Deckfläche sind - im Vergleich zu den Mindelsedimenten - kompliziert. Vorstoßschotter, die im Mindel dominierten, treten im risszeitlichen Vereisungsgebiet weitgehend zurück; dort beginnt die Schichtensequenz meist direkt mit Moränensedimenten. Vorstoßschotter in nennenswerter Verbreitung und Mächtigkeit scheinen im Arbeitsgebiet auf den Raum südlich des Hechen-, Eschl- und Willhartsberges bis zurück zum Riss-3-Endmoränenwall (s.unten) beschränkt.

Ein wichtiger Befund unserer Untersuchungen ist, daß der Chiemseegletscher erstmalig im Riss im Arbeitsgebiet erkennbar wird (Doppler; Hintermaier): Er zwängt sich zwischen Inn- und Salzachgletscher, engt letzteren beträchtlich ein und drängt ihn um fast 10 km nach Osten bis in den Bereich jenseits des Brunnthals zurück. Beim Maximalvorstoß (Riss 1, s.unten) machte sich die Differenzierung räumlich noch kaum bemerkbar; beide Gletscher hingen noch mit einheitlicher Stirn über das Brunnthal hinweg zusammen. Erst während

der Rückzugsphasen klafften Salzach- und Chiemseegletscher auseinander und trennten sich zunehmend an der Naht des Brunnthales. Dadurch kam der Bereich des späteren Alztales zwischen Altenmarkt und Garching, der im Mindel noch voll zum Areal des Salzachgletschers gehörte, nun gänzlich in den Bereich des Chiemseegletschers zu liegen. Die Endmoränenzüge des Chiemseegletschers sind - vielleicht mangels Eismasse - wesentlich flacher und aufgelöster als die des Salzachgletschers und lassen sich daher nur schwer in ein System von Staffeln bringen.

Ebers et al. (1966, Kartenbeilage) zeichnen als weitest nördliche Moränenablagerung am Nordwestrand des Salzachgletschers den Wall von Kirchweidach - Laimgruben b.Burghausen ein. Im zugehörigen Text (S.53) aber erwägt Ebers, ob die auf dem Deckenschotterrücken Wald - Margarethenberg - Edhof liegenden "alten Grundmoränen" einem "kurzen, weitreichenden Alt-Riß-Vorstoß" zuzuordnen sind. Eichler & Sinn (1974, S. 146/147) bestätigten die Zuordnung zum Riss und bewiesen durch geröll- und geschiebepetrographische Analysen, daß die Hangendmoräne auf der Margarethenberger Höhe das Ergebnis einer Oszillation "derselben Vereisung ist, die schließlich bei längerem stationärem Verweilen weiter im S die mehr oder weniger geschlossenen Endmoränenwälle von Freutsmoos - Tyrlaching ... hinterließ".

Unsere gleichzeitigen Untersuchungen (Just 1972; Lobinger 1973; Pakzad 1973) führten zum selben Befund und erbrachten darüber hinaus wesentlich erweiterte und differenzierte Vorstellungen zu den Schüttungsmechanismen und zur Paläogeographie im Riss. Wir unterscheiden 4 Rissphasen, die durch Endmoränenwälle im Salzach- und Chiemseegletschergebiet gekennzeichnet sind, und eine weitere, 5. Phase, die sich durch die Terrassengliederung im Brunnthal zu erkennen gibt.

Riss 1 : Die am Nordrand des Höhenrückens Wald - Margarethenberg dem Mindelsockel auflagernden jüngeren Moränen sowie die östlich gelegenen Moränenkuppen bei Willhartsberg und Voketsberg sind dem Maximalstand des risszeitlichen Gletschervorstoßes zuzuordnen (Just; Lobinger; Muzavor; Pakzad). Die Riss-1-Endmoräne setzt sich in gerader Front westlich der Alz fort und bildet dort - z.B. bei Engelsberg - Moränenkuppen, die nach ihrem Stoffbestand bereits dem Einflußgebiet des Chiemseegletschers angehören (Bläsig; Hintermaier).

Riss 2 : Das vom Maximalstand zurückschmelzende Gletschereis schuf wenige Kilometer weit südlich am Südrand des Höhenzuges Wald - Margarethenberg einen weiteren, nur schwach ausgeprägten Endmoränenwall (Pakzad). Er staute vor sich einen Eisrandsee auf, der fast 10 km weit von Kronposthub im Westen bis Willhartsberg im Osten durch limnische Sedimente, Kamesbildungen und Quellaustritte nachzuweisen ist (Just; Lobinger). Im Westen bog der Riss-2-Eisrand des Salzachgletschers scharf ins Brunnthal zurück und vereinigte sich erst südlich von Feichten mit dem Lobus des Chiemseegletschers. Dessen Riss-2-Stirn zog von Gloneck über Feichten und Edelham, sodann über die Alz nach Nordwesten und bog dann über Peterskirchen zurück, um sich im Mörntal mit dem Inngletscher zusammenzuschließen. Damals wurde also die Anlage des Brunnthales geschaffen durch die Schmelzwässer aus dem Zwickel zwischen Salzach- und Chiemseegletscher, die Anlage des Mörntales durch die Schmelzwässer von Chiemsee- und Inngletscher.

Eine Sonderentwicklung erfolgte während des Riss 1 und 2 im Raum östlich des Höhenrückens Wald - Margarethenberg, nämlich zwischen Eschl- und Hechenberg im Norden und dem Moränenzug Kirchweidach - Laimgruben (Riss 3) im Süden. Dort hatten die Schmelzwässer beim Vorrücken des Rissgletschers eine tiefe und breite Rinne ausgefurcht und wieder flach mit Vorstoßschottern plombiert. Über die eigenen Vorstoßschotter hinweg schob sich dann der rißzeitliche Gletscher nach Norden vor; während er im Bereich des Höhenrückens Wald - Margarethenberg auf Widerstand traf und sich nur langsam über den Bergrücken bis in Höhen über 530 m ü.NN. aufstaute, fand er in der östlich angrenzenden Niederung nur Höhen bis ca. 450 m ü.NN. und drang dadurch beschleunigt vor. Ergebnis ist dort eine geringmächtige, aber durchgängige Grundmoränenlage, die nach Norden ausdünnt und vermutlich den Sockel des Eschlberges und Hechenberges nicht mehr erreicht hat. Später haben sich die Ablagerungen der Hochterrasse (wohl Riss 4) über die dünne Moränenschicht gebreitet (Fakhrai; Lobinger; Ludewig).

Riss 3 : Aus der Riss-2-Position schmolz der Salzachgletscher zurück. Der nächste Stillstand führte zur Bildung des Riss-3-Endmoränenzuges (von Ebers et al. "Riß I" benannt), der mit Unterbrechungen von Oberösterreich (Oberkriebach) über Laimgruben / Marienberg nach Kirchweidach zieht und von dort scharf nach Süden zum Pallinger Rücken zurückbiegt. Dabei wurden weitere, südlichere Abschnitte des Brunnthals

eisfrei und konnten die Schmelzwässer vom Westteil des Salzachgletschers und vom Ostteil des Chiemseegletschers aufnehmen.

Die Riss-3-Moräne des Chiemseegletschers spaltet sich im Bereich des Pallinger Rückens ab, springt weit nach Nordwesten vor über Engertsham und Geberting, sodann über das jetzige Alztal bis Brandstätt, um dann wieder nach Südwesten über Schweinberg und Emertsham zurückzuweichen. Dort vereinigt sie sich mit der Riss-3-Moräne des Inngletschers. In dieser Phase, in der die Gletscherloben weit auseinander klafften, sind auch die Entwässerungssysteme des Salzach- und Chiemseegletschers deutlich gesondert (Doppler; Hintermaier).

Riss 4 : Der Riss-4-Wall des Salzachgletschers (von Ebers et al. 1966 als "Riss II" bezeichnet) bildet die höchsten und prägnantesten Moränenerhebungen des Riss im Arbeitsgebiet. Der Wall kommt von Oberösterreich, wo er auch im Nordteil des Oberen Weilhartforstes durchspießend durch die jüngere Würmschotterflur erkannt wurde (Ludewig). Er quert dann nördlich von Nonnreit die Salzach, auch dort noch weitgehend von Würmmoränenmaterial und Niederterrassenschotter abgeschürft und bedeckt (Fakhrai). Erst weiter südwestlich, bei Oberried, taucht er voll auf und zieht sodann nach Tyrlaching (Kiechle), von dort nach Freutsmoos und über Grafetsstetten hinaus nach Süden (Doppler).

Moränenhügel, die der Riss-4-Phase des Chiemseegletschers zuzuordnen sein mögen, finden sich östlich der Alz bei Engertsham, westlich der Alz in einem südlichen Moränengebiet zwischen Trostberg/Schwarzau und Kienberg und in einem nördlichen, breiten und undeutlichen Moränenareal bei Trostberg, Nunbichl, Feldkirchen, Biburg und Eberting, dort den Mindel-Moränenwall Trostberg - Oberbrunnham kreuzend. Im einzelnen sind die Zuordnungen schwierig, da die Fortsetzung nach Osten durch das Alztal, die nach Westen durch das Schabinger Tal nachträglich unterbrochen wurden.

Die Schmelzwässer, die vom Riss-4-Wall des Salzachgletschers ausgingen, durchbrachen örtlich den vorgelagerten Riss-3-Wall und strömten auf einer Terrassenflur, die bei Freutsmoos wurzelt, innen am Riss-2-Wall, sodann am Höhenzug Wald - Margarethenberg, weiter am Willhartsberg, Eschlberg und Hechenberg entlang nach Burghausen. Sie umfluteten dabei die Riss-3-Moränen und ummantelten sie mit Geröllschutt ("ertrunkene Moränen"). Das weiter südlich bei Traunwalchen wurzelnde Hochterrassenfeld, das Ebers et al. (1966, S. 61) einer "peripheren Hochterrassenflucht Traunwalchen - Burghausen" zuordnen, gehört nicht zu diesem großen Schmelzwassersystem Freutsmoos - Burghausen, sondern entwässerte durch das Brunnthal (Doppler; Hintermaier).

Während der Riss-3-Phase und zu Beginn der Riss-4-Phase dürfte die Hochterrassenflur Freutsmoos - Burghausen noch durch den intakten Deckenschotterzug Wald - Margarethenberg - Höresham - Willhartsberg - Eschlberg - Hechenberg von der Schmelzwasserflur des Inntals getrennt gewesen sein, so daß sich beide Stromsysteme erst östlich von Burghausen zwischen Hechenberg und Adenberg vereinigt haben. Im Verlauf des Riss 4 wurde der Deckenschotterrücken aber von beiden Seiten her zunehmend erodiert. Es kam schrittweise zu Durchbrüchen und zum Kurzschluß der beiden großen Stromsysteme immer weiter im Westen: Zuerst wurde der Durchbruch zwischen Hechenberg und Eschlberg geschaffen, sodann die Verbindung zwischen Willhartsberg und Eschlberg durchstoßen, so daß sich die Schmelzwasserfluten bereits im Raum Burgkirchen vereinigten.

Die Schmelzwasserbahnen des Chiemseegletschers während der Riss-4-Phase sind - wie auch die Endmoränenzüge - nur undeutlich zu erkennen. Die Entwässerung erfolgte durch das Mörntal, das Haitzinger und Schabinger Tal, möglicherweise zwischen Gramsham und Mankham auch bereits durch das Alztal, das damit erstmalig im späten Hauptriss als Talrinne in Aktion tritt (Hintermaier).

Riss 5 : Eine jüngste Rissphase - von Ebers et al. (1966) "Riß III" genannt - ist dokumentiert durch die unterste Rissterrasse im Brunnthal, die in die dortige Riss-4-Terrasse ca. 20 - 25 m tief eingeschnitten ist und sich vom Raum Palling (Doppler) nach Norden in reliktischen Teilstücken bis zum Ausgang des Brunnthals ins Alztal durchverfolgen läßt (Hintermaier; Pakzad). Die Riss-5-Terrasse kann rückwärtig nicht mehr mit dem zugehörigen Moränenwall verknüpft werden, da dieser weit im Hinterland unter Würmmoränen begraben liegen dürfte.

Nach Norden scheinen die Schmelzwässer des Riss 5 vom Brunnthal aus in direkter Fortsetzung durch den Deckenschotterzug bei Garching durchgebrochen sein zur tiefergelegenen Inn-Hochterrassenflur (Grimm; Muzavor; vgl. auch Eichler & Sinn 1974, Fig.6). Der Durchbruch ist heute verschlossen; die Alz hat inzwischen - belegt durch Niederterrassenstufen - wieder einen längeren Weg zum Inntal eingeschlagen, der vom Alzknie bei Garching nach Nordosten an der Außenseite der distalen Günz-, Mindel- und Riss-Endmoränen entlang führt.

Von anderen, benachbarten Gletschergebieten wurde eine Zweiteilung des Riss in Riss I und II infolge einer zwischengeschalteten, weitreichenden Oszillation beschrieben. Eine solche Oszillation kann auch im Salzach-Vorlandgletscher zwischen Riss 3 und 4 vermutet werden und sollte, wenn sie gesichert ist, als "Kirchweidach -Tyrlachinger Oszillation" bezeichnet werden (Grimm). Anzeichen für eine solche Oszillation sind

- die starken Stauchungen des Riss-3-Moränenwalles vor allem bei Kirchweidach (vgl. Ebers et al. 1966, S. 68), die durch den Rückzug und neuerlichen Vorstoß des Eises zwischen Riss 3 und 4 bedingt sein könnten;
- die Bifurkation des Risswalles bei Gilgenberg / Oö., wo sich der einzige Wall im Osten ("vereinigter Rißwall" nach Ebers et al. 1966, S. 161) in zwei Wälle im Westen gabelt. Grund für die Gabelung bzw. Vereinigung beider Wälle könnte eine geänderte Vorstoßrichtung des Gletschers nach der Kirchweidach-Tyrlachinger Oszillation sein;
- die starke Ausräumung und anschließende Verschüttung des Riss-3-Walles durch die späteren Schmelzwasserströme und -sedimente, die auf besonders langdauernde Abschmelzvorgänge nach der Riss-3-Phase hindeuten.

Sollte die Annahme einer Oszillation zutreffen und dieser Oszillation das Ausmaß eines Interstadials zuzumessen sein, so wäre das Riss im Salzachgletschergebiet wie folgt zu gliedern:

Stadium	Phase	Typische Lokalität
Riss I	Riss 1	Racherting
	Riss 2	Schmidtstadt
	Riss 3	Kirchweidach
Kirchweidach-Tyrlachinger Oszillation		
Riss II	Riss 4	Tyrlaching
	Riss 5	Brunnthal (tiefe Hochterrasse).

3.6 Würm

Die von Ebers et al. (1966, S. 84 ff.) als überfahrene Moräne des Frühwürms ("Tenglinger Stadium") gedeutete Hügelkette, die von der Westseite des Waginger Sees über Tengling und Wiesmühl gegen Kay und Tittmoning zieht, konnte durch unsere Untersuchungen nicht als eigener Frühwürm-Vorstoß bestätigt werden (Kiechle): Die Hügel stellen keine altwürmzeitlichen Moränenwälle dar, sondern bestehen in ihrem Kern aus fluviatilen bis fluvioglazialen Kiesen und basalen Seesedimenten des Hauptwürm. Die abdeckende Grundmoränenschicht gehört nicht - wie Ebers et al. meinen - zum Maximalvorstoß (Nunreuter Phase), sondern zur späteren "Radegund-Lanzinger Oszillation" (s.unten).

Die würmzeitliche Vereisung im Arbeitsgebiet beschränkt sich somit auf das Hauptwürm. Der Salzachgletscher ist bei seinem Vorstoß rund 5 Kilometer hinter dem Maximalstand des Günz-, Mindel- und Riss-Gletschers zurückgeblieben und reichte nur im Südosten ins Arbeitsgebiet. Der Chiemseegletscher gar, der im Würm deutlich eigenständig und abgesetzt vom Salzachgletscher getrennt ist, liegt etwa 10 Kilometer hinter dem risseiszeitlichen Maximalstand zurück und bleibt gänzlich außerhalb des Arbeitsgebietes.

Die von Ebers et al. (1966, S.121 ff.) beschriebene morphologische Gliederung des Jungmoränensystems im Salzach-Vorlandgletscher und die Zuordnung der Endmoränenzüge zur Unter-Weißenkirchner, Nunreuter, Radegunder und Lanzinger Phase wurden durch unsere Arbeiten bestätigt. Doch stellen die Moränenwälle keine einfachen Rückzugsstaffeln dar, wie Ebers et al. annahmen, sondern belegen ein komplizierteres paläogeographisches Geschehen infolge einer zwischengeschalteten Oszillation (Kiechle; ergänzt durch Goroncek, Grimm, Schwarz). Im folgenden werden bezeichnet

- die Unter-Weißenkirchner Phase als	Würm 1 a;
- die Nunreuter Phase mit Rückzugsbildungen als	Würm 1 b, c;
- die Radegunder Phase mit Rückzugsbildungen als	Würm 2 a, b;
- die Oszillation zwischen Würm 2 und Würm 3 als	Radegund-Lanzinger Oszillation;
- die Lanzinger Phase mit Rückzugsbildungen als	Würm 3 a, b, c.

Die von Süden ins Arbeitsgebiet hineinreichenden Wälle und Kuppen der Unter-Weißenkirchner Phase sind als distale Würm-Endmoränen nach Norden über Palling hinaus bis nach Tyrlaching erkennbar. Da sie in der übrigen Umrandung des Salzachgletschers zu fehlen scheinen und somit Lokalbildungen darstellen, wurden sie als Würm 1 a mit der Nunreuter Phase (Würm 1 b und c) gekoppelt.

Das anschließende paläogeographische Geschehen verlief wie folgt (Kiechle):

Der Maximalstand des Würmgletschers wird nördlich von Tyrlaching durch den Würm-1-b-Moränenzug der Nunreuter Phase dokumentiert. Gegen Ende dieser Phase begannen der Rückzug und die Zerlappung des Eisrandes, z.B. bei Wiesenzart (Würm 1 c). Der Gletscher schmolz nun ruckweise bis in seine Zungenbecken ab. Unterbrochen wurde dieser Rückzug durch kürzere Stillstandshalte während der Radegunder Phase, die durch zwei hintereinander gelagerte, geringmächtige und zum Teil fluvioglazial überprägte und zerstörte Moränenzüge gekennzeichnet sind (Würm 2 a, b).

Im Verlauf des weiteren Gletscherrückzuges (Beginn der Radegund-Lanzinger Oszillation) wurden Teile des Tittmoninger Zungenbeckens eisfrei. Dadurch war Raum für den "Ersten Tittmoninger Eisrandsee" (Kiechle) geschaffen, der sich im Norden des Arbeitsgebietes an den Moränenwällen der Nunreuter und Radegunder Phase staute. Die zugehörigen Seesedimente wurden vielerorts in einer Höhenlage zwischen 425 und 440 m ü.NN. nachgewiesen, gelegentlich auflagernd auf der Grundmoräne der Würmphasen 1 und 2 und auf den darüber sedimentierten Rückzugsschottern.

Während des erneuten Gletschervorstoßes zur Lanzinger Phase kam es zum Überlauf des ersten Tittmoninger Eisrandsees und zur Schüttung von Vorstoßschottern. Am Rand des Zungenbeckens wurden die Seesedimente und Schotter durch den vorstoßenden Gletscher überschliffen und liegen heute morphologisch als drumloide "überfahrene Hügelkuppen" (von Ebers et al. 1966 als überfahrene Frühwürm-Moränen gedeutet, s. oben) vor. Der Eisvorstoß führte weithin zu großartigen Stauchungen der vorgegebenen Seesedimente und Schotter und zu deren Einquetschung in die Moränen der Würm-3-Phase (z.B. Aufschlüsse bei Unteröd, Hechenberg, Haus und Mönchspoint). Die Eisstirn erreichte wieder den Innenrand der Radegunder Phase, wie die dort angelagerte Grundmoräne zeigt. Mit diesem Vorstoß ist die Radegund-Lanzinger Oszillation abgeschlossen; ob sie regionale oder nur lokale Bedeutung hat, wird durch künftige Arbeiten zu überprüfen sein.

Die kurzen Stillstände des Eises während der folgenden Lanzinger Phase sind durch die Moränenkuppen des Würm 3 a (Wälle bei Leitgering - Enzelsberg und bei Biering), des Würm 3 b (Wall südlich von Diepling) und des Würm 3 c (Wälle von Lanzing, Unteröd, Hechenberg, Haus und Mönchspoint) gekennzeichnet. Beim weiteren Rückzug wurden örtlich und abschließend die Schmelzwasserablagerungen der Niederterrasse an die Endmoränen der Lanzinger Phase angelagert und auf die beckenwärts folgende Grundmoränenlage aufgelagert.

Bezüglich des finalen Gletscherrückzuges mit verschiedenen spätglazialen Eisständen bis zum Eiszerfall im Salzburger Stammbecken und zum Zurückweichen des Vorlandgletschers in die Alpentäler (Grenze Spät-/Postglazial) wird auf die Arbeit Ziegler (1977) verwiesen.

Die durch die Gletscherablagerungen belegte paläogeographische Entwicklung des Hauptwürms konnte durch die Rekonstruktion der Schmelzwassersysteme bestätigt werden (Kiechle):

Zur Zeit der Nunreuter Phase wurden die Schmelzwässer vom peripheren Schnitzinger Eisrandtal gesammelt und, nachdem sie zwischen Oberschnitzing und Kraham den vorgelagerten Riss-4-Wall und die Hochterrasse durchbrochen hatten, durch das untere Halsbachtal dem Vorfluter zugeführt.

Während der Radegunder Phase sammelten die periphere Entwässerungssenke von Zaiselham - Wiesenzart und die Niederterrassenflur von Stockham - Zageln die zunächst zentrifugal und flächenhaft abfließenden Schmelzwässer und führten sie ebenfalls dem Schnitzinger Eisrandtal zu. Dagegen herrschte weiter nordöstlich, bei Asten, direkte zentrifugale Entwässerung nach Norden über die Rissmoränen und -hochterrasse vor.

Die Schmelzwässer der Lanzinger Phase sammelten sich zunächst (Würm 3 a) in der peripheren Rinne von Dürnberg und flossen im Bereich des heutigen Salzachtales ins Vorland ab. In der Folge aber, beim Rückschmelzen des Eises im Tittmoninger Bekken, staute der Moränenzug von Leitgering - Enzelsberg (Würm 3 a) den spätglazialen "Zweiten Tittmoninger Eisrandsee" auf. Seither strömten die peripheren Schmelzwässer aus dem Ollerdinger Eisrandtal in diesen See und mündeten anfänglich (Würm 3 b) im Norden über die Senke von Wallmoning - Moos (topset beds des Deltas östlich von Moos in ca. 470 m ü.NN.), dann im Würm 3 c weiter südlich bei Alm (ca. 465 m ü.NN.) und schließlich bei Dandlberg (ca. 460 m ü.NN.). Die Deltaschüttung von Dandlberg ist zur Zeit prächtig aufgeschlossen mit weitgespannter Schrägschichtung, kamesartigen Verstürzungen und Grundmoränenbändern, die in die Schrägschichtung eingefügt sind.

Die zunehmende Verlagerung des Deltas nach Süden entspricht der Vergrößerung des Tittmoninger Sees durch das allmähliche Abschmelzen des Gletschers. Die gleichzeitig immer tiefere Lage der Delta-Oberflächen beweist, daß die stauende Moränenschwelle im Norden zunehmend ero-

diert wurde und der Spiegel des Eisrandsees dadurch stetig fiel. Dieses Absinken setzte sich, wie die spätglazialen Terrassen des unteren Salzachtales zeigen, stufenförmig fort. Von einem katastrophalen Durchbruch - wie früher gelegentlich angenommen - kann keine Rede sein.

Die spätglazialen Ablagerungen des Zweiten Tittmoninger Eisrandsees finden sich bis hinauf in ca. 410 - 420 m ü.NN., d.h. deutlich unterhalb der Sedimente des Ersten Eisrandsees aus der Radegund-Lanzinger Oszillation (425 - 440 m ü.NN.; s. oben).

Im Arbeitsgebiet nordwestlich des beschriebenen Jungmoränenareals zeugen nur die zumeist tief eingeschnittenen Schmelzwassertäler mit Niederterrassenfüllung von der würmzeitlichen Vereisung im Rückland. Von Osten nach Westen sind dies

für den Salzach-Vorlandgletscher
- das untere Salzachtal, das wohl schon vor dem Würm 3 in der Radegunder Phase und dann als Überlauf des Ersten Tittmoninger Eisrandsees angelegt wurde und nach der Radegund-Lanzinger Oszillation die Wässer des Zweiten Tittmoninger Eisrandsees nach Norden ableitete;
- das periphere Schnitzinger Eisrandtal und seine Fortsetzung nach Norden im unteren Halsbachtal, die vor allem die Schmelzwässer des Würm 1 und 2 aufnahmen;
- das periphere Ollerdinger Eisrandtal, das nach Osten in den Zweiten Tittmoninger See entwässerte;
- das Brunnthal, das die Schmelzwässer vom Westrand des Salzachgletschers während des frühen Hauptwürms aufnahm;

für den Salzach- und den Chiemsee-Vorlandgletscher
- das Alztal nördlich von Altenmarkt, das die Schmelzwässer vom Ostrand des Chiemseegletschers während des frühen Hauptwürms sammelte, im späten Hauptwürm auf einer tiefer gelegenen Niederterrasse zusätzlich auch die Schmelzwässer des Salzachgletschers;

für den Chiemsee-Vorlandgletscher
- wohl das Schabinger Tal;
- das Mörntal, das an der Würm-Endmoräne des Chiemseegletschers bei Kienberg - Sibolding wurzelt und in die würmzeitliche Ampfinger Stufe des Inntals einmündet.

Die im Arbeitsgebiet auf den günstig exponierten Präwürm-Ablagerungen verbreiteten Lösslehme und die seltenen Lösse sind überwiegend der letzten Eiszeit zuzuordnen und als äolische Ausblasungsprodukte aus den bloßliegenden Würmschotterfluren zu deuten. Interne Verwitterungshorizonte, wie sie im Nordosten des Salzachgletschergebietes auftreten können und dort ältere Lössgenerationen belegen, fehlen im Arbeitsgebiet fast völlig. (Doppler hat im Rahmen unserer Arbeiten eingehend die Lösslehmgrube von Sonnleithen südlich von Mauerkirchen / Oö. untersucht und 4 (-6) Verwitterungshorizonte festgestellt und pedologisch gekennzeichnet.)

Der würmzeitliche Lösslehm im Arbeitsgebiet wurde wohl schon synsedimentär entkalkt, da das Areal in Brunnackers (1957) Südlicher Faziesprovinz mit mehr als 900 mm Jahresniederschlag bzw. in der Staublehmlandschaft Finks (1976) mit mehr als 800 mm Niederschlag liegt.

3.7 Spät- und Postglazial

Der mit Stillständen und Oszillationen verbundene Rückzug des Würmgletschers im Spät- und Postglazial bis auf seine Wurzeln im Alpengebirge und die nachfolgenden Klimaschwankungen führten zum Einschnitt weiterer, tieferer Terrassen in den großen Niederterrassentälern Salzach, Alz und Inn. Zumeist handelt es sich um Akkumulationsterrassen, deren Schotterstränge nach vorheriger Erosion neu abgelagert wurden und sich im Geröllbestand von der Niederterrasse unterscheiden. Daneben kommen Abrasionsterrassen mit nur geringer selbständiger Aufschotterung vor. Eine Trennung von spät- und postglazialen Terrassen kann vom begrenzten Arbeitsgebiet aus nicht erfolgen.

Die Kartierung der Terrassenstufen zwischen würmzeitlicher Niederterrasse und subrezenter Auwaldstufe ergab übereinstimmend für das Salzach-, Alz- und Inntal 4 - 6 Terrassenstufen. Wir bemühen uns z.Z., die Terrassenrelikte nach Höhenlage, Böschungssprung und Bodenbildung an den Flüssen entlang zu verbinden und von Tal zu Tal zu korrelieren.

4 ANGEFÜHRTE LITERATUR

4.1 Publikationen

Brückner, E. 1886, Die Vergletscherung des Salzachgebietes, Geogr.Abh. 1 : 183, Wien.

Brunnacker, K. 1957, Die Geschichte der Böden im jüngeren Pleistozän in Bayern, Geologica Bavarica 34 : 95, München.

Ebers, E., Weinberger, L. & Del Negro, W. 1966, Der pleistozäne Salzachvorlandgletscher, Veröff.Ges.Bayer.Landesk. 19-22 : 216, München.

Eichler, H. & Sinn, P. 1974, Zur Gliederung der Altmoränen im westlichen Salzachgletschergebiet, Z.Geomorph.N.F. 18 : 132-158, Berlin,Stuttgart.

Fink, J. et al. 1976, Exkursionen durch den österreichischen Teil des nördlichen Alpenvorlandes und den Donauraum zwischen Krems und Wiener Pforte, Mitt. Kommiss.Quartärforsch.Öst.Akad.Wiss. 1 : 113, Wien.

Penck, A. & Brückner, E. 1909, Die Alpen im Eiszeitalter, 1 : 75-80, 151-166, Leipzig.

Traub, F. 1953, Quartärgeologische Beobachtungen zwischen Alz und Salzach, Geologica Bavarica 19 : 105-113, München.

Traub, F. & Jerz, H. 1975, Ein Lößprofil von Duttendorf (Oberösterreich) gegenüber Burghausen an der Salzach, Z.Gletscherk.Glazialgeol. 11 : 175-193, Innsbruck.

Troll, C. 1924, Der diluviale Inn-Chiemsee-Gletscher, Forsch.dt.Landes- u. Volksk. 23 : 121, Stuttgart.

Weinberger, L. 1950, Gliederung der Altmoränen des Salzach-Gletschers östlich der Salzach, Z.Gletscherk.Glazialgeol. 1-2 : 176-186, Innsbruck.

Weinberger, L. 1955, Exkursion durch das österreichische Salzachgletschergebiet und die Moränengürtel der Irrsee- und Atterseezweige des Traungletschers, Verh.Geol.Bundesanst., S.-H.D., Wien.

Ziegler, J.H. 1977, Spätglaziale Rückzugsstadien des Salzach-Vorlandgletschers in Bayern, Internat.Geol.Correlat.Programme, Project 73/1/24 : 116-125, Prag.

4.2 Unveröffentlichte Diplomarbeiten

Bläsig, H. 1973, Geologische und hydrogeologische Untersuchungen im Westteil des Gradabteilungsblattes 7841 Garching an der Alz in Oberbayern, Dipl.-Arbeit Inst.Allg.Angew.Geol.Univ. : 147, München.

Doppler, G. 1978, Geologische und hydrogeologische Untersuchungen im Mittelteil des Gradabteilungsblattes 7941 Trostberg an der Alz sowie bodenkundliche und sedimentpetrographische Untersuchungen in der Lösslehmgrube Sonnleithen bei Mauerkirchen / Oberösterreich, Dipl.-Arbeit Inst.Allg.Angew.Geol.Univ. : 147, München.

Fakhrai, M. 1972, Geologische und hydrogeologische Untersuchungen am Nordwestrand des pleistozänen Salzachvorlandgletschers, Dipl.-Arbeit Inst.Allg. Angew.Geol.Univ. : 87, München.

Hintermaier, G. 1978, Geologische, pedologische und hydrogeologische Untersuchungen zwischen Inn-Gletscher und Salzach-Gletscher (Oberbayern) auf den Gradabteilungsblättern 7940 Obing (Nordostteil) und 7941 Trostberg (Nordteil), Dipl.-Arbeit Inst.Allg.Angew.Geol.Univ. : 106+19, München.

Just, J. 1972, Geologische und hydrogeologische Untersuchungen im Ostteil des Gradabteilungsblattes 7841 Garching an der Alz in Oberbayern, Dipl.-Arbeit Inst. Allg.Angew.Geol.Univ. : 71, München.

Kiechle, W. 1978, Stratigraphische, sedimentpetrographische, hydrogeologische und baugrundgeologische Untersuchungen im Nordwestteil des Gradabteilungsblattes 7942/43 Tittmoning, Dipl.-Arbeit Inst.Allg.Angew.Geol.Univ. : 125, München.

Lobinger, W.H. 1973, Geologische und hydrogeologische Untersuchungen im Westteil der Gradabteilungsblätter 7742 Altötting und 7842 Burghausen, Dipl.-Arbeit Inst.Allg.Angew.Geol.Univ. : 120, München.

Ludewig, H. 1972, Sedimentpetrographische und hydrogeologische Untersuchungen auf Blatt 7742 Altötting und 7842/43 Burghausen, Dipl.-Arbeit Inst.Allg.Angew. Geol.Univ. : 118, München.

Muzavor, S. 1973, Geologische und hydro geologische Untersuchungen im Südteil des Gradabteilungsblattes 7741 Mühldorf am Inn, Dipl.-Arbeit Inst.Allg.Angew. Geol.Univ. : 72, München.

Pakzad, M. 1973, Geologische, sedimentpetrographische und hydrogeologische Untersuchungen im mittleren Teil des Gradabteilungsblattes 7841 Garching an der Alz (Oberbayern), Dipl.-Arbeit Inst. Allg.Angew.Geol.Univ. : 120, München.

Granulometrische Untersuchungen zur Genese von Moränen im Salzkammergut, Österreich

D. VAN HUSEN
Technische Universität Wien, Wien, Österreich

ABSTRACT

Sedimentological investigations of typical groundmoraines were carried out. Just after a short distance of transportation a granulometric composition appears indicated by an excess of fine material (< 0,063 mm). The cumulative curve on semilogarithmic paper appears as a straight line.

This granulometric composition represents a state of equilibrium that does not change over the whole length of the glacier until it reaches its terminal moraines.

In vertical sections of the valley slopes also no alternation of the granulometric composition was found.

The groundmoraines of Late Glacial stades, although they were deposited by large valley glaciers, show a significantly smaller proportion of fine material. These differences in regard to Hochglazial may probably be due to a shorter time and not to the smaller thickness of ice. This opinion is supported by observations of compactness.

Roundness of larger boulders (2-6,3 mm increases slowly in longitudinal distance up to values due to ice transportation. In the terminal area there are suddenly many boulders much better rounded that probably go together with fluviatile rearrangement in the region of melting.

Calculation of sediment parameters after Folk & Ward (1957) as well as their graphic connections show no significant development neither in vertical nor in horizontal sections although further references are given to the environment of the investigated sediments.

EINLEITUNG

Bei einer sich über das Einzugsgebiet der oberösterreichischen Traun erstreckenden, großmaßstäbigen Kartierung (1:10 000) der quartären Sedimente war es möglich die räumliche Entwicklung des Traungletschers und seiner Eisströme während des Hoch- und Spätglazials zu rekonstruieren (van Husen 1977). Im Zuge dieser Kartierung konnten Beobachtungen über den Aufbau der Grundmoräne gesammelt werden, die gewisse Gesetzmäßigkeiten in der Zusammensetzung der hochglazialen Grundmoräne, und einen deutlichen Unterschied zu jener der spätglazialen Stände erkennen ließen. Sie bezogen sich hauptsächlich auf Lagerungsdichte, Gehalt an Feinmaterial und Form der Geschiebe. Um diese bereits im Gelände erkannten Unterschiede exakter zu erfassen wurden sedimentologische Untersuchungen angestellt.

Gleiche Untersuchungen wurden während der Weiterführung der quartärgeologischen Kartierung im Lammertal (Abtenau, Salzburg) durchgeführt, da aus der geologisch-morphologischen Situation gute Vergleichsmöglichkeiten mit dem Trauntal gegeben sind.

GEOLOGISCHER UNTERGRUND

Das Untersuchungsgebiet liegt in den Nördlichen Kalkalpen und ist von den Dachsteinkalkplateaus (Dachstein, Totes Gebirge, Tennengebirge) geprägt, die als Einzugsgebiet der Gletscherströme dienten. Diese kamen in den dazwischen verlaufenden Talzonen (im Bereich der vielfältigen Gestei-

ne der Hallstätter Entwicklung) zum Abfluß. Bedeutendere Abweichungen dieses generellen Baues sind nur im Bereich von Gosau und im Becken von Abtenau gegeben. Die hier auftretenden großflächigen Vorkommen sehr feinkörniger Nierentaler Mergel (Gosau) und des Haselgebirges (Abtenau) haben Einfluß auf die granulometrische Zusammensetzung der Grundmoränen.

Diesen Untergrundverhältnissen entsprechend sind die Grundmoränen, abgesehen von einem untergeordneten Anteil von Quarz, aus Sandsteinen (Kornklassen < 0,063 mm) als rein karbonatisch anzusehen. Durch die prozentuelle Erfassung verschiedener Karbonate konnten die Eisabflußverhältnisse rekonstruiert werden, die teilweise völlig dem heutigen Talverlauf entgegengerichtet waren.

METHODEN

In den Grundmoränenaufschlüssen (großflächige Bachanrisse und Baugruben) wurden aus dem frischen, unverwitterten Material händisch Proben zwischen 10 und 20 kg entnommen. Dabei wurde besonders darauf geachtet, daß das Sediment der Entnahmestelle für den allgemeinen Aufbau charakteristisch war. Trotz der Abhängigkeit der Probestellen von zufälligen Aufschlüssen konnten einerseits die Grundmoräne des Hochglazials über den ganzen Talverlauf, andererseits alle wichtigen Phasen in der Entwicklung der Eisströme im Spätglazial erfaßt werden.

Das Material wurde primär auf seine Kornverteilung untersucht, wobei die Fraktionen von 63-0,063 mm durch Trockensiebung und die < 0,063 mm mit Hilfe der Aräometermethode nach Casagrande (1934) bestimmt wurden. Die Beschränkung auf ein Größtkorn von 63 mm ergab sich aus der Probenmenge. Als Ergänzung der durch die kombinierte Sieb- Schlämmanalyse gewonnenen Summenkurven wurden die Geschiebe der drei größten Fraktionen (63-20/ 20-6,3/ 6,3-2 mm) auf ihren Rundungsgrad hin untersucht um möglicherweise weitere Hinweise auf die Materialbeanspruchung während des Transportes zu erhalten.

Der Rundungsgrad wurde nach den von Pettijohn (1957) vorgeschlagenen fünf Rundungsgraden ausgezählt und fraktionsweise in die Diagramme eingetragen (Abb.1).

KORNGRÖSSENVERTEILUNG

Der Vergleich der Kornsummenkurven als wertfreie erste Darstellung der Zusammensetzung des Korngemisches in den Grundmoränen des Hochglazials von den Einzugsgebieten bis zu den Endmoränen ergab zwei gleichbleibende Aspekte.

So stellt sich bereits nach kurzer Transportstrecke ein Verhältnis von Feinteilen zu gröberem Material ein, das durch ein Überangebot der feinen Kornklassen charakterisiert wird. Dieses Verhältnis kommt sehr deutlich durch den geraden Verlauf der Kornsummenkurve in halblogarithmischer Darstellung zum Ausdruck (Abb.1). Darüber hinaus ist ein Wert von ca. 10 % für den Gehalt an Tonfraktion für die Grundmoräne charakteristisch, der unbeschadet kleinerer Areale mit mergeligen oder tonigen Gesteinen im Bereich der Hallstätter Zone weitgehend konstant bleibt.

Abb. 1.

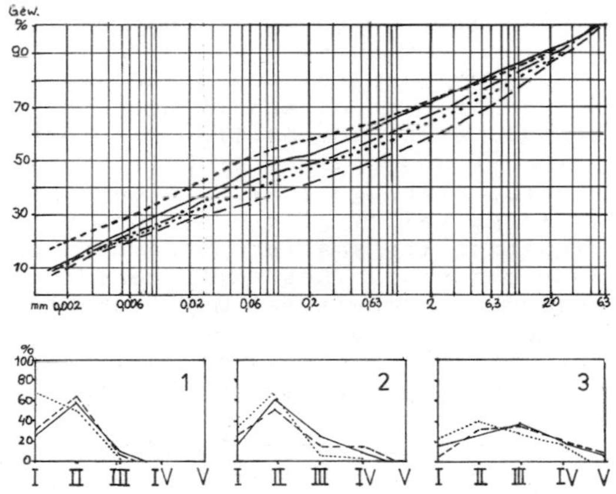

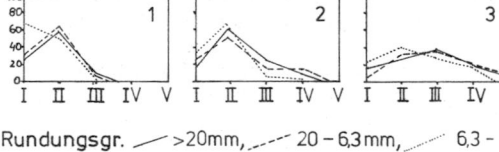

Rundungsgr. >20mm, 20 – 6,3 mm, 6,3 - 2 mm

Trauntal

+ 1 nach ca. 5 km Transportstrecke

+ 2 nach ca. 25 km

+ 3 nach ca. 60 km

Abtenau

Talbereich

Hanglage

Es stellt nach den Untersuchungen im Trauntal (van Husen 1977) einen Gleichgewichtszustand in der Kornverteilung dar, der sich bereits nach der kurzen Wegstrecke über die Plateaus und den Abstieg in die Täler einstellt und dann bis vor die Endmoräne gleich bleibt.

Nur in Gebieten mit weit verbreiteten, extrem leicht erodierbaren Gesteinen wie Nierentaler Mergel (Gosau), Haselgebirge und Werfener Schiefer (Abtenau) kommt es zu einer Erhöhung auf 15-20 % der Tonfraktion und einem höheren Angebot an Schluffen (Abb.1). Auf eine ähnliche Beeinflussung des Gehaltes an Feinanteilen in Grundmoränen weist schon Dreimanis (1975:96 f) im Gebiet der Großen Seen in Kanada hin.

Dieser erhöhte Gehalt an Tonfraktion findet sich in allen Proben aus dem Abtenauer Becken, die im engeren Talbereich meist 5-10 m über der Auflage auf dem Untergrund entnommen wurden.

Nun war es möglich in einem Seitengraben, der zur Gänze in Grundmoräne verläuft, über eine Höhe von ca. 200 m eine Serie von Proben zu entnehmen. In diesen konnte in der vertikalen Entwicklung am Talhang der abnehmende Einfluß der feinkörnigen Gesteine (Werfener Schiefer) des Talgrundes in einer gleichbleibenden Abnahme von 20 auf 12 % der Tonfraktion festgestellt werden (Abb.1). Abgesehen von dieser Beeinflussung des Kornaufbaues in den feinsten Kornklassen tritt keine Änderung im Verhältnis der untersuchten Kornklassen zueinander auf. Da sich auch im Abtenauer Becken nach einer annähernd gleich kurzen Wegstrecke der selbe Gleichgewichtszustand im Aufbau der Grundmoräne des Hochglazials einstellt, könnte es sich hiebei um eine Zusammensetzung handeln, die für Gletscher dieser Größenordnung charakteristisch ist. Ob eine Abhängigkeit vom generellen Aufbau (Kalkalpen, Kristallin) des Untergrundes und der Größe des Gletschers besteht, werden weitere Untersuchungen zeigen müssen.

Eine davon gänzlich abweichende Zusammensetzung zeigen die Kornsummenkurven der Grundmoränen der spätglazialen Gletscherstände im Trauntal.

Aufgrund der detaillierten Kartierung war es möglich die spätglazialen Gletscherstände in Ausdehnung und Sedimentausbildung zu umreißen (van Husen 1977). So war es möglich aus sicher diesen Gletscherständen zurechenbaren Grundmoränen Proben zu entnehmen. Dabei zeigte es sich, daß diese Grundmoränen einen wesentlich geringeren Anteil der feinen Kornklassen aufweisen (Abb.2), obwohl die ersten spätglazialen Stände beachtliche Gletscherzungen zwischen 10 und 20 km Länge in den Tälern zwischen den Kalkplateaus aufwiesen.

Abb. 2.

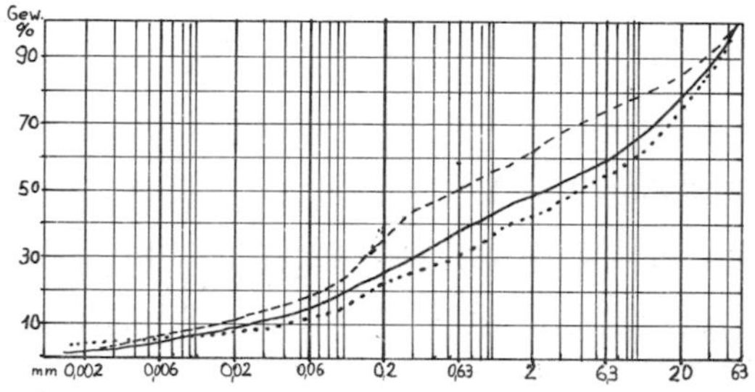

Dabei auftretende Unterschiede in der Kornverteilung zwischen den einzelnen Gletscherständen deuten darauf hin, daß sich auch in der Entwicklung der Grundmoräne die Charakteristik des Gletscherstandes (kurzfristige Wiederbelebung einer vorhandenen oder rascher, weitgehender Neuaufbau einer weit zurückgeschmolzenen Gletscherzunge) widerspiegeln könnte.

Daß die Unterschiede im Kornaufbau der Grundmoränen des Hoch- und Spätglazials keinen Effekt des weiteren Zungenbereiches oder der geringeren Eisüberlagerung darstellen, zeigen die Kornsummenkurven aus dem Gebiet der hochglazialen Gletscherzunge um den Traun See (Abb.1, Kurve 3).

Die verschiedenen Zusammensetzungen im Kornaufbau der Grundmoräne dürften demnach wahrscheinlich hauptsächlich von der Dauer des Ereignisses abhängen. Diese Meinung läßt sich noch durch folgende Beobachtungen weiter untermauern.

In allen Grundmoränenaufschlüssen des Hochglazials war eine sehr hohe Lagerungsdichte zu beobachten, die gemeinsam mit dem hohen Feinstoffgehalt zu einer fest verbackenen, steifen Ausbildung des Sedimentes

führt.

Im Gegensatz dazu weisen die Grundmoränen der spätglazialen Stände eine wesentlich geringere Lagerungsdichte auf, die das schon durch den geringeren Ton- und Schluffgehalt bedingte sandigere Aussehen noch verstärkt. Diese Unterschiede treten nicht nur in Gebieten mit unterschiedlich hoher Eisüberlagerung im Talbereich auf, sondern auch in den Zungenbereichen bei einer genau rekonstruierbaren Eisüberlagerung von jeweils 100-150 m.Der Grund ist hier meiner Meinung nach in der längeren Dauer der Eisüberlagerung und der damit verbundenen Zeitspanne der Wasserauspressung zu suchen, die in dermaßen feinkörnigen Sedimenten nur sehr langsam erfolgen kann.

RUNDUNGSGRAD

In der Rundung der gröberen Kornklassen der Geschiebe konnte eine deutliche Entwicklung rekonstruiert werden (Abb.1).

Die Zurundung der Geschiebe, die sich im Vorfeld der heutigen Gletscher des Dachsteinstockes nur auf wenige der größeren Geschiebe in Form von Kantenrundung auswirkt, ist aber bereits nach kurzem Transportweg in den Grundmoränen am Fuß der Plateaus so weit fortgeschritten, daß die gröberen Komponenten schon mehrheitlich dem Rundungsgrad II zuzuordnen sind.

Ab dieser Position nimmt die Zurundung der Gerölle - bis auf plötzliche Änderungen durch Aufarbeitung älterer Sedimente (Gosaukonglomerate) mit gut gerundeten Komponenten oder kleinstückig zerfallender Dolomitareale - in Fließrichtung des Eises langsam zu und erfaßt auch die kleineren Kornklassen (Flint 1971:167 f).

Eine auffällige Änderung ergibt sich im Umkreis des Traun Sees (Abb. 1). Es treten hier plötzlich verhältnismäßig viele Geschiebe der Rundungsgrade III, IV und selten V auf, die sich aber nicht auf die Aufarbeitung aus älteren Sedimenten zurückführen lassen (van Husen 1977).

Dieses plötzlich vermehrte Auftreten von Geschieben mit Rundungsgraden, wie sie beim Eistransport nicht entstehen (Flint 1971:165), wird wohl auf fluviatilen Transport innerhalb des vergletscherten Areals und auch auf nachträgliche Einarbeitung in die Grundmoräne zu suchen sein und somit im Zehrgebiet eine nicht unbedeutende Rolle spielen.

GRAIN-SIZE PARAMETER

Bei der Berechnung der grain-size Parameter nach dem Vorschlag von Folk & Ward (1957) konnten einige Aspekte gewonnen werden, die die oben gegebene Deutung der Kornzusammensetzung der Grundmoränen im untersuchten Gebiet weiter untermauern, indem sie manche Unsicherheitsfaktoren ausschalten. Die Durchschnittswerte mit ihren Min- und Maximalwerten sind in der Abb.3 zusammengefaßt.

Abb.3. Durchschnittswerte der Sedimentparameter (Max-/Min Werte)

Hochglazial	
M_z	2,73 (4,13/ 1,53)
σ_1	5,31 (5,88/ 4,03)
Sk_1	0,06 (0,30/ -0,28)
K_G	0,71 (0,87/ 0,60)
Spätglazial	
M_z	-0,41 (0,13/ -1,03)
σ_1	4,48 (4,70/ 4,16)
Sk_1	0,14 (0,31/ -0,06)
K_G	0,79 (0,92/ 0,71)

Eine deutliche Trennung der hoch- und spätglazialen Grundmoränen kann bei der durchschnittlichen Korngröße M_z festgestellt werden. Wesentlich geringer sind bereits die Unterschiede in der Sortierung σ_1, wogegen in den beiden übrigen Parametern Skewness Sk_1 und Kurtosis K_G kein nennenswerter Unterschied mehr feststellbar ist.
Der Unterschied in den σ_1-Werten zwischen hoch- und spätglazialen Grundmoränen mag ebenso wie der der M_z-Werte auf den verschiedenen Aufbereitungsgrad zurückzuführen sein. Die recht gleichmäßig schlechten σ_1-Werte können aber als Hinweis auf einen alleinigen Eistransport gewertet werden. Einen weiteren Hinweis auf alleinige glaziale Bildungsbedingungen sind in den niedrigen Sk_1-Werten zu finden, da hohe Sk_1-Werte eine Überschneidung verschiedener Bildungsbeding-

ungen (Folk 1974:86) anzeigen.

Die niedrigen Kurtosiswerte $< 0{,}8$ sind nach Folk (1974) auch dahingehend zu interpretieren, daß in den untersuchten Grundmoränen keine älteren Lockersedimente aufgearbeitet wurden, was die Interpretation der Kornsummenkurven unterstützt.

Keiner der vier Parameter oder ihre graphische Kombinationen zueinander - wie sie Braun (1973) und Schlüchter (1977) zur Trennung verschiedener glazialer Sedimente verwendeten - läßt eine Entwicklungstendenz erkennen, die mit der Lage der Probenpunkte in horizontaler oder vertikaler Entwicklung des Eisstromes in Einklang gebracht werden könnte.

ZUSAMMENFASSUNG

Im Rahmen einer Detailkartierung der quartären Sedimente im Trauntal wurden auch sedimentologische Untersuchungen an in ihrem Habitus typischen Grundmoränen durchgeführt.

Dabei zeigte es sich, daß sich bereits nach einer kurzen Transportstrecke eine Kornzusammensetzung einstellt, die von einem starken Überangebot der feinen Kornklassen ($<0{,}063$ mm) geprägt wird. Die Kornsummenkurve bildet sich in halblogarithmischer Darstellung als Gerade ab.

Diese Kornzusammensetzung stellt einen Gleichgewichtszustand dar, der über die gesamte Länge des Gletschers bis vor seine Endmoränen gleich bleibt. Diese Ergebnisse wurden in fortführenden Untersuchungen im Lammertal voll bestätigt. Hier konnte auch nachgewiesen werden, daß diese Kornzusammensetzung auch in vertikaler Erstreckung am Talhang unverändert bleibt.

Einen stark abweichenden Kornaufbau mit wesentlich geringerem Feinanteil zeigen die Grundmoränen der spätglazialen Stände, obwohl sie von ausgedehnten Talgletschern abgelagert wurden.

Diese Unterschiede können sehr wahrscheinlich auf die kürzere Dauer gegenüber dem Hochglazial und nicht auf die geringere Eismächtigkeit zurückgeführt werden. Die Meinung wird noch durch Beobachtungen über die Lagerungsdichte unterstützt.

Der Rundungsgrad der größeren Geschiebe (2-63 mm) nimmt in Längserstreckung des Gletschers langsam bis zu den durch Eistransport erreichbaren Werten zu. Im Zungenbereich treten aber plötzlich viele, wesentlich besser gerollte Geschiebe auf, was wahrscheinlich mit fluviatiler Umlagerung im Bereich des Zehrgebietes zusammenhängt.

Die Berechnung der Sedimentparameter nach Folk & Ward (1957) und ihre graphische Gegenüberstellung ergab keine signifikante Entwicklung in vertikaler sowie horizontaler Erstreckung, obwohl weitere Hinweise auf das Bildungsmilieu der beprobten Sedimente zu erkennen waren.

LITERATUR

Braun, A.F. 1973, Klassische Methoden der Einzelkornanalyse an Lokkersedimenten, Zbl.Geol.Paläont. 137-149

Casagrande, A. 1934, Die Aräometermethode zur Bestimmung der Kornverteilung von Böden und anderen Materialien. Berlin

Dreimanis, A.& V.J.Vagners 1975, Effect of Lithologyupon Texture of Till. In R.P.Goldthwait (ed.) Glacial Deposits, p.86-101. Benchmark Papers Strandsburg,Penns.

Flint, R.F. 1971, Glacial and Quaternary Geology. John Wiley & Sons inc., New York

Folk, R.L. 1966, A Review of Grain-Size Parameters. Sedimentology, p.73-93

Folk, R.L. 1974, Petrology of Sedimentary Rocks. Hemphill Publishing Co., Austin, Texas

Folk, R.L. & W.C. Ward 1957, Brazos River Bar: A study in the significance of grain size parameters. J.Sed.Petrology, 27:3-26

Husen, D.van, 1977, Zur Fazies und Stratigraphie der jungpleistozänen Ablagerungen im Trauntal. Jb. Geol.B.-A. 120:1-130

Schlüchter, Ch. 1977, Grundmoräne versus Schlammoräne- two types of lodgement till in the Alpine Foreland of Switzerland. Boreas 6: 181-188

Considerations on the relations between landforms, sediments and genesis at ice margins of the Würm Maximum

Based on geomorphological mapping in the area of the former Iller Glacier (Allgäu/Bavaria)

KARL ALBERT HABBE
Universität Erlangen-Nürnberg, Germany

I.

Research work on the Quaternary of the Alpine Foreland has been piloted since about 80 years by a model, a base concept, which explains the specific landforms and sediments of that region as the result of a few dominating processes within a morphogenetic system: the "glacial series" of A. Penck (1901/09). Penck argued as follows: because every glacier advance is connected with three processes, namely 1. glacial erosion, 2. glacial transport and glacial accumulation, and 3. meltwater transport and the accumulation of meltwater deposits, in any case a certain sequence of landforms and sediments will come into existence: 1. in the inner region of prevailing glacial erosion the terminal basin, 2. at the margin of the glacier the mainly rampart shaped terminal moraines and before them 3. the area of meltwater deposits. This sequence he called a "glacial complex". Because every glaciation caused several glacier advances, there must be as much glacial complexes, too, which all together Penck called a "glacial series".

The model of the glacial series has been verified in numerous cases since Pencks times, it has been improved and completed by several authors (Troll 1924, 1926; Moser 1958; German 1968), but no one called it in question fundamentally. Only when R. German published his paper on "Sediments and Landforms of the Glacial Series" (1973), a new discussion on the validity of the Penck model was opened.

German criticizes first of all the nonreflected conclusion from the landform to the forming process. Sedimentological investigations only could give sufficient evidence of the processes which caused the relief forms. Especially moraines, above all so called ground moraine covers should be investigated carefully concerning their sediments. In most cases they would turn out to be meltwater deposits, not tills. Accordingly he demands a largely altered glacial series - altered, too, as against his own presentation of the glacial series of 1968 - in which meltwater sediments have to be taken into consideration much more than before. In other words: German demands a glacial series, which is ruled by the forming processes of glacier retreat, not of glacier advance as one was accustomed since Penck.

Surely German is right, when he emphasizes the importance of meltwater deposits in the inner area of a former glacier bed, in the intramarginal zone, as he calls it. On the other hand Germans base of his criticism are investigations in the eastern Rhine Glacier area. There, however, existed atypical conditions for meltwater runoff in glacial times, because the glacier of the last glaciation for the most time lay behind the European main watershed between Rhine and Danube and so meltwater runoff was hampered. Meltwater deposits therefore must be abnormally frequent in that region.

But be that as it may, Germans demand for a revision of the glacial series remains harassing. For if the glacial series ought to be able to accomplish its function as the central model for glacimorphological investigations, then the connexion between landform, sediments and form genesis should be reliably cleared. This is the reason, why my students and I, when we began geomorphological mapping in the area of the former Iller Glacier, not only had to map the landforms and to look for the sediments, which they are consisting of, but had to consider the model of the glacial series, too.

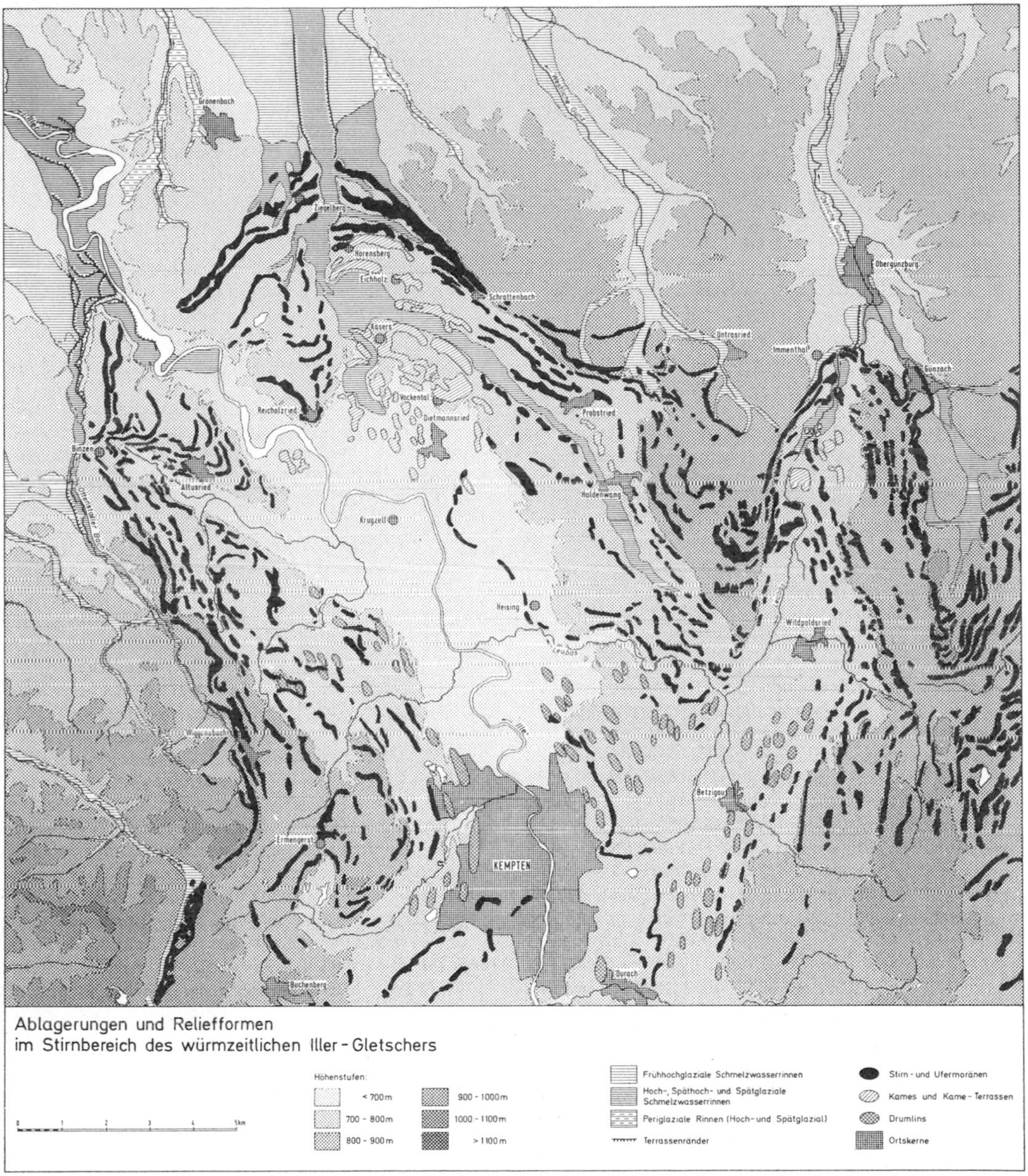

fig. 1. Landforms and deposits of the Würm glaciation in the Iller Glacier area (after Brutscher 1975).

II.

The Iller Glacier has been the smallest one in the German Alpine Foreland. Its extension was limited, it just reached the mountain border. Nevertheless, the Iller valley down-valley of Kempten is a broad basin with flat slopes, on which the glacial and glacifluvial deposits of the last glaciation could be accumulated as well as being preserved (fig. 1).

At first sight the Iller area seems to be a classical example of the glacial series. All its members are present: the terminal basin, the belt of the terminal moraines, and the glacifluvial accumula-

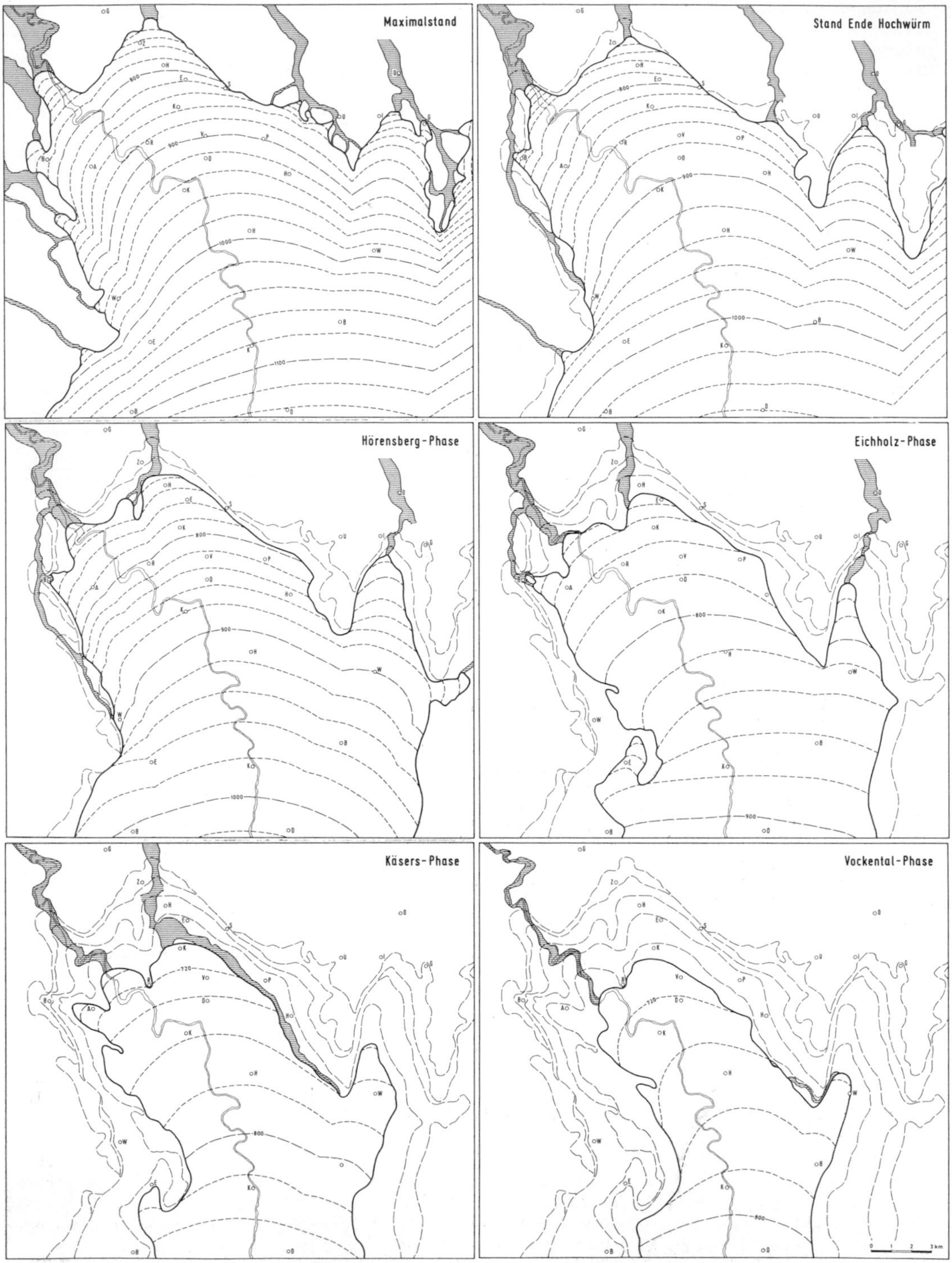

fig. 2. Ice margins and glacier surfaces of the Würm glaciation in the Iller Glacier area.

tion complex of the lower terrace. One is struck by the large number of terminal moraines, which are situated at different altitude levels and reveal a remarkably differentiated behaviour of the former Iller Glacier. And it is to be seen at once, that the Iller river of today does not follow the way it went until the last glaciation towards the north into the "Memminger Trockental" (Memmingen dry valley), but has made its own way through a narrow canyon towards the north-west. The changing meltwater run-off of the Iller glacier has been a main subject of investigation in the 1950s (Schaefer 1940, Graul & Schaefer 1953, Stepp 1953, German 1959), but in general, too, the existing literature on the Quaternary of the Iller area since the times of Penck has dealt mainly with glacifluvial landforms and sediments of the Iller Glacier foreland.

Contrary to that former research work our studies were concentrated on the glacier area itself, the glacial landforms and their deposits. It turned out, that the Iller Glacier area is in fact an area of mainly glacier deposits sensu stricto, i.e. of moraines in a broader sense, as it already has been shown on the only existing geological map of the whole area (Geologische Übersichtskarte ... 1956), which bases on surveys of the 1920s. They are but partly terminal moraines, there are vast areas of ground moraine - of real till - between the terminal moraines, too, especially on higher ground at the outer margins and on the flanks of the terminal basin. Even in the north-eastern area, where meltwater deposits are more frequent, till can be traced in many places. This can be said, although till outcrops are not very frequent and mostly bad, because till was used but for making the farm ways passable in former times. Larger outcrops nowadays are in gravel pits only. But within the Iller glacier area one will find very often these gravels capped by a thick layer of till,- mainly real lodgment-till, in the uppermost parts sometimes ablation-till. We know it from many borings, too. There is - that must be emphasized - a till cover, indeed. The tills are typically bouldery, often sandy - i.e. somewhat meltwater influenced - but clearly to be distinguished from the well stratified gravelly meltwater deposits by the nearly complete lack of sorting or orientation of its particles.

The terminal moraines at the front of the moraine belt as well as on its flanks may look rather different. Real vallums are frequent on elevated but flat underground and on gently rising slopes as e.g. at the outermost rim of the terminal moraine belt and in parts of the western and eastern flanks. Where slopes rise steeper, occasionally only an aggradation of ablation till, frequently mixed up with slope debris and some meltwater deposits, between ice and neighbouring higher slopes took place and formed a moraine-terrace (e.g. at Eschachberg near Wiggensbach). In some places - mainly at the outermost margin and at the angles between the glacier lobes - where the glacier readvances were hampered by older moraine deposits or solid bedrock, a superposition or a juxtaposition of moraine material without formation of separate landforms is not unfrequent. Where the pre-Würm relief is sloping outwards (as above Probstried) the glacier snout has oscillated so frequently, that - notwithstanding the otherwise favourable conditions - moraine landforms of what kind ever could not come into existence at all. The terminal moraines form long-ranging files, frequently interrupted, but always traceable, gently ascending upvalley on the neighbouring slopes of solid bedrock,- thus representing the margin of a glacier advance - or at least a longer lasting steady state of the glacier.

In any case the glacial, the moraine member of several glacial complexes sensu Penck can be frequently traced. But the glacifluvial, the meltwater members are present, too. There are, of course, thick meltwater deposits in front of the moraine belt, the thickest ones in the old outlet towards the north, the Memmingen dry valley, but also in the valleys of the Bettrich meltwater channel, the Iller canyon and the two valleys of the western and eastern Günz river. But meltwater work outside the moraine belt is not only represented by meltwater deposits, but by erosional activity, too, - often, but not always testified by glacifluvial terraces - mainly in the Iller canyon, in the trumpet valley near Ziegelberg, but in the small autochthonous valleys of the Adelegg mountains in the west, too, the low watersheds of which were overridden by the advancing glacier of early high glacial times. Meltwater spillways formed the valley of the Weitenau channel, the Kimratshofen valley, the unnamed valley between Wiggensbach and Ursulers, the Kürnach valley and, above all, the first predecessors of the Iller canyon and the valley of the Ottenstaller Bach.

When after the Würm maximum, meltwater had to run off within the moraine belt, the run-off was concentrated to only a few outlets. Meltwater had to flow the later the more below the glacier as well as at its margins, mainly at the rim between living ice and neighbouring higher slopes.

Best example of such a marginal meltwater stream is the channel traceable between Börwang and Käsers for over 10 kilometers, which must have been active for a long time according to its thick outwash gravel layers. But there were numerous other meltwater courses, the deposits of which can be seen in many gravel pits, mainly at the north-eastern flank of the terminal basin. These gravels partly have been deposited in open channels, often as kame-terraces, sometimes on dead-ice. Thus in some cases kame-terraces may substitute the terminal moraines of a glacier phase, and this the more the nearer you come to the bottom of the terminal basin. But there are not only meltwater deposits of open channels. There can be found delta gravels and sands, too, which have been deposited in small marginal lakes (e.g. near Altusried), and also the silty sediments of those marginal lakes. Eventually there are kame deposits, frequently in the area of Eichholz and Käsers and near Sellthüren in the Wildpoldsried lobe area. Here during certain phases large dead-ice masses must have rested, which were overflown by meltwater heading towards the north for the trumpet valley and the eastern Günz valley, and so gave rise to the kame landscapes of today in those regions.

But meltwater sediments of different type came into existence not only during glacier phases, which we can reconstruct today, but also during older phases, which later on were overridden by glacier readvances. This e.g. can be seen in a gravel pit near Haldenwang, where the gravels are overlain by lake silts and tills, which both have been heavily disturbed by pressing and pushing of a readvancing glacier. The Haldenwang gravel pit is one of many proofs, that the Iller Glacier of late glacial times - at least during one certain period - retreated and readvanced over longer distances.

III.

After all, the glacifluvial members as well as the glacial members of several glacial complexes sensu Penck are present in the Iller Glacier area. Both of them can be reliably connected on the base of their altitude levels, and thus the glacial complexes themselves - or glacier phases - can be reconstructed(fig. 2). That means: the model of the glacial series sensu Penck and Moser as a sequence of glacial complexes can be claimed as valid still today. But is it fit to answer the more comprehensive question for the processes, that brought these glacial complexes into existence? Were they readvances or mere stops of the glacier during a general retreat?

Proofs, that there were readvances, indeed, are not only the wide-spread till covers, which often cap older meltwater deposits, or the contortion of the Haldenwang marginal lake sediments. Another argument for the readvance character of at least one of the distinguished glacier phases are the drumlins of the Iller glacier area. They are situated at different altitude levels between 680 and 800 meter on the flanks of the terminal basin near Kempten. Thus their position is not behind the front of the Würm maximum, but somewhat retreated, - just as it is in the drumlin type areas of the Alpine Forland, e.g. in the Rhine glacier area behind the vallums of the "Innere Jungendmoräne" (Inner terminal moraine). This means, that they, too, came into existence not when the glacier advanced to its maximum extent, but during a later readvance, presumably during the Eichholz phase, which thus most likely represents the same readvance as does the Innere Jungendmoräne of the Rhine glacier area.

But were all phases of the Iller glacier, which can be traced today, readvances? It seems to be so. The glacier surfaces of the distinguished phases of the Iller glacier, which can be reconstructed according to the altitude level of their frontal and lateral terminal moraines, have so different shapes, and especially so different gradients of their longitudinal profile, that it can be hardly imagined, that the younger phases should be mere decomposition phases of the older ones. They must have got their shape by glacier readvances of their own.

Thus on the base of the experiences got in the Iller Glacier area the statement can be given, that Pencks glacial series as a sequence of glacial complexes, consisting of terminal moraines as well as of the respective meltwater deposits and landforms, and - in terms of a more process orientated geomorphology - as a sequence of glacier readvances (and not of retreats)- is still valid and useful as a model for the explanation of the landforms and sediments of an Alpine Foreland glaciation.

REFERENCES

Brutscher, P. (1975): Glazialmorphologische Probleme im Bereich des würmzeitlichen Illergletschers. - Staatsexamensarbeit Erlangen (Mscr.).

Geologische Übersichtskarte von Baden-Württemberg 1:200 000, Blatt 4, ed. by Geologisches Landesamt in Baden-Württemberg, 2. ed. 1956.

German, R. (1959): Die Würmvereisung an Rhein- und Illergletscher zwischen Federseebecken und Günztal. - Geologica Bavarica 43: 3-73.

German, R. (1968): Moraines. - In: The Encyclopedia of Geomorphology, ed. by R.W. Fairbridge. New York: 710-716..

German, R. (1973): Sedimente und Formen der glazialen Serie. - Eiszeitalter und Gegenwart 23/24: 5-15.

Graul, H. & I. Schaefer (1953): Zur Gliederung der Würmeiszeit im Illergebiet. - Geologica Bavarica 18.

Jerz, H., W. Stephan, R. Streit & H. Weinig (1975): Zur Geologie des Iller-Mindel-Gebietes. - Geologica Bavarica 74: 99-130.

Moser, S. (1958): Studien zur Geomorphologie des zentralen Aargaus. - Diss. phil. Basel.

Penck, A. & E. Brückner (1901/09): Die Alpen im Eiszeitalter. - 3 Bände. Leipzig.

Schaefer, I. (1940): Die Würmeiszeit im Alpenvorland zwischen Riß und Günz. - Abhandl. Naturkunde- u. Tiergartenver.f. Schwaben 2.

Stepp, R. (1953): Zur Talgeschichte der mittleren Iller. - Geologica Bavarica 19: 168-185.

Troll, C. (1924): Der diluviale Inn-Chiemseegletscher - Das geographische Bild eines typischen Alpenvorlandgletschers.- Forschungen zur deutschen Landes- und Volkskunde 23: 1-121.

Troll, C. (1926): Die jungglazialen Schotterfluren im Umkreis der deutschen Alpen. - Ihre Oberflächengestalt, ihre Vegetation und ihr Landschaftscharakter. - Forschungen zur deutschen Landes- und Volkskunde 24: 157-256.

Glacigenic and glaciofluvial sediments, typification and sediment parameters

R. GERMAN
Bezirksstelle für Naturschutz und Landschaftspflege, Tübingen, Germany

M. MADER
Kirchheim/Teck, Germany

B. KILGER
Tübingen/Kressbach, Germany

Translated by C. Boone

SUMMARY

The constantly changing geological activity on the margin of the ice, the reworking and redepositing of sediments is demonstrated in actuogeological terms (Roseg Glacier, Switzerland) and for Quaternary sediments (sections 2, 4 and 6). The various Pleistocene and recent glacigenic, glacifluvial and glacilimnic deposits are described in section 5 as sediment types. In section 3 their parameters are given, inasfar as findings to date permit. From the sediment parameters emerge both the importance of as uniform an investigation and evaluation of sediments as possible, and the necessity of a data bank (Braun et al. 1976), at least for the comparison of sediments and for the reconstruction of the processes involved, with their significance for the local history of the earth (e.g. natural monuments). In the case of surface sediments, a sufficiently precise designation is frequently possible during field work by the naming of the sediment types.

The different sediments at and near the margin of the ice have been summarised in the "Glaziale Serie" model since the work of Penck & Brückner (1909). In the further development of this model by German (1968b and 1973), the state of knowledge of that time has been given. Details of the geological activity (re-sedimentation, dead ice effects, etc.) may require separate treatment in the illustration. Sections 1 and 7 deal with the significance of the sediment facies and its importance to stratigraphy.

This model is an aid to the general classification of glacigenic sediments at the margin of the ice. The various sediment types (section 5) are characterised by sediment parameters (section 3). The examination of exposures and their sediments often reveals breaks (hiatuses) in the sediments (section 6). This shows once again that the Quaternary on the continent was a period of episodic, locally greatly differing accumulation and erosion, as is the case with the formation of terrestrial sediments. But these conditions, too, can - carefully examined - lead to an accurate picture of the history of the earth.

1 INTRODUCTION

The reconstruction of the stratigraphy is widely considered today to be the principal aim of quaternary research. But unfortunately, in terrestrial sediments suitable and far-reaching key horizons are in the geological-stratigraphic sense, as a rule, absent. Penck and his contemporaries found in "terrace stratigraphy" a means adequate for their time to establish a provisional stratigraphy of the Quaternary for the Alpine region. Generations of scholars have now been trying for almost a century to make this classification more precise in the type-area and to extend it throughout the world. Numerous "oszillations" of the margin of the ice varying greatly in size (phases, stages, etc.) have been observed in the area of the Alps, but nevertheless the basic framework of four glacial epochs by Penck and Brückner has been retained. An increasing number of climatic variations in the non-glaciated regions of the world (e.g. deep-sea sediments, limnic sediments) have been become known, so that in the future we need to ask ourselves critically

whether the criteria used hitherto for the stratigraphic classification of glacial and glacifluvial sediments can still be maintained. It should also be kept in mind that demands for a more differentiated classification of the Quaternary are contradicted by important palaeontological findings (e.g. Adam 1964,1969). Since palaeontological standpoints have been used for geological-stratigraphic classification throughout the history of geology, we cannot depart from this principle without very good reasons with regard to the last system of the earth's history, the Quaternary. Furthermore, since Penck & Brückner (1909), it has emerged that "gravel stratigraphy" is of only local significance, and that cold-climatic bodies of gravel can merge at lower levels of valleys with warm-climatic ones (Graul & Groschopf 1953, Woldstedt 1954, fig. 124, German 1965). Thus the way of further work on the Quaternary was indicated many years ago: the replacement of morphological studies, and comparison of terraces with bodies of gravel presumed to be uniform, by an examination of the various strata as a three-dimensional sediment unit (e.g. gravel complexes, not terraces only). It has become clear that even over small distances gravel complexes change their various characteristics and that breaks in the sedimentation (section 6) appear. From the palaeogeographical-geomorphological-climatically orientated "terrace stratigraphy" of the Quaternary, a sedimentological method has emerged leading to direct facies interpretations. The profiles of descent of the terraces were supplanted by examinations of the three-dimensional gravel body (time and place) with the further variables climate and facies: these dimensions have to be taken into account in work in Quaternary geology. This procedure has increasingly been supported by a number of ancillary disciplines.

The authors present findings from the application of this method. Detailed geological work, particularly careful investigation of exposures (cf. Braun et al. 1976), has made it possible to obtain considerably more precise results than with older methods. Methods of sediment investigation are important as basic research for correct lithostratigraphic and palaeogeographic interpretations (cf. Geyer 1974). In our investigations the following schema has served as guidelines :

1. Observation of repeated sedimentation, especially at the margin of the ice (see sections 2 and 4);

2. Observation of breaks in the sedimentation (discontinuous deposition, cf. section 6);

3. the localisation of key markers in the sediment (boulder layers, weathering phenomena, organic enclosures, etc., cf. section 6);

4. Establishing of sediment types (according to texture, structure, grain size, cf. section 5);

5. the characterisation of sediments by sediment parameters (see section 3).

The symposium on "Moraines and Varves" is not the first occasion to discuss these problems in Switzerland: 79 years ago, at the Glacier Conference, August 20-25, 1899, held at Gletsch in the Valais with 18 participants (Richter 1900) moraine sediments were under discussions too. As early as 1901, Böhm provided a detailed and critical analysis of the results of the conference in his "Geschichte der Moränenkunde" with a "Moränenregister" (moraine index) as appendix containing over 100 titles. It is indicated in German (1968) that such a large number of mainly geomorphological terms is not necessary and that the problem can be more easily presented in terms of genetic sedimentology. In the meantime, this line of investigation is carried further by Braun, Hermann, Holzmann, Kilger, Mader, Tsiakiris and Weinhold (see bibliography).

2 SEDIMENTS AT THE MARGIN OF THE ROSEG GLACIER

The Roseg Glacier is located on map sheet No. 1277, "Landeskarte der Schweiz", "Piz Bernina", 1:25'000. Approximately 18 sq.km of the Roseg valley are covered by the glacier, most of it belonging to the two major tributaries: the Tschierva Glacier with ca. 6.5 km^2, and the Roseg Glacier with ca. 8.5 km^2 respectively.

Like many Alpine glaciers, both of these glaciers have retreated considerably in the last hundred years. The more unfavourable relation of the area of ablation to that of accumulation has led to a much greater loss of length in the case of the Roseg Glacier than in that of the Tschierva Glacier. As a result, since 1935, the Roseg Glacier

has remained well behind the tongue of the Tschierva Glacier that runs into the valley from the east (Campell 1930 to 1935). The uppermost part of the Roseg valley has been closed off by the medial moraine, which has once been common to both glaciers, and by accumulations of marginal moraine material. Since 1935, the Roseg Glacier has retreated another 1500 m. Behind the medial moraine, a rather flat glacier forefield has arisen, containing a lake which at present has an area of ca. 22 ha, a maximum depth of 22.5 m and a content of over 1.5 million cubic meters of water.

For the high mountains, these are quite atypically flat topographic conditions more comparable with the relief of foreland glaciation.

This means that the Roseg Glacier provides the opportunity to observe both conditions of outflow and the formation of sediments which are close to those found at the ice front of the ice-age glaciers in the Alpine Foreland. The relationship between the sediments and the forms at the ice margin of the Roseg Glacier will be described below as at three different times:

10th August 1976: Sediment-laden melt water emerges from the glacier cave on the west side of the glacier tongue, and deposits part of its burden towards the upper end of the glacier lake in the form of a delta ca. 0.5 ha in size. Crater-shaped holes on the west side of the delta suggest that there is ice beneath this area. Fault-like structures at the same place show that this ice is still in motion and is evidently still connected with the active part of the glacier. The edge of the visible glacier tongue takes the shape of a series of conical hills. They ablate from a slightly inclined layer of sediment which spreads out far across the ice, covering it with a thin coating. This surface moraine material is surrounded by a few smallish supra-glacial melt water courses, and thus incorporated into the delta.

2nd July 1977: The level of the lake is virtually unchanged. On the other hand, the whole of the delta is raised above the lake level on this occasion, because it was deposited on glacier ice which has now broken free from the bottom. The central portion has been raised highest, over 2 m and is traversed by several longitudinal faults. The glacial stream had used its old bed until the change. The dome-shaped root of the delta, deposited upon the glacier ice, has collapsed (ca. 500 m^2). Since then the stream has been flowing through beneath its own delta, reaching the lake by an invisible route. The ice beneath the delta is now visible in places, and in parts is only about 1 m beneath the surface of the sediment.

10th August 1977: The portion of the delta which has collapsed has now considerably increased. The glacier tongue above the old glacier cave has also collapsed. Fresh breaks in the west side and in the part of the delta towards the lake show that ice lies beneath the surface here, too.

The east side of the delta has risen further. Near the old glacier cave, about 1000 m^2 in area has emerged from the water again. Fresh holes indicate that there is ice underneath too. The visible glacier tongue is not rising with it. The melting of the surface of the ice and the creation of fresh ice cones are clearly to be seen.

In July 1978, the re-sedimentation process of the raised delta and at the ice margin continued.

These observation indicate that although the glacier provides large amounts of sediment, it is the melt water that is responsible for the final deposition of the sediments in front of the ice. Together with dead ice melting away, meltwater is continually involved in loosening, "shaping" and re-depositing the sediments (cf. section 4).

3 SEDIMENT PARAMETERS

The classification of sediments by grain size parameters is an important key to their genetic interpretation. Braun (1972) has demonstrated this of certain ice margin sediments and (in 1977) for volcano-clastics. Following Braun (1972) and expanding his findings sediment parameters of the ice margin sediments are given here.

The Roseg Glacier provides an excellent opportunity to study the process. The grain size composition of the deposits can be related to direct observation of the processes at the tongue of the glacier. This facilitates the interpretation of the Pleistocene sediments: because of repeated covering with ice (several glaciations and/ or oscillating ice margins) the present land surface at Roseg Glacier and in the

Alpine foreland indicates only the most recent activity. Huge amounts of sediment are often concealed further down (e.g. German et al. 1967b).

The classification of grain sizes is carried out here after Braun (1972), using the ϕ-scale. For conversion to metrical values, the following formula obtains:

$$(1) \quad \phi = -\log_2 x = 1.4427 \ln \frac{1}{x}$$

where x represents the grain size diameter in millimetres, and ϕ the corresponding value on the ϕ-scale.

The characteristic differences between the cumulative curves of typical melt water sediments and moraine were demonstrated in German (1964) and expanded in German et al. (1967a).

Fig. 1 shows typical cumulative curves of a limnic sediment (bottom-set beds) according to German (1967) and of a ground moraine according to Schlüchter (1973).

4 characteristic statistical parameters can be derived from the cumulative grain-size distribution curve by graphic calculation. Braun (1972b) has discussed the existing literature. The four sediment-parameters (M_Z, σ_I, SK_I, K_G) are calculated after the following formulas:

$$(2) \quad M_Z = \frac{\phi16 + \phi50 + \phi84}{3}$$

(after Folk & Ward 1957) or

$$M_Z = \frac{\phi10 + \phi30 + \phi50 + \phi70 + \phi90}{5}$$

(after McCammon 1962)

The formula for the Inclusive Graphic Standard Deviation is

$$(3) \quad \sigma_I = \frac{\phi84 - \phi16}{4} + \frac{\phi95 - \phi5}{6.6}$$

(after Folk & Ward 1957)

The farther the value departs from zero, the less well-sorted was the sample.

The formula for the Inclusive Graphic Skewness (SK_I) is

$$(4) \quad SK_I = \frac{\phi84+\phi16-2\phi50}{2(\phi84 - \phi16)} + \frac{\phi95+\phi5-2\phi50}{2(\phi95 - \phi5)}$$

(after Folk & Ward 1957)

The value zero indicates a symmetrical grain size frequency distribution. Positive values of SK_I indicate that the samples have a tail of fines; negative values indicate a tail of coarser grains by comparison with normal distribution.

The formula for the Kurtosis is

$$(5) \quad K_G = \frac{\phi95 - \phi5}{2.44\ (\phi75 - \phi25)}$$

(after Folk & Ward 1957)

According to Braun et al. (1976), ground-, marginal- and surface moraines as well as material of outwash plains and "Eisrandschwemmkegel" display characteristic sediment parameters. It emerges from the values given in table 1 that the ground moraine contains very large quantities of fine-grained material (positive graphic mean). This is not surprising, if one considers the physical conditions at the base of a glacier. The surface moraine forms the logical counterpart to this (negative graphic mean), especially where finer portions of sediment that may be present are removed from the surface of the glacier by melt water. The values of graphic mean for the melt water sediments reflect the hydrological environment.

The degree of sorting also expresses a clear difference between moraine material and melt water deposits. Moraines are poorly sorted, melt water sediments well sorted. Outwash and delta sediments form a link between the two types of sediment. Their degree of sorting is apparently often quite poor. The sampling procedure may lead to inadequate results as often several layers of unimodel sediments each, are "combined" in the same sample.

Considering the skewness values it is characteristic that the ground moraine is very close to the value zero (=symmetrical). It does not have the propensity to sort according to grain size, and thus creates sediments that come close to a mathematical chance distribution. In particular, the marginal moraine, but also surface moraine and outwash diverge considerably from this chance distribution of grain size because the spectrum of grain sizes which the water has sorted from the sediment provided by the glacier is still very narrow.

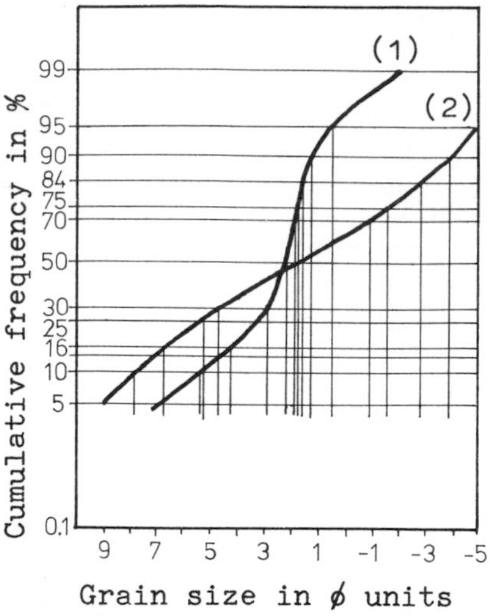

Fig. 1 Typical cumulative curve of a melt water sediment (1) according to German (1967) and of ground moraine (2) according to Schlüchter (1973).
The reverse arrangement of the grain size diameters (positive values to the left) implies the "unusual" parameters in Table 1, as the formulas for M_Z, σ_I, SK_I and K_G after FOLK & WARD have not been rearranged by the authors (Editor).

Kurtosis, except from delta sediment, where it diverges considerably from the value 1, offers no clear help in distinguishing between moraine sediments and melt water sediments.

The values presented in Table 1 are "ideal" ones. The results of sediment analysis will diverge from them to greater or lesser extent. Large numbers of samples only will give results clustered close to those mentioned.

The analysis of the sediments from the marginal area of the Roseg Glacier has shown that the results given in Table 1 after Braun et al.(1976) are not adequate for the genetic interpretation of glacial deposits from flat terrain. A large proportion of the sediment samples can be brought into close connection with outwash-plains and "Eisrandschwemmkegel"; similarly, the values between melt water deposits and marginal or surface moraines are demonstrated. The rest of the results, however, shows a considerable divergence from the values mentioned: towards

1. smaller mean sizes
2. a better degree of sorting
3. a skewness that remains evenly small, and
4. a kurtosis which indicates either strictly unimodal or very complex grain size composition

In the area of the Roseg Lake the following 3 types of sediment are also involved: top-set beds, bottom-set beds and shore deposits

Top-set beds:

Sediment samples taken close to each other from a delta can be quite different. Frequently, even a single sample will contain a large number of different layers. This is not surprising: direct observation of the formation of a delta shows that sedimentation and erosion often alternate within millimeters of each other, and that melt-water flows from the high energy glacial stream to the fine threads of water are to be found. The kurtosis of the top-set beds therefore diverges very strongly from the value 1, and shows how much these sediments reflect mixed hydrological regimes.

Geomorphologically speaking, an ice margin delta is hardly recognisable as such quite soon after its formation. A more regular sedimentation, as on firm ground (Fürbringer 1976), is not always possible, on top of melting ice. Strata tilted in the direction of the lake are destroyed, and inclinations indicating former directions of flow can be altered.

Sediment analysis and calculation of the sediment parameters are therefore critical for identifying a fossil delta or its reworked or re-deposited sediments. Together with the kurtosis, the degree of sorting provides a good possibility of differentiation, for instance from a moraine. Sorting is much better in a delta deposit, even if one sample contains parts from various layers of sediment. The values of Graphic Mean almost always are between +3ϕ and -3ϕ. Very large and very small grain sizes are absent.

Bottom-set beds:

Bottom-set beds and ground moraines are not always easy to distinguish from each other by macroscopic methods. Both sediments may have a wide horizontal spread,

Table 1. Parameters of glacigenic and glacifluvial sediments. - Upper part of the table and numerical values after Braun et al. (1976), also suggestion for placing the limnic sediments (bottomset beds). - Lower part of the table, for delta deposits (top-set beds) and shore sediments after results obtained at the Roseg Glacier in 1977 and 1978.

	Mean	Sorting	Skewness	Kurtosis
Ground moraine	+3.6	-2.4	-0.13	1.10
Marginal moraine	-1.4	-2.8	+1.13	0.97
Surface moraine	-2.2	-2.3	+0.24	0.98
Sed. of outwash plain	-2.1	-2.4	+0.20	0.92
"Eisrandschwemmkegel"	-0.1	-1.8	-0.08	0.93
Top-set beds (delta deposits)	+3 and coarser	0 to -2.5	-	widely deviant from 1
Shore sediments	+2 to -2	0 to -0.5	0	close to 1
Bottom-set deposits (limnic sed.)	+2 and finer	approx. -1	0	1 to 1.5

a more or less definite stratification, and a very high proportion of fine-grained material (German et al.1967a).

Small grain sizes predominate in the limnic sediments (bottom-set beds): mean size can, therefore, be in the clay fraction. In the case of banded clay, it is of the order of +9ϕ (summar varves) or +12ϕ (winter varves) (Ashley 1975). Closer to the glacier, it can also be found in the sand-fraction, e.g. at +2ϕ.

The water in the lake in front of the Roseg Glacier is replaced every ten days in summer, roughly calculated. This time is too short for sediment particles smaller than +9ϕ to settle down. They are therefore carried out of the lake in suspension ("Gletschermilch"). In winter, too, when the replacement of the water is slower, no winter varves are formed, because the northern parts of the lake are very shallow, and water freezes to the bottom. Conditions may have been similar in the old lake, which drained in 1954 (Bisaz 1955, Töndury 1954), and whose level was ca. 150 cm higher. Among the sediments of the former lake bed which are immediately accessible at the surface today, only sediments of mean size down to +7ϕ are to be found for this reason.

Coarser components (boulders) are sometimes conveyed to the lake bed by surface moraine material falling from the ice margin into the lake. In contrast to ground moraine, however, there are no grain sizes present between the coarse and the fine, so that in this case the mixing can be clearly identified.

The degree of sorting is much better in limnic sediments than in moraine material. It is seldom poorer than corresponds to the value -1, and is better the smaller the sample. Within individual varves it approaches to the value zero. The skewness-values, too, are small, thus indicating symmetrical granule distribution. The kurtosis is close to 1.

Shore sediments:

On the eastern shore of the lake, remains of the Roseg Glacier project from the water in form of a wall of ice ca. 500 m long and up to 10 m high. Large pieces of ice are continually breaking off, causing calving waves. For this reason, the opposite west shore is marked by their erosive power and has a narrow shelf at low-water levels. This shelf is formed from material above the level of the lake. The wave action draws from the shore all sediment up to a certain size, and tranports the finer particles to the lake. On the shelf itself there remains a residual sediment. Its grain size composition approaches an "ideal" set of values.

With the exception of Graphic Mean, all the sediment parameters can be defined: The sorting degree is excellent (-0.5 or better); hardly any skewness can be observed (nearly 0); the kurtosis is close to 1.

The grain size distribution alternates on a broader scale and could be considered as a function of exposure to formerly prevailing winds or the position relative to possibly existing walls of ice (fossil shore-lines, cf. cliff at the former Federsee, Upper Swabia).

4 RE-SEDIMENTATION AT THE MARGIN OF THE ICE

Sedimentation at the glacier margin is by no means a single action of deposition as it is assumed to be for the deep oceans but it is an extremely detailed "flow of processes" of repeated erosion and redeposition of the same particle associations (cf. on this also Gripp 1964 and earlier). This changing pattern at the ice margin is influenced by the following factors: movement of the glacier, meltwater regime, changing insolation, changing of meltwater currents, collapsing of moraines and frontal parts of the glacier, dead-ice phenomena, temporarily formed lakes, freezing and thawing etc. (cf. Table 2).

Recent observations at ice margins, in the Arctic (Alaska, the North-West Territories of Canada), in the high mountains (the Alps, the Himalaya) indicate, particularly if the observations are carried out for months or comparisons made year by year, that re-sedimentation is normal and everyday at the ice margin everywhere. German (since 1962) has repeatedly demonstrated in the most varied landscapes and on numerous glaciers that melt water sediments predominate at the ice margin because of continued re-sedimentation, since very little moraine sediment can remain unaltered there. Mapping by workers in Upper Swabia has shown that the predominance of melt water sediments is equally general and widespread (German et al. 1965, 1967, 1968). The glacifluvial sediments, particularly the kames deposits (Richmond 1953) are of a rich facies-variety as can be observed today at recent ice margins in the high mountains and the Arctic. The extent of this facies-variety becomes clear when evaluating the sediments three-dimensionally. For example, the genesis of a sediment type can still be reconstructed with the parameters even after re-working. Our investigations have shown, by studying the distribution of the sediments, the genesis of the types and the sediment parameters, that there is no longer any valid reason to adhere to the resolution of the Glacier Conference of 1899, inasfar as it distinguishes between moved (Recent) and deposited (Pleistocene) moraines. It becomes clear that modern sediment research produces useful and profitable results if performed in conjunction with actuo-geology, as by our study group. The retreat of glaciers in the present century offers an excellent opportunity to make comparisons between the actual processes and those of the Riss or late Würm glacial period.

In order to demonstrate the importance of the re-sedimentation at the ice margin, the respective causes and processes are listed below:

Table 2: Causes and processes of re-sedimentation in ice-margin areas

- Thawing of partially frozen sediment particles (e.g. by insolation or air temperature),
- Englacial transport of sediment,
- Melting-down of the ice, with settling or re-depositing of the sediment on it (Gripp 1964),
- Melting of "basins" of dead ice (see section 2),
- Collapse of caves in the ice (section 2),
- Deep thawing of dead ice (Gripp 1964),
- Ice-pushing (forming "contorted" moraines, Gripp 1964, Weinhold 1973),
- Oscillation of the ice margin,
- Waves, caused by calving avalanches or land-slides into an ice-margin lake, and the influence of the wind, surface moraine breaking off and falling in or on to the margin of the ice,
- Alternation of the course of melt water,
- Erosion by changes in water-courses at different times of day or year (by raising of underlying ice section 2),
- Side erosion with a break-through to a deeper basis of erosion,
- Dwindling of the ice in the case of kames formation,
- Retrograde erosion,
- Meandering of the melt water stream,
- Variations in the lake level,
- Outflow from the lake with flooding of the foreland (or with slipping of the limnic sediments),
- Transport on ice-floes or icebergs (drift of limnic or ocean sediments),
- Avalanches, landslides etc. on the glacier surface (cf. surface moraine, earthquake in Alaska 1964).

5 SEDIMENT TYPES AND THEIR FORMS

Mapping in Upper Swabia (Braun 1972, German since 1958, esp. 1968, 1970, 1973, Hermann 1973, Holzmann 1970, Mader 1970, 1971, 1972, 1976, Schiftah 1970, Tsiakiris 1972, Weinhold 1973) have demonstrated that the glacial sediments of the former Rhine-Foreland Glacier can be grouped genetically into a small number of sediment types too. Their facies depends essentially on the activity of the melt-water, and the topographical shape of the glacier foreland. This means that the geomorphological forms at the terrain surface may not necessarily correspond with the sediment content. Investigating the sediment becomes even more important in the case of massive buried quaternary deposits (German 1965, 1967, 1968), where the geological history has to be derived from the sediments alone.

Moraine sediments

Surface moraine sediment is the debris that covers the surface of a glacier. It consists mostly of components of every grain size from silt to the largest block, which are not rounded at all and are seldom or never striated. Grain size analyses of recent surface moraine material have been published by Braun (1972b, 1974).

In Upper Swabia, no surface moraine material is found as a rule, because this has been sorted by melt water while still on the glacier surface (cf. to "Eisrandschwemmkegel"). On rare occasions, individual large blocks are found directly on ground moraine sediment; these blocks, being the coarsest fraction of the former surface moraine, remain longest on the glacier, before rolling off the glacier snout. They can be clearly seen in the foreland of the Upper Theodul Glacier near Zermatt, where they lie on stretches of ground moraine sediment (German 1972a) and in the foreland of the Biferten Glacier (German, Hantke & Mader, in press). Where the ice margin remains for a long period, the blocks can pile up to form a ridge-like structure, such as can be observed particularly well at the Tschierva Glacier (Roseg valley) or in the foreland of the Vadret da Porchabello on Piz Kesch (Grisons, Switzerland).

Marginal moraine sediment is mostly created by pushing, when the advancing glacier pushes material in the foreland together (advance phase). Melt water is not involved in the formation of this sediment, if one excepts the superficial sorting of the material by water from crevasses in the glacier, which often appear at the glacier snout (e.g. German 1972, fig. 2).

The ice pushing causes the sediment to be constantly redistributed and mixed in front of the glacier snout, so that it may display neither texture nor structure (grain size analyses in German 1964, 1965, 1967, Mader 1970, Holzmann 1970, Braun 1972). In this case, the sediment consists mainly of unrounded components of every grain size, as in the case of the surface moraine. It is, however, easy to distinguish from the latter because of the presence of striations on the stones caused by the mixing movement. Marginal moraine sediment of the composition described occurs frequently in ridge zones between different confluent ice-streams (marginal basins), and in the "walls" of the respective maximum glacial advances of the Mindel, Riss and Würm glaciations (Mader 1976). Marginal moraine sediment re-distributed by melt water occurs in kames terraces, etc.

"Contorted" moraines are a unique feature of ice marginal deposits. They are mostly formed when retreating glaciers exhibit a readvance. This process may cause the piling up of more or less stratified melt water sediments in front of the glacier snout, partly in a frozen state, so that the original texture is still clearly recognisable. If the advance is more than about 100 m, these sediments are usually so intermingled with new material of marginal moraine that no earlier sedimentary texture can be discerned any more (Mader 1972).

Medial moraines, as geomorphological forms with marginal moraine sediment, have been preserved in Upper Swabia only in sill areas between two glacier lobes. They are to be discerned in the southward continuation of the Ziegelbacher Rücken (Ziegelbach Ridge), a molasse ridge beneath sediments from the Würm glacial period that runs from Arnach S Wurzach to the Wolfegger Ach S Kisslegg (German et al.1967, Holzmann 1970, Schiftah 1970; cf. German et al.1972; Figure 2).

Ground moraine sediment is formed by friction and pressure of glacier ice on bedrock. Apart from the fine-grained matrix as product of the pulverisation of bedrock, it contains pebbles and boulders of all sizes, which are scored or polished, according to

the type of rock, but always well rounded and striated. By contrast to the components of marginal moraine sediments which are in general loosely distributed, the striations of the stones of ground moraines run straight and usually in concentration (cf. Weinhold 1973, 1974). These striations are caused by the individual boulders being firmly embedded in the matrix. If a boulder changes position, a new set of striations will be caused on its surface.

The amount of clay contained in ground moraine is determined by the clay content of the primary material, because the weathering process during its transport in the ice, or after its deposition, appears without significance.

In the colour of the ground moraine, too, the components of the reworked material are reflected: Alpine rocks rich in limestone principally produce a grey colour and the sediments of the Upper Freshwater Molasse in the Alpine Foreland produce a yellow tone in the matrix. Thus, the colour is only indicative of local conditions, and can change over the shortest distances (cf. Gasser & Nabholz 1969).

This implies that from the mode of origin of the ground moraine the "basal rock" gradually merges upwards into ground moraine sediment, as long as it does not consist of hard rock. By contrast, the contact between the ground moraines and overlying sediments is always definite. Towards the glacier snout, ground moraine material yields to marginal moraine material.

In Upper Swabia, ground moraine material occurs at the surface extensively on ridges only, whereas in basins it is covered by massive amounts of melt water sediments (Fig. 2). A recent example is the almost level area immediately in front of the Upper Theodul Glacier near Zermatt, where ground moraine occurs extensively (German 1972). By contrast, no ground moraine at all is to be found in the basin of the Roseg Glacier (section 2).

Glacifluvial sediments

In Upper Swabia "Vorstossschotter" is the term for the deposits of glacial melt water streams that have been overridden by the advancing glacier. In profile, an idealized "Vorstossschotter" is composed as follows: at the base, layers of silt and sand, which were deposited at some distance from the glacier. As the "delivery area" approaches (advancing glacier), the material becomes increasingly coarser, whilst the degree of roundness of the pebbles decreases. In the upper layers, individual pebbles can have a diameter of 50 cm and more. This material, deposited near the ice margin, merges further upwards into ground moraine. Grain size analysis of "Vorstossschotter" are to be found in Mader (1970).

Directionof transport is deduced from the imbricated arrangement of the pebbles. The longitudinal axes are generally transverse to the direction of transport. In a profile in the direction of transport a gravel-body displays stratification persisting over long distances, whilst a profile transverse to the direction of transport shows lentiform sedimentation.

"Outwash plains" are usually cone-shaped deposits. Their sediments consist mainly of coarse, poorly rounded material, deposited in front of the glacier. In accordance with the definition, this sediment type occurs in the upper layers of "Vorstosschotter" (grain size analyses,e.g. in Mader 1970). Towards the glacier cave, the outwash sediment can merge into marginal moraine sediments.

Gravels ("Fluss-Schotter")are well sorted and stratified (grain size nalysis, e.g. in Mader 1970). The layering is caused by the hydrographic regime or by changes in the direction of flow. At some places graded bedding may occur.

The maximum grain size of gravels remains more or less constant in vertical sections. Glacifluvial gravel bodies, by contrast, display increasing grain size from bottom to top, indicating the approach of the glacier. The grain size in the uppermost layers is often decreasing again. There, fine-grained water-meadow sediments may be enclosed, or other traces indicate reworking and/or re-sedimentation (see below), which indicates the late-glacial epochs or even the beginning of the post-glacial (cf. German et al.1972).

Stratification and pebble imbrication are similar in river gravels as in "Vorstossschotter" (cf. German 1973).

"Eisrandschwemmkegel" are sediment cones formed at the ice margin by melt water from the surface of the glacier. They consist mainly of the finer components of surface moraine material. Occasionally a larger block falls from the glacier surface and

remains between the sand and pebble layers. The material of the "Eisrandschwemmkegel" is thus silt, sand and small stones. The components are unrounded and rarely striated. For the sedimentological modelling of "Eisrandschwemmkegel" see Mader (1970) and Braun (1972 , 1973, 1974).

Geomorphologically, the "Eisrandschwemmkegel" mainly take the form of small hillocks, which are superimposed one on top of the other. In the frontal basins, mainly from the Würm glacial period, particularly on sheet 8225 Kisslegg (topographic map 1:25'000), "Eisrandschwemmkegel" are to be found piled up behind each other. This landscape was hitherto generally referred to as a "kuppige Grundmoränenlandschaft" (dome-shaped ground moraine landscape), although ground moraine does not come to the surface at all there (cf. German 1971).

Kames sediments (ice margin formations): the stratification and sorting of sediments at the ice margin occurs in Upper Swabia in places where glacier tongues have been confined to marginal basins (German 1958, Mader 1970, 1971). Because of the topography, widely varying sediments have been deposited between the ice margin and the rim of the basin rising away from the glacier. To some extent, these sediments must be assigned to the moraine facies, but mostly to melt-water deposits. Melt-water emerging everywhere along the ice margin, collects marginal moraine sediments in front of the glacier with subsequent sorting and re-sedimentation. In this way, kames terraces are formed (for instance in the Wurzach area, Mader 1971; the Frohnhofen area, Hermann 1973; or the Zussdorf basin; Tsiakiris 1972). In depressions, and because of increased melt-water activity, delta-shaped structures can form within such terraces - the individual components still being quite angular. This sorting effect is to be seen in the south-east part of sheet 8125 Diepoldshofen and the north-east part of sheet 8225 Kisslegg (Holzmann 1970, Schiftah 1970). In the Simmers gravel pit in the north-west part of sheet 8025 Wurzach, the marginal moraine sediment is unsorted only where it projects above the surface of an ice-dammed lake formerly present in the Füramoos frontal basin (Mader 1972).

Former "Eisrandschwemmkegel" can be located (on sheet 8225 Kisslegg and sheet 8122 Weingarten) where "contorted" moraines consist of "Eisrandschwemmkegel"-material. It is the first sediment to be affected by an advance of the glacier (cf. Schiftah 1970, Hermann 1970).

Delta deposits can easily be distinguished from "Eisrandschwemmkegel": In contrast to the latter, they display bottom-, fore- and top-set bedding. The pebbles are usually rounded. The top-set layers contain in particular pebbles and sand, the bottom-set layers silt and clay (varved).

Varves are typical bottom-set layers occurring in Upper Swabia, e.g. in the Federsee or Wurzach Basins (German et al.1965, 1967, 1968). Before a stratigraphic evaluation of the varves is possible, it needs to be checked whether they do in fact display annual layering. In varves, hiatuses can exist which destroy their usefulness for stratigraphic purposes.

6 STRATIGRAPHIC BREAKS (HIATUSES) AND MARKER HORIZONS

Breaks within an apparently continuously deposited lithological unit are of equal importance as the sediments themselves for stratigraphic reconstructions. For example, boulder layers are enriched in particles whose size and weight correspond to or exceed the maximum energy level of a river. They may mark the base of a gravel unit. Some of the formations described by Graul (1953) as periglacial basal facies are to be interpreted as boulder layers. Different units of gravel on top of each other, are often separated by a boulder layer ("Kondensationshorizonte"). Examples are: Gravel-pit near Scholterhaus N Biberach/Riss (Mader 1976, Braun, German & Mader 1976), gravel units in the Aare valley S of Berne (Schlüchter 1973), gravels in the Danube valley (Löscher 1972, 1974) and on

Fig. 2 Basin Formation and Moraine Sediments in Upper Swabia (South-West Germany). The contour lines indicate the upper edge of the Upper Miocene-Lower Pliocene "Obere Süsswassermolasse (OSM)". In the shaded parts, sediments of ground moraines and marginal moraines are widespread. A ground moraine layer only about a meter thick lies over the highest molasse ridges in the Biberach and Ochsenhausen area. Ground moraine sediment - together with marginal sediment - is widespread particularly in the area Kisslegg-Wurzach-Waldsee.

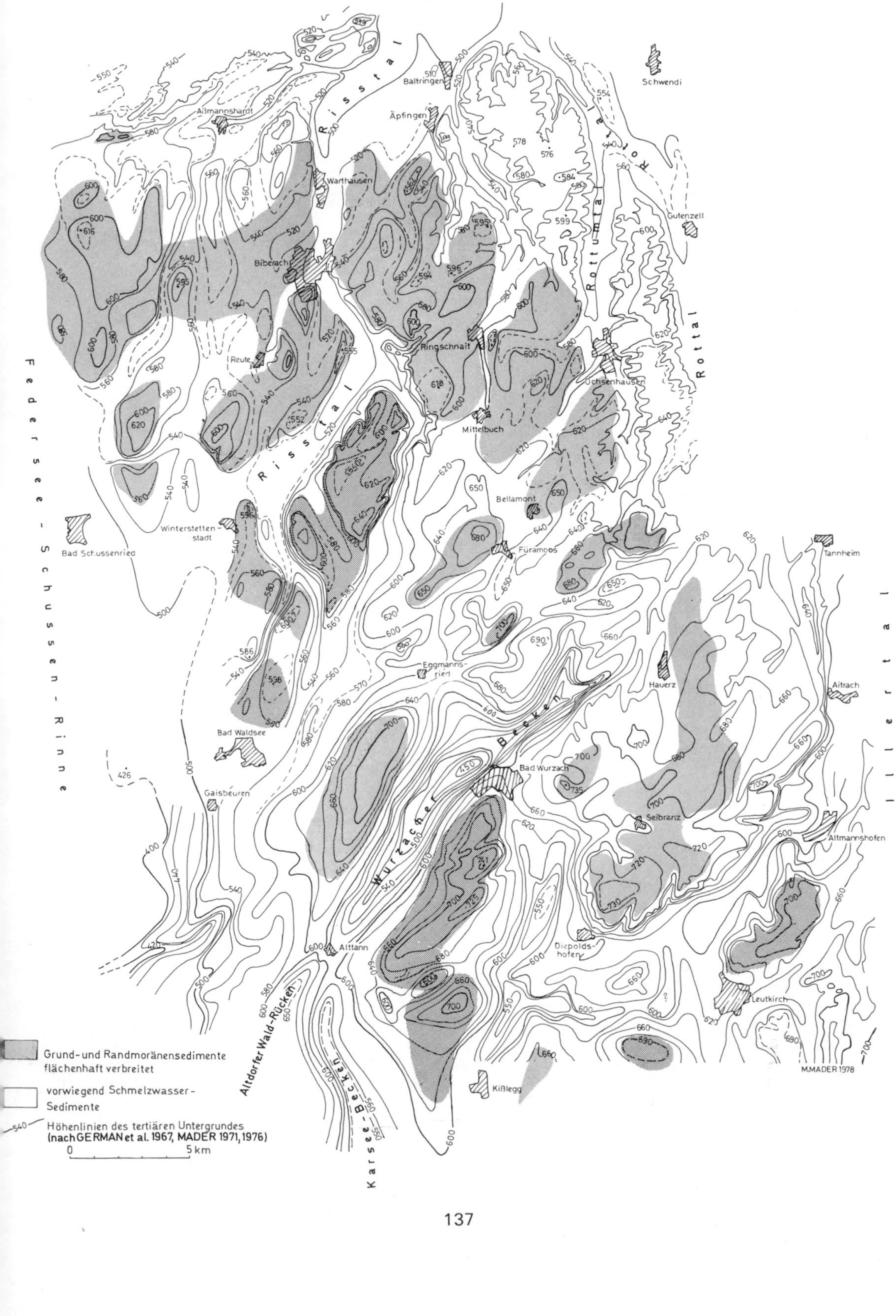
Federsee - Schussen - Rinne
Karsee - Becken
Altdorfer Wald-Rücken
Wurzacher Becken
Risstal
Rottumtal
Rottal
Illertal
Aßmannshardt
Baltringen
Äpfingen
Schwendi
Warthausen
Gutenzell
Biberach
Ringschnait
Ochsenhausen
Reute
Mittelbuch
Bellamont
Winterstetten-stadt
Bad Schussenried
Füramoos
Tannheim
Eggmanns-ried
Hauerz
Aitrach
Bad Waldsee
Bad Wurzach
Gaisbeuren
Seibranz
Altmannshofen
Alttann
Dizpolds-hofen
Leutkirch
Kißlegg
M. MADER 1978
Grund- und Randmoränensedimente flächenhaft verbreitet
vorwiegend Schmelzwasser-Sedimente
Höhenlinien des tertiären Untergrundes (nach GERMAN et al. 1967, MADER 1971, 1976)
0 5 km

the Iller-Lech plateau (Scheuenpflug 1972, Löscher 1972, 1974), Oligocene Silvana limestone boulders in the gravel-pit Queck SE Zwiefaltendorf-on-Danube (Zöbelein 1973, Schreiner in Geyer 1977, Mader 1976), gravel-pit Baur "beim Vogelwäldle" S Riedlingen-on-Danube (Schädel, Weidenbach & Werner 1961, Schädel & Werner 1963, Graul 1962, Mader 1976).

In the marginal area of a glaciation erratic blocks within a boulder layer are an indication of reworked and washed ground moraine or marginal moraine material. The original sediment does not need to be preserved to provide adequate proof, provided that scratches (glacial striae) are still to be found on the blocks. Example: angular and striated boulders in a boulder layer of the A+B gravel-pit at Ingoldingen S Biberach, in front of the so-called "Aeussere Jungendmoräne", the geomorphologically outermost Würm glacial moraine (Mader 1976, German & Mader 1976).

Sudden decrease in the maximum grain sizes within a gravel unit can be a sign of a stratigraphic break, even if the petrographic composition of the detritus remains unchanged and if the maximum clast size increases again above the surface of discontinuity. For example, in front of the Aeussere Jungendmoräne between Bad Schussenried and Bad Waldsee, the deposits of an older, more powerful ice advance are covered in places by those of a more recent and weaker one (Mader 1976).

Changes in the petrographic composition may indicate a stratigraphic break, even with no change in the maximum grain size, if still water deposits, such as water meadow sediments, occur in places just below the surface of discontinuity (cf. Schlüchter 1973). A sudden change in the petrographic composition of the detritus, together with smaller maximum grain sizes, always indicate a stratigraphic break. Example: the gravel of Inneberg-Fürbuch east of the Iller (Schädel 1952, Sinn 1972, Mader 1976b).

Ice polishing on the surface of glacifluvial conglomerates. Examples are: gravel-pit near Ertingen (Schädel, Weidenbach & Werner 1961, Schädel & Werner 1963), gravel-pits near Uttenweiler E Dieterskirch (Mader1976) and in the Riss valley (Mader 1976).

Fossil soils. Examples are: in the Danube valley at Riedlingen (Graul 1952, 1962, Schädel, Weidenbach & Werner 1961, Brunnacker in Graul 1962, Mader 1976), gravel-pit in the Kirchen valley W Ehingen-on-Danube (Mader 1976), Hochgelände S Biberach-on-Riss (Weidenbach 1937, Eichler 1970,etc.).

Weathering phenomena (often called, geologically inappropriately, "weathering horizons") are alterations in the sediment at the surface and at suitable places within the sediment body (cf. Brunnacker in Graul 1962, Kaiser 1963, Fezer 1969). Examples are: gravel-pit at Scholterhaus N Biberach-on-Riss (Mader 1976, Braun, German & Mader 1976), Hochgelände (Weidenbach 1937, Eichler 1970).

Intercalation of tree trunks, peat, calcareous mud, diatomite, molluscs, etc. Examples are: Iller alluvial cone (Graul & Groschopf 1953, Becker 1972, Brunnacker 1960), Ziegelberg S Bad Wurzach (Göttlich & Werner 1967), Unterpfauzenwald (Göttlich & Werner 1968), Karrestobel (Hermann 1973), gravel-pit near Rudenweiler (Weinhold 1973), core drilling Wurzacher Becken 1 (DFG) (German et al.1968, German & Filzer 1964).

7 THE "GLAZIALE SERIE" (Penck & Brückner 1909)

Geomorphological and sedimentological processes by the ice and at the ice margin were schematically modelled as the "Glaziale Serie" by Penck & Brückner (1909). Originally, this model was intended for valley-glaciers only and developed on the basis of observations in the Alpine region. Investigations at the margin of Recent and Pleistocene inland ice indicate that this model can be used for ice-margin formations in general. Instructive and clear as it is, it has required to be modified (Schäfer 1950, Moser 1958). The decisive improvement was achieved by German (1968), with a more differentiated account of the sediments and a conceptual separation of sediments and forms (German 1970, 1975, 3.15, p.63, and German et al.1972, Fig. 1). The application of Penck & Brückner's general model combined with geomorphological investigations occasionally caused the neglection of geological details such as sediment inclusions, alterations and breaks within gravel accumulations. Therefore, we prefer to encourage sedimentological and geologi-

cal investigations because more detailed information on the minute processes of deposition and erosion can be obtained (cf. Braun et al.1976, Vossmer-Bäumer 1976).

The results of physical dating seem to indicate that the breaks within the Quaternary sediment sequences both in the Alps and in the Alpine Foreland, are greater than the time indicated by the sedimentation. The breaks in some cases embrace the whole of a cold or warm period (or several). The importance of stratigraphic breaks relates not only to the elevated ground on either side of the river systems, where - apart from possible periglacial processes - breaks are more easily to be recognised. These breaks also apply to the river systems and their gravel bodies below the valley floors, both in the case of the rivers taking their rise in the Alps (autochthonous rivers), and those taking their rise in the foreland (allochthonous rivers). The complex alternation between erosion and accumulation even in late-glacial and post-glacial time alone, with fluvial re-depositing in the river meadows has been known since Brunnacker (1960) and German & Filzer (1964). Hiatuses can be demonstrated both horizontally and vertically in gravel units, which externally appear unbroken (German et al.1972, figs. 3a and b). Where such profiles of different glacial epochs are simply juxtaposed (cf. Göttlich & Werner 1976), they do not provide an overall profile of the Quaternary. Under favourable conditions breaks can be excluded in the case of a lake, provided it has not been overridden by the glacier (glacial abrasion). The alternation of erosion and accumulation (and resedimentation) should be investigated more carefully.
Without a careful geological and sedimentological exposure analysis, we may miss essential observations. In field-work, the designation of sediment types is an important aid. In cases of doubt, they can be evaluated by determing the sediment parameters.

Quaternary research in the Alpine region has often dealt with formations near the surface. Considerable quaternary accumulations in bedrock bassins provided by drillings and field-work (incl. German 1963, German et al. 1965a and b, German et al. 1967a and b) are proof that much stratigraphic evidence is stored in depth: The same is true for the marginal area of the northern European glaciations (Cepek 1967). The 60 m of sediments, presumably mainly of the Würm glacial epoch, at the margin of the Argen valley (Schiftah 1970) or the 144 m thick strata of the "Ur-Federsee" basin have been evaluated from the geological-sedimentological and pollen points of view. In these profiles stratigraphic breaks appear too.

As more and more geological evidence of the earth's history is being destroyed in the cause of "progress" (e.g. by material extraction), it is essential for us to obtain and preserve samples or finding-data of this evidence. This can, for instance, be done by the aid of a data bank (Braun et al.1976). Furthermore, the world-wide typification of sediments, with the development of the sediment parameters, is thereby promoted.

In English-speaking contexts, the term "till" is frequently used. Here, the significance of the word often diverges widely comprising glacial and glacifluvial sediments. For this reason, the meaning of "till" would appear to correspond to the meaning of the word "moraine" in the Alps 150-200 years ago (cf. German 1968). For the purposes of a modern nomenclature of sediments it appears necessary to avoid such general notions in future, and rather to reveal the findings as precisely as possible by means of the sediment type (section 5). The difficulty which besets the quaternary geologist of having to work in at least 5 dimensions (3-dimensional space, time, climate and facies), is alleviated by the designation of sediment types and their respective parameters. Even if it means a certain simplification sometimes (neglecting one dimension or another), it would be shortsighted to overlook this difficulty in the case of wide-ranging comparisons. The world-wide existence of similar types of sediments and their nomenclature in German, English and French (German 1968) permit a uniform designation of glacial and glacigenous sediments in the future. A practicable way to realise this is offered by the suggestion of Braun et al.(1976) with the establishing of a data bank. In this context the localities Hurifluh, Wässerifluh (Schlüchter 1973) and Uznach, could be put under protection on account of their importance for quaternary geology. Their naming as natural monuments would be appropriate for this purpose, together

with other geological natural monuments (e.g. in Upper Swabia). Either for methodical or stratigraphic reasons, reference localities can be preserved for posterity (German 1974). For the "Glaziale Serie" model, the gravel-pit "Mosertal" near Waldburg (E. Ravensburg) would be a possibility, because it displays a comprehensive sequence of strata. It is important to know reference localities with characteristic sediment types and their parameters, in order to be able to carry out comparative investigations and to establish a data bank, and in order to characterise the "Glaziale Serie" with its various sediments more completely.

8 REFERENCES

Adam, K.D. 1964, Die Grossgliederung des Pleistozäns in Mitteleuropa, Stuttgarter Beitr. Naturkde., No. 132, 12 p., Stuttgart.

- 1969, Zur Grossgliederung der Altsteinzeit Europas, Stuttgarter Beitrg. Naturkde, No. 207, 16 p., Stuttgart.

Ashley, G.M. 1975, Rhythmic sedimentation in glacial lake Hitchcock, Massachusetts, Connecticut. In: Glaciofluvial and Glaciolacustrine Sedimentation, Soc. Econ. Paleontologists and Mineralogists, Spec. Publ. No. 23: 304-320. Tulsa (Oklahoma).

Bisaz, O. 1955, Jährliche Mitteilungen des Kreisforstamtes 29 Oberengadin-Bergell über den Stand des Roseggletschers an die Schweizer Gletscherkommission, Celerina.

v. Böhm, A. 1901, Geschichte der Moränenkunde, Abh. geogr. Ges. Wien, vol. 3, No. 4, 334 p., 4 tab., 2 fig., Vienna.

Braun, A.F. 1972a, Statistik in der Sedimentology, ein Hilfsmittel zur genetischen Deutung, Zbl. Geol. Paläont. Teil I, 1972, 1/2, 1-9, Stuttgart.

- 1972b, Rezente Gletschergebiete in den Alpen als Modell für die eiszeitlichen Verhältnisse in Oberschwaben. Zur Problematik von Sedimenten und Formen, Diplomarbeit Tübingen.

- 1973a, Klassische Methoden der Einzelkornanalyse an Lockersedimenten, Zbl. Geol. Paläont, Teil I, 1972/5, 6, 257-269, 137-149, Stuttgart.

- 1973b, Einfaches sedimentologisches Modell zur Gliederung der von Gletschern abgelagerten Sedimente. Zur Problematik von Eisvorstoss- und Eisabschmelzbildungen, N.Jb. Geol. Paläont., Mh.6, 315-326, Stuttgart.

- 1974, Eine sedimentologische Ableitung der Eisrandschwemmkegel aus der Obermoräne. (Untersuchungen an rezenten Gletschern in den Schweizer Alpen), Eclogae geol. Helv. 67, 155-161, Basel.

- 1977, Pyroklastika im Interandinen Längstal von Ecuador. - Ein Beitrag zur Systematik der Pyroklastika, Dissertation, Tübingen.

Braun, A.F., German, R. & Mader, M. 1976, Der Beitrag der Sedimentanalyse zur Quartärstratigraphie, Bezirksstelle Natursch. Landschaftspfl. Tübingen, Mitt. Nr. 4, 28 p., Tübingen.

Brosse, P., Filzer, P. & German, R. 1965, Neues zur Geologie der Umgebung von Bad Wurzach (Württ. Oberschw.), N.Jb. Geol. Paläont., Mh. 1965, 255-275, Stuttgart.

Brunnacker, K. 1960, Zur Kenntnis des Spät- und Postglazials in Bayern, Geologica Bavarica 43, 74-150, München.

Campell, 1930-1935, Jährliche Mitteilungen des Kreisforstamtes 29 Oberengadin-Bergell über den Stand des Roseggletschers an die Schweizer Gletscherkommission, Celerina.

Cepek, A.G. 1967, Stand und Probleme der Quartärstratigraphie im Nordteil der DDR, Ber.deutsch. Ges. geol. Wiss., A, Geol. Paläont., 12, 3/4, 375-407, Berlin.

Charlesworth, J.K. 1957, The Quaternary Era, 2 vol., Arnold, London.

Eichler, H. 1970, Das Präwürmzeitliche Pleistozän zwischen Riss und oberer Rottum. Ein Beitrag zur Stratigraphie des nördlichen Rheingletschers. Heidelberger Geogr. Arb. 30, 128 p., Heidelberg.

Fezer, F. 1969, Tieferverwitterung circumalpiner Pleistozänschotter, Heidelberger Geogr. Arb. 24, Heidelberg.

Folk, R.L. & Ward, W.C. 1957, Brazos River Bar: A study in the significance of grain size parameters, J. Sediment.Petrol. 27, 3-26.

Fürbringer, W. 1976, Zur Sedimentologie eines arktischen Deltas (Corville Delta, Nordalaska), Geol. Rundschau 65, 577-614, Stuttgart.

Gasser, U. & Nabholz, W. 1969, Zur Sedimentologie der Sandfraktion im Pleistozän des schweizerischen Mittellandes, Eclogae Geol. Helv. 62, 67-516, Basel.

German, R. 1958, Zur Feinmorphologie letzteiszeitlicher Ablagerungen des Rheingletschers in Württemberg, Württ.Jh. 113, 78-90, Stuttgart.

- 1962a, Zur Geologie des Lechvorlandgletschers, Jber.Mitt.oberrhein.geol. Ver., N.F. 44, 61-83, Stuttgart.

- 1962b, Zur Deutung pleistozäner Sedimente mit rezenten Gletschergebieten (Grosser Aletschgletscher), Württ.J. 117, 122-141, Stuttgart.
- 1963, Der Ur-Federsee, Jber.Mitt.oberrhein.geol.Ver., N.F.45, 61-86, Stuttgart.
- 1964, Korngrössenuntersuchungen an glazigenen und glazifluvialen Sedimenten, N.Jb.Geol.Paläont., Mh. 1964/7, 388-390, Stuttgart.
- 1965a, Glazial oder interglazial? Gedanken zur zeitlichen Einstufung der Südostabdachung, Mitt. geogr.Ges.Wien 104, 1-19, Wien.
- 1965b, Neue Ergebnisse zur quartären Landschaftsgeschichte Oberschwabens, Württ.Jh. 120, 124-125, Stuttgart.
- 1968a, Halbtagsexkursion Biberach-Bad Buchau.-Beitr. zu den Exk. anlässlich der DEUQUA-Tagung August 1868 in Biberach an der Riss, Heidelberger geogr. Arb. 20, 9-28, Heidelberg.
- 1968b, Moraines. - Encyclopedia of Geomorphology, 710-717, Rheinhold Book Corporation New York.
- 1970a, Die Unterscheidung von Grundmoräne und Schmelzwassersedimenten am Beispiel des württembergischen Allgäus, N.Jb. Geol.Palöont., Mh.1970/2, 69-76, Stuttgart.
- 1970b, Rand und Vorland von Bernina-Gletschern und ihre Bedeutung für Oberschwaben, Württ.Jh.125, 76-87, Stuttgart.
- 1970c, Studienbuch Geologie, Eine Einführung unter besonderer Berücksichtigung der exogenen Dynamik, Ernst Klett Verlag, Stuttgart.
- 1971a, Die wichtigsten Sedimente am Randes des Eises. Ein aktuogeologischer Bericht von der Stirn des Kiagtut sermia bei Narssarssuag (Süd-Grönland). N.Jb. Geol. Paläont., Abh.138, 1-14, Stuttgart.
- 1971b, Sediment-Umlagerung am Rande des Inlandeises. Ein aktuogeologischer Bericht über die Bedeutung von Windarbeit und Toteis an der Stirn des Kiagtut sermia bei Nassarssuag (Süd-Grönland), Württ.Jh. 126, 125-136, Stuttgart.
- 1971c, Gibt es Grundmoränenlandschaft im Umkreis der Alpen? Ein Beispiel geomorphologischer Grundlagenforschung, Regio Basiliensis, H. 12/2, 362-376, Basel.
- 1972a, Die Sedimente am Rand des Oberen Theodul-Gletschers bei Zermatt, Württ.Jh. 127, 52-59, Stuttgart.
- 1972b, Die Bedeutung der Schmelzwasserarbeit in früher eisbedeckten Gebieten, Jber.Mitt.oberrhein.geol.Ver., N.F.54, 53-57, Stuttgart.
- 1973, Sedimente und Formen der glazialen Serie, Eiszeitalter und Gegenwart 23/24, 5-15, Oehringen/Württ.
- 1974a, Das mittelfristige Programm zum Schutz geologisch wichtiger Naturdenkmale in Baden-Württemberg. Veröff. Landesstelle Naturschutz u. Landschaftspfl. Bad.-Württ. 42, 85-92, Ludwigsburg.
- 1974b, Erdwissenschaftliche Beiträge zur Naturschutzarbeit in Vergangenheit und Zukunft, Naturschutz u. Naturparke, H.74, 58-63, Stuttgart.
- in printing, Sediment-Umlagerung am Tschierva-Gletscher im Rosegtal bei Pontresina (Graubünden, Schweiz).
- in printing, Veränderungen an der Stirn und im Vorland des Morteratsch-Gletschers (Graubünden, Schweiz).

German, R. & Filzer, P. 1964, Beiträge zur Kenntnis spät- und postglazialer Akkumulation im nördlichen Alpenvorland, Eiszeitalter und Gegenwart 15, 108-122, Oehringen/Württ.

German, R., Dehm, R., Ernst, W., Filzer, P., Käss, W., Müller, G. & Witt, W. 1965, Ergebnisse der wissenschaftlichen Kernbohrung Ur-Federsee 1,, Oberrhein.Geol. Abh. 14, 97-139, Karlsruhre.

German, R., Borneff, H., Brunnacker, K., Dehm, R., Filzer, P., Käss, W., Kunte, H., Müller, G. & Witt, W. 1967a, Ergebnisse der wissenschaftlichen Kernbohrung Ur-Federsee 2, Oberrhein.Geol.Abh. 16, 45-110, Karlsruhe.

German, R., Lohr, P., Wittman, D., & Brosse, P. 1967b, Die Höhenlage der Schichtengrenze Tertiär-Quartär im mittleren Oberschwaben, Eiszeitalter und Gegenwart 18, 104-109, Oehringen/Württ.

German, R. & Filzer, P. 1967c, The Upper Pleistocene Stratigraphy of Core Ur-Federsee 2, Southern Germany, Quaternary Paleoecology, 341-347, New Haven and London.

German, R., Filzer, P., Dehm, R., Freud, H., Jung, W. & Witt, W. 1968, Ergebnisse der wissenschaftlichen Kernbohrung Wurzacher Becken 1 (DFG), Württ.Jh. 123, 33-8, Stuttgart.

German, R. unter Mitwirkung von Braun, A.F., Hermann, R. & Mader, M. 1972, Quartäre Sedimente im Alpenvorland, Bezirkst. Naturschutz u. Landschaftspfl. Tübingen, Mitt.nr. 1, Tübingen.

German, R. & Mader, M. 1976, Die Aeussere Jungendmoräne bei Bad Waldsee und das Riedtal, Württ.Jh.131, 39-49, Stuttgart.

German, R., Hantke, R. & Mader, M. in printing, Der subrezente Drumlin im Zungenbecken des Biferten-Gletschers (Kanton Glarus, Schweiz), Eiszeitalter und Gegenwart, Oehringen/Württ.

Geyer, O. 1977, Grundzüge der Stratigraphie und Fazieskunde, Bd. 2, Paläontologische Grundlagen II, Paläographie, Fazieskunde, Schweizerbart, Stuttgart.

Göttlich, K.H. & Werner, J. 1967, Ein Pleistozänprofil im östlichen Rheingletschergebiet, N.Jb.Geol.Paläont., Mh. 1967/4, 202-216, Stuttgart.

- 1968, Ein vorletztinterglaziales Torfvorkommen bei Hauerz (Landkreis Wangen i.Allg.), Jh.Geol.L.A.Bad.-Württ. 10, 73-78, Freiburg i.Br.

Graul, H. 1952, Die Gliederung der mittelpleistozänen Ablagerungen in Oberschwaben, Eiszeitalter und Gegenwart 2, 133-146, Oehringen/Württ.

- 1953, Ueber die quartären Geröllfazien im deutschen Alpenvorland, Geologica Bavarica 19, 266-280, München.

- 1972, Eine Revision der pleistozänen Stratigraphie des schwäbischen Alpenvorlandes (Mit einem bodenkundlichen Beitrag von K. Brunnacker), Petermanns geogr. Mitt.106, 253-271.

Graul, H. & Groschopf, P. 1952, Geologische und morphologische Betrachtungen zum Iller-Schwemmkegel bei Ulm, Naturf. Ges. Augsburg, 5. Ber., Augsburg.

Gripp, K. 1929, Glaziologische und geologische Ergebnisse der Hamburgischen Spitzbergenexpedition 1927, Abh. naturwiss. Ver.Hamburg 22, 2.-4.H. 145-249, Hamburg.

- 1964, Erdgeschichte von Schleswig-Holstein, Karl Wachholtz, Neumünster.

Holzmann, H. 1972, Sediment-Untersuchungen an altpleistozänen Ablagerungen in der Umrandung des Wurzacher und Arnacher Beckens (Württembergisches Alpenvorland), Jber.Mitt.oberrhein.geol.Ver., N.F.54, 45-52, Stuttgart.

Inman, D.L. 1952, Measures for describing the size distribution of sediments, J.Sediment.Petrol. 22, 125-145, Tulsa/Okl.

Kaiser, K.H. 1963, Zur Frage der Würm-Gliederung durch einen "Mittelwürm-Boden" im nördlichen Alpenvorland bei Murnau, Eiszeitalter und Gegenwart 14, 208-215, Oehringen/Württ.

Kilger, B. 1979, Sediment-Untersuchungen am Rand des Roseg-Gletschers (Graubünden, Schweiz), Dissertation, Tübingen (in preparation).

Löscher, M. 1972, Der altpleistzäne Donaulauf in der Zusamplatte und seine Bedeutung für Schotterstratigraphie, Führer zu den Exkursionen anlässlich der DEUQUA-Tagung 1972 in Hohenheim, 42-44, Stuttgart-Hohenheim.

- 1974, Die präwürmzeitlichen Schotterablagerungen in der nördlichen Iller-Lech-Platte, Dissertation, Heidelberg.

Mader, M. 1970, Das Quartär zwischen Adelegg und Hochgelände, Diplomarbeit, Tübingen.

- 1971, Das Quartär zwischen Adelegg und Hochgelände. - Bildungsweise und Stratigraphie, Württ.Jh. 126, 177-205, Stuttgart.

- 1972, Stratigraphie und Bildungsablauf der quartären Ablagerungen im Raum Füramoos, Führer zu den Exkursionen anlässlich der DEUQUA-Tagung 1972 in Hohenheim, 93-96, Stuttgart-Hohenheim.

- 1973, Geologische Karte des nördlichen württ.Oberschwaben, in: Der Kreis Biberach, Konrad Theiss Verlag, Stuttgart und Aalen.

- 1976, Schichtenfolge und Geschehensablauf im Bereich des Schussenlobus des pleistozänen Rhein-Vorlandgletschers, Dissertation, Tübingen.

- 1976b, Die paläogeographisch-tektonische Entwicklung des süddeutschen Molassetroges, Manuskript.

- 1976c, Geologische Karte des südlichen württ. Oberschwaben, in: Der Kreis Ravensburg, Konrad Theiss Verlag, Stuttgart und Aalen.

McCammon, R.B. 1962, Efficiencies of particle measures for describing the mean size and sorting of the sediment particles, J.Geol.70, 443-465.

Moser, S. 1958, Studien zur Geomorphologie des zentralen Aargaus, Mitt.geogr.-ethnol. Ges.Basel 10, 1-98, Basel.

Penck, A. & Brückner, E. 1909, Die Alpen im Eiszeitalter, 3 Vol, Tauchnitz, Leipzig.

Richmond, G.M. 1953, Surficial Geology of the Georgiaville Quadrangle Rhode Island, US-geol. Survey, geol. Quadrangle maps of the US, Washington DC.

Richter, E. 1900, Die Gletscherkonferenz im August 1899, Petermanns geogr.Mitt. 46/4, 77-81, Gotha.

Schädel, K. 1952, Die Stratigraphie des Altdiluviums im Rheingletschergebiet, Jber.Mitt.oberrhein.geol.Ver., N.F. 36, 1-20, Stuttgart.

Schädel, K., Weidenbach, F. & Werner, J. 1961a, Exkursion in das Pleistozän von Oberschwaben, Führer zu den Exkursionen anlässlich der 82. Tagung des Oberrheinischen Geologischen Vereins in Ulm, Arb. Geol.Paläont.Inst.TH Stuttgart, N.F.30, 46-51, Stuttgart.

- 1961b, Exkursion in das Pleistozän von Oberschwaben. Bericht über die 82. Tagung des Oberrheinischen Geologischen Vereins vom 4. April bis 8. April 1961 in Ulm (Donau), Exkursion D, Jber.Mitt.oberrhein. geol.Ver., N.F. 43, XII-XIV, Stuttgart.

Schädel, K. & Werner, J. 1963, Neue Gesichtspunkte zur Stratigraphie des mittleren und älteren Pleistozäns im Rheingletschergebiet, Eiszeitalter und Gegenwart 14, 5-26, Oehringen/Württ.

Schaefer, I. 1950, Die diluviale Erosion und Akkumulation. Forsch. deutsch. Landeskde. 49, Landshut.

Scheuenpflug, L. 1972. Die Zusamplatte und Schweinsberg bei Dinkelscherben, Führer zu den Exkursionen anlässlich der DEUQUA-Tagung 1972 in Hohenheim, 35-36, Stuttgart-Hohenheim.

Schiftah, S. 1970, Quartärgeologische Untersuchungen auf Blatt Kisslegg Nr. 8225, Diplomarbeit, München.

Schlüchter, Chr. 1973, Geologische Untersuchungen im Quartär des Aaretals südlich von Bern (Stratigraphie, Paläontologie, Sedimentologie). - 307 Seiten, mit 42 Fig., Karten und Tabellen und 112 Fotos. - Dissertation, Philosophisch-Naturwiss. Fakultät, Universität Bern; Manuskript.

Schreiner, A. 1974, in: Gwinner, M.P. et al., Geologische Karte von Bad.-Württ., 1:25'000, Erläuterungen zu Blatt 7723 Munderkingen, Stuttgart.

Sinn, P. 1972, Zur Stratigraphie und Paläogeographie des Präwürm im mittleren und südlichen Illergletscher-Vorland, Heidelberger Geogr. Arb. 37, Heidelberg.

Töndury, G.A. 1954, Ursachen und Bekämpfungsmöglichkeiten der zunehmenden Hochwassergefahr im Engadin, Wasser- und Energiewirtschaft 12, 308-323.

Vossmer-Bäumer, H. 1976, Allgemeine Geologie. Ein Kompendium, Schweizerbart, Stuttgart.

Weidenbach, F. 1937, Bildungsweise und Stratigraphie der diluvialen Ablagerungen Oberschwabens, N.Jb. Geol.Paläont. 78, 66-108, Stuttgart.

Weinhold, H. 1973, Beiträge zur Kenntnis des Quartärs im württembergischen Allgäu zwischen östlichem Bodensee und Altdorfer Wald, Dissertation, Tübingen.

Woldstedt, P. 1954, Das Eiszeitalter, I. Vol., Die allgemeinen Erscheinungen des Eiszeitalters, Enke, Stuttgart.

Zöbelein, H.K. 1973, Ueber das Pleistzän um Zwiefaltendorf/Donau (Bad.-Württ.), Jh.Geol.L.A. Bad.-Württ. 15, 251-302, Freiburg i.Br.

B. General aspects and lowland environments

Composition and dispersal of debris by modern glaciers, Bylot Island, Canada

R. N. W. DILABIO & W. W. SHILTS
Geological Survey of Canada, Ottawa, Canada

1 INTRODUCTION

1.1 Location and geography

Bylot Island lies opposite the settlement of Pond Inlet, off the northeastern tip of Baffin Island, about 400 km southwest of Thule, Greenland and 430 km east of Resolute Bay (Fig. 1). The central, northwest-trending spine of the 180 km long by 100 km wide island is mountainous, with peaks averaging 1400 m a.s.l. up to a maximum altitude of about 2000 m a.s.l. in the Byam Martin Mountains. The highlands (Fig. 2) are aptly described by Jackson and Davidson (1975:1) as - "...mostly ice-covered, and the bedrock, much of which is deeply weathered, is exposed in a myriad of jagged nunataks, arêtes, côls, tors, and cirques which appear like islands in a white sea. Innumerable glaciers flow outward from the backbone of the Byam Martin Mountains; ...".

In 1977, the authors chose several glaciers on the southwest side of Bylot Island as sites to study the composition of debris in lateral moraines and debris entrained in ice. Bylot Island was selected for these studies specifically because of its bedrock geology — highly metamorphosed, crystalline, Precambrian-age terrane in the rugged, high-altitude accumulation areas is surrounded by a gently rolling apron of unmetamorphosed, late Proterozoic sediments and poorly consolidated, coal-bearing, Cretaceous-Tertiary sediments in the dispersal area (Jackson and Davidson, 1975). Glaciers chosen for sampling cross at least two of these lithologically distinct bedrock terranes.

1.2 Objectives and methods

The principal objectives of the study are: (1) to determine the proportions of "far-travelled" and "local" components in various size fractions of the debris; (2) to determine the nature of entrainment and transport processes; and (3) to determine how compositions of debris change according to vertical or areal position within or around a glacier. To achieve these objectives the lateral moraines of five contiguous glaciers (Fig. 1) from two major drainage basins have been sampled systematically, and four vertical profiles of samples have been collected from two glaciers, the largest of the five, located less than 15 km apart. (In 1978 two additional profiles were collected from the upper reaches of Aktineq glacier and profiles or samples of englacial debris were collected from glacier C-55 and from a glacier located wholly on crystalline terrane. Additional ice samples were also collected from several glaciers. None of these 1978 materials have been analyzed at this writing.)

The lateral moraines of the five glaciers (numbered or named with reference to map area 46201, Bylot Island, Inland Waters Branch, 1969) were sampled by collecting 2 kg samples of diamicton from the crest of the moraines at approximately 1.5 km intervals around each glacier.

At Aktineq and "Camp" glaciers (B-17 and B-7, Fig. 2) debris bands were sampled at sections that exposed debris (accessible without special climbing equipment) from near the base of the glacier to a height of several metres. The fragments of ice and debris were stored in 1000 ml polypropylene bottles where they remained during shipment and at room temperature

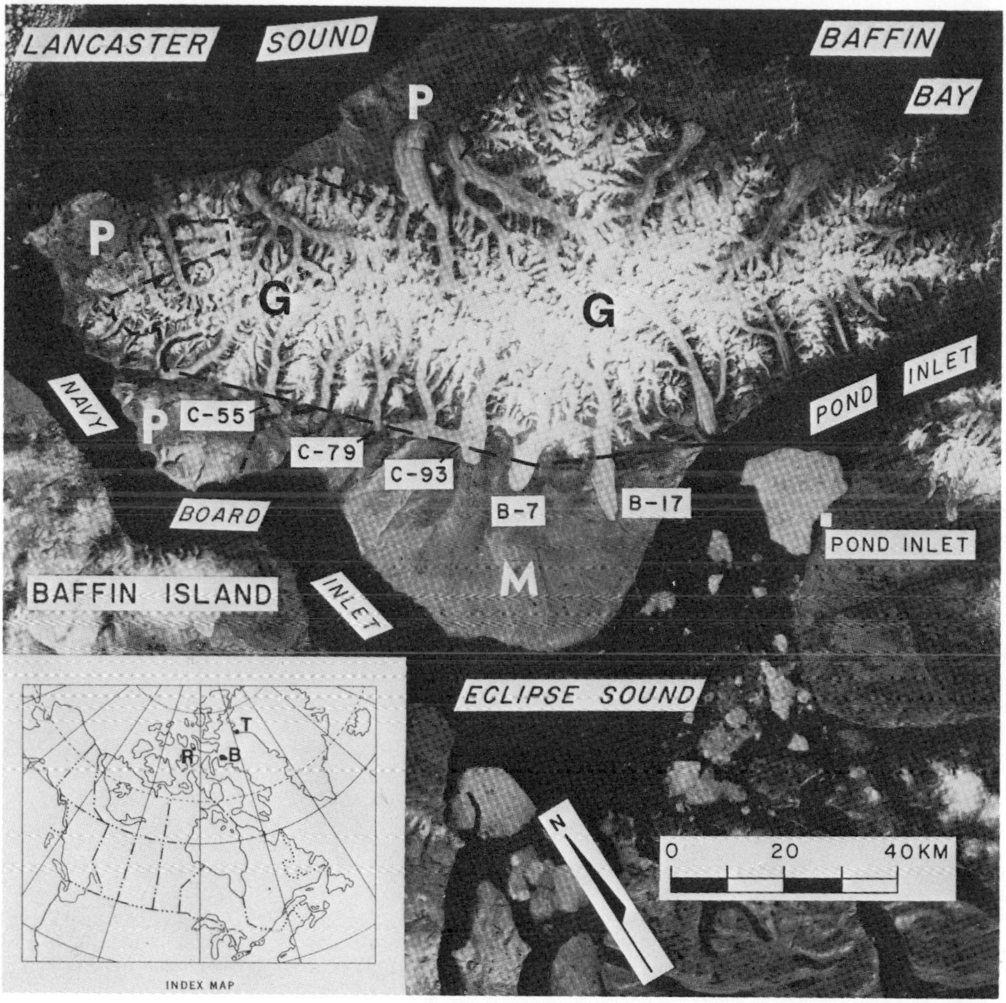

Figure 1. Satellite image of Bylot Island. P = Proterozoic sedimentary rocks; G = Precambrian gneisses and related rocks of high metamorphic grade; M = Mesozoic and Tertiary-age, poorly consolidated sandstones and coal measures. Glacier numbers are explained in text. R = Resolute Bay; B = Bylot Island; T = Thule, Greenland; Glaciers are numbered according to Glacier Atlas of Canada, Sheet 46201, Inland Waters Branch (1969). (LANDSAT image E-1748-16544, band 7, August 14, 1974)

for an average of two months in Ottawa. The supernatant water in each sample bottle was filtered and analyzed for trace and major elements by atomic absorption methods. It is thought to have had ample time to equilibrate with the sediment in the bottles.

The <4μm fraction was removed from all mineral sediment samples by centrifugation techniques. This fraction was analyzed by atomic absorption for several trace and minor elements after a hot $HCl-HNO_3$ bath leach.

Suspensions containing the <4μm fractions were also mounted on glass slides which were analyzed by X-ray diffraction first at room temperature and humidity and later after saturation with ethylene glycol. Because significant data were derived from these treatments, no further treatment of the slides was performed. Thus, the identifications of the 17 Å, 14 Å, 10 Å, and 7 Å peaks can be no more precise than montmorillonite group, montmorillonite-chlorite, micas, and chlorite-kaolinite, respectively. For the purpose of this paper, a more

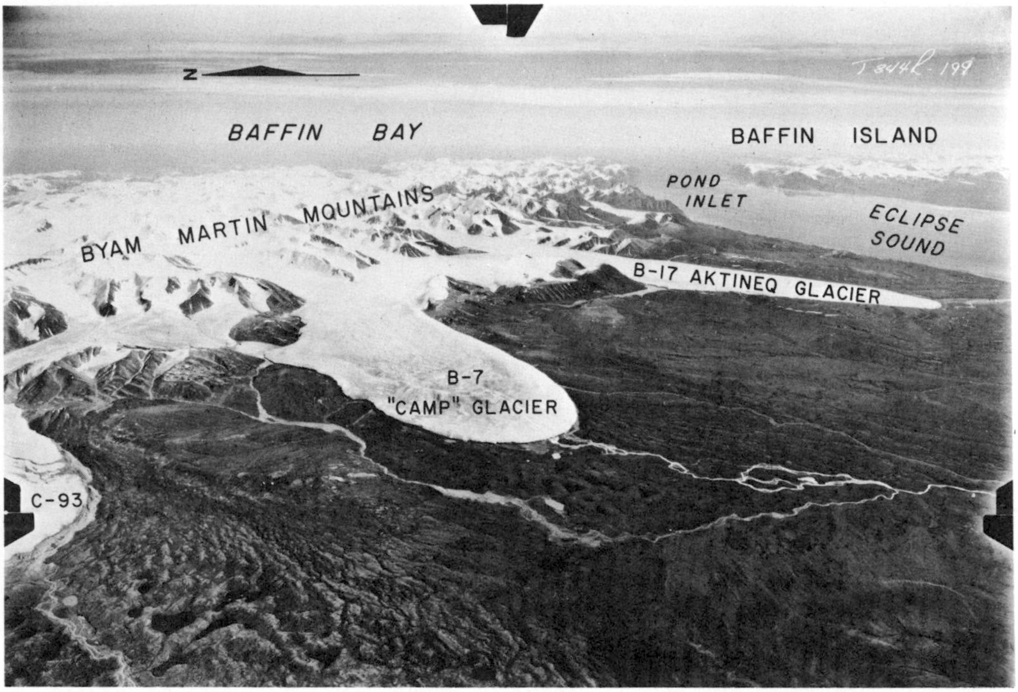

Figure 2. View southeast of Bylot Island. Contact of Gneissic and Mesozoic-Tertiary terrain marked by prominent escarpment in central part of photo. Hummocky moraine in foreground and in front of "Camp" Glacier probably greater than 8,000 years old. Byam Martin mountains include peaks over 2,000 m a.s.l. (EMR T344R-198).

sophisticated mineralogical breakdown is not required.

Heavy minerals (s.g. >3.3) were separated from the fine sand (64 to 250μm) fraction of all moraine samples. Weight percentages of heavy minerals (weight heavy minerals/ weight heavy plus light minerals times 100) were calculated.

1.3 Outline of glacial history

Bylot Island is located at or near the presumed northeastern edge of the Laurentide ice sheet. The complex temporal and spatial relationships between local highland ice bodies and the continental ice sheet are presently in the initial stages of study by R.A. Klassen. From the reconnaissance studies of Klassen, DiLabio and Shilts (1978), and Hodgson and Haselton (1974), it is apparent that at least the lower parts of the island were covered by continental ice during the Wisconsinan Stage. Several "old" dates (>30,000 yr. B.P.) on shells from undisturbed raised marine deltaic beds at Pond Inlet, only 25 km from Bylot Island (Hodgson and Haselton, 1974), suggest that the last episode of continental glaciation affected Bylot Island several tens of thousands of years ago, probably in the early to middle Wisconsinan. On terrain underlain by Cretaceous-Tertiary bedrock drift deposited by continental ice differs from drift deposited by glaciers originating on Bylot Island in its higher content of erratics with lithologies found in late Proterozoic and Paleozoic formations of northern Bylot Island and Baffin Island.

As noted by Hodgson and Haselton (1974:8) the glaciers of Bylot Island appear to have advanced recently and many have begun to retreat, leaving massive, ice-cored moraines covered by lichen-free bouldery detritus. The outermost lichen-free boulders of the modern moraine that surrounds Aktineq glacier lie on a section exposing wind-deposited sands bearing peaty layers as young as 450±70 ^{14}C years B.P. (GSC-2597). About 2 m lower in the same section, a peat in growth position lies on a till-like diamicton and associated fluvial(?) gravels; this peat is 7860±100 ^{14}C years old (GSC-2541),

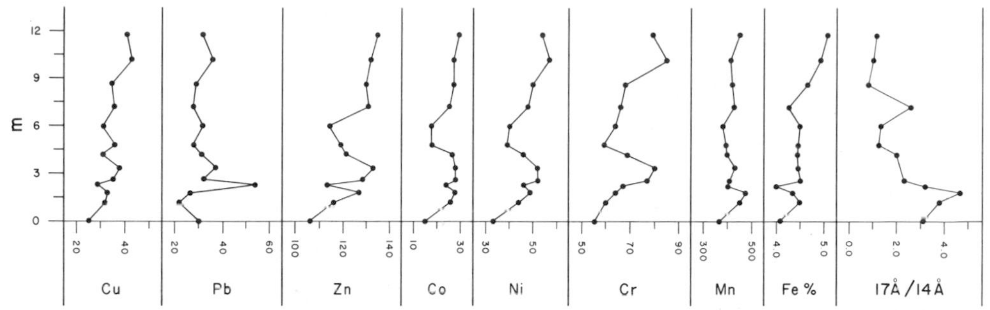

Figure 3. Trace elements in clay and relative amount of expansible clay in profile 2, Aktineq glacier. Concentrations in ppm, higher 17 Å/14 Å ratios equal higher amounts of expansible clay.

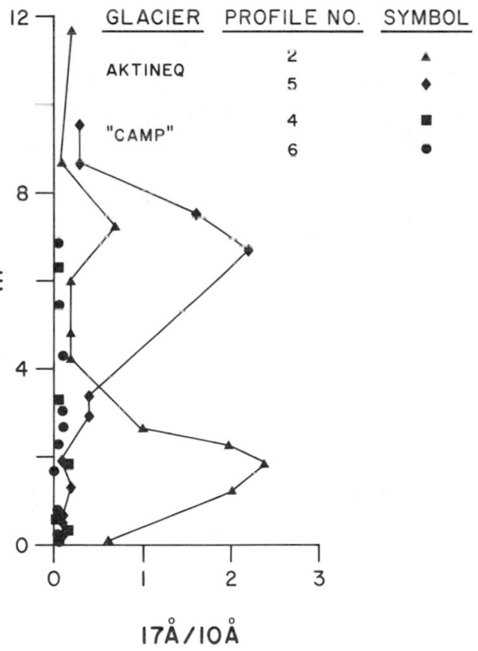

Figure 4. Relative amounts of expansible clay in four ice profiles studied. Higher 17 Å/10 Å ratios equivalent to higher concentrations of expansible clay.

suggesting that Aktineq glacier advanced to its present position within the last 400-500 years and is as far advanced as it has been in the past 7000 to 8000 years, or longer.

2 RESULTS

2.1 Clay mineralogy

2.1.1. Expansible minerals

The clay-sized detritus of the debris in transport in Aktineq and "Camp" glaciers is a mixture of expansible, montmorillonite group clay minerals and well crystallized 10 Å micas and chlorite-kaolinite. In general, Aktineq glacier seems to be richer in expansible minerals than "Camp" glacier. The probable source for the expansible minerals is the Cretaceous-Tertiary bedrock that underlies the sample sites at "Camp" and Aktineq glaciers. The well-crystallized chlorites and micas are typical for till derived from crystalline rocks of the Canadian Shield and are most likely derived by glacial comminution of phyllosilicates eroded from the crystalline Precambrian substrate in the upper reaches of "Camp" and Aktineq glaciers.

In profile 2 (Fig. 3) on Aktineq glacier, the apparent concentration of montmorillonite group minerals is greatest near the base of the profile (an undetermined distance above the base of the glacier) and decreases upward in a 12 metre vertical distance. In other profiles, however, the amount of montmorillonite group minerals is less obviously related to height of sample above the base of the glacier (Fig. 4). All but two debris samples from "Camp" and Aktineq glaciers contain some expansible clay, but the sites sampled on Aktineq are greatly enriched in expansible minerals, if the ratio of 17 Å to 10 Å peaks is considered to be a valid measure of relative concentrations (Fig. 4).

2.1.2 Green layers

In "Camp" glacier, several distinctly green debris bands were noted as high as 30 m above the base of the glacier.

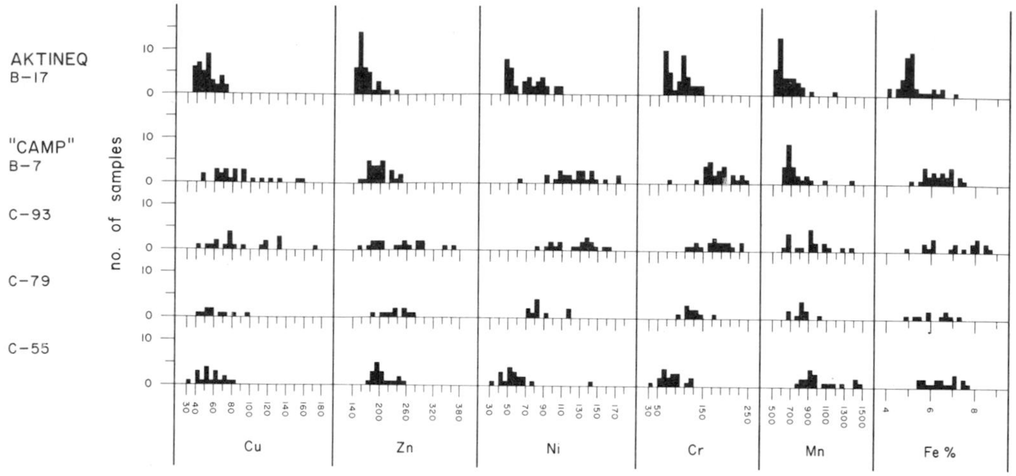

Figure 5. Histograms of trace element concentrations (ppm) in clay fractions of samples from lateral moraines of 5 glaciers indicated on Figure 1.

In the field, they seemed to be associated with pink gneissic erratics containing abundant epidote that apparently occurs as fracture fillings or as coatings on fracture surfaces. Numerous angular erratics of this type were lying on the glacier and on the lateral moraines, and pink and grey gneisses are common about 5 km up-ice from where the green bands were sampled. The clay fraction of the green bands is predominantly chlorite with little or no montmorillonite and relatively small amounts of 10 Å mica, which is the dominant component of the clay suite in all other samples. The chlorite may occur in the pink gneisses as a product of retrograde metamorphism associated with their epidotization.

More than 95% of the coarse clasts from the green bands are fragments of epidotized pink gneiss. The gneissic debris was either picked up from outcrops up glacier or comprises detritus derived from individual blocks that were transported on the surface of the glacier, dropped from the sides or snout, and ultimately picked up again at the base. The peculiar mineralogy and monolithologic nature of the debris indicates that a large block (or blocks) of gneiss was crushed and sheared into place with little or no dilution by superadjacent debris. The lack of other types of debris supports an origin within the outcrop area.

2.2 Trace element variations in clay fractions

2.2.1 Comparison among glaciers

Figure 5 indicates that the clay fractions of lateral moraines around "Camp" glacier and glacier C-93 are similar in their trace element content, but that they differ markedly from the other three glaciers which, in turn, differ among themselves. The moraines of "Camp" glacier and glacier C-93 are markedly enriched in Cu, Ni, and Cr, elements usually enriched in basic or ultrabasic intrusive or extrusive rocks or in their metamorphic equivalents. A volcanic-sedimentary bedrock unit that may include such rocks is mapped by Jackson and Davidson (1975) between these adjacent glaciers, about 10 km up-ice from their termini. Numerous sulphide-bearing gneissic boulders with significant amounts of Cu, Ni, and Cr were found on and adjacent to the moraines of "Camp" glacier (DiLabio and Shilts, 1978). Whether the boulders are derived from the volcanic-sedimentary unit or are related to unknown zones of sulphide mineralization is not known at present.

Because the Cretaceous-Tertiary bedrock is impoverished in trace elements (averages: Cu= 7 ppm, Pb=14 ppm, Zn=37 ppm, Co=3 ppm, Ni=11 ppm, Cr=47 ppm, Mn=96 ppm) it is assumed that most of these metals in the clays in the moraines are derived ultimately from the Precambrian terrane, either as cations leached from labile minerals and sorbed onto clay-sized phyllosilicates, as part of the silicate structures of clay-sized particles, or as very finely ground non-silicate minerals.

2.2.2 Variations within lateral moraines of individual glaciers

A complex set of factors affects the composition of till at a given location within a lateral moraine. These include (1) the amount of supraglacial debris falling onto the moraine, (2) the amount of proglacial sediments sheared upwards in debris bands within ice adjacent to the moraine, (3) the amount of meltwater erosion of local bedrock and contribution of water-eroded bedrock to outwash, which could then be incorporated into the glacier, (4) the degree of weathering or age of the moraine, (5) the nature and degree of reworking of morainic material (slump, collapse, washing, inundation, etc.), and, most importantly, (6) the chemical, mineralogical, and physical nature of the bedrock that supplies debris to the ice and the distance of a sample site from specific source rocks. In this study, it is the latter influence on the composition of lateral moraines or debris in transport that the authors are attempting to distinguish from the other, secondary factors.

Two types of element dispersal patterns are illustrated by Figure 6. Glacier C-93 is intersected by a tributary glacier, the lateral moraines of which outline effectively the part of the main lobe that it comprises. In this figure, it can be seen that the southeastern part of the glacier has lateral moraines with Zn, Ni, Fe, and Cr compositions that are significantly higher than concentrations within the snout and northwestern parts of the glacier. In effect, these trace elements are distributed in the moraine of this glacier in bimodal fashion, the higher metal contents being associated with the southeastern tributary glacier. That tributary drains bedrock terrain near "Camp" glacier, the debris of which has already been shown to be enriched in these metals.

2.2.3 Variations in long profile of trace elements, lateral moraines of Aktineq glacier

The moraines of Aktineq glacier also have a bimodal metal distribution in the clay sizes. In this case, the metal values are highest in the moraines within the Precambrian terrain and seem to decrease in a regular manner toward the snout of the glacier (Fig. 7). This appears to reflect dilution by increasing contents of metal-poor clays derived from the Cretaceous-Tertiary rocks on which the lower 12 km of the glacier lie. These examples serve well to show the advantage of studying composition on the contrasting bedrock terranes of Bylot Island. They also show how studies of modern glacier debris can point up the complexities liable to be detected in ancient glacial deposits, particularly in areas of high relief.

2.2.4 Variations in vertical profiles of Aktineq glacier

Although the profiles of trace-element variations in Aktineq glacier show discrete zones of enrichment or dilution that are inversely related to amounts of expansible clays, the range of variation in a given profile is within the range of variation of the adjacent lateral moraine. Thus, the average composition of in-ice debris is as characteristic of a glacier as is the average composition of its lateral moraines.

By comparing the chromium and zinc contents of clays from the profiles to the amount of montmorillonite group minerals within the clay fraction (as represented by 17 Å/14 Å ratios) (Figs. 3, 8) one can see that trace element content generally decreases with increasing amounts of montmorillonite. If the same relationship holds for samples from lateral moraines, the regular down-ice decrease in trace elements noted above (Sec. 2.2.3) can be similarly related to dilution by metal poor debris from bedrock in the dispersal area.

2.3 Mineralogy of sand-sized fractions

Compared to tills formed elsewhere on the Canadian Shield, the sand-sized debris of these glaciers is greatly enriched in heavy minerals, reflecting the large amounts of garnet and magnetite in the high-grade metamorphic rocks that form the Precambrian "core" of the island. It is rare, in the authors' experience, to encounter more than 5 weight per cent heavy minerals (s.g. >3.3) in the sand fractions of Canadian tills, but these tills range from 5.5% to 24.6% heavy minerals by weight. The fact that heavy mineral percentages are high reflects derivation from the Precambrian terrain. The observation that the percentages do not seem to decrease down-ice in a manner similar to the trace elements in clay, reflects the fact that much of the

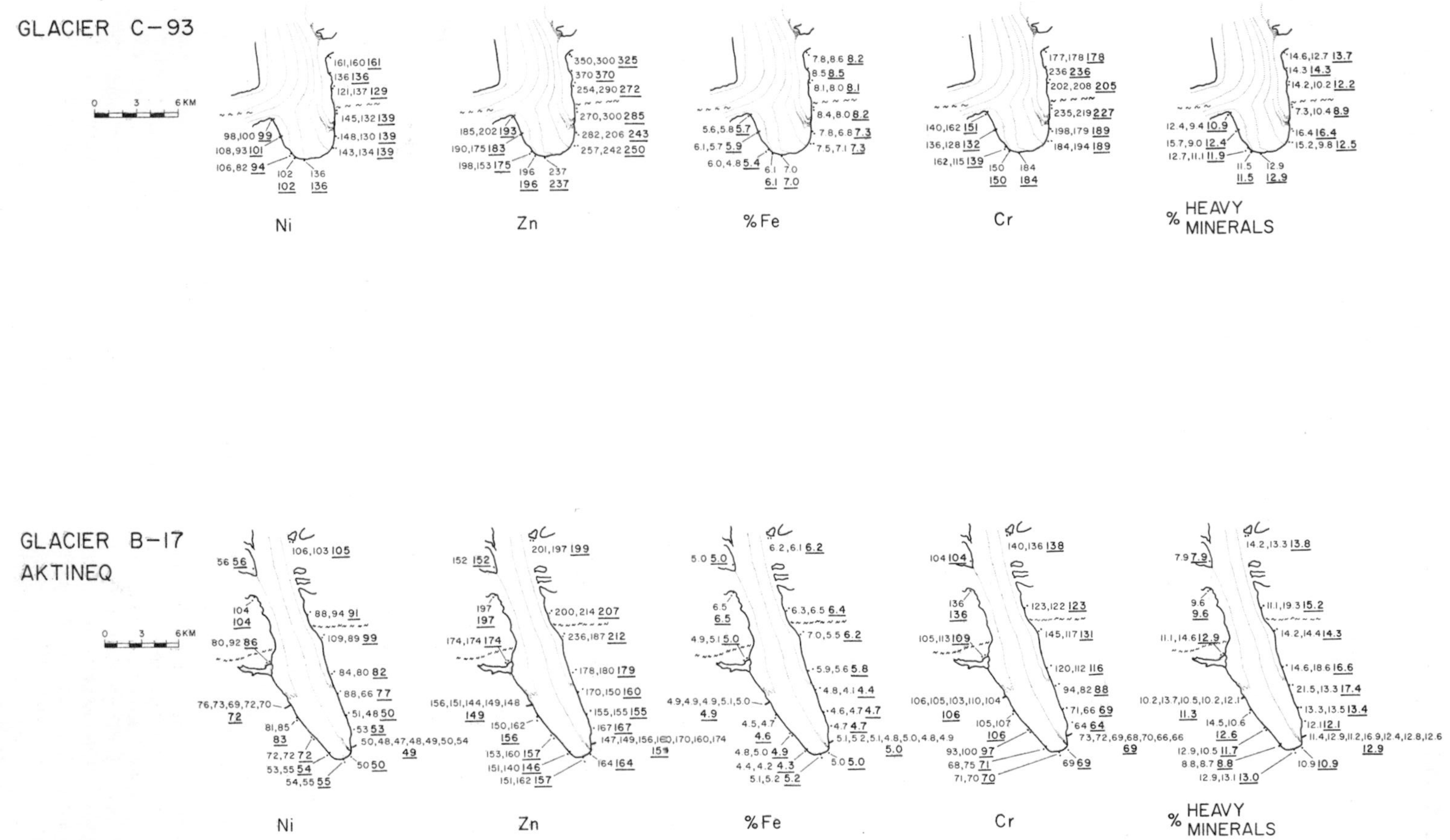

Figure 6. Comparison of trace element contents (ppm) of clay from lateral moraines on Aktineq glacier and glacier C-93. Average values at each sample site are underlined. Dashed line marks contact between Mesozoic-Tertiary and Precambrian crystalline rocks.

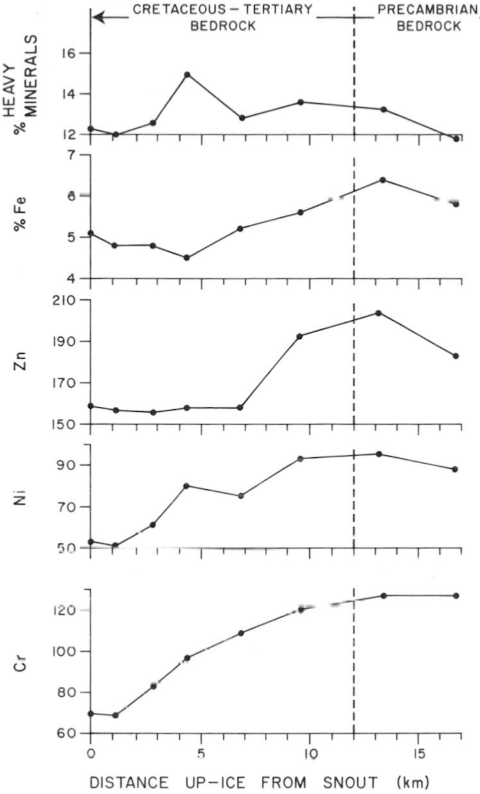

Figure 7. Long profiles of average trace element concentrations (ppm) in clays and percentage of heavy minerals (s.g. >3.3) from lateral moraines of Aktineq glacier.

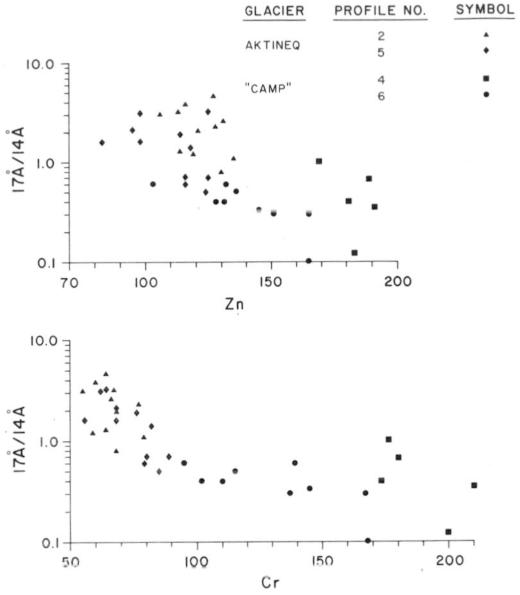

Figure 8. Relationship of Cr and Zn (Precambrian source) to amount of expansible clay (17 Å/14 Å ratio directly proportional to amount of expansible clay) derived from Mesozoic-Tertiary rocks.

Cretaceous-Tertiary sediment is very immature and contains large amounts (5-6 weight per cent) of angular heavy minerals. Thus, dilution by incorporation of sand from the younger "rocks" does not affect the concentration of heavy minerals in the sand sizes noticeably. It is interesting to note that the fluvial beds of the Tertiary-Cretaceous formations are nearly as immature as the modern glacial debris.

2.4 Rock types in coarse fractions

The pebble and boulder fractions of the debris in transport and in morainal accumulations are composed almost exclusively of Precambrian rock types. The lack of apparent influence of the Cretaceous-Tertiary substrate on the coarser fractions at Aktineq is at least partially due to the unconsolidated nature and fine grain size of the younger rocks — they simply do not have enough strength to survive as large fragments. The rock types in moraines and in the ice are mostly light-coloured gneisses and garnetiferous gneisses. Fragments of late Proterozoic sedimentary rocks are present but rare in modern glacial sediment of all glaciers except for C-55, which lies on Proterozoic rock.

2.5 Composition of clean and debris-rich ice

Although the data are limited, some comparison may be made of the major element content of clean glacier ice vs. melted ice allowed to equilibrate with its included debris. Water filtered from debris-rich ice is markedly enriched in dissolved major elements compared to water derived from clean ice in a nearby glacier. For example, Ca ranges from 4 to 35 ppm in melted debris-rich ice from Aktineq glacier whereas clean ice from an adjacent glacier contains 1/5 to 1/10 these amounts (pers. comm. L. Johnston and G. Holdsworth, Inland Waters Directorate). A sample of meltwater from a stream draining a glacier near Aktineq is enriched in Ca and other elements to the same order of magnitude

as water allowed to equilibrate with its included debris, suggesting that cations are readily exchanged between sediment and ice on melting.

3 DISCUSSION

3.1 Textural fractions

From data presented in the previous section and from field observations, it is clear that the debris of the glaciers studied comprises three textural fractions that are fundamentally different in composition. The compositional differences can be used to derive basic information about the nature of entrainment and distance of transport of detritus. The three fractions, to be discussed from coarsest to finest, will be called the (1) Coarse fraction, comprising poly-mineralic (rock) fragments in the coarse sand, granule, gravel, and boulder size ranges (>250μm); (2) Sand fraction, comprising monomineralic particles dominated by quartz-feldspar with significant amounts of heavy minerals in the sand, fine sand, and silt sizes (250μm-4μm); and (3) Clay fraction, comprising dominantly phyllosilicate minerals (micas and clays) in the very fine silt and clay sizes (<4μm). These are essentially the same textural/compositional classes long recognized in the study of till and are slightly modified from those recently discussed by Shilts (1971, 1975).

3.1.1 Coarse fraction

The coarse fractions of all lateral moraine and glacier samples consist almost entirely of crystalline rock from the Precambrian highlands. The rare exceptions to the highland source are occassional (<<1%) fragments of Proterozoic sediments from northwestern Bylot or from Baffin Island and occasional pieces of coal from the Cretaceous beds.

The coarse fragments vary from very angular to well rounded. The most angular fragments seem to have fallen onto the ice in the highlands and are scattered about on the surface of the ice tongues. They fall continuously onto the lateral moraines, often splitting further or splitting and scarring the other rocks.

The more rounded particles are mostly reworked proglacial outwash which is incorporated as individual particles or as cross-bedded "rafts" as the glacier advances. Many undeformed, boulder-sized "rafts" of outwash were observed near the snouts of the glaciers, suggesting that slabs of sediment, frozen onto the base of the glacier in its outer portions, are dragged up to the surface of the glacier along shear planes. One undeformed "raft" of cross-bedded gravel, covered with peat 120±120 years old (GSC-2529, uncorrected), was found over 30 m above the base of "Camp" glacier. This "raft" must have been lifted from the glacier bed within a kilometer of the site where it was found because such vegetated outwash does not occur much further than that up ice. Similar "rafts" within 10 m of the one that was dated were deformed and disrupted, being smeared out by shearing.

That there is almost no contribution to the coarse fraction by the Cretaceous and Tertiary beds is not surprising in view of the fact that these sediments are generally very poorly consolidated and are not likely to produce coarse rubble under any condition. Also, the younger beds lie beneath the sole of the glaciers in broad valleys so that no debris from them could fall onto the glacier.

3.1.2 Sand fraction

The sand fraction contains one component, the heavy mineral fraction, that can be linked to a source within the Precambrian highlands. The metamorphic grade of the highlands is such that garnet and magnetite are abundant in the Precambrian gneisses and granulitic rocks. The few samples of Cretaceous-Tertiary sandstones studied, however, also contain abundant heavy minerals of >3.3 specific gravity. Their heavy mineral suite is remarkably immature, consisting of a fairly wide compositional range of very angular heavy grains, reminiscent of the same size grains in the superadjacent glacial deposits.

Thus, in the case of Aktineq glacier, the unusually high concentration of heavy minerals in the Cretaceous-Tertiary rocks does not allow any firm conclusion to be drawn as to the ultimate source of the sand fraction. The fact that a significant proportion of the clay fraction comes from the younger rocks of the dispersal area (see below) does suggest, however, that the sand portion of the moraine and in-ice debris does contain a Cretaceous-Tertiary component.

3.1.3 Clay fraction

The clay fraction shows systematic variations in trace element chemistry and in mineralogy that lead to several important conclusions relating to entrainment and transport of debris. Data from scanning electron microscopy and X-Ray Diffraction indicated that the particles in the clay fraction from ice profile samples of "Camp" and Aktineq glaciers are almost all (>95%) phyllosilicates. The profiles of clay mineralogy from sites directly over montmorillonite-bearing bedrock on Aktineq glacier show significant amounts of montmorillonite group clays in debris bands, indicating that the material in the bands is at least partially derived from the sole of the glacier, regardless of the height of the bands above the base. Almost all debris bands in "Camp" glacier also contain montmorillonite, but they are much less enriched in this component than those in Aktineq glacier (Fig. 4). Although in one profile on Aktineq glacier the bands closest to the base are most enriched in the montmorillonite group minerals, indicating that they are most greatly diluted by the underlying montmorillonite-bearing Cretaceous bedrock, the vertical location of greatest dilution is more erratic in other profiles, sometimes occuring in the uppermost samples.

The trace element concentration of profile number 2 on Aktineq glacier shows a trend opposite to that of the clay minerals (Fig. 3). Because most of the trace element component is derived from the Precambrian highlands this reflects either increasing dilution by Cretaceous bedrock toward the base of the glacier or increasing amount of Precambrian components upward in the glacier, or both.

The long profile of trace elements in the clays of the lateral moraines of Aktineq glacier (Fig. 7) shows a regular decrease in element levels from the Precambrian-Cretaceous contact out onto the Cretaceous terrane. This is taken as an indication of down-ice dilution by debris derived basally from the Cretaceous rocks. In the case of Aktineq glacier, the clay fraction reflects this dilution much better than the sand fraction, as discussed above.

In general, the dilution of Precambrian debris is best shown by the negative correlation of montmorillonite and trace elements in the clay (Fig. 8). This relationship is presumed to hold for debris from the lateral moraines as well as for the debris collected directly from the ice. Finally, the comparison of trace element variations in the clays of the moraines of the five glaciers sampled shows that each one is influenced by a distinctive chemistry within its catchment area (Fig. 5). Around glacier C-93 (Fig. 6), in fact, one can clearly detect the lateral morainic debris of one of its major tributaries by the strong contrast in trace element chemistry with the rest of its moraines. The tributary, which drains an area near "Camp" glacier, shows much more chemical affinity for the moraines of "Camp" glacier than for adjacent morainic debris on the northwest side of its snout.

These latter observations have important practical implications for mineral prospecting in regions presently or formerly covered by Alpine-type glaciers. Sampling of lateral moraines might be used in the same way as stream sediment geochemistry is used in mountainous regions to detect potential ore bodies lying within the drainage basin.

4 CONCLUSIONS

Based on the preceding discussion, a number of preliminary conclusions can be proposed:

(1) The composition of coarse englacial or morainic debris cannot necessarily be used as an indicator of entrainment, transport, or depositional processes as has been attempted elsewhere (Boulton, 1970; Souchez, 1971). The contrasts in apparent sources among the coarse, sand-sized, and clay-sized fractions of debris in and around Aktineq glacier clearly demonstrate this (i.e., examination of only the coarse fraction would lead to the false conclusion that all debris was derived from the Precambrian highlands).

(2) Material in the debris bands in the Bylot Island glaciers, although including a high proportion of components ultimately derived from several kilometers up-ice, includes significant amounts of debris dragged up from the base of the glacier from sites a few hundred metres or less from the profiles studied. On Bylot Island, the most distinctive component of the "local" debris is montmorillonite group clay from the Cretaceous-Tertiary beds that underlie the lower portions of the glaciers.

(3) There are significant vertical variations in the physical and chemical nature of debris bands. Some variation is systematic with respect to height above the glacier base and some is

erratic, probably due to the presence of bands containing debris eroded from unusual lithologies at or near the glacier base some distance up ice from the sites of study. The "green bands" of chlorite-rich debris in "Camp" glacier are examples of the latter situation. These vertical variations may be related to similar ones noted in ancient tills deposited by continental glaciers (Shilts, 1978; Podolak and Shilts, 1978).

(4) The finest debris carried by Aktineq glacier is diluted in a regular way in the down-ice direction, a feature best shown by the trace element content of the clay fraction.

(5) The chemical composition of glacial debris can be used, under conditions on Bylot Island, to characterize the drainage basin of individual glaciers or even to distinguish among the morainic materials of tributary ice streams. Apart from the sedimentological and glaciological uses of such information, these contrasts may be useful in mineral exploration in Alpine areas.

(6) Detailed compositional studies of modern glaciers, located in areas where distinctive differences exist among bedrock types in their accumulation and dispersal areas, can shed much light on the nature of glacial sedimentation, erosion, and transportation processes.

5 ACKNOWLEDGMENTS

Logistical support in the field was provided by the Polar Continental Shelf Project. Field work was done in conjunction with projects carried out by H.R. Balkwill and R.A. Klassen, who provided valuable technical and logistical assistance. Field assistance was provided by M. Rupners and S. Lavender.

L.M. Johnston (Inland Waters Directorate) carried out major element analyses of ice and water and G. Holdsworth of the same branch provided data on ice and meltwater from samples collected by H. Steltner.

Dr. R.J. Fulton read the manuscript and suggested changes that greatly improved its clarity.

6 REFERENCES

Boulton, G.S. 1970, On the origin and transport of englacial debris in Svalbard glaciers, J. Glaciol. 9:213-228.

DiLabio, R.N.W. & W.W. Shilts 1978, Compositional variation of debris in glaciers, Bylot Island, District of Franklin. In Current Research, Part B, Geol. Surv. Can., Paper 78-1B:91-94.

Hodgson, D.A. & G.M. Haselton 1974, Reconnaissance glacial geology, north-eastern Baffin Island, Geol. Surv. Can., Paper 74-20:10p.

Inland Waters Branch 1969, Bylot Island, Glacier Inventory Area 46201, Glacier Atlas of Canada, Dept. of Energy, Mines and Resources, Canada.

Jackson, G.D. & A. Davidson 1975, Bylot Island map-area, District of Franklin, Geol. Surv. Can., Paper 74-29:12p.

Podolak, W.E. & W.W. Shilts 1978, Some physical and chemical properties of till derived from the Meguma Group, southeast Nova Scotia. In Current Research, Part A, Geol. Surv. Can., Paper 78-1A:459-464.

Shilts, W.W. 1971, Till studies and their application to regional drift prospecting, Can. Mining Jour. 92:45-50.

Shilts, W.W. 1975, Principles of geochemical exploration for sulphide deposits using shallow samples of glacial drift, Can. Inst. Mining Metall. 68(no. 757):73-80.

Shilts, W.W. 1978, Detailed sedimentological study of till sheets in a stratigraphic section, Samson River, Quebec. Geol. Surv. Can., Bull. 285:27p.

Souchez, R.A. 1971, Ice-cored moraines in southwestern Ellesmere Island, N.W.T., Canada, J. Glaciol. 10:245-254.

Glazigene Press-Schuppen, frontal und lateral

K. GRIPP
Lübeck, Bundesrepublik Deutschland

Zusammenfassung

Press-Schuppen entstehen

1. frontal
   a) im Eis
   b) in Stauchmoränen
   c) bei erneutem Eis-Vorstoss in niedergeschmolzenem, von Schmelzwasser-Sanden überdecktem Rest-Eis in Zungenbecken

2. lateral
Bei anschwellendem Eis werden Seiten-Moränen mit Eis-Kern und Hangschutt
   a) aufwärts gepresst
   b) über die Trogwand hinaus gepresst und als Zunge horizontal verschleppt
   c) die Trogwand mit anschliessendem, noch von Rest-Eis bedecktem Gelände örtlich zu einer Stauchmoräne gepresst. Es entstehen Press-Schuppen, die oben Eis und darunter Grundmoräne und Schmelzwasser-Ablagerungen enthalten.

Abstract

PRESSED SCALES IN ICE, IN PUSH-MORAINES, IN LATERAL MORAINES

If a down melted ice-tongue is reswelling the lateral parts (debris, ice-cored lateral-moraines) are increasingly pressed and pushed upward in form of scales.

Passing the uppermost of the lateral wall

a) this scales are laid down horizontally by overflowing upper glacier-ice

b) parts of the wall are sheared and carried off.

A push-moraine (compressed moraine) with dikes lowering to the ice side is explained as two fold pressed. The second push forward of the glacier compressed scales of ice covered by a thick layer of meltwater-deposits. The inside scale, pressed the highest and containing the most of ice, sank the deapest.

1 BEOBACHTUNGEN, DIE NICHT GEDEUTED WERDEN KONNTEN

Der Penck-Gletscher, im Van Keulen Fjord, in Spitzbergen, war 1927 steil und hoch am Widerlager einer Stauchmoräne aufgestiegen. Diese verschwand an der Ostseite der Stirn der Eiszunge. Hier ragte eine Eiswand steil auf. An deren Basis war im Eis eine liegende Falte zu erkennen (Fig. 1 und K. Gripp 1929 Taf. 18, Abb. 3).

An der Stirn des gleichen Gletschers fielen Schermoränen mit petrographisch verschiedenem Gestein auf, (K. Gripp 1929 Taf. 19 Abb. 2). Auch die Schermoränen an der Flanke des Eidem-Gletschers (ebenda Taf. 7 Abb. 1+2) konnten nicht hinreichend gedeutet werden. Polnische Forscher zeigten auf, dass in der Stauchmoräne des Penck-Gletschers Molluskenschalen führende Meeresabsätze enthalten sind. Damit wurde ersichtlich, dass jener Gletscher in einen zuvor vom Meere erfüllten Fjord vorgedrungen war. Am Widerlager der Stauchmoräne hatte das Eis eine Höhe erreicht, die es ermöglichte, dass randliches Eis auf die benachbarte Fels-Oberfläche vordrang. (K. Gripp 1929 Taf. 18 Abb. 1+2).

Ferner: Im Winter 1952/53 war am Steilufer der Ostsee bei Marienfelde unweit von Stohl, also in der Moränen-Gabel = Press-Zone zwischen den Eis-Zungen der Kieler und der Eckernförder Bucht eine zunächst nicht zu deutende, liegende Falte aufgeschlossen (Abb. 2).

Vergleichbare flache Störungen wurden 1957 an der Westküste der äusseren Lübecker Bucht bei Kraksdorf angetroffen (siehe unter 4.1). Zum Verständnis dieser Beobachtungen erschien eine vergleichende Untersuchung über die Rolle der Scherflächen im Gesamtgeschehen am Eisrand erwünscht.

2 DAS ABSCHEREN VON SCHUPPEN IN DER RANDZONE EINER EISZUNGE

Bewegungen im Eise finden überwiegend auf Scherflächen statt. Das Fliessen des Eises durch Schub geschieht auf Gleitflächen, die dem Untergrund parallel verlaufen = liegende Scherflächen.

Beim Ausweichen des Eises vor einem Widerlager aus Eis reissen steigende Scherflächen auf. Sie queren die flachliegenden älteren.

2.1 Scherflächen und Schuppen

Eine durch Pressung entstandene Schuppe wird zumeist von steigenden Scherflächen begrenzt. Dies ist auch an steiler Felswand der Fall. Hier laufen diese Scherflächen nahezu parallel zu den aufgerichteten "liegenden" Scherflächen.

Bei den Schuppen unterscheiden wir frontale, die an einem nur gering weichenden Widerlager in Stauchmoränen und im Eise selber entstehen. Laterale Press-Schuppen treten an stellen, [illegible] festen Widerlagor der Flanken einer Eiszunge auf. in solche kann, bei erneutem Anstieg der Eis-Oberfläche, Hangschutt, Seitenmoräne und deren Eiskern aufwärts gepresst werden.

2.2 Schermoränen

Im Eis selber entstandene Press-Schuppen sind nicht selten. Sie werden bislang nur erkannt, wenn basaler Moränen-Schutt von der Schuppe emporgetragen wurde. Solche Schermoränen bezeugen zumeist eine vom Grunde des Eises aufgestiegene Trennfläche zwischen ruhendem oder geringer bewegtem und aufwärts ausweichendem Eis.

2.3 Steigende Scherflächen in schwindendem Eis

Wenn eine noch fliessende Eiszunge flächenhaft niederschmilzt, kann eintreten, dass der Widerstand für ein Fliessen über lange Strecken, dem Untergrund parallel, mehr Druck erfordert als ein kürzerer Weg aufwärts. Hierbei wird verstärktes Schmelzen des gehobenen Eises diesen Weg erleichtern. An der Grenze von ruhendem und fliessendem Eis kann Basis-Schutt bis auf die Oberfläche des Eises gelangen und als Obermoräne auf der Oberfläche des ruhenden Eises verschwemmt werden. Je nach Dauer entsprechender Gegebenheiten kann sich der Vorgang weiter oberhalb wiederholen. Dies dürfte allerdings in grösseren räumlichen Abständen geschehen, nicht so häufig wie Sigurd Hansen (1932) annahm.

Fig. 1 Am Ostrand des Penck-Gletschers (Spitzbergen) ist eine Eis-Schuppe seitlich über die Trogwand hinausgepresst und bleibt als liegende "Falte" die Basis des zur Seite abfliessenden Eises (nach Gripp 1929, Tafel 18, Abb. 3).

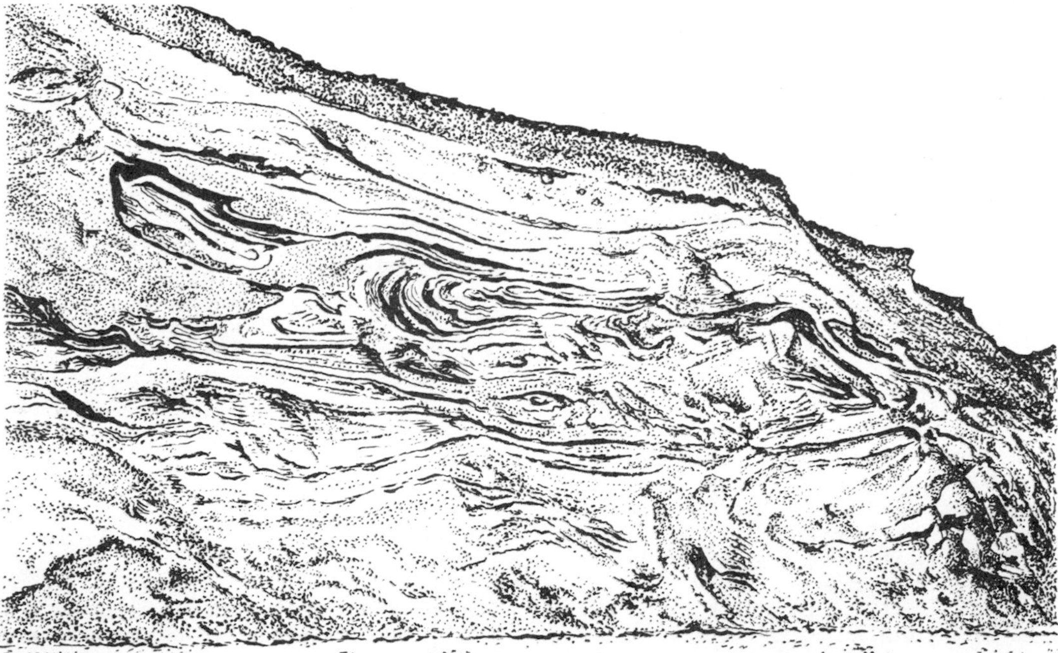

Fig. 2 Steilküste Südseite Ausgang Eckernförder Bucht bei Marienfelde - Stohl. - Liegende "Falten" bezeugen, dass am Rande der Gabel zwischen Kieler und Eckenförder Eiszunge Press-Schuppen von überlaufendem Eis verschleppt wurden (nach Foto von A. Möller).

Fig. 3 Die jüngste Press-Schuppe in der Stauchmoräne des Grönfjord-Gletschers Is-Fjord, Spitzbergen. - Das Eis des Grönfjord-Gletschers war hoch aufgestiegen. Der Druck des Eises hatte die Stauch-Schuppen weiter zusammen und höher hinauf gepresst. Dadurch war aussen an der Grenze von Stauchmoräne und Sander eine neue Schuppe aufgestiegen (nach Foto in Gripp 1929, Tafel 4, Abb. 4).

2.4 Steigende Scherflächen bei zunehmendem Eisfluss

Gegen stagnierendes Eis erneut vorrückendes Eis dürfte gleichfalls bald den Weg aufwärts nehmen. Dies dürfte in Abständen eintreten, falls das ruhende Resteis von grösserer Ausdehnung ist. Es entstehen Schuppen, die bald in den Eisfluss einbezogen werden.

3 DAS GESCHEHEN AN DER STIRN DER EISZUNGE

Je länger der Weg, umso schmaler und steiler gestellt dürften die Schuppen werden. Soweit wie deren Basis-Schutt nicht vorher als Oberflächen-Moräne verschwemmt wird, wird er schliesslich der Grundmoräne auf dem Innenhang der Stauchmoräne angelagert werden.

Falls das Eis am Widerlager der Stauchmoräne aufwärts fliessend hoch genug ansteigt, werden

a) die Schuppen der Stauchmoräne unter Verschmälerung höher hinaufgepresst

b) sobald wie der Schuppen-Komplex der Stauchmoräne auf die Basis am Aussenrand verstärkt drückt, tritt hier eine zunächst kleine weitere Stauch-Schuppe hervor, auch aus gefrorenem fein- bis mittelkörnigem Sediment (Fig. 3). Diese wird von zwei sich spitzwinklig treffenden Scherflächen begrenzt.

Allgemein besteht eine Stauch-Endmoräne aus einer Reihe von Press-Schuppen, die von aussen nach innen an Höhe zunehmen. Bei dem erheblich niedergeschmolzenen Grönfjord-Gletscher (früher Greenbay-Gletscher genannt) in Spitzbergen (K. Gripp und E. Todtmann 1926) war eine an Geschieben reiche Satzmoräne der aus sandigen Meeres-Absätzen bestehenden Stauchmoräne innen hoch hinauf angelagert. Hier wie bei den an Press-Schuppen reichen Stauch-Endmoränen von Penck-, Holmström-und Usher-Gletscher auf Spitzbergen liegen die höchsten Teile nahe am Innenrand.

Jedoch bei der Duvenstedter Stauch-Endmoräne, nördlich von Rendsburg am Nord-Ostsee-Kanal gelegen, liegt die höchste Zone aussen und der Innen-Hang fällt mit zahlreichen Stauch-Rücken allmählich bis an den Rand des von Wasser erfüllten Zungenbeckens, Wittensee genannt, ab. Dieser ist heute noch, trotz alluvialer

Sedimentation, bis 17 m tief.

Es erscheint ausgeschlossen, dass die zum Zungenbecken hin an Höhe abnehmenden Press-Schuppen in der heutigen Höhenlage entstanden sind. Sie müssen - auch wenn diese Stauchmoräne ersichtlich durch mehrfache Eisvorstösse zusammengepresst ist - auf der Eis-Seite am höchsten gewesen sein.

Um die heutige Gestalt zu erlangen, muss ein Teil des ehemals in der Stauchmoräne vorhanden gewesenen Materials geschwunden sein. Anders ausgedrückt: Die Stauchmoräne bestand zum Teil aus Eis. Jede dieser Press-Schuppen muss aus zwei Teilen bestanden haben, und zwar einem sandig-kiesigen oben, der von erheblicher Mächtigkeit war, und einem aus Eis bestehenden Teil unten. Die Mächtigkeit des Eises war gegen aussen am geringsten, nahe dem Zungenbecken am grössten. In dem Masse wie das verschieden mächtige Eis der Press-Schuppen schwand, sank der sedimentäre obere Teil der einzelnen Schuppen schliesslich in die heute noch vorhandene Lage.

Es erhebt sich die Frage nach der Herkunft des später gepressten Eises und ob einmalige oder mehrfache Stauchung vorliegt.

Bei erstmaligem Vorrücken einer Eiszunge wird zunächst, wie unter 3 dargelegt, gefrorener Lockerboden vor dem Eise gefaltet oder aufgerollt, (R. Köster 1957, K. Gripp 1975). Bei längerer Andauer des Vorganges werden diese zu Schuppen, den neugebildeten gleichend. Bei geradem Verlauf der Eisstirn bilden diese Schuppen lange gerade Rücken. In der Biegung zur Flanke der Eiszunge sind die Press-Schuppen kurz. Wenn von solchen mehrere Reihen vorhanden sind, stehen die Schuppen häufig "auf Lücke", also in Quincunx-Stellung zueinander. An die Stauchmoräne legt sich innen Gletschereis an, das an ihr aufwärts fliesst und in den tieferen Lagen mit Resten von Eis-Schuppen durchsetzt ist, die basalen Schutt führen.

Eine eiswärts an Höhe abnehmende Serie von aus Sediment bestehenden Schuppen kann bei dem vorstehend beschriebenen Vorgang nicht entstehen.

Eine Schuppen-Serie wie in der Duvenstedter Stauchmoräne setzt Eis mit einer ansehnlichen Auflage von Sand und Schottern voraus. Eine solche konnte nur als Sander im Zungenbecken eiswärts eines Stauchmoränen-Zuges entstehen.

Das Geschehen jünger als diese erste Stauchmoräne umfasst

1. Aussetzen des Eis-Zuflusses
2. Niederschmelzen des Eises im Zungenbecken
3. Bedeckung des Rest-Eises im Zungenbecken durch einen Sander, dessen Wässer am Innen-Fuss der Stauchmoräne seitlich abliefen. Diese Sediment-Masse war von erheblicher Mächtigkeit, dies nicht nur durch Niederschmelzen der Eis-Oberfläche sondern auch durch zunehmend höhere Aufschüttung
4. Erneutes kräftiges Vorrücken der Eis-Front
5. Dadurch Aufschuppung der Füllung des Zungenbeckens, die aus Eis unten und Sander-Ablagerungen oben bestand. Dies geschah bis zu einer die ältere Stauchmoräne an Höhe übertreffenden, mit dieser eine Einheit bildenden Stauchmoräne
6. Nach Beendigung des Eis-Vorstosses erfolgte lansames Schwinden des Eises unten im jüngeren Teil der Stauchmoräne. Dabei sackten die Schuppen entsprechend der Höhe ihres Anteils an Eis unten in die Tiefe. Das Zungenbecken wurde zum See.

Die Formen der Duvenstedter-Stauchmoräne bezeugen:

Mehrfache Stauchung durch wiederholtes Vorrücken des Eises,

Stauchung der von Sander-Ablagerungen überdeckten Eisfüllung des Zungenbeckens.

4 PRESSUNG DES EISES AN DER FLANKE EINER EISZUNGE

Bei abnehmender Mächtigkeit des Eises nimmt der Druck auf die Flanke einer Eiszunge ab. Allein die geringe Fliessgeschwindigkeit in den steil stehenden randlichen = tiefsten Eislagen bestimmt das Geschehen. - Neben und auf diesen randlichen Eislagen können Schutt der Seitenmoränen, Hangschutt, lagern und unter der Schuttdecke Rest-Eis (Eis-Kerne) erhalten bleiben.

Steigt die Oberfläche des Eises infolge verstärkten Zuflusses auf-

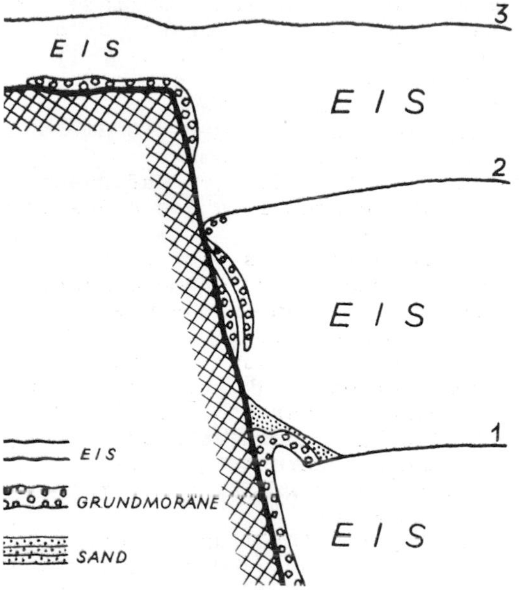

Fig. 4 Querschnitt durch die Wand eines Gletschertroges: (a + b) bei verstärktem Eiszufluss werden mit Steigen der Eisoberfläche Schuppen aufwärts gepress, die Hangschutt und die Seitenmoräne mit ihrem Eis-Kern enthalten. (c) wenn solche Schuppen die Oberkante der Trogwand erreichen, werden sie von schneller fliessenden oberen Eislagen rechtwinklig abgebogen und verschleppt. (d) die Oberkante der Trogwand dürfte nicht selten abgepresst und gleichfalls verschleppt werden.

wärts, so werden die randlichen Schuttmassen mitsamt dem Rest-Eis in die steil aufsteigenden basalen Eislagen aufgenommen. Hierbei werden Unregelmässigkeiten im losen Material und die Form des Widerlagers zu Abscherungen und Bildung von Schuppen führen. Diese weichen vor zunehmendem Druck unter Verschmälerung nach oben aus und zwar bei gleichzeitiger langsamer Fliessbewegung beckenauswärts.

Dies dürfte sich nicht nur an einer Wand am Fels sondern auch in Eis-Betten abgespielt haben, die in gefrorenem Lockerboden ausgefurcht waren.

4.1 Schuppen aus dem Gletscherbett zur Seite herausgepresst und verschleppt

Wo eine Eiszunge höher als die flankierende Wand anschwillt, breitet sich das obere, schneller fliessende schuttfreie Eis über das benachbarte Gelände aus. Von diesem Eis werden die an der Flanke hochgepressten Schuppen erfasst und an der Oberkante der Talwand im rechten Winkel abgebogen und unter spitzem Winkel zur Trogwand mehr oder weniger weit verfrachtet. Bei diesem Vorgang dürften oberste Teile der Trogwand bisweilen abgeschert und verschleppt werden (siehe Fig. 4 und 5).

Derartige abgebogene Press-Schuppen an der Basis dieser sekundären flankierenden Eiszungen liegen unter Eis, das überwiegend den obersten Lagen der Eiszunge angehört.

Als Belege für derartiges seitliches Ueberlaufen einer Eiszunge werden angesehen:

1. faltenähnlicher Verlauf der Eis-Bänderung an der Basis des ausgelaufenen Eises auf der Ost-Seite des Penck-Gletschers (Fig. 1),
2. der waagrechte Schichten-Komplex mit runden Stirnen, der im Steilufer der Ostsee zu Marienfelde bei Stohl, also in dem Winkel zwischen Kieler- und Eckernförder Bucht, 1952 aufgeschlossen war (Fig. 2),
3. topographisch und durch Profile werden abgebogene Press-Schuppen belegt im Gebiet zwischen dem Kieler und dem Lübecker Eislobus. Das Gelände liegt auf +10 m NN, also 30 m höher als der Boden jener Eiszungen. Im Westen und Norden traten an dem Trogrand laterale Pressungen auf, wodurch die Wandelwitzer- und Heiligenhafener-Stauchmoränen entstanden. Auf der Ostseite jenes interlobaten Gebietes, bei Kraksdorf, war der obere Teil der Trogwand im Jahre 1957 durch winterliche Brandung gut aufgeschlossen. Fig. 6 lässt rechts und links liegende "Falten" und dazwischen hochgepresste Grundmoräne erkennen. Vermutlich liegt der Grenzbezirk zweier in die Bildtiefe hineingeschleppten Schuppen vor.

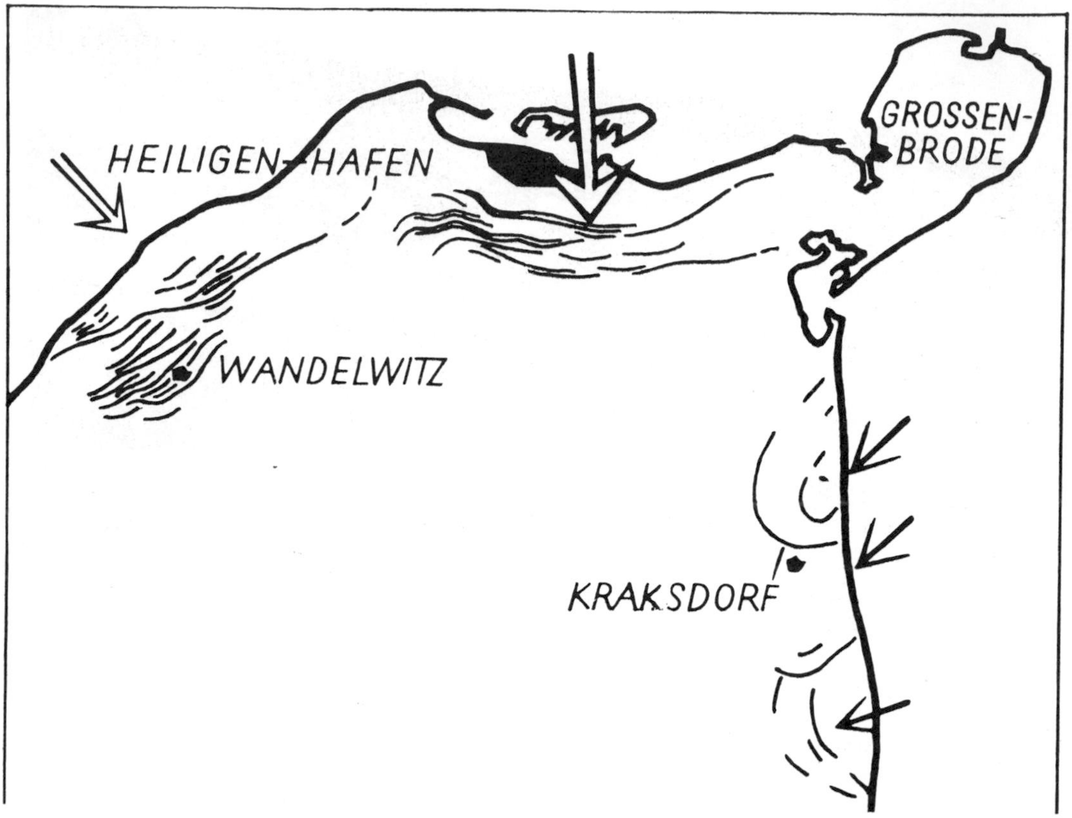

Fig. 5 Das Land zwischen Lübecker und Kieler Bucht war noch weitgehend von stagnierendem Eis bedeckt, als die Trogwand örtlich zur Wandelwitzer- und Heiligenhafener-Stauchmoräne aufgepresst wurde. - Bei Kraksdorf an der Lübecker Eiszunge wurden Press-Schuppen unter Eis, das über die Trogwand erneut hinaufstieg, horizontal verfrachtet (vgl. Fig. 6, 7). Neu gedeutet in Anlehnung an Seifert 1954.

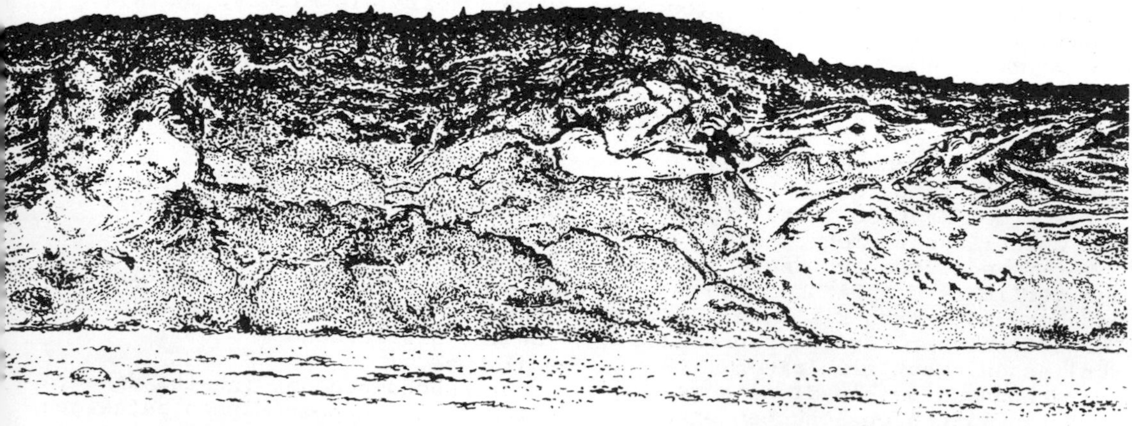

Fig. 6 Steilufer an der Lübecker Bucht bei Kraksdorf (Ost - Holstein). - Die Flanken zweier verschleppter Schuppen mit gewölbten Rändern liegen einander in Grundmoräne gebettet gegenüber (Gripp fot. 1957).

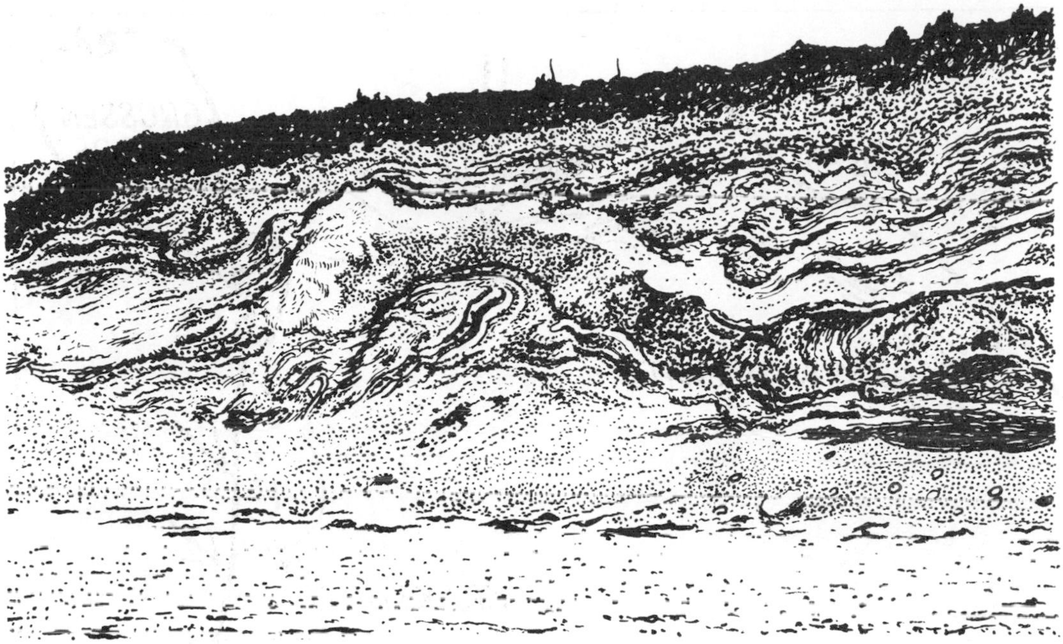

Fig. 7 Steilufer an der Lübecker Bucht bei Kraksdorf (Ost - Holstein). - Rechts: heller Sand überlagert eine Schicht von Grundmoräne. Von links erfolgter Druck durch den Rand einer abgelagerten Prss-Schuppe hat eine helle Sandschicht zur Seite hochgepresst und zu über 12 aufeinanderliegenden Falten eingeengt (Gripp fot. 1957).

Figur 7 lässt rechts die Schichtenfolge
- obere Grundmoräne
- helle Sande
- untere Grundmoräne
- ältere kiesige Sande

erkennen. Diese an sich unruhig gelagerte Schichtenfolge ist von links (Süden) durch die Flanke einer liegenden Schuppe hochgepresst, wobei das helle Sandband randlich 12-fach gefaltet wurde. Hier dürfte ein Schnitt durch die Flanke einer verschleppten Schuppe vorliegen. Dies hat schon G. Seifert (1954, Taf. 1) bei Kraksdorf und Siggen aus den Oberflächenformen erkannt.

Vorstehende Beobachtungen lassen auf örtliches Ueberlaufen von basalen und oberen Eislagen als Folge örtlicher lateraler Pressung schliessen. Nach bisherigen Beobachtungen weisen diese abgelagerten lateralen Schuppen eine ihnen eigene Tektonik auf.

Ferner hat W. Prange 1975 an Steilufern der westlichen Ostsee, also an den durch Meeresabtrag rückverlagerten Trogwänden der Eiszungen, Schichten-Folge und Schichten-Lagerung über Jahre hin festgehalten. Zwischen Grundmoräne oben und unten liegen Schmelzwasser-Sande und Becken-Absätze. Diese werden teils als Falten, teils als Schuppen angesprochen. Deren Achsen liegen horizontal. Die Vergenz der Falten ist vom Trog weg gerichtet. Für Stohl wird angegeben, dass der Eis-Schub aus der Eckernförder Bucht gekommen sei.

Diese Angaben lassen sich nicht auf frontale Pressung beziehen. Hingegen stimmen sie weitgehend überein mit dem Auftreten lokaler lateraler Schuppen, die eine anschwellende Eiszunge unter seitlich von ihr abfliessendem Eis mitschleppte.

Untersuchungen des Inhalts dieser Falten und Schuppen müssen aufzeigen, wie weit sie anderer Herkunft als aus Seitenmoränen sind.

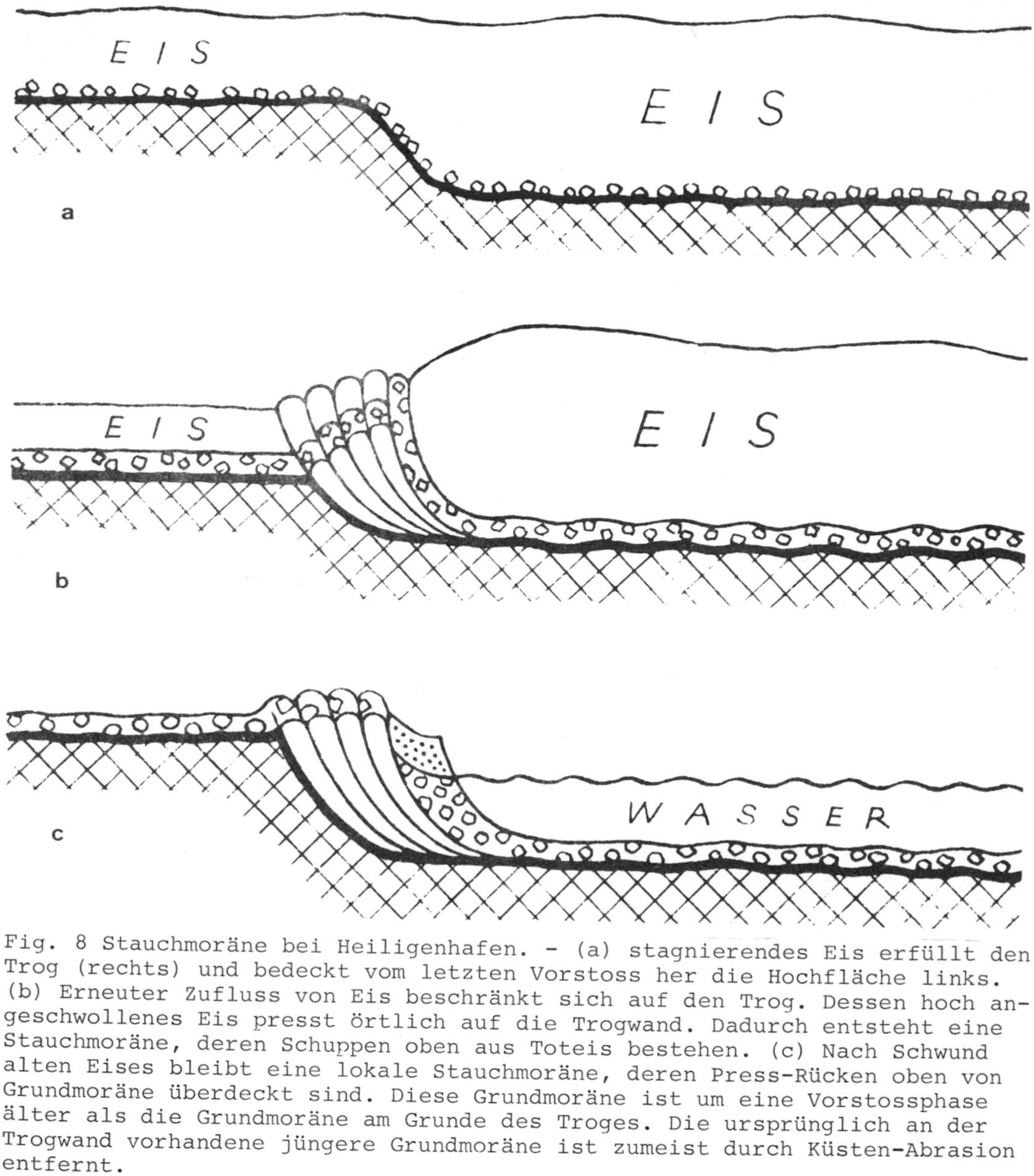

Fig. 8 Stauchmoräne bei Heiligenhafen. - (a) stagnierendes Eis erfüllt den Trog (rechts) und bedeckt vom letzten Vorstoss her die Hochfläche links. (b) Erneuter Zufluss von Eis beschränkt sich auf den Trog. Dessen hoch angeschwollenes Eis presst örtlich auf die Trogwand. Dadurch entsteht eine Stauchmoräne, deren Schuppen oben aus Toteis bestehen. (c) Nach Schwund alten Eises bleibt eine lokale Stauchmoräne, deren Press-Rücken oben von Grundmoräne überdeckt sind. Diese Grundmoräne ist um eine Vorstossphase älter als die Grundmoräne am Grunde des Troges. Die ursprünglich an der Trogwand vorhandene jüngere Grundmoräne ist zumeist durch Küsten-Abrasion entfernt.

5 LATERALE PRESS-SCHUPPEN, STAGNIERENDES EIS ENTHALTEND

Bei durch Schub fliessendem Eis ist nicht nur langfristig, sondern auch kurzfristig und lokal ein Wechsel in der hinzufliessenden Menge eingetreten. Die Trogwände waren daher wechselndem Druck ausgesetzt.

In dem Winkel zwischen Kieler und Lübecker Eiszunge des westbaltischen Eises ist lokaler Druck auf die Trogwand durch die Wandelwitzer- und Heiligenhafener-Stauchmoränen belegt (Fig. 5, nach G. Seifert 1954, Taf. 1). Die erstgenannte erreicht über 60 m an Höhe über NN, die Heilighafener 50 m. Da das stauchende Eis in Höchstlage noch höher aufgesteigen war und das benachbarte Meer bis 20 m tief ist, können wir für das Höchststadium

mit zeitweiser Mächtigkeit des Eises von reichlich 100 m rechnen. Dieses Eis hat örtlich beschränkt auf die Trogwand gedrückt und in Schuppen aufwärts gepresst. H.J. Stephan 1971 hat für die Heilighafener Stauchmoräne aufgezeigt, dass Eis oberhalb der Trogwand, zu dieser parallel in O-W Richtung geflossen ist. Er besteht darauf, dass Eis die Stauchmoräne in Längsrichtung überfahren habe. Aus Gründen, die in Gripp 1975 (S. 51/52) dargelegt sind und weil der 10 km nordöstlich gelegene, 2,5 km lange Wulfener Berg zeigt, wie ein echter Drumlin aussieht, vermag ich der Deutung als drumlinisierte Stauchmoräne nicht zuzustimmen (Fig. 8). Zutreffender erscheint für Heiligenhafen die Auffassung: die Trogwand einer von restlichem stagnierendem Eis bedeckten Oberfläche sei örtlich lateral gepresst. Es entstanden Stauch-Schuppen, wie folgt aufgebaut

- stagnierendes Eis
- Grundmoräne
- ältere Ablagerungen.

Es dürften Stauch-Schuppen vorliegen, die oben Eis enthielten.

6 ZUSAMMENFASSUNG

Wenn ein niedergetauter Gletscher erneut anschwillt, werden dessen laterale Partien an der Wand des Troges als Press-Schuppen dem zunehmenden Seitendruck folgend nach oben ausweichen. Schutt und Eis-Kerne der Seitenmoränen nehmen daran teil. Falls diese Schuppen den Oberrand der Flanke des Troges erreichen, werden sie umgebogen und am Grunde des sich schneller ausbreitenden Eises verschleppt und abgelagert. Von der oberen Trogwand kann dabei Gestein abgeschert und mitgeschleppt werden.

Eine Stauchendmoräne, deren Wälle heute eiswärts zunehmend niedriger sind, wird als mehrfach gestaucht gedeutet. Der zweite Eis-Vorstoss erfasste die Füllung des Zungenbeckens, die aus Rest-Eis unten und mächtigen Schmelzwasser-Sanden oben bestand.

Hingegen entstanden Press-Schuppen, Eis oben - Grundmoräne unten, als die Trogwand des westbaltischen Eises bei Heiligenhafen lateral gepresst wurde, während stagnierendes Eis noch im Gebiet neben dem Trog lag.

7 DANK

Herrn A. Möller, Graphiker in Kiel, danke ich für die Umzeichnung der Fotografien.

8 SCHRIFTTUM

Gripp, K. 1929, Glaciologische und geologische Ergebnisse der Hamburgischen Spitzbergen Expedition 1927 - Abh.Nat.Verein Hamburg, 22: 154-249, 32 Taf.

Gripp, K. 1964: Erdgeschichte von Schleswig-Holstein - 412 Seiten 57 Taf. Neumünster i.H.

Gripp, K. 1975: 100 Jahre Untersuchungen über das Geschehen am Rande des nordeuropäischen Inlandeises - Eiszeitalter und Gegenwart, 26: 31-73, 8 Taf.

Gripp, K. und Todtmann, E. 1926: Die Endmoräne des Green-Bay Gletscher, eine Studie zum Verständnis norddeutscher diluvialer Gebilde - Mitt.Geogr.Ges. Hamburg, 37: 45-75, 14 Taf.

Hansen, Sigurd 1932: Nye opfattelser af Bewegelsermekaniken for Gletscheris belegte ved Jagttagelser paa Frederikshaab Isblink - Naturens Verden 357, Kobenhavn.

Köster, R. 1957: Experimente zur glazialen Schuppung - Neues Jb.Geologie, Paläontologie Mh. 1957, 11: 510-517.

Prange, W. 1975: Gefügekundliche Untersuchungen zur Entstehung weichselzeitlicher Ablagerungen an Steilufern der Ostseeküste Schleswig-Holsteins - Meyniana, 27: 41-54.

Stephan, H.-J. 1971: Glaziologische Untersuchungen im Raume Heiligenhafen (Ost-Holstein) - Meyniana, 21: 67-86.

Seifert, G. 1954: Das mikroskopische Korngefüge des Geschiebemergels als Abbild der Eisbewegung, zugleich Geschichte des Eisabbaues in Fehmarn, Ost-Wagrien und dem Dänischen Wohld, Meyniana, 2: 124-184.

The problems of waterlain tills

ALEKSIS DREIMANIS
University of Western Ontario, London, Ont., Canada

INTRODUCTION

The term 'waterlain till' will be used here as it was proposed for 'waterlaid till' in 1967 (Dreimanis, 1969, Table 1):
"Glacial drift deposited as till in lake or sea", with following alterations: "waterlaid" changed to "waterlain" and "lake or sea" to be replaced by "glaciolacustrine, glaciomarine or glaciofluvial environments". (Note: "waterlain" is linguistically more appropriate than "waterlaid", as "waterlaid" means "laid down by water", while 'waterlain' - "lying in water", according to Francis, 1975: 57.)

The above difference between "waterlain" and "waterlaid" points at both the genetic and the terminologic problems concerning these tills. G e n e t i c a l l y waterlain tills occupy the boundary area between unquestionably glacial and dominantly water-deposited (glaciolacustrine, glaciomarine, glaciofluvial) sediments (Table 1).

TABLE 1. Interrelationship of waterlain till and other glacigenic deposits.

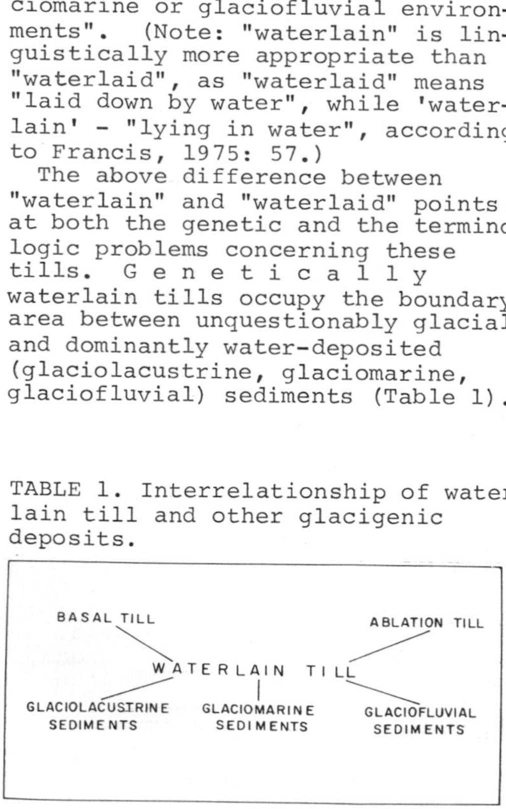

As ice thaws to water during deposition of most tills (Carey and Ahmad, 1961; Lavrushin, 1968; Boulton, 1971; Goldthwait, 1971; Muller, 1977; just to mention a few discussions), the presence of water during deposition of till is not embarrassing, unless the water causes noticeable disaggregation and sorting of glacial debris. Also, water has been more and more commonly reported from test drillings reaching the bottom of present day terrestrial glaciers, e.g. in the "active subsole drift" of Blue Glacier, USA, (Kamb et al., 1978) and as lakes under the Antarctic ice sheet (Unpublished reports on new work by D. Drewry and I. Whillans at the 1978 Symposium on glacier beds, Ottawa, Canada). Therefore we may expect more recognition of participation of water during the depositional process of till. Whatever are the differences in opinions how much water may be allowed in formation of till for a diamicton to be called till, I would like to stress the beginning of the original definition of waterlain till: "glacial drift deposited as till", thus emphasizing that it is essentially a glacial, not water deposit. This concept will be elaborated in the following paragraphs.

Prior to specific discussions of waterlain till, I better first clarify some past misunderstandings, that have arisen probably from my comparison (Dreimanis, 1969, 1976) of waterlaid till with the older term para-till of Harland et al (1966). Like para-till, the waterlain till is deposited in water,

but the concept of waterlain(formerly waterlaid) till was introduced as being narrower than para-till: the waterlain till is either deposited in direct contact with glacier ice or by deriving glacial debris from glacier ice without substantial disaggregation or sorting, what is possible only in close vicinity of glacier ice. Therefore, I would never call a till-like sediment a waterlain till, if it was deposited 20 km away from glaciers, here referring to an example mentioned by Boulton (1976: p. 71-72) as an illustration of the "problem introduced by the descriptive concept of "waterlaid tills" (ibid: 71). Also, waterlain (formerly waterlaid) till was never meant a descriptive concept, and it has not been called till "because it looks like till", as suggested by Boulton (ibid); most waterlain tills, because of their stratification, do not look like conventional till "commonly defined as nonstratified sediment" (Flint, 1971: 154).

Waterlain till may be still considered a genetically controversial sediment, but our present knowledge of it places it in the same group as terrestrial flow till and lee-side till - diamictons, that have been generally accepted as true glacial deposits. In order to be certain about the correct genetic position of waterlain till, its distinction from non-glacial diamictons should be both t h e o r e t i c a l l y s o u n d and p r a c t i c a l l y i d e n t i f y a b l e by field and laboratory criteria.

THEORETICAL CONSIDERATIONS

Theoretically, waterlain till should be essentially a glacial deposit that has derived from glacially transported debris and has not undergone subsequent significant disaggregation by nonglacial agencies. This theoretical concept is similar to Boulton's (1972: 378-379) definition of till. Being in full agreement with Boulton's (1972: 378) suggestion "that a very narrow definition is often difficult to apply" and with Goldthwait's (1971: 5) very thoughtful discussion "When is till not a till?" I have added the word "significant" to "disaggregation" in the first sentence of this paragraph. The final wording of the definition of waterlain till should be postponed until a decision is made by the commission on genesis and lithology of quaternary deposits or by any other international body: what diamictons should be called tills, and after the definition of till will be carefully worded to encompass all these specific diamictons.

PRACTICAL CRITERIA

Considering the p r a c t i c a l side for recognition of waterlain till, in combination with its theoretical origin, two general criteria are given here, followed by more specific comments afterwards.

1. Waterlain tills are bedded, containing thin interbeds, laminae or lenses of aquatic sediment, though poorly sorted till dominates in them. Flow structures are common; primary fabric of clasts is of variable strength, but unrelated to glacial movement.

2. Waterlain tills are associated laterally or in depositional sequence both with unquestionable terrestrial tills and with sediment units deposited by water.

Photographs of waterlain tills illustrate the type of stratification of the criterion (1) better than any lengthy descriptions, e.g. middle unit of Fig. 38 and the central thin-laminated layer in Fig. 28a of MacClintock and Stewart (1965), Photo 14 in Karrow (1967), Fig. 11 in Dreimanis (1976), Figs. 5,6,8,10,11,13-16 in Evenson et al. (1977), Figs. 2 and 3 in May (1977). Flow structures are present in some types of waterlain till (Figs. 13-17 in Evenson et al., 1977; Fig. 3 in May, 1977), and some flow folds have been measured as long as 10 m (Hicock and Dreimanis, 1978). In some waterlain tills, however, flow structures are hardly noticeable. Primary orientation of clasts usually has not been found related to the known glacial movement, even though many clasts show striae and other glacial abrasion features. The clast fabric is either random, particularly if they are emplaced as dropstones (Dreimanis 1976, Fig. 13; Gibbard 1977), or their long axes are mainly parallel and transverse to the local mass movement

directions (Evenson et al., 1977; Hicock and Dreimanis 1979).

The sorting, determined as QDø, is similar in waterlain tills and the related basal tills: while in some instances the waterlain tills are slightly better sorted (Evenson et al., 1976: 128), in others (Zubrzycki 1972: 11) the opposite is true. Those waterlain tills that appear less stratified are as poorly sorted as the related lodgement tills; with increased abundance of sorted laminae, the sorting of the adjoining till laminae increases slightly (Evenson et al., 1977). The silty and clayey waterlain tills investigated by Dreimanis (1970,1976) Zubrzycki (1972) and Evenson et al. (1977) from Lake Erie basin usually contain more silt than related basal tills, while the more sandy waterlain tills deposited in sea at Victoria, British Columbia (Hicock and Dreimanis, 1978) are of the same granulometric composition as the overlying and the underlying lodgement tills. The skewness (SK_G) of waterlain Catfish Creek Till (Evenson et al., 1977: 129) was found to be similar to that of typical tills, and quite different from mudflows. Though present laboratory investigation do not suggest any significant sorting of waterlain tills by water action, except for possible incorporation of silt in some cases, further laboratory investigations, application of various statistical parameters and integration of laboratory investigations with field observations, particularly on structures, are still desireable. Geochemical investigations may assist in deciphering the influence of water in the process of formation of waterlain tills.

WATERLAIN TILL VERSUS WATER-DEPOSITED SEDIMENTS

Though waterlain tills have been found to be associated laterally or in depositional sequence with water-deposited sediments (criterion 2), the grain size composition of them differ markedly. The glaciomarine sediments have been found much finer textured than the associated waterlain tills in Spitsbergen (Lavrushin, 1968) and along the west coast of Canada (J.E. Armstrong, 1977; Hicock and Dreimanis, 1978). This difference is a very important criterion for distinguishing waterlain tills from glaciomarine sediments which often look like tills because of their massive or just faintly stratified structure and the presence of dropstones in various abundance. Freshwater glacial lake deposits being more distinctly stratified than waterlain tills, can be easier differentiated from them by their structure. Sorting of glaciofluvial and glaciolacustrine gravels, sands and silts is much better than that of texturally similar waterlain tills, while glaciolacustrine and glaciomarine silty clays (Dreimanis, 1976, Fig. 1; Hicock and Dreimanis, 1978) are as poorly sorted as tills.

RELATIONSHIP OF WATERLAIN TILL TO BASAL TILL

The spatial relationship of waterlain and basal tills (other part of criterion 2) may not be noticeable readily in small sections, particularly if the waterlain till deposits are thick or laterally extensive. However, this is a very important criterion to prove that a diamicton which looks like waterlain till, according to the criteria listed under (1), was indeed deposited by or from glacier ice. Lateral contacts, by massive lodgement till layer grading proximally through interfingering into stratified waterlain till have been observed by the author in the Catfish Creek and Southwold Drifts in several Lake Erie bluff sections northeast of those described in Evenson et al. (1977). Upward grading from undeformed waterlain till through glaciodynamically sheared waterlain till into basal fissile lodgement till consisting of glacially reworked local waterlain till was observed in 1978 in the upper 2 m of Catfish Creek Drift right above the fabric analysis sites 9 and 10 in the Lake Erie bluff shown in Figs. 2 and 4 of Evenson et al. (1977). May (1977, Figs. 2-4) depicts sharp boundaries separating lodgement till from two different facies of overlying waterlain till in Alberta. A variety of contacts is noticeable along the about 50 km long Port Stanley drift sections

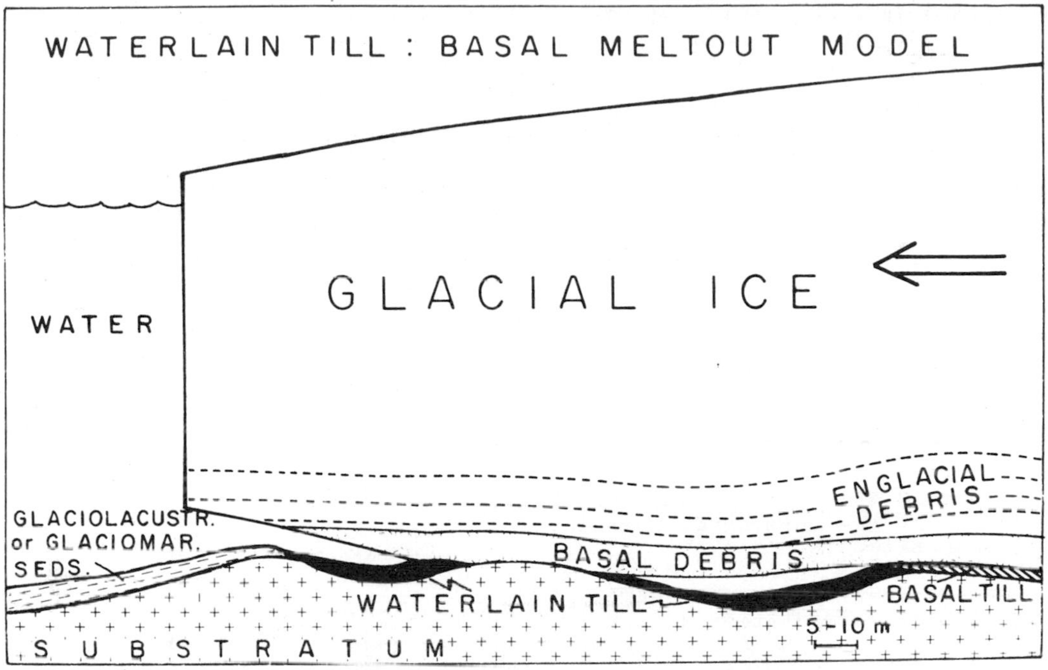

FIGURE 1. Waterlain till: basal meltout model, underneath floating terminus of glacier (left side) or in a subglacial water pocket of grounded glacier (right side).

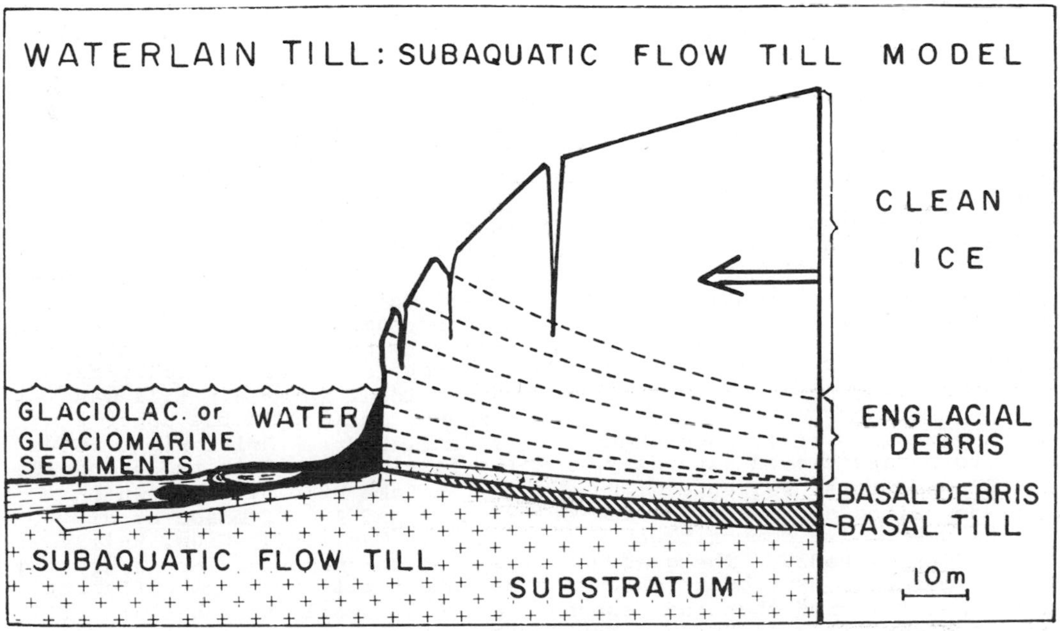

FIGURE 2. Waterlain till: subaquatic flow till model, along the glacial margin grounded in lake or sea.

between Tyrconnell and Port Burwell, Ontario. Here waterlain till is very common either at the top or at the base of some of the Port Stanley Till members, as these tills were deposited by the Erie lobe oscillating in a large proglacial lake. The contacts of the Port Stanley lodgement and waterlain till facies vary along the same stratigraphic position: some are sharp, others are transitional. The transitional contact zones are often rich in subrounded silt pebbles. If the waterlain till underlies lodgement till, it is often glaciotectonically deformed, and some of the deformed waterlain till has been dragged into the basal part of the lodgement till either along discrete shear planes, or both facies of till are so intermixed, that no sharp boundary can be drawn between them. Where it is difficult to define sharp laterally traceable contacts between the lodgement and the waterlain facies of a till unit, the genetic interrelationship of the waterlain till and the basal lodgement till become most obvious.

MECHANISMS OF DEPOSITION OF WATERLAIN TILL

Following specific mechanisms have been suggested for the deposition of waterlain tills:

A. Deposition underneath floating glacial termini, shelf ice or in subglacial basins (Fig. 1), essentially as waterlain basal meltout till (Carey and Ahmad, 1961; Lavrushin, 1968; Dreimanis, 1969 and 1976; Gibbard, 1977; May, 1977).

B. Deposition as subaquatic flow till along the edge of glacial margins (Fig. 2) grounded in lakes or seas (Evenson et al., 1977, May, 1977, Hicock and Dreimanis, 1978).

C. Deposition by grounded icebergs (Fig. 3). Fecht and Tallman (1978) have recently described them as "bergmounds", up to 50 m diameter and 4 m high, consisting of glacial drift; they have been deposited probably by current driven icebergs grounded in slackwaters of Lake Missoula floods along Columbia River, USA. I have observed lenses of till up to 20 m diameter in glaciolacustrine deposits in the Late Wisconsin deposits of Lake Erie basin and west of Steeprock

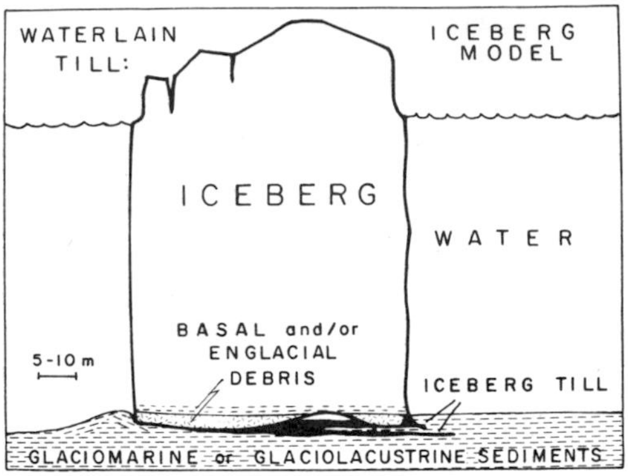

FIGURE 3. Waterlain till: iceberg model, underneath and along grounded iceberg.

Lake, Ontario. The iceberg moraines of various USSR classifications include both deposits of grounded icebergs, and glaciomarine deposits (here not considered as waterlain till) rich in glacial debris dropped from icebergs (e.g. Lavrushin, 1968).

Deposition by mechanisms (A) and (B) may be accompanied by strong participation of meltwater streams, meaning additional interaction of glacial and glaciofluvial processes, as suggested by Evenson et al. (1977), and Hicock and Dreimanis (1978) for the model (A). As Kalix till (see Lundqvist, 1977 for its brief description and further references) contains sorted sediments, particularly sand, interbedded with till, its origin as waterlain till, with participation of glaciolacustrine and glaciofluvial processes has been considered (Dreimanis, 1976 and references therein).

The criteria for distinguishing different mechanisms of waterlain tills still require further studies, some of them being currently in progress. The dispute of Evenson (1977) and Gibbard (1977) on several possible mechanisms responsible for the deposition of the Catfish Creek waterlain till unit at Plum Point, Ontario, suggests an overlap of some criteria and/or a combination of several mechanisms.

TERMINOLOGY

The names used for waterlain till are listed in Table 2. Some of them have originated as English terms, e.g. "lacustrotill" (Odynsky and Newton, 1950; see May, 1977,

for its discussion), some are translations from other languages, e.g. 'submarine moraine' from the Russian, appearing in the English translation in Lavrushin (1968). Many of them are not synonyms of 'waterlain till', but they may include it as one of their possible constituents, e.g. the already mentioned 'para-till' (Harland et al., 1966) and the Russian term 'basin moraine' and its subtypes 'shelf moraine' and 'iceberg moraine', judging by the discussion of Rukhina (1973). More clarification of the original meaning of all the terms used, and the variance of their meanings by subsequent users is needed.

Some confusion in the existing terminology has been created by May's (1977) proposal for the usage of the terms 'waterlaid till' and 'lacustrotill'. It is difficult to agree with May's (ibid: 177) statement, "It (lacustrotill, A.D.) . . . is the waterlaid till of Dreimanis (1969, 1976)" for the following reasons: lacustrotill is deposited in lacustrine environment only (lacustro- derives from the Latin 'lacus' = lake), while the waterlaid till is a broader term including also deposits of glaciomarine and glaciofluvial environments (Dreimanis, 1969: 16; 1976: 39-41). Therefore the term lacustrotill, though introduced earlier than waterlaid till, can not "take preference over the term waterlaid till", as proposed by May (ibid). May's (ibid: 178-179) proposal that the term "waterlaid till" is used restrictively "to describe essentially massive diamictons overlying basal tills" contradicts the previous descriptions of waterlaid till which state that it is "crudely stratified" (Dreimanis, 1976: 39), and its position is not only overlying, but also underlying basal tills (Zubrzycki, 1972).

TABLE 2. Terms of Waterlain Tills or Sediments that May Include Waterlain Till as Their Constituent, in alphabetical order; some names (*) are translations from the Russian into the English; published sources are preferred, but only a few of them indicate the first publication of the term listed; most are references used for discussions in this paper:

*Aquatic moraine, aquatil (E.V. Shantser, person. communic.)
*Aqueo-glacial deposits (E.V. Rukhina, personal communic.)
*Ablation submarine moraine (Lavrushin, 1968)
*Basin moraine (Rukhina, 1973)
Boulder beds (H. Ahmad, person. communic.)
*Fluvio-ablation moraine (Rukhina, 1973)
Glaciolacustrine till (Francis, 1975)
Glaciomarine till (Francis, 1975)
*Iceberg moraine or till (Rukhina, 1973)
Kalix till (Lundqvist, 1977)
Lacustrotill (Odynsky and Newton, 1950; May, 1977)
*Marine moraine (Lavrushin, 1968)
Para-till (Harland et al., 1966)
*Shelf moraine (Rukhina, 1973)
Subaquatic or subaqueous flow till (Evenson et al., 1977)
Subaquatic ablation till (Drozdowski, 1975)
Subaquatic ablation flow till (Drozdowski, 1974)
Subaquatic melt-out till (Dreimanis, 1977)
Subaquatic moraine deposits (Rzechowski, 1969)
- fluviotropic
- limonotropic
- thalassotropic

*Subwater or underwater moraine or till (E.V. Shantser, person. communic.)
Waterlaid till (Dreimanis, 1969)
Waterlain till (Francis, 1975)

SUMMARY

The term 'waterlain till' will be used here as a general group term, as it was origianlly proposed by the author at our Commission's symposium in Poland in 1967: "Waterlaid till: Glacial drift deposited as till in lake or sea", with following minor alterations: 'waterlaid' changed to the linguistically more appropriate term 'waterlain', and 'lake or sea' replaced by the broader 'glaciolacustrine, glaciomarine or glaciofluvial environments'.

In order to be classified as till, waterlain till has to be essentially a glacial deposit that has derived

from glacial transport and has not undergone subsequent major disaggregation and sorting by nonglacial agencies. Genetically waterlain till occupies the boundary area between glacial and water deposited (glaciolacustrine, glaciomarine and glaciofluvial) deposits. Some authors have favoured their inclusion in the latter group, by giving as examples those till-like deposits which look like till but truly belong to the water deposited group. To avoid such confusion, criteria should be established, both theoretically sound and practically determinable by field and/or laboratory methods, which permit to distinguish waterlain tills from till-like aquatic diamictons.

Genetically, waterlain till, though deposited in water, is essentially a glacial deposit. This interrelationship may be recognized by the stratification of waterlain till, though texturally its till laminae resemble the related basal till; flow structures vary in abundance depending upon the mechanism of deposition of waterlain till; the clast fabric is usually local and unrelated to the direction of glacial movement. Waterlain till may be found either about or below related basal till, or it may grade laterally into basal till.

Waterlain till may be distinguished from the related glaciomarine till-like deposits by its much coarser texture; from related glaciolacustrine and glaciofluvial deposits it differs usually by its poorer sorting, though glaciolacustrine clays may be equally poorly sorted. A few other criteria are listed, but further investigations are still required to establish what field and laboratory criteria are most useful for distinguishing waterlain till from other diamictons.

Three main mechanisms have been proposed for the deposition of waterlain till: as waterlain basal meltout till, as waterlain flow till, and as a deposit of stranded iceberg. Though a list of terms, either more or less equivalent to waterlain till, or including merely some varieties of waterlain till, is compiled in Table 2, the exact meaning of some of the terms is not entirely clear.

REFERENCES

Armstrong, J.E. 1977, Terminology and genesis of glaciomarine sediments, based upon studies along the Pacific coast of Canada. In A. Dreimanis (ed.) Commission on genesis and lithology of Quaternary deposits. INQUA X Congress Programme and abstracts of commission meetings Aug. 16, 1977, p. 3. London, Canada.

Boulton, G.S. 1971, Till genesis and fabric in Svalbard, Spitsbergen. In R.P. Goldthwait (ed.) Till: a symposium, p. 41-72. Columbus, Ohio State Univ. Press.

Boulton, G.S. 1972, Modern arctic glaciers as depositional models for former ice sheets. Jour. Geol. Soc. London 128: 361-393.

Boulton, G.S. 1976, A genetic classification of tills and criteria for distinguishing tills of different origin. In W. Stankowski (ed.) Till, its genesis and diagenesis, Univ. im. A. Mickiewicza w Poznaniu, Ser. Geografia 12: 65-80.

Carey, S.W. & Ahmad, N. 1961, Glacial marine sedimentation. 1st internat. symp. arctic geol. Proc. 2: 865-894.

Dreimanis, A. 1969, Selection of genetically significant parameters for investigation of tills. Zesz. Nauk. Univ. im. A. Mickiewicza w Poznaniu, Geografia 8: 15-29.

Dreimanis, A. 1970, Last ice-age deposits in the Port Stanley map-area, Ontario (40/1/11), Report of activities, Geol. Survey Canada Paper 70-1, pt. A: 167-169.

Dreimanis, A. 1976, Tills, their origin and properties. In. R.F. Legget (ed.) Glacial till. Roy. Soc. Canada Spec. Publ. 12: 11-49.

Dreimanis, A. 1977, Terminology and development of genetic classification of materials transported and deposited by glaciers. In A. Dreimanis (ed.) Commission on genesis and lithology of Quaternary deposits. INQUA X Congress, Programme and abstracts of commission meetings Aug. 16,1977, 7-10, London, Canada.

Drozdowski, E. 1974, Facial variations of tills in the profile of the second moraine stratum at Sartowice (Lower Vistula Valley), Zesz. Nauk. Universit. im. A. Mickiewicza w Poznaniu, Geografia

:121-136 (Polish, with English summary).

Evenson, E.B. 1977, Subaquatic flow tills: their characteristics and recognition. In A. Dreimanis (ed.) Commission on genesis and lithology of Quaternary deposits. INQUA X Congress, Programme and abstracts of commission meetings Aug. 16, 1977, p. 11, London, Canada.

Evenson, E.G., Dreimanis, A., Newsome, J.W. 1977, Subaquatic flow tills: a new interpretation for the genesis of some laminated till deposits, Boreas 6: 115-133.

Fecht, K.R. & Tallman, A.M. 1977, Bergmounds along the western margin of the channeled scablands, south-central Washington, Abstract with programs, 1978 annual meetings, Geol. Soc. America 10, No. 7: 400.

Flint, R.F. 1971, Glacial and Quaternary geology. New York, John Wiley and Sons, Inc.

Francis, E. 1975, Glacial sediments: a selective review. In A.E. Wright and F. Moseley (eds.) Ice Ages: Ancient and modern. Geol. Jour. Spec. Issue 6: 43-68.

Gibbard, P.L. 1977, The origin of stratified Catfish Creek till by basal melting (north shore of Lake Erie, south-western Ontario, Canada). In A. Dreimanis (ed.) Commission on genesis and lithology of Quaternary deposits. INQUA X Congress, Programme and abstracts of commission meetings Aug. 16, 1977, p. 12. London, Ontario.

Goldthwait, R.P. 1971, Introduction to till, today. In R.P. Goldthwait (ed.) Till: a symposium, p. 3-26, Columbus, Ohio State Univ. Press.

Harland, W.B., Herod, K.N. & Krinsley, D.H. 1966, The definition and identification of tills and tillites. Earth Sci. Reviews 2: 225-256.

Hicock, S.R. & Dreimanis A. 1978, Late Wisconsinan marine subaquatic flow tills, Victoria, B.C., Abstracts with programs, 1978 annual meetings, Geol. Soc. America, 10, No. 7: 421.

Kamb, B., Engelhardt, H.F. & Harrison, W.D. 1978, The ice-rock inter face and basal sliding process as revealed by direct observation in boreholes and tunnels. Volume of Abstracts, Symposium on glacier beds: the ice-rock interface, p. 11-12, Ottawa, Canada.

Karrow, P.F., 1967, Pleistocene geology of the Scarborough area. Ontario Dept. Mines Geological Report 46.

Lavrushin, Yu. A. 1968, Features of deposition and structure of the glacial-marine deposits under conditions of a fjord coast (based on the example of Spitsbergen) (Translated from the Russian). Litologia i poleznije iskopae myl 3: 63-74.

Lundqvist, J. 1977, Till in Sweden. Boreas 6: 73-85.

MacClintock, P. & Stewart, D.P., 1966. Pleistocene geology of the St. Lawrence Lowland. New York State Mus. and Sci. Service Bull. 394.

May, R.W. 1977, Facies model for sedimentation in the glaciolacustrine environment. Boreas 6: 175-180.

Muller, E.H. 1977, Dewatering during lodgement of till. In A. Dreimanis (ed.) Commission on genesis and lithology of Quaternary deposits. INQUA X Congress, Programme and abstracts of commission meetings Aug. 16, 1977, p. 18, London, Canada.

Odynsky, W. and Newton, J.D., 1950, Soil survey of the Rycroft and Watino sheets. Alberta Soil Survey Rept. 15.

Rukhina, E.V. 1973, Litologiya lednikovikh otcozhenii. Leningrad, Nedra.

Zubrzycki, A.J. 1972, Granulometric composition of some waterlaid tills. Unpublished B.Sc. thesis, Dept. Geology, Univ. Western Ontario, London, Canada.

ACKNOWLEDGMENTS

The terminology of waterlain tills (Table 2 and text) and several ideas expressed in this report have derived from discussions and correspondence on genetic varieties of tills at the INQUA Commission on genesis and lithology of Quaternary deposits, and the author is grateful to all those members of this Commission who have given their thoughts on the problems of waterlain tills. The preparation of this report was made possible by the National Research Council of Canada grant A 4215.

APPENDIX

The following illustrations are examples of waterlain till (photos by A. Dreimanis):

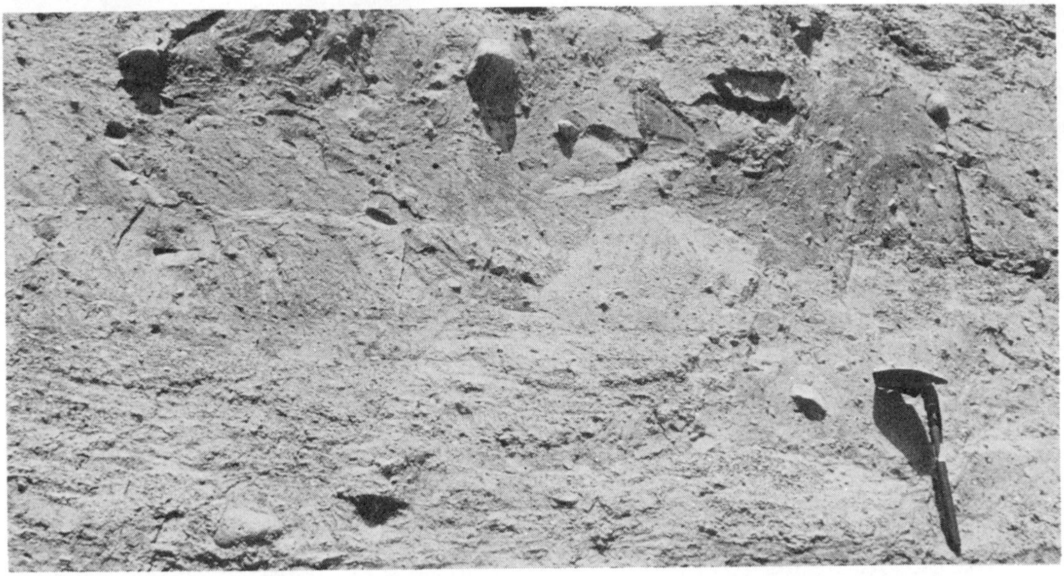

Figure 4. Coarse textured stratified waterlain till, under the "pocket knife", overlain by silty basal till (sheared by glacial movement from right to left); both are Catfish Creek Tills, Lake Erie cliff NE of Plum Point, Ontario, at 470 m of Fig. 4, Evenson et al., 1977.

Figure 5. Waterlain clayey silt till, with white streaks of silt, partly deformed by glacial movement from right to left; it is overlain and underlain by basal Port Stanley Till. Lake Erie shore 10 km E. of Port Stanley, Ontario.

Figure 6. Pebbly and silty well stratified waterlain till of Catfish Creek Formation, at 90 m of Fig. 4, Evenson et al., 1977.

Figure 7. Flow Lobes in coarse textured waterlain till. Sea cliff at Clover Point, Victoria, British Columbia.

Figure 8. Pebbly silty waterlain till, crudely stratified, with dropstones. The contact with overlying basal till is about 10 cm above the photo. St. Mary's Cement Co. old quarry, W. end, Ontario.

The origin of preconsolidated and normally consolidated tills in eastern Wisconsin, USA

D.M.MICKELSON, L.J.ACOMB & T.B.EDIL
University of Wisconsin, Madison, Wis., USA

1 ABSTRACT

This paper is based on a geological and geotechnical analysis of tills along 174 kilometers of Wisconsin's Lake Michigan shoreline. The study involved the identification, vertical profiling and mapping of eight tills and several lacustrine units, exposed in the eroding coastal bluffs, and the determination of the tills' geotechnical properties in a drilling and sampling program. Standard geologic laboratory determinations were made. Standard engineering index tests were performed on split-spoon and 7 1/2 cm, thin-wall tube samples. In addition, compression characteristics and preconsolidation stresses were investigated by standard consolidation tests and effective strength parameters, ϕ' and c' were obtained by conducting consolidated undrained triaxial compression tests with pore water pressure measurements.

Each till was remarkably uniform in its engineering properties throughout the length of exposure. The tills deposited before about 12,800 ± 300 years B.P. are normally consolidated or only slightly overconsolidated. Tills with otherwise similar properties deposited after this time are definitely preconsolidated. There seems to be no differences in drainage, stratigraphic position, erosion, or other factors that might affect this property; therefore, a difference in the nature of till deposition is postulated. One possible explanation is that prior to 12,800, the tills were deposited under permafrost conditions (which prevented drainage of pore fluids) whereas in the warming climate after 12,800 non-permafrost conditions existed. Evidence suggests that the change from tundra or shrub vegetation to spruce (Picea) corresponds to about the time of the change in the consolidation properties.

2 INTRODUCTION

The sciences of geology and geotechnical engineering have evolved along separate paths as a consequence of the requirements each has been asked to fill. Classically, geologists have studied the earth's surficial materials in an attempt to interpret the processes and history producing the mass whereas geotechnical engineers have tried to assess the present characteristics of a material with respect to a specific purpose. However, this artificial separation of the sciences can be detrimental to the complete understanding of a material, from both the engineers' and geologists' point of view. A more beneficial and scientifically sound "approach" can be obtained from an interdisciplinary study of earth materials. Although some studies of this type have been undertaken, especially in Europe, very little use has been made of engineering studies to develop ideas on the genesis of till. In this paper we shall try to hypothesize on the environment and mode of deposition of a stratigraphic unit based on the mechanical characteristics of that unit.

The information presented in this paper is derived from a study of Wisconsin's Lake Michigan Shoreline, conducted during the summer of 1976 for the Wisconsin Coastal Zone Management Program (Mickelson et al. 1977). The area of study extends from the Illinois-Wisconsin state border north, a distance of approximately 174 km. The generally north-south trending shore includes a wide variety of shoreline types. The dominant coastal form is a series of wave cut bluffs (up to 43 m high) consisting of interbedded lacustrine clays, silts and fine sands and tills. Other coastal forms include low bluffs of interbedded lacustrine and glacial material; low bluffs of

only lacustrine material; sand terraces of ancestrial Great Lakes; and sand dune areas. The high and low bluffs of interbedded glacial and lacustrine materials are best suited to fulfill the objectives of this paper, hence discussion will be limited to these shore types.

3 GEOLOGICAL AND GEOTECHNICAL INVESTIGATION

Detailed field surveys and measurements were made during the summers of 1976-78 and the preliminary results were presented by Mickelson (et al. 1977) and Edil, Mickelson and Acomb (1977). This paper is a revision of the latter presentation. Lab analysis of samples collected by the geologic field parties included textural determination using a standard hydrometer analysis and a calgon dispersing agent. Clay minerals were determined by X-ray diffraction by the method outlined by Fricke (1976) which is modified from that used by the Illinois Geological Survey. Samples for this analysis were taken from bluff exposures and at least 50 power auger borings.

In order to determine the geotechnical properties of the glacial deposits, a boring and sampling program was undertaken. Two drill rig units were used in the sampling of subsurface materials in 25 bore holes located at selected sites over five counties along the Lake Michigan shoreline. In each borehole representative samples of materials were obtained by intermittent split spoon sampling, and the standard penetration resistance was recorded with depth. Pocket penetrometer readings were taken on the surface of the split spoon samples. When cohesive strata were encountered, 7 1/2 cm, thin-wall tube samples were taken for the purpose of "undisturbed" sampling at approximately every 5 m or when a new recognizable layer was reached. Engineering index tests (Atterberg limits, mechanical sieving and hydrometer analysis for grain size distribution) were conducted in the laboratory on practically all of the several hundred samples obtained. Natural water content, unit dry weights were also determined for all thin-wall tube samples, and many split spoon samples.

Consolidated undrained triaxial compression tests with pore water pressure measurements were conducted on all undisturbed samples in order to determine the effective strength parameters, ϕ' and c'. Cylindrical specimens 3.6 cm in diameter and 7.8 cm high were trimmed and subjected to consolidation pressures in the range of 100 to 600 kN/m^2. Back pressures of 300 to 500 kN/m^2 were generally required to achieve

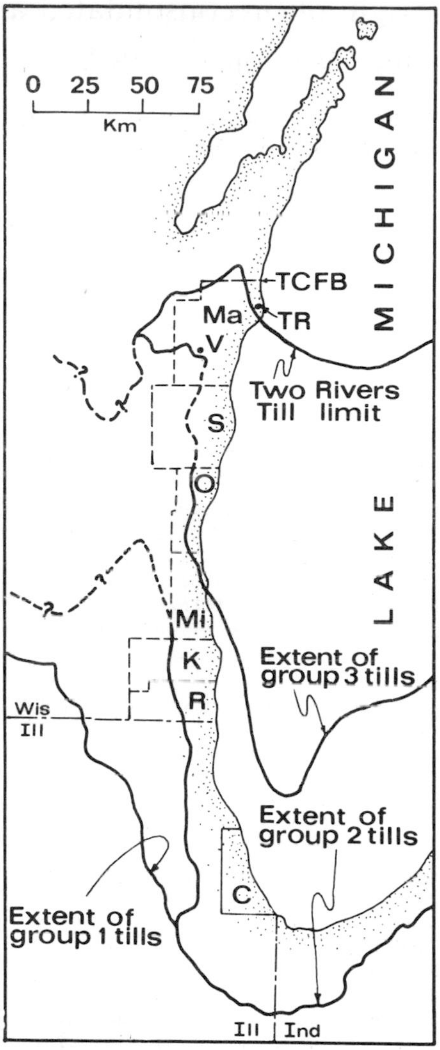

Figure 1. Extent of till units in eastern Wisconsin and estimated extent in Lake Michigan based on Lineback (et al. 1974). Modified from Evenson (et al. 1976). Key to counties: K = Kenosha, R = Racine, Mi = Milwaukee, O = Ozaukee, S = Sheboygan, Ma = Manitowoc, C = Chicago, TR = Two Rivers, TCFB = Two Creeks Forest Bed, V = Valders.

saturation prior to shearing which was carried at a rate of about 0.13 percent axial strain per minute, and the pore water pressure was measured by an integrated circuit pressure transducer. Failure was defined at the maximum principal stress difference. The effective strength parameters, ϕ' and

TIME - DISTANCE DIAGRAM

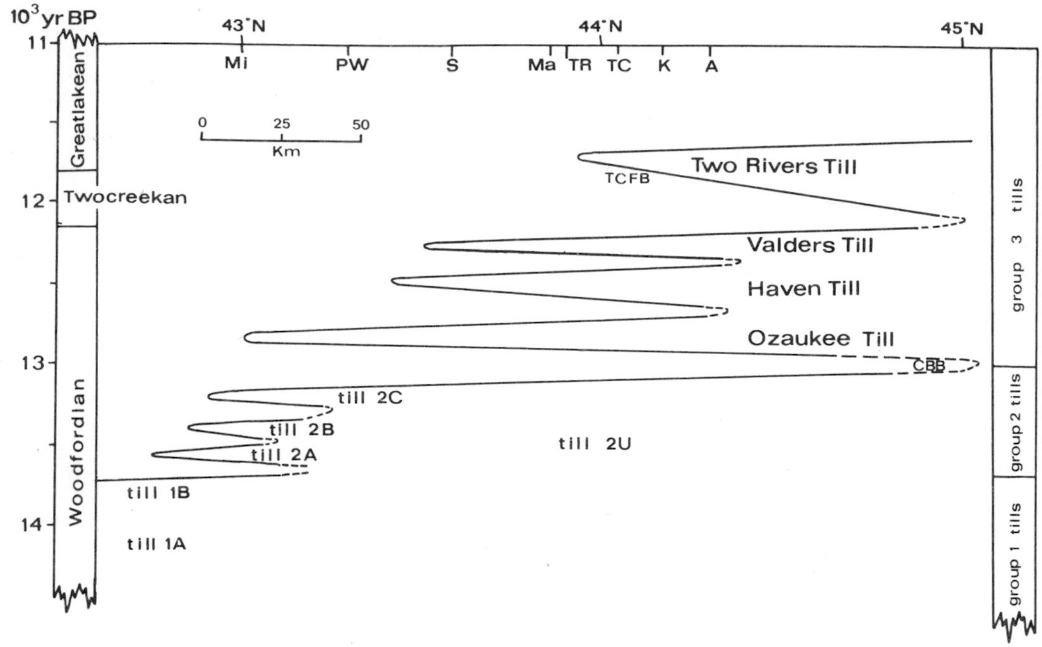

Figure 2. Time-distance for the Lake Michigan basin. Vertical scale represents time and horizontal scale is approximate distance along the glacial path of Lake Michigan. Numbers on the units are those used in this paper. TCFB = Two Creeks; C.B.B. = Cheboygan Bryophyte Bed projected to axis of lake. Mi = Milwaukee, PW = Port Washington, S = Sheboygan, Ma = Manitowoc, TR = Two Rivers, K = Kewaunee, A = Algoma.

c' were computed using the least squares fit of the failure data of the specimens from a sample. This testing program involved the determination of the shear strength parameters for some 90 samples. Conventional consolidation tests were also performed on a number of selected samples from certain till units in order to determine the stress history characteristics.

4 GLACIAL STRATIGRAPHY

The Lake Michigan basin has been deepened and widened during several glacial phases which took place over the last one to two million years. However, only the last few advances (late Wisconsin or Woodfordian age) have left a record of deposits which are now exposed in the eroding bluff. Sandwiched between the glacial till units are lacustrine sands, silts, and clays deposited in a fluctuating facies environment as the water level of the lake rose and fell with each ice advance and retreat. Because of their similarities, tills have been grouped informally into three major categories (1, 2 and 3) and then subdivided into members. The tills are not formally named although they are used informally by Acomb (1978).

The oldest tills exposed along the shoreline are group 1 (Fig. 1,2). They are coarse sandy tills with abundant cobbles and many large boulders (which are their most distinguishing feature). These tills are not cohesive and consolidation information is not available for them so they are not discussed in detail here but characteristics of the tills are given in Table 1.

A subsequent ice advance, probably about 13,500 years B.P. incorporated some of the previously deposited lacustrine sediment from the basin, reworked it and deposited it as a till along the Wisconsin and Illinois shorelines. These tills (group 2) may be correlated with the Wadsworth till of Illinois. In southern Milwaukee and Racine counties (Fig. 1) three separate layers of till 2 can be distinguished, however, north of Milwaukee and in Kenosha County only one layer is present and due to similarities between the 3 units it is impossible to differentiate which unit is represented

Table 1. Average grainsize, relative clay mineral percentages of tills in eastern Wisconsin

Till	% Sand	%[1] Silt >2μ	%[1] Clay <2μ	%[2] Expandable	%[2] Illite	%[2] Kaolinite and Chlorite
1A	35±	49±	16±	7.1±	78.5±	14.7±
1B	30±	50±	20±	15.7±7.3	72.6±5.8	11.7±4.1
2A	4±	69±	27±	19.7±	64.0±	16.2±
2B	7.6±6.5	50.7±7.2	41.5±11.3	17.7±5.8	66.3±4.1	15.9±2.8
2C	14.2±4.8	55.4±5.1	30.4±4.5	28.62±43	62.8±3.7	8.4±3.3
2U	8.1±3.8	42.1±12.9	49.6±13.2	17.1±	64.6±	18.2±
Ozaukee	13.0±3.9	47.3±5.0	39.6±6.5	19.7±5.8	60.3±5.1	20.0±2.3
Haven	15.5±7.1	56.2±6.2	27.7±6.6	24.5±4.5	56.2±5.0	19.0±3.2
Valders	30.2±8.6	52.0±8.1	17.8±7.9	45.5±5.0	42.1±3.7	12.4±2.7

1. Percent of < 2 mm fraction. Boundaries used are 2 mm, 0.0625 mm, .002 mm. ± is one standard deviation; not calculated where less than 10 samples.
2. Method modified from Glass (1977, pers. comm.). Percentages are relative amounts of clay minerals analyzed (total always adds to 100 percent). Standard deviation not calculated when less than 4 samples.

north of Milwaukee and which is present in Kenosha. These are referred to as till 2 undifferentiated (2U). At the Milwaukee, Racine County line the till reaches a maximum thickness of 21.3 m.

Till 2A and till 2B are characteristically fine grained (>90% silt and clay) and contain frequent shale pebbles. Their field color is dull gray to brown gray and their suspended color is a true 10 YR 6/3. The percentage of expandables is similar in both (~18%) as are the amounts of illite (~65%) and kaolinite (~16%) (Table 1). The uppermost unit of the second grouping of tills, till 2C, is slightly more sandy and has a more yellow field color. Pebbles and cobbles are more frequent and it differs from 2A and 2B in that kaolinite represents only 8% of the clays considered.

The next major ice advance (approximately 13,000 yrs. B.P.) again incorporated lacustrine sediments and redeposited them as tills of group 3 (Figure 2). The lower till unit of this series, the Ozaukee till extends from the City of Milwaukee north through Ozaukee County and presumably continues northward below lake level through Manitowoc and Sheboygan Counties (Fig. 2). Fluctuations of this ice mass, following the deposition of the Ozaukee till resulted in the deposition of the Haven and Valders tills. These tills extend from central Sheboygan county north at least to Kewaunee County. All of these units have a distinct red color in the field and a 7.5 YR suspended color. This parameter is useful in differentiating the 3rd series tills from other groups but is of no help in differentiating between members of this 3rd grouping. Mechanical analysis of Ozaukee, Haven and Valders tills shows an increase in the sand percentage and a marked decrease in clay content upward in the section (Table 1) although the youngest red till, the Two Rivers has grain size characteristics of the Valders till. The clay mineral analysis (Table 1) also shows a distinct difference between several of the tills of the third grouping. The Two Creeks Forest Bed, dated at 11,850 years B.P. (Broecker and Farrand, 1963) lies between the Valders and Two Rivers tills. Sufficient engineering tests have not yet been performed on the Two Rivers till to provide for discussion here.

5 GEOTECHNICAL PROPERTIES

The soils along the shoreline vary considerably and as discussed can be broadly grouped as glacial tills, lacustrine silts and clays and sandy outwash deposits based on their genesis and geotechnical behavior. The first two are generally cohesive and the last one is cohesionless. The properties and behavior of only the tills are considered herein. However, it is recognized that the other groups also form an important part of the coastal bluff materials in Wisconsin. The classification and nomenclature of the till units encountered was presented in the preceding section based on geological observations. The

Table 2. Average geotechnical properties of tills in eastern Wisconsin

Till Unit	γ_d* (kN/m^3)	w (%)	w_L (%)	I_p (%)	ϕ (degrees)	c' (kN/m^2)
1A	19.6	4.8	Non-plastic		34.6	0
2A	18.6±0.6	15.8±3.1	24.5±2.9	9.9±1.9	31.1±0.1	0
2B	18.5±0.6	15.2±1.4	25.1±2.8	10.0±2.3	31.4±	0
2U	18.7±	15.7±	23.4±2.1	10.2±2.0	30.5±	0
Ozaukee	17.9±0.4	17.6±1.2	30.6±2.6	14.0±2.5	31.4±0.8	0
Haven	18.6±0.9	16.5±2.5	30.3±6.0	14.4±4.7	31.2±0.5	23.8±5.6
Valders	17.7±1.2	17.4±4.1	28.4±5.0	13.1±3.4	29.3±0.6	28.3±6.9

*Unit Dry Weight

Note: Standard deviation not calculated when less than 4 samples.

geotechnical properties of these units are considered here and variation of these properties within and between these units is investigated. Because of the small number of samples available, tills 1A and 1B are not discussed.

5.1 Index Properties

The index properties of the till units are summarized in Table 2 in terms of means (plus or minus one-standard deviation). Size of the sample population varies from one till unit to another, being as few as 3 or 4 and as many as 28 samples. The liquid and plastic limits for all samples, including the lacustrine deposits, varied in a relatively narrow range and resulted in the classification of the cohesive soils along the shoreline as either low-plasticity silts or clays (CL according to the Unified Classification System). This is even more striking when the average values of the Atterberg limits are considered (Table 2) because practically all of the averages for liquid limits vary between 23 and 31, with corresponding plasticity indices between 10 and 14 per cent. Nevertheless, distinct differences exist between group 2 and group 3 tills in terms of the Atterberg limits. This type of differentiation is also apparent for the natural water content. However, differences in these properties within these two till groupings are not significant. Mean activity numbers, obtained by averaging the ratios of plasticity index to per cent clay fraction (less than 2 micron size), do not vary significantly from one till unit to another and the average is around 0.35. Mean unit dry weights also vary in a relatively small range (17 to 19 kN/m^3).

5.2 Shear Strength

The effective strength parameters (Table 2) ϕ' and c', which relate the shearing resistance of a soil to the normal effective stress acting on the failure surface, are parameters normalized with respect to the influence of consolidation pressure and corresponding equilibrium void ratio. Thus, they are quite fundamental parameters and basically are functions of the composition, texture, fabric and stress history of the soils. The sample population is again variable: one to six data points for group 2 tills, twelve to sixteen data points for group 3 tills. Based on all data it becomes apparent that the shear strength parameters vary within very narrow limits for a given cohesive till unit in spite of geographic distances involved. The effective angle of internal friction, ϕ', has a standard deviation of less than 1° in each till unit; it varies only from 29° to 31° for all cohesive till units with remarkable consistency. Tills 2A, 2B, 2U and the Ozaukee till exhibited no effective cohesion intercept whereas the Haven and Valders tills had effective cohesion intercepts varying between 19 and 40 kN/m^2 with a combined average value of 26 kN/m^2 for the two units. A generalization cannot be drawn relating the Atterberg limits of a particular sample and its effective strength parameters. Presence of the cohesion intercept in the two northern tills (Haven and Valders) indicates the overconsolidation of the tills in the range of test consolidation pressures (100 to 600 kN/m^2) and this is also supported by the

Table 3. Compressibility and preconsolidation of group 3 tills

Unit	C_c	$\sigma'_c(kN/m^2)$	$\bar{\sigma}_o(kN/m^2)$
Ozaukee	0.19	440	135
Haven	0.14	700	190
Valders	0.19	640	160

relationships between the undrained strength and the laboratory consolidation pressure for these tills. This overconsolidation can possibly be traced back to the processes which take place during deposition and/or post depositionally and this is discussed in the following section.

5.3 Compressibility and Preconsolidation

The conventional consolidation tests performed on selected samples from group 3 tills provides information regarding the compressibility and stress history of the soils (Table 3). Compression index, C_c, is the slope of the straight line (virgin) part of the compression curve (void ratio versus log of effective stress) and it represents the compressibility of the soil. When the vertical effective stress has been decreased, the preconsolidation pressure, σ'_c (the maximum pressure under which the soil was consolidated) exceeds the existing overburden stress, $\bar{\sigma}_o$. In such a condition, the soil is considered to be overconsolidated. The preconsolidation pressure can be estimated from a laboratory compression curve by observing the stress at which a change in the slope of the compression curve occurs from recompression to virgin compression. Since this transition is gradual, it may not be very easy to identify the preconsolidation pressure. Certain procedures such as the one suggested by Casagrande (1936) can be used. The preconsolidation pressure defined by the Casagrande empirical procedure is considered to represent the most probable value. Representative values of these parameters are summarized in Table 3 for group 3 tills. These tills, in general, are relatively stiff with compression indices less than 0.2. The preconsolidation pressures clearly indicate that the Haven and Valders tills are more heavily overconsolidated than the Ozaukee till, i.e., they have a value of σ'_c about one and a half times as the σ'_c for the Ozaukee till.

6 DISCUSSION

The Characteristics of tills deposited by various glacial advances are relatively consistent over extended distances (Fig. 2) roughly along the direction of transport.

Natural water content, liquid limit and plasticity index (Table 2) show minor but clear differences between the till groupings 2 and 3. These groups are usually fairly easily distinguishable by color and the so-called "red tills" (group 3) have long been recognized as distinct from the "grey tills" (group 2) (Alden 1918; Evenson 1973). Although differences in clay percentage, (Table 1) exist within the major groupings of tills (especially between the Ozaukee and Haven tills these do not seem to be reflected in the index properties (Table 2) as they are in other studies (e.g. Roderick 1975; Seed et al. 1964). For example, the Ozaukee Till has a clay content more than 12% higher than either the Haven or Valders tills (Table 2), yet its liquid limits fall within the same range as those of the Haven and Valders tills.

Another striking difference in the till characteristics is in the consolidation stress history. As pointed out previously (Tables 2 and 3) the Haven and Valders tills show distinct overconsolidation and the older tills (2A, 2B, 2U, Ozaukee) are only slightly overconsolidated (or not at all). Several recent workers, especially Boulton (1975) have tried to explain the presence or absence of overconsolidation in tills. Several early papers (e.g. Harrison 1958) suggested that the weight of glacial ice was responsible for creating the preconsolidation often found in tills. Others have suggested that desiccation, fluctuation of water table, erosion of overlying materials and ability of pore water to drain under loading are also ways of explaining preconsolidation (Dreimanis 1976). Boulton (1975) has concluded that the state of consolidation "depends almost entirely on depositional and postdepositional changes." He also concludes that lodgement tills are generally overconsolidated, flow tills often have low preconsolidation pressures, and melt out (ablation) tills seem likely to be normally consolidated. In the case investigated it seems likely that all of the tills are basal in origin although washed or flow tills have been noted in local areas but not included in our sampling (Mickelson et al. 1977). In some localities, the lower parts of the till units are probably water-laid but these were also avoided for the geotechnical sampling. Thus we must explain why all of the tills in group 2 and the Ozaukee till are slightly overconsolidated

or not at all where the two younger tills are distinctly overconsolidated. It seems unlikely that post glacial desiccation could explain the difference. All of the tills occur at different heights above Lake Michigan in different places along the bluff and samples of the tills were taken at various depths below the bluff top. It also seems unlikely that the mode of deposition or general source of materials was different. All of the units are fine-grained compared to the earlier group 1 tills shown in Figure 2. All seem to have been derived from fine-grained lake sediments deposited in the Lake Michigan basin. Lake level during the time of deposition of the tills was probably at the same level (Glenwood, or about 18 m above present level) during the deposition of all of these tills. There is no consistent relationship between the presence or lack of overconsolidation and the presence of permeable sandy units above or below the till. We can see no stratigraphic reason why the Haven and Valders tills should have drained more easily than the others.

The consolidation tests (Table 3) indicate that, in the case of the Haven and Valders tills, the excess above the overburden stress is less than the total stress the glacier ice would be expected to exert, i.e. 3000 kN/m^2 or more, if tills were deposited under fairly thick ice or during ice advance and if pore pressures were low.

The degree of overconsolidation exhibited by a soil is not only related to ice thickness, but is also related to the duration of loading and the ability to drain. As discussed here, the duration of loading (100's of years) is sufficient for 100% consolidation if adequate drainage is provided, and therefore, the ability to drain is the factor controlling the amount of overconsolidation. It is expected that near the margin of a stagnant glacier (not advancing) complete drainage could occur; however, farther under the ice mass, incomplete drainage would be the rule rather than the exception. This effect would be due to the long distance over which the head must be dissipated. Studies (e.g. Hodge 1976) have indicated that the pore water pressure at the bed of modern temperate glaciers can have a head on the order of two-thirds of the ice thickness. In a situation similar to this the soil would be supporting only a portion of the weight of the ice mass in terms of effective stresses. This probably partially explains the relatively low overconsolidation values of even the Haven and Valders tills.

Another possible explanation of the relatively low overconsolidation values and of the differences seen between the tills is a difference in load due to ice thickness. This explanation would require thicker ice during or after the deposition of the Haven and Valders tills than previous tills. There seems to be no reason for this. In addition, we have situations where overconsolidated till lies stratigraphically above, in the same section, a till showing less overconsolidation. Clearly, factors other than ice thickness are most important in determining the values we measure today.

Differences in materials cannot be used to explain differences in soil drainage. A more likely control is the temperature regime and resulting distribution of frozen or melted bed and sub-bed.

Birks (1976) has shown that in Minnesota, adjacent to Wisconsin, tundra vegetation was present until about 14,700 years ago when it was replaced by shrubs lasting until about 13,600 when a spruce dominated woodland developed. Today, the northern limit of the boreal forest nearly corresponds with southern boundary of continuous permafrost from the area. No dated pollen records containing the tundra zone are available in Wisconsin although the tundra/spruce boundary may occur at about the same time in central Wisconsin.

The best minimum date available on the advance of the Ozaukee till is about 13,000 years B.P. in the northern part of the lower peninsula of Michigan (Farrand et al. 1969) (Fig. 2, C.B.B.). Minor retreat after deposition of this till, advance and deposition of the Haven and Valders tills are necessary before about 11,800 years B.P., the date of the overlying Two Creeks Forest Bed. Vegetation at the Cheboygan Bryophyte Bed is Dryas, a typical tundra plant and the Two Creeks Forest Bed contains the remains of a boreal forest (Broecker and Farrand, 1963). Thus the change from deposition of tills which are now normally consolidated to those which are now overconsolidated is bracketed by the tundra-spruce and possibly the permafrost-nonpermafrost transitions.

When a polar glacier advances over permafrost, presumably no consolidation can occur for a number of related reasons. The major factor to consider is that the entire soil mass is in the solid phase with essentially no void volume. Secondly, even if melting under the ice takes place, the impermeable barrier of overlying ice and permafrost beyond the margin would probably prohibit drainage and the release of pore pressure, again rendering consolidation ineffective. In this situation the till is in a frozen state until the ice thins and disappears. It is this type of environment that is hypothesized during the ice advances that

deposited till 2 and the Ozaukee till.

Between the deposition of the Ozaukee till and the deposition of the Haven till, we suggest, the climatic environment may have changed and continuous permafrost disappeared. This change could have caused a corresponding change in the temperature of the glacier ice and the basal depositional environment. Under these new conditions drainage in front of the ice margin and even along the base of the ice or through the ice could occur, which would decrease the neutral stress (increase the effective stress) and allow consolidation under the load of the ice.

Similar mechanisms to explain the distribution of normally consolidated and overconsolidated sediments have been suggested by Aario (1971) for clays in Finland in a similar topographic situation. Mathews and MacKay (1960) have considered the development of permafrost beneath large water bodies (like Lake Michigan) by the advance of polar ice as would be expected in tundra conditions and our mechanism seems to work even adjacent to and under Lake Michigan.

We conclude, therefore, that the difference in preconsolidation of the tills may be due to the temperature of the ice mass and the landscape across which it advanced. Certainly other factors can, at times, also cause preconsolidation but in this case we see no valid options to our hypothesis. We hope that with continued testing on these tills and others, especially tills being deposited now, that further documentation of this or another hypothesis might develop. We believe that by combining information from geology and geotechnical engineering much progress can be made in solving problems in both of the disciplines.

7 ACKNOWLEDGEMENTS

Financial assistance for the original field and laboratory work was provided by the Coastal Zone Management Act of 1972 administered by the Federal Office of Coastal Zone Management, National Oceanic and Administration through the Office of State Planning and Energy. Contributions of Drs. D. Hadley of the Wisconsin Geological and Natural History Survey, A. F. Schneider of the University of Wisconsin-Parkside, N. Lasca and Mr. R. Klauk of the University of Wisconsin-Milwaukee and C. Fricke are recognized in the geologic field work. Efforts of Messrs. B. J. Haas, N. H. Severson, T. Wolf, G. H. Bahmanyar and D. Bennett in the laboratory tests for geotechnical properties are greatly appreciated. The University of Wisconsin Sea Grant Program provided additional funds during the latter part of the project.

8 REFERENCES

Aario, R. 1971, Consolidation of Finnish sediments by loading ice sheets, Bull. Geol. Soc. Finland 43:55-63.

Acomb, L. 1978, Stratigraphic relations and extent of Wisconsin's Lake Michigan Lobe red tills, M.S. Thesis, University of Wisconsin-Madison.

Alden, W.C. 1918, The Quaternary Geology of Southeastern Wisconsin, U.S. Geol. Surv. Prof. Paper 106.

Birks, H.J.B. 1976, Late-Wisconsinan vegetational history at Wolf Creek, Central Minnesota, Ecol. Monographs 46:395-429.

Boulton, G.S. 1975, The genesis of glacial tills - A framework for geotechnical interpretation. In Proc. of Symposium "The Engineering Behavior of Glacial Materials". Midland Soil Mechanics and Foundation Engineering Society: 53-59.

Broecker, W.S. & W.F. Farrand 1963, Radiocarbon age of the Two Creeks forest bed, Wisconsin, Geol. Soc. Amer. Bull. 74:795-802.

Casagrande, A. 1936, The determination of the preconsolidation load and its practical significance. First Intn. Conf. on Soil Mechanics and Foundation Engineering 3:60.

Dreimanis, A. 1976, Tills: their origin and properties. In Legget (ed.), Glacial Till, Royal Soc. Can. Spec. Publ. 12: 11-49.

Edil, T.B., D.M. Mickelson & L. Acomb, 1977, Relationship of geotechnical properties to glacial stratigraphic units along Wisconsin's Lake Michigan shoreline, Saskatoon, Proc. 30th Canadian Geotechnical Conference: 36-54.

Evenson, E.B. 1973, Late Pleistocene shorelines and stratigraphic relations in the Lake Michigan basin, Geol. Soc. Amer. Bull. 84:2281-2298.

Evenson, E.B., W.R. Farrand, D.F. Eschman, D.M. Mickelson, & L.J. Maher 1976, Greatlakean substage: A replacement for Valderan substage in the Lake Michigan basin, Quat. Res. 6:411-424.

Farrand, W.R., R. Zahner, & W.S. Benninghoff 1969, Cary - Port Huron Interstade: Evidence from a buried bryophyte bed, Cheboygan County, Michigan, Geol. Soc. Amer. Spec. Paper 123:249-262.

Fricke, C.A.P. 1976, The Pleistocene geology and geomorphology of a portion of central-southern Wisconsin, M.S. Thesis, University of Wisconsin-Madison.

Harrison, W. 1958, Marginal zones of vanished glaciers reconstructed from the pre-consolidation-pressure values of overridden silts, Jour. Geol. 66:72-95.

Hodge, S.M. 1976, Direct measurement of basal water pressures: A pilot study, Jour. Glaciol. 16:205-218.

Hunt, C.B. 1974, Natural Regions of the United States and Canada. W.H. Freeman, San Francisco.

Lineback, J.A., D.L. Gross, and R.P. Meyer 1974, Glacial tills under Lake Michigan. Ill. Geol. Surv. Envir. Geol. Note 69.

Mathews, W.H. & J.R. MacKay 1960, Deformation of soils by glacier ice and the influence of pore pressures and permafrost, Trans. Royal Soc. Can. 54, ser. 3:27-36.

Mickelson, D.M., L. Acomb, N. Brouwer, T. Edil, C. Fricke, B. Haas, D. Hadley, C. Hess, R. Klauk, N. Lasca, & A.F. Schneider 1977, Shoreline erosion and bluff stability along Lake Michigan and Lake Superior shorelines of Wisconsin. Shore Erosion study Technical Report, Madison, Wisconsin, Office of State Planning and Energy.

Roderick, G.L. 1975, Properties of some glacial soils in Wisconsin. In Proceedings of a Symposium, "The Engineering Behavior of Glacial Materials," Midland Soil Mechanics and Foundation Engineering Society: 75-82.

Seed, H.B., R.J. Woodward, & R. Lundgren 1964, Fundamental aspects of the Atterberg limits, Jour. Soil Mechanics and Foundations Div., Proc. ASCE 90, SM6: 75-105.

La genèse des drumlins

G.SERET
Université de Louvain, Belgique

1 INTRODUCTION

1.1 Définition et description des drumlins

Les drumlins sont des collines allongées, hautes de 20 à 50 mètres, longues de quelques centaines de mètres et de largeur 3 à 4 fois inférieure à leur longueur. Les formes sont molles, écrasées, arrondies. Gravenor C.P. (1953) propose une synthèse de leurs caractères descriptifs, qui peut être résumée en quelques points :

1. La composition des drumlins est variable. La plupart correspondent à une colline de moraine de fond. Ce sont les "till drumlins". D'autres, moins nombreux, sont sculptés dans le substratum : les "rock drumlins". Plusieurs, enfin, sont constitués d'un placage morainique adossé à un noyau de substratum : les "rock and till drumlins".
2. Le matériel morainique peut présenter un litage, des plis, des failles.
3. Les drumlins sont allongés parallèlement à la progression des glaciers.
4. Ils se répartissent en champs, au sein desquels ils sont nombreux, et à peu près tous de mêmes dimensions. Dans ces champs, les trois types de drumlins peuvent être présents, mais les till drumlins sont en général les plus nombreux.
5. Les drumlins sont dissymétriques, leur versant amont étant plus raide que leur versant aval. Notons cependant que cette dissymétrie assez générale n'est pas vraiment absolue. Par exemple, elle ne s'observe guère à l'est de la Clew Bay (Irlande), surtout au centre du champ de drumlins.

1.2 Aperçu bibliographique

Plusieurs centaines de publications se rapportent à l'étude des drumlins, à leur forme, à leur répartition, à leur constitution. Il est exclu d'en donner ici l'inventaire (Voir notamment Charlesworth J.K. 1957). Les problèmes les plus généralement rencontrés sont l'asymétrie longitudinale des drumlins et leur uniformité dimensionnelle relative dans un même champ, où leur équidistance est à peu près constante. Leur constitution soulève un autre problème : les drumlins sont-ils des formes d'érosion, comme l'indiquent les rock-drumlins, ou des formes d'accumulation, tels les till-drumlins ? Comment expliquer dans un même champ de drumlins le voisinage de formes identiques, les unes d'érosion, les autres d'accumulation ?

Deux articles récents, de Muller E.H., 1974 et de Gillberg G., 1976 abordent l'ensemble de ces problèmes à la suite d'analyses très soigneuses, respectivement aux Etats-Unis et en Suède. Parmi leurs nombreuses conclusions, il faut mentionner :

1. Les drumlins s'établissent dans des zones où les glaciers sont actifs ;
2. La glace se trouvait à son point de fusion ;
3. La granulométrie du matériel des drumlins peut être très variable ;
4. Des courants différentiels dans l'écoulement de la glace peuvent être à l'origine de cavités basales où les drumlins s'établiront.

En 1968, Andrews J.T. and King C.A ont conclu qu'au Yorkshire, l'analyse statistique de la position des éléments suggère une accumulation correspondant à des tensions latérales apparues au sein de la glace.

2 CHOIX DES REGIONS ETUDIEES

Sur la côte occidentale de l'Irlande, face à l'Atlantique, l'érosion côtière affecte le vaste champ de drumlins de la Clew Bay, y ménageant de multiples coupes naturelles.

La composition pétrographique des constituants morainiques n'est pas uniforme. Au centre de la baie, le matériel est riche en galets calcaires noyés dans une matrice argilo-limoneuse. Au sud, les constituants sont des galets de gneiss et des granites dans une matrice sablo-limoneuse. Dans le nord, les éléments de l'Old Red Sandstone, souvent conglomératiques, sont mélangés à des débris cristallins de teinte claire, surtout des gneiss, des migmatites et des micaschistes. La matrice est sableuse. Or, au sein des drumlins, plus la matrice est sableuse, mieux les structures sédimentaires sont conservées et identifiables.

Les variations de composition reflètent la nature du substratum local, mais elles indiquent aussi l'origine des courants de glace responsables de l'édification des drumlins. Ceux-ci se sont édifiés sous l'action de courants de glace poussés vers l'ouest par l'inlandsis, mais localement déviés par les collines de Nephin (2646 ft) au nord, de Croaghmoyle (1412 ft) à l'est et celles de Croagh Patrick (2510 ft) au sud . A la suite de cette déviation, un mouvement local de convergence des glaces vers la dépression de la Clew Bay s'est établi. La disposition en éventail des drumlins le confirme, ainsi que la nature des constituants morainiques.

L'abondance de coupes fraîches et de grandes dimensions, les variations de lithologie du substratum et des constituants, ainsi que les modifications granulométriques de la matrice rassemblent dans la Clew Bay un maximum de conditions propices à l'étude des drumlins. Ailleurs en Irlande, en Ecosse, en Suède, en Finlande, les conditions d'étude sont moins favorables. Nous n'y ferons référence qu'occasionnellement.

3 ORIENTATIONS DES PRESSIONS EXERCEES AU SEIN DES DRUMLINS

3.1 Disposition des galets

Au centre du champ, les drumlins ont une majorité de galets calcaires aux arêtes arrondies et à faces planes finement striées. La figure 1 illustre graphiquement l'orientation du grand axe des cailloux, respectivement au centre des drumlins (206 cailloux), sur leur flanc nord (146 cailloux) et leur flanc sud (75 cailloux). Une majorité de cailloux se disposent grosso-modo parallèlement à l'axe des drumlins (265° W) dans leur partie centrale. Sur les flancs nord et sud, un maximum se dessine à 45° environ. Il est

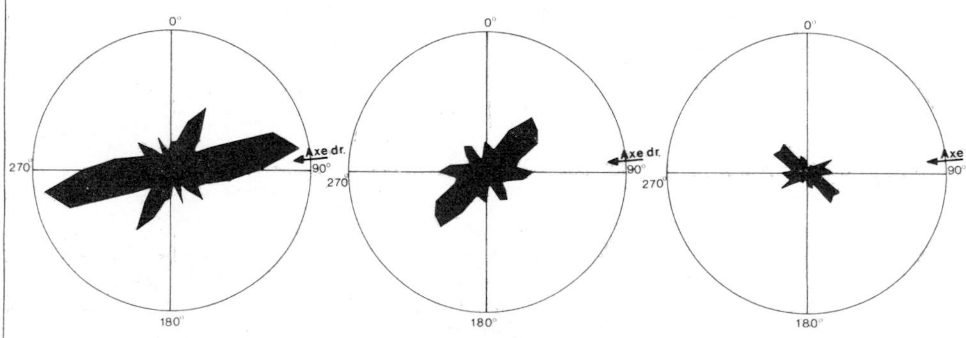

Fig 1 ; Orientation des cailloux, Axe et flancs des drumlins de INISHGOWLA, INISHOO, INISHBEE.

particulièrement net sur le flanc nord. 70,5 % des galets sont relevants vers l'aval, c'est-à-dire que leur grand axe décrit avec l'horizontale un angle positif dans le sens de progression des glaces.

Au sud de la Clew Bay (drumlin de Tully Lo au sud de Louisbourgh) et au nord (drumlins de Furnace Lough, Rosturk et Rosmurrevagh), les mêmes analyses ont livré des graphiques moins nets, bien que s'y perçoivent encore les maxima latéraux à environ 45° de l'axe des drumlins.

3.2 Litage, plis et failles

La matrice sableuse des drumlins au nord de la Clew Bay a favorisé l'apparition et la préservation d'un litage fin, parfois lenticulaire, indiquant une mise en place par l'eau courante. Souvent aussi, le litage résulte d'une sédimentation en eau calme avec graded bedding, foreset beds, et présence possible de dropstones. Le passage latéral d'un faciès d'eau courante au faciès d'eau calme peut être occasionnellement observé (drumlin de Altapheebera, au SW de Mulrany).

Ces ensembles lités ne s'observent que sporadiquement. Ils ne sont jamais épais : quelques dm au maximum. Ils sont intercalés entre des séquences plus grossières dépourvues de structures apparentes, à faciès de moraine de fond. Les conditions sous-glaciaires favorables à leur mise en place ont donc toujours été provisoires et occasionnelles.

Les structures litées sont affectées par deux types de perturbation : des déformations sont dues au tassement, d'autres résultent de poussées latérales. Les séquences de moraine de fond contenaient certainement une importante proportion de glace dont la fusion a engendré une perte de volume appréciable. Un tassement du matériel s'est opéré. (Boulton 1970 a décrit au Spitzberg des tranches de moraine de fond contenant de 10 à 80 % de glace de regel. Au glacier de Jostedalsbreen (langue de Brattebakk, Norvège) et dans plusieurs glaciers de Maurienne (Savoie, France) nous avons pu mesurer des teneurs en glace atteignant environ 40 à 60 % dans des moraines vives de cisaillement). Dans le matériel lité, un réseau serré de petites failles subverticales traduit ce tassement de la moraine sous-jacente. Il est important de remarquer que les lits superposés ne sont pas tous affectés d'un même rejeu. Les lits supérieurs sont moins affectés, voire même épargnés par les structures d'affaissement. Il y a donc eu d'abord mise en place d'un placage de moraine de fond riche de glace. La fusion de cette glace est ensuite survenue, concomitante de la sédimentation de formations litées, d'eaux courantes et d'eaux calmes.

Des plis déversés ou couchés, parfois des plis laminés ou même des plis-failles affectent les séquences litées. Il s'agit cette fois de poussées exercées latéralement, dans des directions indiquées par la disposition du plan axial des plis.
A proximité de l'axe longitudinal des drumlins, les poussées se sont exercées vers l'aval, parallèlement à cet axe, mais avec une composante ascendante d'au moins 30°. Sur le flanc des drumlins, les poussées étaient aussi ascendantes, mais exercées obliquement, suivant une résultante à la fois aval et latérale (fig.2).

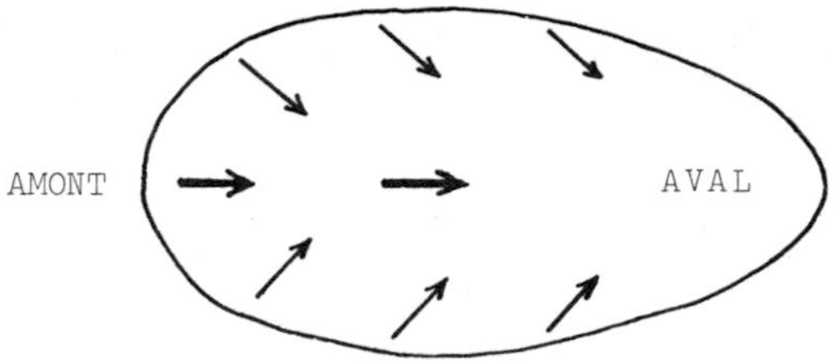

Fig. 2 - Sens des poussées exercées lors de l'édification d'un drumlin.

3.4 Etalement des débris de blocs éclatés par gélivation sous-glaciaire.

La variété lithologique de la charge graveleuse des drumlins, au nord et au sud de la Clew Bay, permet d'identifier les débris provenant d'un même bloc que la gélivation sous-glaciaire a préalablement débité en gélifracts. Pour quelques rock-and-till drumlins, il a même été possible de reconstituer le trajet de quelques mètres, parcouru par des blocs, depuis leur prise en charge dans le substratum et leur incorporation dans la moraine.

La prise en charge dans le substratum commence par un débitage dû à la gélivation sous-glaciaire. Elle provoque un écartement des joints de la roche : joints de schistosité, de stratification, diaclases.

L'ouverture de ces joints est due au gel d'une boue intersticielle injectée sous pressions par feuillets successifs. Au moyen de résines, nous avons induré in situ le matériel de remplissage des joints et des lames minces ont ensuite pu être taillées au laboratoire. Pour les micaschistes du nord de la Clew Bay, l'examen microscopique montre des feuillets de mica arrachés à la paroi des joints, que la pression a pliés et entrainés vers l'intérieur de ces joints.

Photo 1. - Boue injectée et feuilletée, dans un joint du substratum.

Aux éléments du substrat local se mélangent des débris allochtones, notamment des particules d'Old Red Sandstone et de calcaire carbonifère. (Echantillons prélevés au SW de Corraun Hill, environ 400 m à l'W du Whiteheather Cott.). La boue intersticielle injectée à l'état liquide puis gelée dans les joints, par feuillets successifs provient donc en partie d'apports du glacier. Il s'y ajoute des éléments arrachés au noyau rocheux du drumlin. (Photo 1.)

Le matériel ainsi préparé par la gélivation est ensuite incorporé à la moraine. Ce sont des glocs encore soudés entre eux par la boue intersticielle gelée. Mais dès cette incorporation à la moraine, les débris s'individualisent progressivement. La fusion de la glace intersticielle facilite cette séparation des débris. Elle libère de l'eau liquide et engendre la perte de volume, dont le litage et les failles de tassement, décrits ci-avant, sont les conséquences. (Photo 2).

Dans les cas favorables, il est possible de reconstituer le bloc débité par gélivation, et de préciser le trajet relatif

Photo 2. - Bloc d'Old Red Sandstone éclaté et débité de la droite vers la gauche.

parcouru par chaque gélifract. Les nombreux cas observés confirment les mouvements décelés par l'étude de l'orientation des galets et la disposition des failles de poussée et plans axiaux des plis couchés et déversés. Les mouvements sont ascendants, à composante aval, parallèles à la direction générale du glacier à proximité de l'axe du drumlin, et obliques sur les flancs. La plus forte valeur angulaire verticale mesurée a été de 54° (Drumlin de Furnace).

3.4 Galets injectés

Des résultats identiques sont livrés par les traces dues à l'injection de galets dans du matériel lité ou contre de fragiles plaquettes de schiste. Le sens des pressions exercées peut en effet se déduire des déformations affectant le litage ou de la disposition des débris de plaquettes.

Photo 3. - Galet injecté dans du matériel lité. Mouvement de droite à gauche

3.5 Conclusion

En bref, plusieurs structures indiquent que lors de l'élaboration des drumlins, les pressions qui s'exercent sont orientées :

- dans l'axe du drumlin, parallèlement à cette axe ;
- sur les flancs du drumlin, obliquement, suivant une résultante à la fois aval et centripète. (fig.2).

Dans tous les cas, les mouvements résultants de ces pressions sont nettement ascendants.

4 COMPORTEMENT DU GLACIER

4.1 Température de la glace basale proche du point de fusion.

Lors de l'élaboration des drumlins de la Clew Bay, l'état physique de la glace à son plancher fluctuait de part et d'autre du point de fusion. L'injection de films de boue liquide dans les joints du substratum et le regel de ces feuillets de boue le confirme (3.3 ci-avant). Les séquences litées concomitantes de la fonte de la glace interne des apports morainiques l'indique aussi (3.2).

D'autres structures sédimentaires révèlent cette fluctuation fréquente eau-glace. Des ensembles sableux finement lités et intensément chiffonnés ont été pris en charge par le glacier, transportés et incorporés en vrac à la moraine des drumlins. Ce transport en vrac implique une cimentation préalable par la glace. Un bloc ainsi gelé a subi dans la moraine de fortes tensions qui l'ont brisé. Une boue silto-argileuse sous pression s'est alors infiltrée dans les fissures. Il y a donc eu présence simultanée de matériel gelé et d'eau liquide sous pression. De telles conditions sont propices au débitage sous-glaciaire (Lliboutry, 1965 p.688).

4.2 Mouvements de la glace basale

Zumberge J. et al. (1960) ont décrit l'apparition d'un réseau de plis anticlinaux et synclinaux sur le shelf de Ross. D'orientation longitudinale, l'axe des plis résulte de la compression latérale de la langue de glace. Rozycki (cité p.612 par Lliboutry L., 1965) a décrit et cartographié un phénomène semblable sur le glacier Denman, comprimé latéralement contre le shelf de Shackleton.

Au glacier des Evettes (Savoie, France), la convergence de courants latéraux de glace comprime une langue centrale, y déterminant une succession de plis longitudinaux. La fonte à peu près totale de la couverture de neige de 1970 et 1971 et l'abondance de crevasses dues à un mouvement extensif à l'amont d'une zone de séracs nous ont permis d'effectuer à la boussole de géologue un lever structural relativement continu. Les anticlinaux et synclinaux se suivent régulièrement et l'axe des plis montre un ennoyage amont. Au plancher du glacier, l'axe des synclinaux constitue le point de départ du mouvement ascendant d'une moraine de cisaillement. Cette moraine contient environ 40 % de glace microgrenue de regel, dont l'axe des cristaux ne montre aucune orientation préférentielle. Le matériel détritique est constitué de gélifracts et de matrice sablo-limoneuse. La nature des cailloux indique une origine latérale double : gneiss à hématite du nord-est et serpentines du sud.

Des levers semblables mais moins précis sur le glacier de Fannaraken (massif de Jotunheimen, Norvège) nous ont aussi montré une succession de plis anticlinaux et synclinaux affectant une langue émissaire s'insinuant entre les deux versants d'une vallée.

5 EDIFICATION DES DRUMLINS

La déviation de la convergence des courants de glace vers la Clew Bay a probablement déterminé l'apparition dans le glacier d'une structure plissée, d'anticlinaux et de synclinaux, dont les axes étaient disposés longitudinalement. Nous venons de voir que cette allure structurale s'établit dans les glaciers actuels subissant des contraintes latérales.

Le plissement de la glace répartit inégalement les pressions exercées sur le plancher du glacier. Si les plis sont droits, c'est dans l'axe des synclinaux que les pressions sont les plus fortes. Sous le coeur des anticlinaux, le plancher glaciaire est au contraire le siège de pressions moindres. Le point de fusion est donc plus fréquemment atteint sous les synclinaux, ce qui y localise un pouvoir d'érosion accru. Ces synclinaux peuvent être à l'origine d'importantes moraines de cisaillement, avec glace interne de regel, que la progression du glacier incorpore et déplace longitudinalement vers l'aval et latéralement vers les anticlinaux. Ainsi s'opérerait sous le glacier une répartition inégale de la charge

morainique : accumulation de moraine sous les anticlinaux et, au contraire, prise en charge sous les synclinaux. D'après cette hypothèse, un champ de drumlins serait donc le reflet de la structure plissée du glacier responsable de son édification.

Les structures sédimentaires analysées dans les drumlins corroborent l'hypothèse .

1. L'orientation des galets et la déformation des lits indiquent un apport centripète, latéral sur les flancs et axial au centre des drumlins. L'étalement des gélifracts issus d'un même bloc le confirme. En outre, ces mouvements ont une composante ascendante.

2. Les failles de tassement pénécontemporaines de l'accumulation révèlent la fonte de la glace interne de regel et montrent que la glace basale était proche du point de fusion. L'engel de bouc liquide injectée dans les fissures le confirme.

L'asymétrie longitudinale des drumlins résulterait aussi du plissement du glacier. L'ennoyage amont de l'axe des plis détermine en effet une répartition inégale des pressions sur la plancher. Une poussée exercée longitudinalement résulte de la progression du glacier vers l'aval. Elle est maximum au point d'ennoyage de l'axe des plis, et notamment à l'amont des anticlinaux qu'occupent les drumlins. La raideur du versant amont des drumlins correspondrait à ce point d'ennoyage et donc de forte poussée longitudinale. En ce point s'exerce en effet une triple contrainte à composante horizontale. A la poussée longitudinale s'ajoutent deux poussées centripètes transmises par le plissement, opposées entre elles et orientées latéralement (Fig.3).

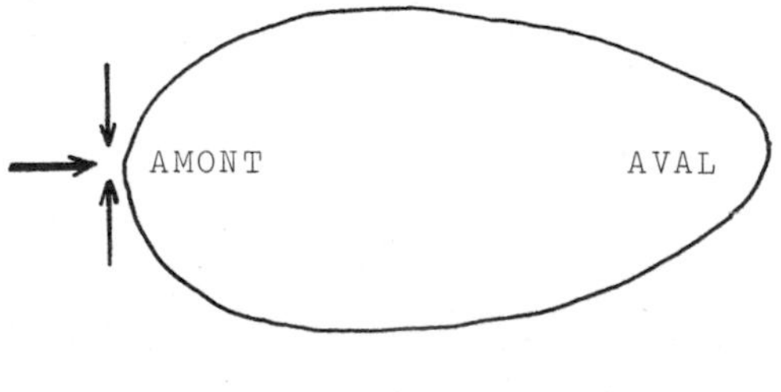

Fig.3 - Convergence de poussées à l'amont des drumlins. Résultantes horizontales.

L'amont des drumlins, zone de pressions élevées, est propice à l'érosion sous-glaciaire. L'accumulation ne pourra s'opérer qu'à l'aval où, grâce à l'ennoyage amont des plis, la glace s'écarte du plancher.

Lorsque les drumlins ne présentent pas l'asymétrie longitudinale, on peut supposer que l'angle d'ennoyage de l'axe des plis était faible. Peut-être passerait-on aux drumlins très étirés longitudinalement, les "drumlinoïds" (Tyrell J.B., 1906 et Dean W.G., 1953), lorsque l'ennoyage est nul et que l'axe des plis est proche de l'horizontale.

La présence simultanée de till-, rock-, et de rock-and-till drumlins, peut aussi se concevoir à la base d'un glacier plissé par des contraintes latérales. Si les conditions locales déterminent une érosion sous-glaciaire importante, celle-ci s'exercera plus activement sous les synclinaux du glacier, et moins intensément sous les anticlinaux. Le site d'un rock-drumlin correspondrait alors à un point d'érosion sous-glaciaire moins intense qu'au voisinage, car exercée sous un pli anticlinal de glace.

Un calibrage dimensionnel caractérise les drumlins d'un même champ, de même qu'une équidistance relative. Cette régularité ne serait que le reflet de l'uniformité des plis qui affectent la nappe de glace.

6 CONCLUSION

L'analyse sédimentaire des drumlins de la Clew Bay montre qu'il s'agit de formes édifiées par apports sous-glaciaires successifs résultant de la convergence de courants de glace. Cette glace était proche du point de fusion. L'apport de matière résulte de mouvements ascendants, à composante aval centripète, parallèles à l'axe des drumlins en leur centre, mais obliques sur les flancs.

Il semble que le plissement d'un glacier par contraintes latérales soit responsable de la naissance d'un champ de drumlins. Une pression plus importante s'exerce sur la plancher glaciaire à l'emplacement des synclinaux. L'érosion y est accrue. La glace se charge de débris incorporés à une moraine de cisaillement qui se concentre au coeur des anticlinaux voisins, vers lesquels convergent les courants sous-glaciaires. Si le glacier plissé présente des conditions d'érosion intense, celle-ci sera moins forte sous les anticlinaux où subsisteront les "rock-drumlins".

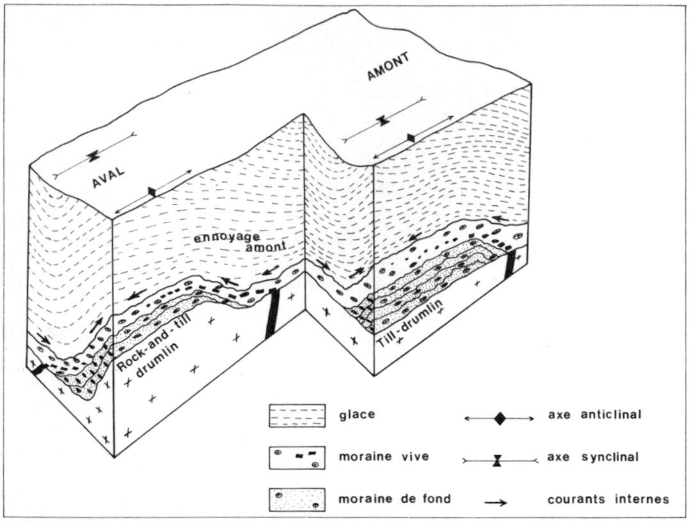

Fig. 4- Schéma de la localisation des drumlins sous les glaciers plissés.

Dans un même champ de drumlins, on observe un calibrage dimensionnel relatif des formes et une équidistance assez constante. Ce calibrage et cette équidistance reflèteraient la régularité des plis du glacier. L'asymétrie longitudinale des drumlins, avec une pente plus forte sur le versant amont, résulterait de l'ennoyage amont de l'axe des plis.

L'ensemble de ces hypothèses est schématisé sur la fig.4.

Deux conditions doivent être réunies pour la formation d'un champ de drumlins :

1. la température de la glace basale doit être légèrement sous le point de fusion. Cette fusion doit pouvoir s'opérer localement par accroissement de la pression ;

2. La nappe de glace doit subir une contrainte latérale, de telle sorte qu'un plissement longitudinal se réalise, répartissant inégalement les pressions sur le plancher glaciaire.

7 REMERCIEMENTS

Un subside du Fonds National de la Recherche Scientifique (F.N.R.S.) a permis les levers en Irlande. Nous remercions vivement cet organisme.

8 BIBLIOGRAPHIE

Andrews, J.T. & King C.A.M. 1968, Comparative till fabric variability in a till sheet and a drumlin : a small scale study. Yorkshire Geol.Soc.Proc. 36 : 435-461.

Andrews, J.T. & Smith D.I. 1970, Statistical analysis of till fabric : methodology, local and regional variability. Quaterly Journ.Geol.Soc. of London - 500 vol 125 : 502-542.

Baranowski S. 1969, Some remarks on the origin of drumlins. Geogr. Polonica - 17: 197-208.

Boulton G.S. 1970, On the origin and transport of englacial debris in Svalbard glaciers. Journ. of Glaciol. Vol 9-56 : 213-229.

Boulton G.S. 1970, On the deposition of subglacial and melt-out tills at the margin of certain Svalbard glaciers. Journ. of Glaciol. vol.9-56 : 231-245.

Boulton G.S. 1974, Processes and patterns of glacial erosion. In D.C.Coates (Ed.), Glacial Geomorphology, Ch 2 : 41-87.

Charlesworth J.K. 1957, The Quaternary Era. E.Arnold - London : 211-509.

Charlesworth J.K. 1967, The drumlins of East Down - Ireland. Journ. of Glaciol. vol.6-48 : 960-961.

Chorley R.J. 1959. The shape of drumlins. Journ. of Glaciol. vol.3-25 : 339-344.

Dean W.G. 1953, The drumlinoïd landforms of the "Barren Grounds". The Canadian Geogr. 19-30.

Gravenor C.P. 1953, The origin of drumlins. Amer.Journ. of Sc., vol.251-9 : 676-681.

Gillberg G. 1976, Drumlins in Southern Sweden. Bull.Geol.Institutions Univ.Uppsala. N.S., vol.6 : 125-189.

Guilcher A. 1962, Morphologie de la Baie de Clew (Comté de Mayo-Irlande). Bull. Assoc.Géogr.Français. N.303-304 :53-65.

Hill A.R. 1971, The internal composition and structures of drumlin in the North Down and South Antrim - Northern Ireland. Geogr.Annaler 53 A : 14-31.

Hillefors A. 1974, The stratigraphy and genesis of stoss-and lee-side moraines - Lund Studies in Geogr. - A - Physical Geogr. - n°56 : 139-154.

Lliboutry L.A. 1965, Traité de Glaciologie. T.II - Masson (Ed.) Paris : 429-1040

Muller E.H. 1974, Origins of drumlins. In D.C.Coates (Ed.), Glacial geomorphology, Ch 7 : 188-204.

Reed B., Galvin C.J. & Miller J.P. 1962, Some aspects of drumlin geometry. Am.Journ. of Sci. - 260 : 200-210.

Rose J. & Letzer J.M. 1977, Superimposed drumlins. Journ. of Glaciol. vol.18,80 : 471-480.

Smalley I.J. & Urwin D.J. 1968, The formation and shape of drumlins and their distribution and orientation in drumlin field. Journ. of Glaciol. vol.7, 51 : 377-390.

Tyrell J.B. 1906, Report on the Dubawnt, Fergusson and Kazan Rivers. Geol.Surv. Canadian Ann.Rept.9 : 3-33.

Some glaciogenic landforms in Glacier Bay National Monument, Southeastern Alaska

GEORGE M. HASELTON
Clemson University, Clemson, S.C., USA

1 INTRODUCTION

Glacier Bay National Monument lies at the north end of Alaska's Panhandle (Figures 1 and 2). It is approximately 130 km northwest of the capitol city of Juneau. The monument covers an area of approximately 8000 square kilometers. It is part of the maturely glaciated Coast Range Cordilleran complex. Highest peaks lie along the western flanks of the monument, dominated by Mount Fairweather, whose altitude is 4670 meters. The highest mountains along the eastern boundary reach altitudes close to 2134 meters. The central interior of the monument is occupied by a deep arm of the sea, Glacier Bay, and branching from it are numerous fjords (Figure 2).

The largest tributary fjord of Glacier Bay is Muir Inlet. It extends northward into the eastern portion of the Monument and is bordered along much of its east side by a large lowland covered by outwash and till. The ice sheet, which once covered all of Glacier Bay National Monument, had retreated to the entrance of Muir Inlet as recently as 1880 where it formed a steep tidal glacier.

There was rapid recession of glacial ice in Muir Inlet due to calving. The decade-by-decade retreat, from 1880 to 1960, can be seen on Figure 3. Unlike the tidal portion of this ice sheet, the ice on the shores of Muir Inlet retreated and downwasted much more slowly resulting in the development of several large detached ice masses some of which still exist today.

One of these stranded or residual ice masses was McBride Remnant Glacier (Figure 3). Its location can also be seen just south of Van Horn Ridge and on the east side of Muir Inlet north of the Klotz Hills (Figure 2).

This separated mass of glacial ice existed until the late 1960's. It became detached from ice in Muir Inlet sometime between 1930 and 1940. From the time of its separation until its disappearance, in the late 1960's, numerous ice marginal erosional and depositional features developed.

Some of the most obvious features that could be seen developing around dead ice blocks include: flow till deposits, cross-valley till ridges, crag-and-tail deposits, eskers, and ice-marginal drainage channels.

2 ICE CONTACT FEATURES

Of all these features, the most interesting are the cross-valley till ridges. The ones that could be seen developing in the field resulted from wet, plastic till that had been squeezed up into basal fractures or shear planes within dead-ice blocks. Some could possibly have formed in open crevasses, but none were ever observed forming this way. The wet till is extruded into these fractures from the base of the ice and may first be observed on the ice surface as tiny protruding "squeeze-up" ridges (Figures 7A and 7B). Once the material reaches the surface of the ice, it may flow down gradient as a

sheet. Please refer to Figure 7A for a typical illustration.

After the enclosing glacial ice melts these till ridges may persist, depending on the volume of till, porosity and permeability, and how much of the till ridge is actually ice-cored.

In the vicinity of the McBride Remnant ice these ridges seldom exceed a height of two meters but can be traced along strike for distances of up to 0.8 km. Most are much shorter having been cut by meltwater channels (Figures 4, 5, and 6). Some are developed at right angles to their neighbors and others could be seen crossing each other. They have been seen to overlap small crag-and-tail features and in one or two places small eskers were seen crossing over the tops of several of these till ridges.

Where crossings occur, the upper set of ridges crosses the lower set at angles varying from 90 to 10 degrees. These ridge dimensions and patterns show up extremely well on the plane table maps of Figures 4, 5, and 6. The highest till ridges measured ranged from 3.6 to 8 meters. Their widths varied from just a few centimeters up to 9 meters.

The average strike of the ridges is N70°E but the range in orientation is considerable, varying from N40°E to E-W. The distance between the ridges is also quite variable, ranging from 3 meters up to 150 meters, yet the average of those measured was 24 meters.

There is nothing about their spacing or morphology that suggests they are annual morainelets; as a matter of fact there seems to be nothing annual about their nature whatsoever. Several of these ridges can form in one summer.

The fact that these till ridges can be traced directly into shear planes is positive evidence for one mode of development.

Many of these ridges are asymmetrical when formed. They are steep on their lee sides and gentle on the stoss side. The angle of repose on the stoss side was found to vary from 50 to 60 degrees, while on the lee side this angle was only 10 to 12 degrees. The angular differences between the stoss and lee sides are a function of the geometry of the shears.

3 ICE DISINTEGRATION RIDGES

Small, circular, oval or irregular ridges unlike those above have formed around isolated ice blocks. They are especially well developed around those blocks that carry an excessive amount of ablation debris. Irregularities in form of these ridges reflect the shape or outline of the ice block around which they were deposited.

4 CRAG-AND-TAIL FEATURES

The crag-and-tail features across the McBride Remnant Valley consist of either a knob of resistant bedrock or an individual boulder with an elongate ridge of till streaming outward on the lee side. These features are extremely common in this area and their trend parallels that of the drumlins. Where they were measured, the till tails in the lee of boulders were 30 to 90 meters in length, from one-half to one meter in height, and one to three meters in width. Please refer to Figure 6. The length, width and height of the tails are a function of the size of the bedrock knob or boulder which provided the obstruction. Some of the crags on mountain sides east of the McBride Remnant Valley are up to one kilometer long.

The entire McBride Remnant Valley is "drumlinized" giving the topography a "roller-coaster-like" surface.

5 ESKERS

Eskers are rather numerous in the Muir Inlet area and are especially well preserved on the Casement Glacier outwash. Many of the esker ridges were still ice-cored in the middle 1960's. The interested reader is referred to the work of R. J. Price (1964).

During downwasting of the McBride Glacier, one large englacial esker system developed in the northeast corner of what is now the McBride Remnant Valley. It extended southward down the gentle gradient of an alluvial fan for at least three kilometers. In 1964, it was still cored by some 20 meters of ice.

Eskers were also seen forming from subglacial meltwater channels within the terminal portions of McBride Glacier north of Van Horn Ridge (see Figure 2).

Small eskers were examined that crossed over the tops of cross-valley till ridges. The smallest eskers measured ranged from 90 to 120 meters in length, from one to three meters in width, and one to two meters in height.

These small discontinuous eskers along the eastern edge of McBride Remnant Valley are in many cases associated with lake sediments or were found crossing small ponds. Their entrance into the small ponds allowed aggradation to take place in the subglacial channels.

An example of englacial or even superglacial esker development was seen on an isolated glacial mass referred to as Muir Remnant (see Figure 3). Collapsed features were well developed in the bedding with many cross-bedded structures slumped as separate blocks.

Perhaps one of the most interesting aspects of Glacier Bay Monument is the rapidity of retreat of its glaciers from their advanced historic positions in the late 1600's or early 1700's. By 1916 glacial ice had already retreated 105 km from the mouth of Glacier Bay to the terminus of the Grand Pacific Glacier of that date. Even today Muir Glacier is approximately 40 km from its 1890 position (Figure 3).

As recently as 1910 there was still approximately 300 meters of glacial ice over the McBride Remnant Valley. By 1935 it had thinned to 150 meters. By 1950 only about half the area was ice covered and its thickness is estimated to have been between 30 and 60 meters. By 1966 only small blocks of ice remained and these were covered by thick masses of till and gravel. Some small ice blocks may still exist today in this region and if so they have been well protected by a thick mantle of overburden.

The most recent Neoglacial ice advance probably was encroaching on the Muir Inlet region between 4000 and 3500 years ago and had thickened to 460 meters above sea level by 2200 years before the present.

At elevations close to sea level, downwasting of glacial ice takes place at rates close to 7 meters per year. These observations have been made by W. O. Field (1947), L. D. Taylor (1962) and the writer (Haselton, 1966).

6 REFERENCES

Field, W. O., Jr., 1924-1926, The Fairweather Range: Mountaineering and Glacier Studies, Appalachia, vol. 16, pp. 460-472.

_______, 1937, Observations on Alaskan Coastal Glaciers in 1935, Geog. Rev., vol. 27, pp. 63-81.

_______, 1942, Glacial Studies in Alaska, Geog. Rev., vol. 32, pp. 154-155.

_______, 1947, Glacier Recession in Muir Inlet, Glacier Bay Alaska, Geog. Rev., vol. 37, pp. 369-399.

_______, Editor, 1975, Mountain Glaciers of the Northern Hemisphere, Corps of Engineers, Tech. Info. Analysis Center, Cold Regions Res. and Eng. Lab., Hanover, N.H., USA; vol. 1, 689 pp., vol. 2, 932 pp., and accompanying Atlas with 48 plates.

Goldthwait, R. P., 1963, Dating the Little Ice Age in Glacier Bay, Alaska, Intern. Geol. Cong. XXI, 1960, pt. XXVII, pp. 37-46.

_______, et al., 1966, Soil Development and Ecological Succession in a Deglaciated Area of Muir Inlet, Southeast Alaska, Ohio State Univ. Res. Foundation, Inst. Polar Studies, Rept. 20, 167 pp.

Haselton, G. M., 1966, Glacial Geology of Muir Inlet, Southeast Alaska, Ohio State Univ. Res. Foundation, Inst. Polar Studies, Rept. 18, 34 pp.

_______, 1967, Glacial Geology of Muir Inlet, Southeast Alaska, Ph.D. Dissertation, The Ohio State Univ., 228 pp.

McKenzie, G. D., 1970, Glacial Geology of Adams Inlet, Southeast Alaska, Ohio State Univ. Res. Foundation, Inst. Polar Studies, Rept. 25, 121 pp.

Price, R. J., 1964, Land Forms Produced by the Wastage of the Casement Glacier, Southeast Alaska, Ohio State Univ. Res. Foundation, Inst. Polar Studies, Rept. 9, 41 pp.

Taylor, L. D., 1962, Ice Structures, Burroughs Glacier, Southeast Alaska, Ohio State Univ. Res. Foundation, Inst. Polar Studies, Rept. 3, 110 pp.

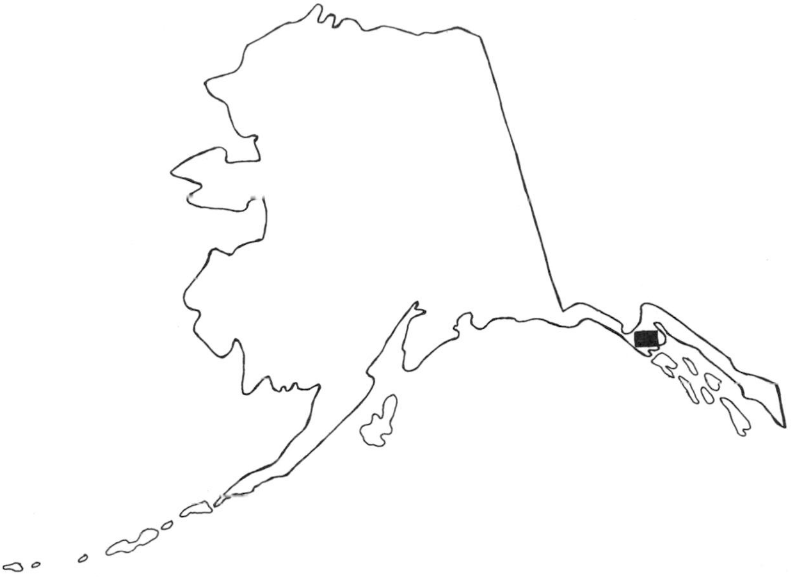

Figure 1. Alaska and location of field area shown by black square.

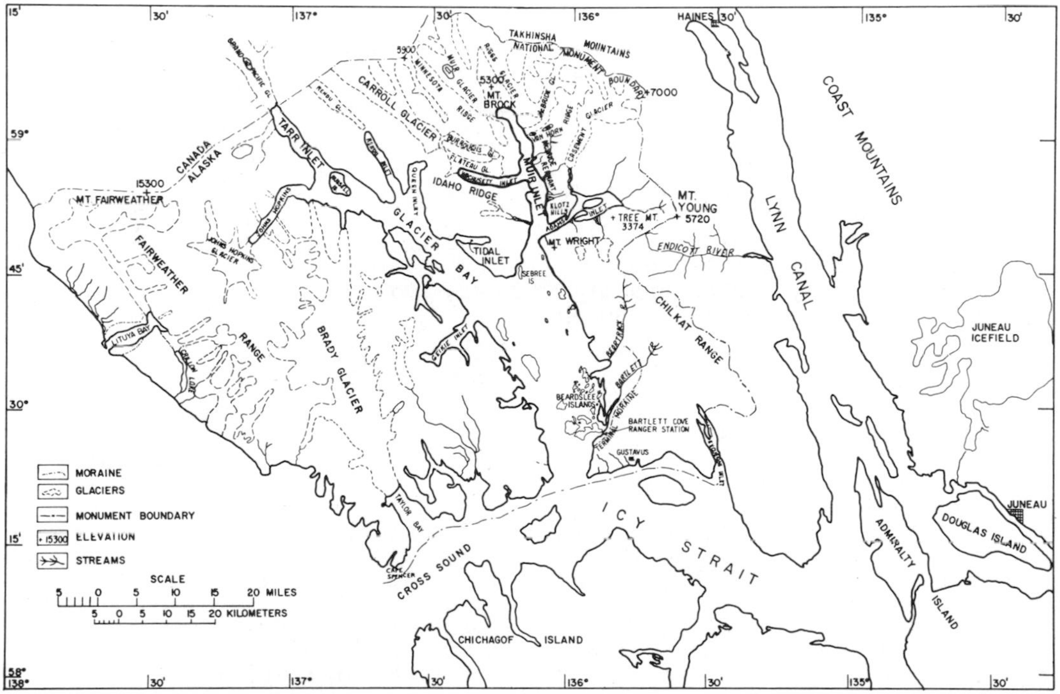

Figure 2. Location map of Glacier Bay National Monument west of Juneau, Southeastern Alaska.

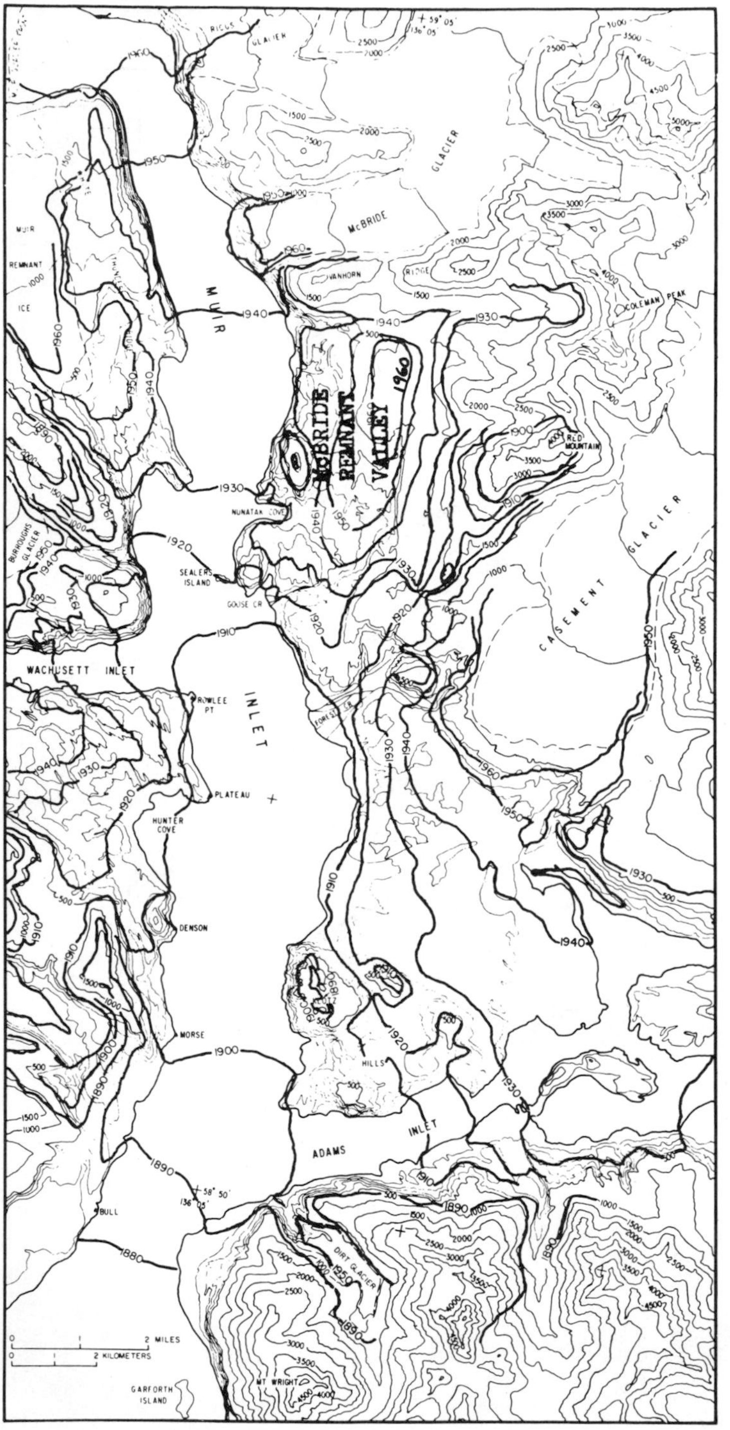

Figure 3. Map of Muir Inlet region showing retreat of glacier fronts by decades since 1880 and location of McBride Remnant Valley.

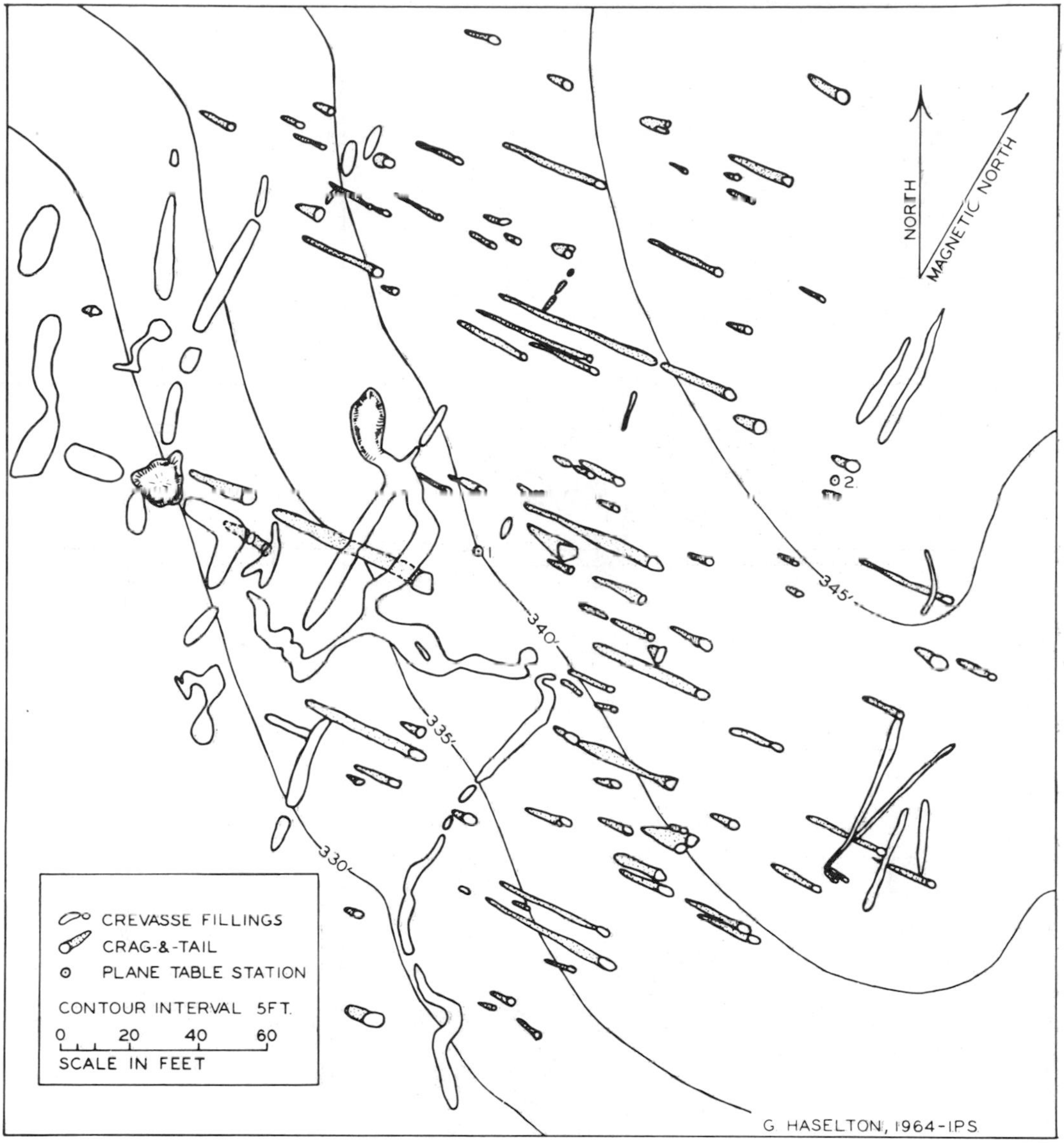

Figure 4. Plane table map of crevasse filling ridges and crag-and-tail ridges at McBride Remnant Valley, east side of Muir Inlet, Alaska.

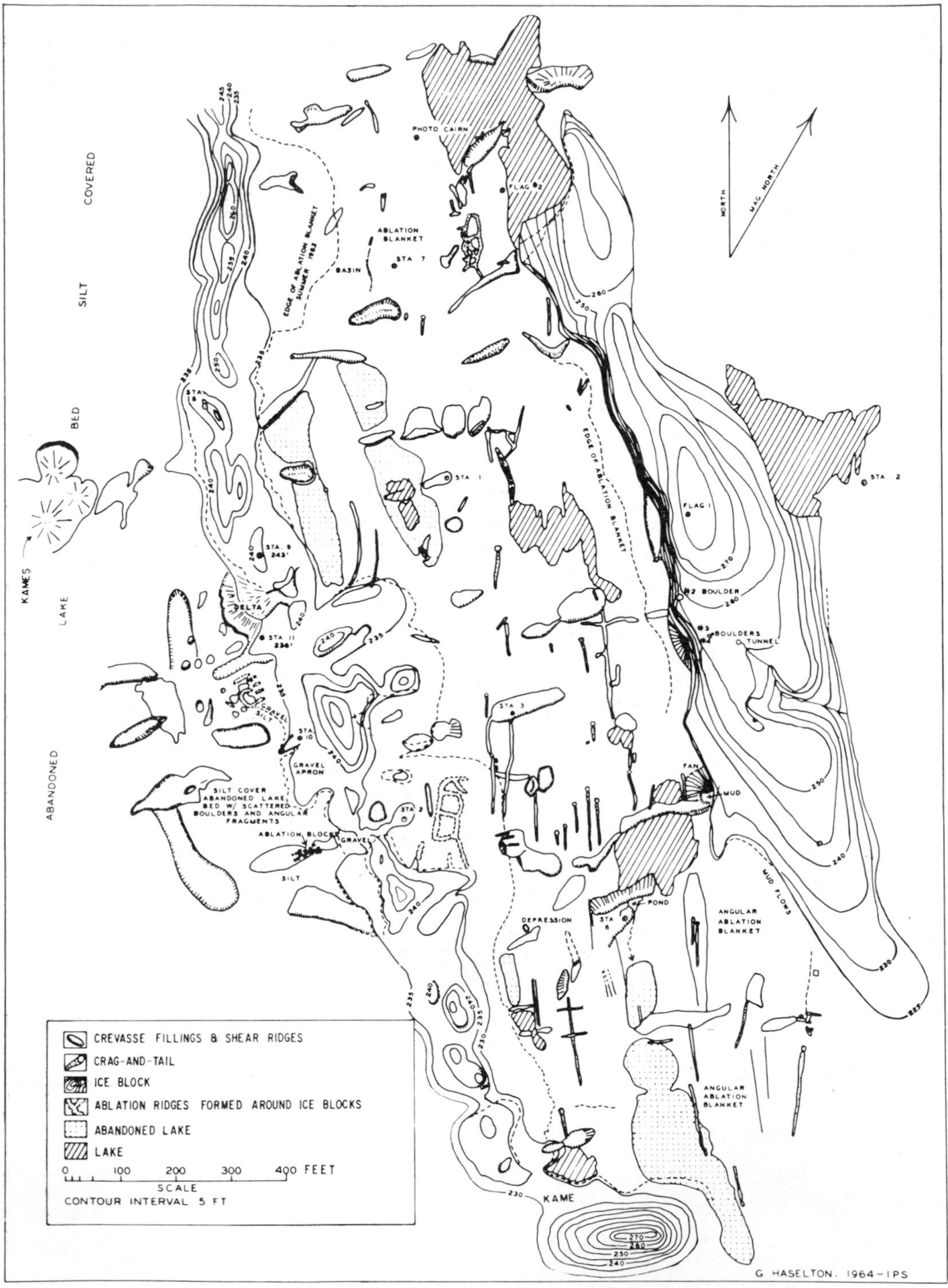

Figure 5. Plane table map of the details of ice contact deposits and residual ice blocks of McBride Remnant Glacier, east side Muir Inlet, Alaska.

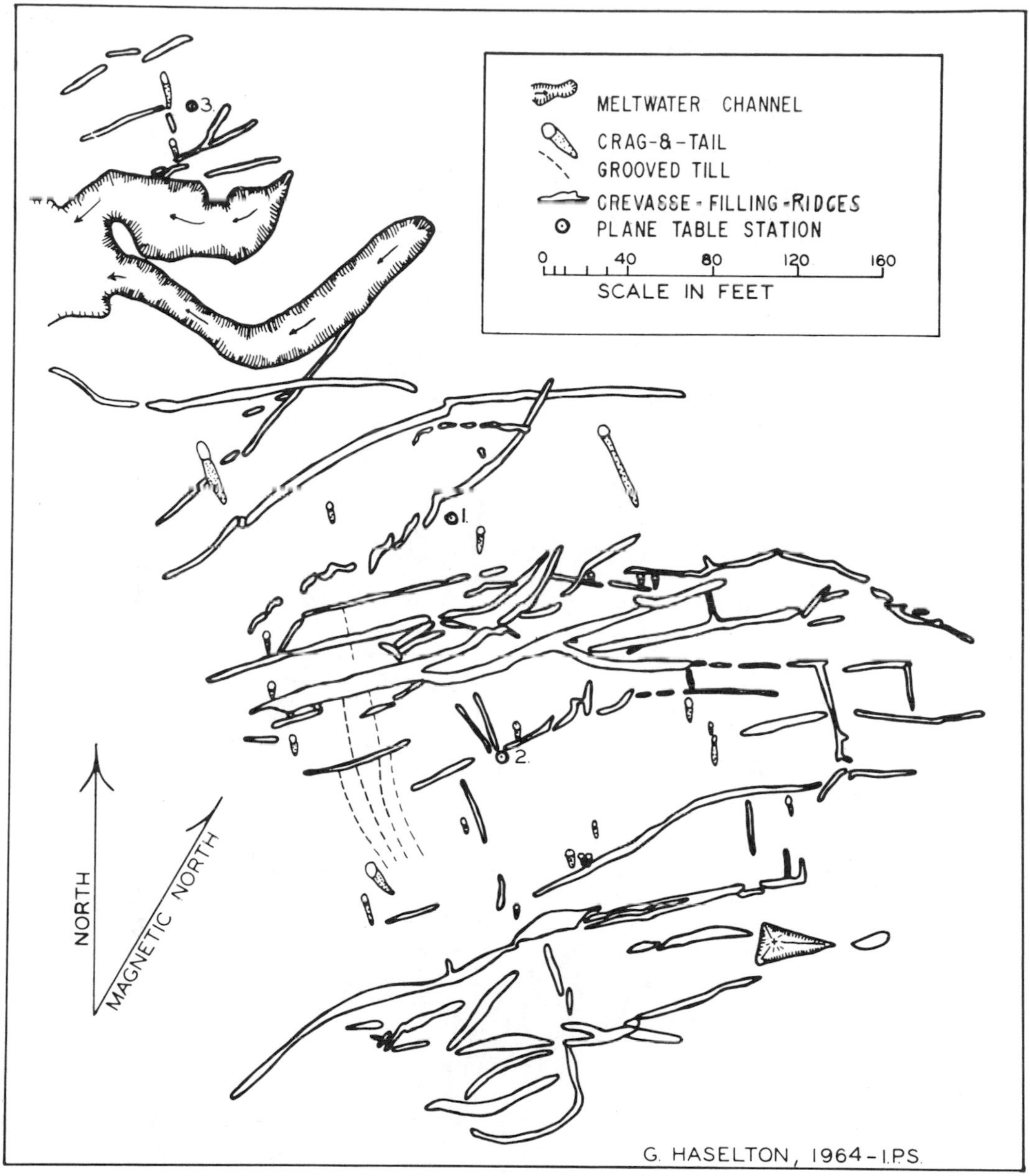

Figure 6. Plane table map of cross-valley till ridges developed on a drumlinized till plain, McBride Remnant Valley, east side of Muir Inlet, Alaska.

(A)

(B)

Figure 7. (A) Water saturated silt-rich till being forced upward into a small fissure or shear plane along the terminus of Plateau Glacier, Wachusett Inlet, Alaska (ice thickness = 6 m).
(B) Shear plane filled with till being released by melt-out and flow till, McBride Remnant Valley, Alaska, Muir Inlet.

Weichselian till stratigraphy in central South-Norway

KARI GARNES
Universitetsbiblioteket, Bergen, Norway

1 INTRODUCTION

There are local thick basal tills with a distinct stratigraphy in the central parts of South-Norway. These deposits are often topographically conditioned and situated in so-called till traps. Some tributaries in Gudbrandsdalen are examples of such traps. In these steep, narrow tributaries recent slope processes have been intensive and the till bodies are laid bare in open ravines. Detailed litho- and chronostratigraphic investigations of the tills throw light on important questions in relation to the development and course of the last Scandinavian inland ice. The studies give many informations about till genesis.

Valley fillings by tills in Norway were discussed by Mangerud (1965) who claimed that the thick tills in the tributaries in Gudbrandsdalen were accumulated only by ice which moved across the deep tributaries.

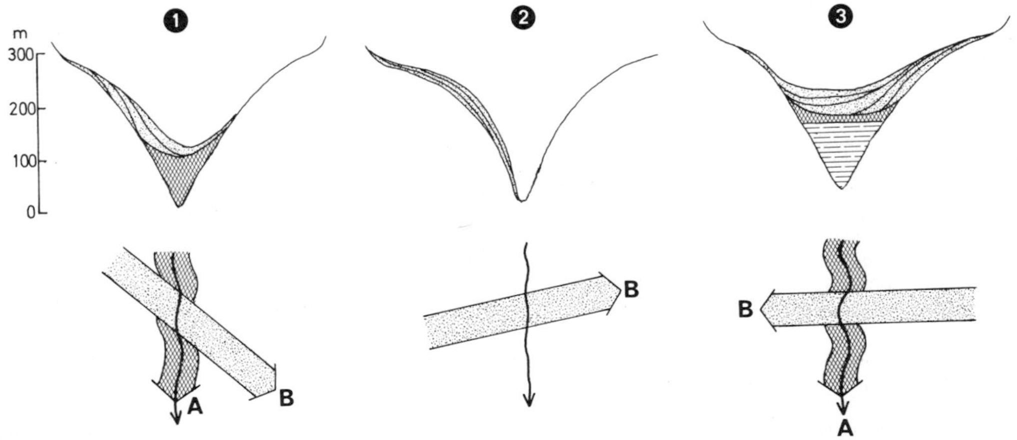

Fig.1. Mode of accumulation of thick basal tills in tributary valleys in the Mid-Gudbrandsdal valley. Type 1 shows a lower basal till deposited by a glacier flowing out the valley (phase A on Fig.3). The upper till is accumulated by glaciers flowing obliquely across the valley (phase B). The accumulation first started as a lee-side moraine on the upper part of the valley side, just below the well-marked valley shoulder. Type 2 lacks the valley-glacier phase, the large tills belonging to phase B only. The accumulation did not reach the valley bottom. Type 3 corresponds to type 1, but here the valley bottom contained thick waterlaid sediments when the deposition of tills started. Due to these old sediments, and also because of glacier flow directly across the valley, the accumulation during phase B reached the opposite valley side. After Garnes & Bergersen (1977).

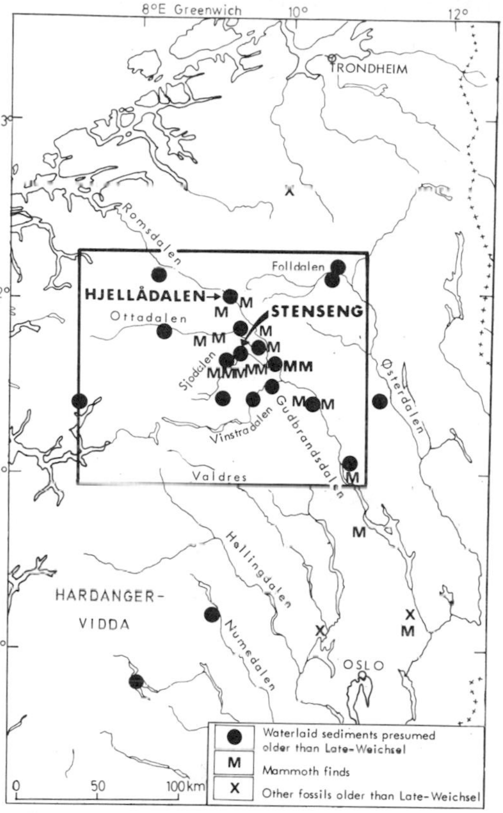

Fig.2. Key map. Localities in central South-Norway with sediments and fossils presumed older than Late-Weichsel.

The author has later in several papers described the till stratigraphy in both the main valleys and their tributaries (principally in areas situated beneath the ice-divides of the inland ice) (Bergersen & Garnes 1972, Garnes 1973, Garnes & Bergersen 1977). Three conclusions are emphasized in particular in these papers:

1. Since the till stratigraphy has been found to be younger than the sub-till sediments dating from an ice-free period (denoted "The Gudbrandsdalen Interstadial"), the reconstructed stratigraphy must belong to the last ice age (Weichsel) (see Garnes & Bergersen 1977).

2. Different basal tills correlate accurately with reconstructed ice movements. The valley infills of tills are result from different modes of accumulation. In tributary valleys three types are distinguished (Fig.1).

3. Most of the Norwegian basal tills seem to be deposited during early phases of the last ice age.

These conclusions are to a great extent built on the existence of sub-till sediments found in a number of places in the investigated area (cf. Fig.2). Regional studies of sediments combined with recent ^{14}C datings of mammoth finds which have been excavated in the sediments, indicates that "The Gudbrandsdalen Interstadial" existed more than 40,000 years ago (Garnes & Heintz 1978, in print). Although the indicates are not conclusive it is highly probable that the existence of sub-till sediments in the Gudbrandsdal Region were all deposited during the same ice-free period. These sediments are the oldest Quaternary deposits hitherto found in the investigated area.

Results from till stratigraphic investigations in a tributary to Gudbrandsdalen, namely the Hjellådal, situated close to the main water divide in the north and thus peripheral in relation to the main ice-divide's southernmost position, will be discussed now, see Fig.2. A genetic interpretation of the till stratigraphic profiles requires a detailed reconstruction of the ice movements in the region. This is the case especially in areas situated underneath migrating ice divides.

2 ICE MOVEMENTS

Based on striae analysis, studies of till stratigraphy, geomorphology, and deglaciation phenomena, the last ice age in East-Jotunheimen and Gudbrandsdalen (a mountain and valley landscape with relief up to 1.500 metres) can be divided into four phases (see Figs. 3 and 4). It is noteworthy that whereas during phases A and B great erosion and accumulation have taken place in the investigated area, little has occured during phase C. However, the latter is supposed to comprise a longer period of time during the maximum of last ice age. In considerable areas around the Mid-Gudbrandsdal there are relatively few traces of the deglaciation, phase D. It is important to bear in mind that the deglaciation of the inland ice in central parts of South-Norway was dominated by spot-wastage so that the last remnants of ice could be found in valleys and depressions. During the wastage dramatic changes of the ice movements were brought about because the melt-water found new and lower-lying overflow passes. The alteration of the main drainage routes brought in many cases an increased ice-gradient with the formation of considerable (not climatically conditioned) lateral moraines and other lateral phenomena (Garnes 1978).

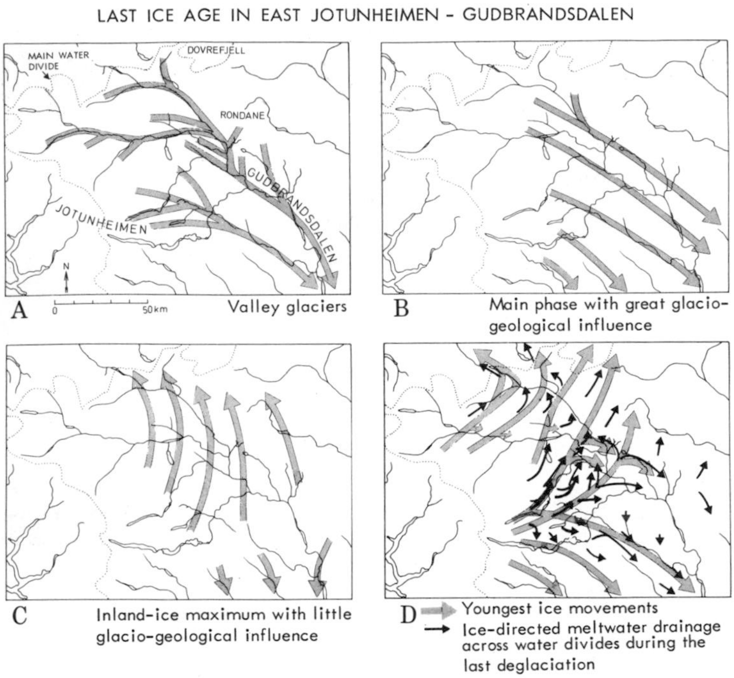

Fig.3. A reconstruction of four phases of ice movements during Weichsel. Location, see Fig.2. The region along the main water divide and to the west and north are not investigated (after Garnes 1975).

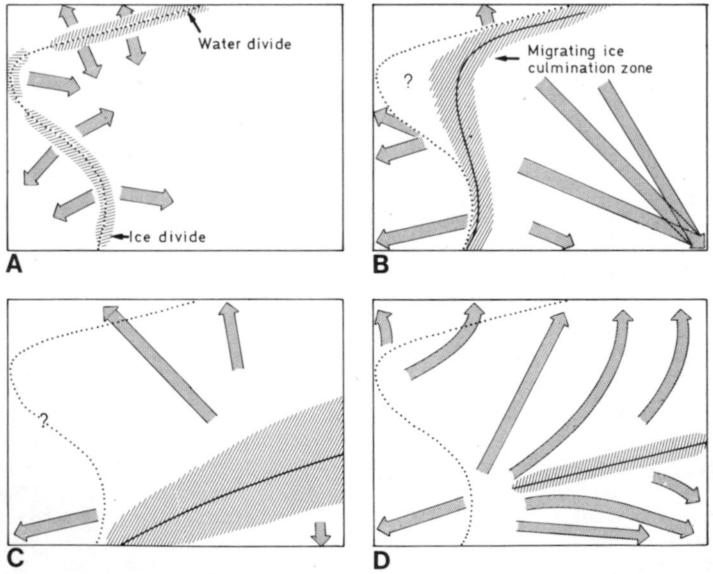

Fig.4. The interpretation of the ice divide's migration in part of central South-Norway during the last ice age (cf. Fig.3): A. Following "The Gudbrandsdalen Interstadial", the glaciation is found to have started in the highest mountain massivs along the main water divide. The glacier streams were guided by the main valley pattern. B. Slowly the ice divide migrated both to the south and to the east. When this ice divide passed an area the sense of flow was reversed. The striae, therefore, can be found having opposite directions. The ice movements towards the west and the north-west crossed the water divide. C. The ice divide approaching its southernmost position. D. At an early stage of the deglaciation period several ice culminations were built up in the vicinity of the water divide. However, the east-west main ice divide still existed. This situation gave north-north-easterly ice movements in the area investigated and lasted until the inland ice had melted to such an extent that it could only be located in the valleys.

3 ICE DIVIDE MIGRATION

While we find in Fig.3 a continuous development from phase A to phase B and from phase C to D, the investigations seem to indicate that there is a discontinuity between phases B and C, especially when the striae are examined. This ought to indicate a sudden alteration from an ice divide position B to position C. Similar observations have been made by others (e.g. Lundqvist 1969: 148-149). Vorren (1977: 255) postulates "that such a migration on Hardangervidda was caused by a glacier surge out of the Hardangerfjord". He also suggests applying this explanation to other parts of the Weichselian ice-sheet (Vorren op.cit.). Such apparently sudden migration of an ice divide from one position to another can in the investigated area be most simply explained as a successive migration of the ice divide zone from the water divide where glaciation presumably started, to a position circa 150 km south to south-east of this. The observations indicate that around the time when the inland ice culminated there was minimal movement underneath the ice in a relatively wide zone. This phenomenon has been pointed out by many authors. Judging from the area where striae are absent or are so faint that they cannot be pointed out in the investigated area, this zone has a width of about ten kilometres or more. When a mobile ice divide zone passes an area almost parallel with itself the migration will presumably be difficult to indicate. Theoretically the change from little erosion and insignificant striae caused by movements in the ice divide effects thin and poorly visible striae twice, viz. just before and after approaching any locality. Therefore, it is difficult to draw accurate conclusions about the direction and age of the striae near the ice divide zone. There is possibly a corresponding gradual transition in the accumulation of tills in this zone.

It is outlined on Fig.4 how a continuing, successive movement of the ice divide can explain the following results that combined analyses have led to, see also Fig.3. If this interpretation is valid the illustration also gives an explanation of regional differences in the distribution of different till bodies. While the till that can be correlated with pase C are totally absent in the area around the ice divide in this phase, tills are found deposited during phase C closer to the water divide, cf. Fig. 5, T_2. When it comes to the tills from the deglaciation period too, it becomes evident that these are generally more extensive in the periphery than near the ice divide zone in phase D. The figure, therefore, offers an explanation of the relationship which the author has emphasized in several papers, namely that the tills in the area around the ice divide zone's southernmost position almost exclusively date from phases A and B.

An important conclusion is that age and extension of the different types of tills depends on the area's location in relation to the ice divide's position.

4 THE HJELLÅDAL

The investigations in the Hjellådal include mapping, analysis of striae, texture- and structure investigations of various types of tills by means of stone counts, analyses of roundness and grain-size distribution, mineralogic investigations and fabric measurements. The mineralogic investigations are carried out by separation with heavy fluid of the fraction 125-250 μ and thin-section microscopy of this heavy mineral fraction. In addition X-ray diffractometric investigations of the fraction more than 8 μ are carried out. The fabric analyses include 50-100 particles in each analysis. In some cases with a very good orientation a smaller number have been employed. Dip measurements of particles with a dip more than 30° are in most cases omitted. The average of the measured dip can, therefore, be less than the real average. The investigations have been undertaken in many sections but the results are only presented from one profile, Einbu, cf. Figs.5 and 8.

The Hjellådal is a three-kilometre, narrow tributary hanging about 100 metres above Gudbrandsdalen which here has a one-kilometre wide floor and an altitude of 500 metres. The Hjellådal declines on an average 190 m/km. Apart from a cover of coarse fluvial sediments (mainly blocks) the Hjellå river is flowing close to the bedrock along the valley floor. In great parts of the valley-slopes the rock is exposed. Circa 800 metres from Gudbrandsdalen the Einbudal valley comes from the north-west into the Hjellådal hanging circa 50 metres. Here the mouth of the Einbudal is completely filled with loose material, Fig.6. The Einbudal is near parallel with Gudbrandsdalen and has acted as a pass from a large valley parallel to the Hjellådal. The Einbudal has played a great role as drainage for ice for a long period during the ice ages because it leads in the same direction as the main valleys like Gudbrandsdalen.

The terrain around the Hjellådal consists

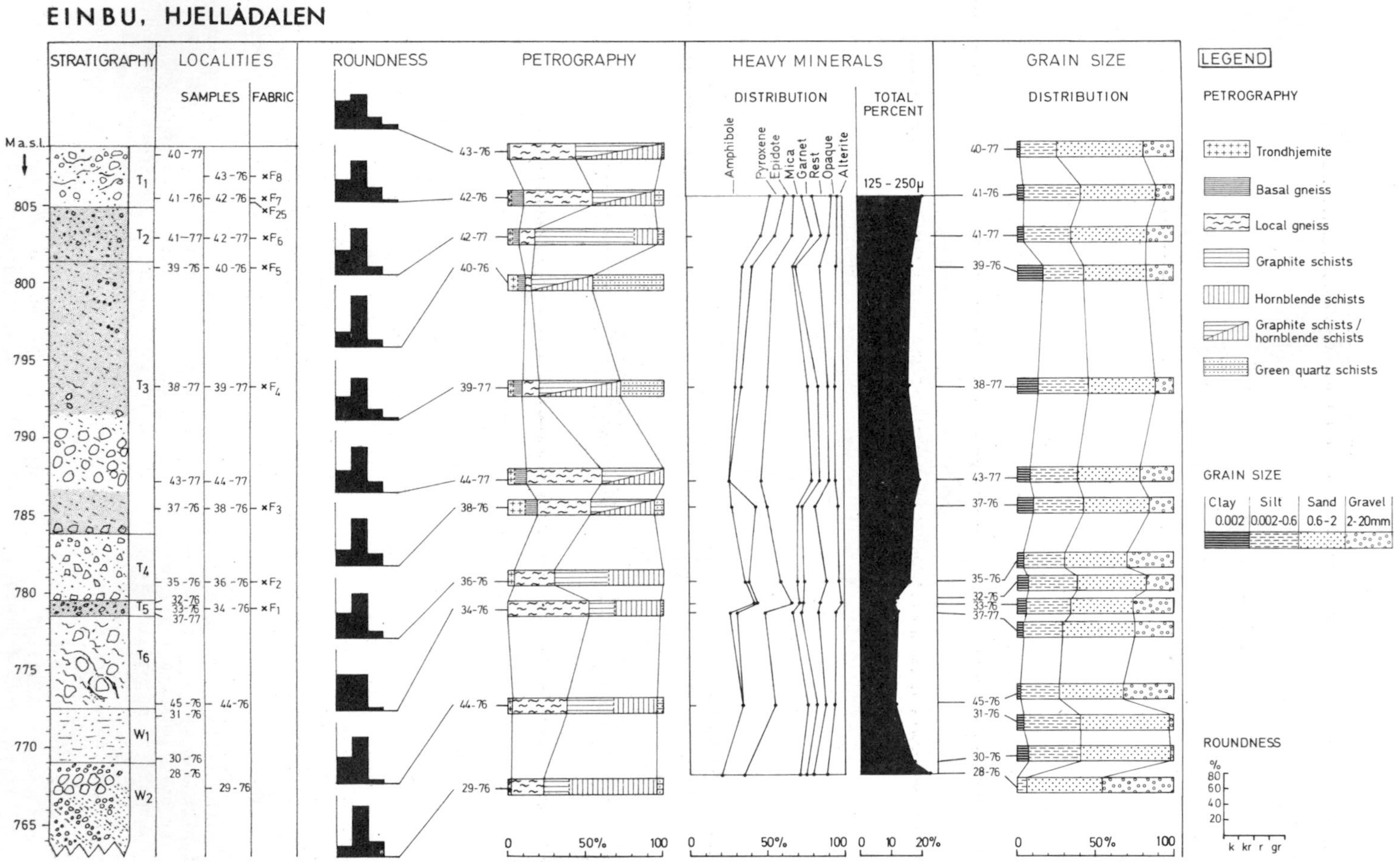

Fig.5. Compound diagram showing the result of different analyses in one section at Einbu in Hjellådalen. The diagram shows the stratigraphy with sample localities plotted. Roundness classes are: angular (k), abraded angles (kr), rounded (r), and well rounded (gr). The method is after Bergersen (1973). The heavy mineral contents are from the fraction 125-250μ separated with liquid with specific gravity 2,82. The distribution of each mineral group is found by microscopy-identification of 300 grains from each sample. The method of heavy mineral identification is after Galehouse (1969). The result of the fabric analyses is shown on Fig.8.

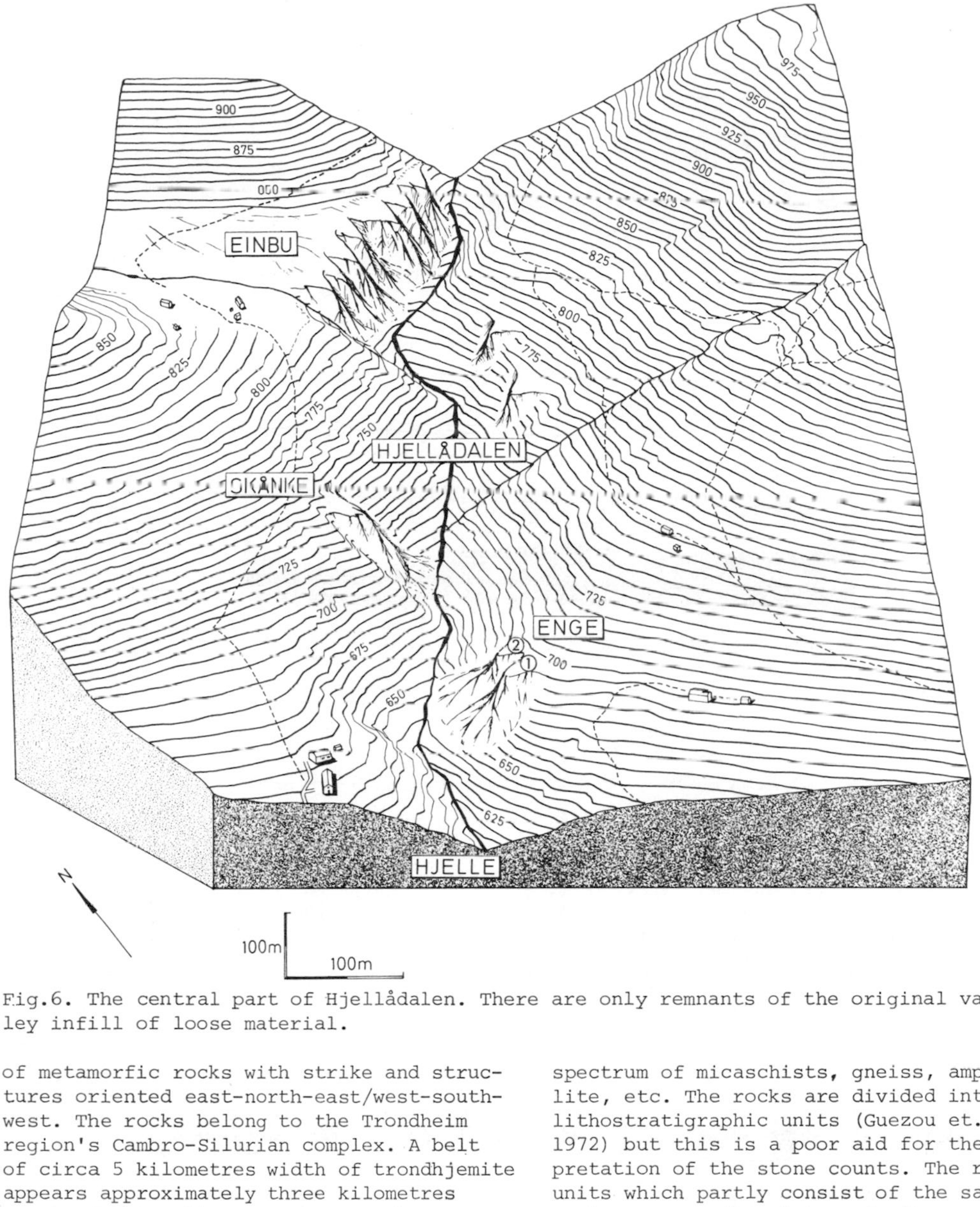

Fig.6. The central part of Hjellådalen. There are only remnants of the original valley infill of loose material.

of metamorfic rocks with strike and structures oriented east-north-east/west-south-west. The rocks belong to the Trondheim region's Cambro-Silurian complex. A belt of circa 5 kilometres width of trondhjemite appears approximately three kilometres north of the valley. Further to the east intrusive gabbro is exposed in an almost equally large area towards the north-east. Stone counts in the tills of the Hjellådal all show insignificant amounts of gabbro (0-2%) which strongly indicate that there has been no transport of material to the Hjellådal from north-east. The scattered gabbro finds can originate from the existence of gabbro along the trondhjemite's border a little north of the Hjellådal, or can be misinterpreted amphibolite.

The metamorphic rocks include a wide spectrum of micaschists, gneiss, amphibolite, etc. The rocks are divided into lithostratigraphic units (Guezou et. al. 1972) but this is a poor aid for the interpretation of the stone counts. The rock units which partly consist of the same rocks are overthrust towards the south-southeast. Each unit have thus got numerous narrow ribbons which are repeated constantly. It is necessary to have a detailed knowledge of the rocks of the area if one is to draw conclusions about the directions of the rock transport based on the petrography of the tills. Seven kilometres towards the north-east the terrain consists of basal gneiss (see Fig.7).

There are appreciable differences in rocks in the till stratigraphy and a great deal

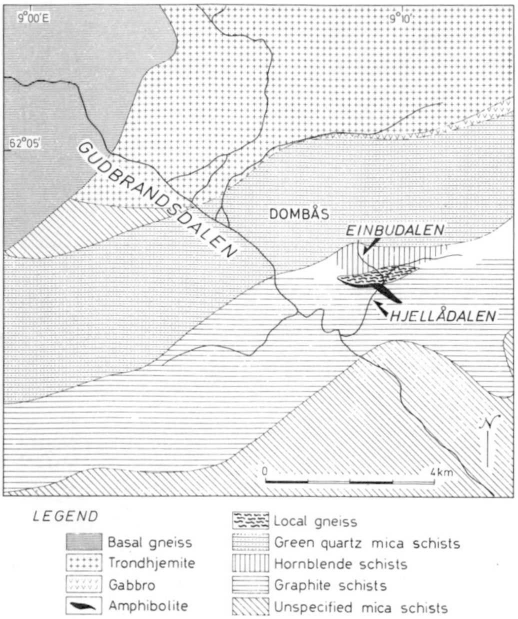

Fig.7. Simplified bedrock map adjusted to the stone-count classification groups, mainly after Guezou et.al.(1972).

of work has been carried out to make use of stone counts to differentiate or correlate the tills. The difficulties in finding a suitable classification are, however, so great that only a rough classification has been chosen. In addition to the above-mentioned intrusives (trondhjemite and gabbro) a zone division east-north-east/west-south-west of graphite-schist containing gneiss, mica-schists, and greenish quartz-mica-schist which largely lodge in order northerly from the Hjellådal to the intrusive border three kilometres north of the valley, has played a great part in these considerations.

In the lower parts of both slopes along the Hjellådal large accumulations of loose material are sporadically deposited and usually display ledges. The inner rims of these ledges reach 75-100 metres above the valley floor near the mouth of the valley but only twenty metres above the valley floor two kilometres up. Here the valley infill ceases to exist. This shows that the valley's gradient is greater than that of the ledge. The ledge is situated circa fifteen metres higher in the north-west than in the south-east. While there is little loose material above the ledge to the north-west, there is on the south-east side an even cover of till above the ledge as well. At all obvious ledges there are open sections and the slope prosesses are very active today. It is obvious that the loose material was much more widely distributed than we can observe today, and there are strong reasons to believe that at the end of the last ice age a valley filling existed which extended at least two kilometres into the valley and can be considered to be of a thickness from fifty to about ten metres along Talweg. Corresponding altitudes, till fabric and extention of ablation tills indicate that the ledge forms are remains from a primary accumulation level, although at the north of the valley at Skånke glacio-fluvial erosion can be traced.

Stratigraphic investigations of the valley filling in seven profiles show that the loose material distinctly consists of tills which to a great degree can be correlated from one section to another, see Fig.8. As can be derived from the figure the youngest till (T_1) is present in all profiles, except at Enge 1, and its thickness increases up the valley.

5 DESCRIPTION AND DISCUSSION OF THE STRATIGRAPHY AT EINBU

The Einbu locality is situated at the mouth of the Einbudal. Where the valleys converge the Hjellådal is wide and here the largest remains from the valley filling are preserved. It is noteworthy that at the outer end of the Einbudal there seems to be an intact valley filling, which must be considered with the fact that the stream flowing through the Einbudal carries little water, cf. Figs.9 and 6. Fig.5 shows the lithostratigraphy and some of the results of analyses made in one of the sections at Einbu. The stratigraphy is founded on a comprehensive use of different methods but are mostly supported on fabric and texture/structure studies.

At the bottom of the profile coarse fluvial sediments are laid bare (W_2). They are at least five metres thick but the sediments are likely to continue under slumped material for another ten metres. The sediments which mainly consist of gravel, pebbles, and blocks, have in this section a dip opposite the slope of the Hjellådal and is distinctly deposited from the Einbudal as a fan. Further into the Hjellådal sediments are observed which can have been deposited along the Hjellådal also.

Above the coarse sediments there is three to four metres of well-sorted sand and silt, nearly wholly without pebbles and little disturbed, W_1. In a section at the stream from the Einbudal corresponding sediments

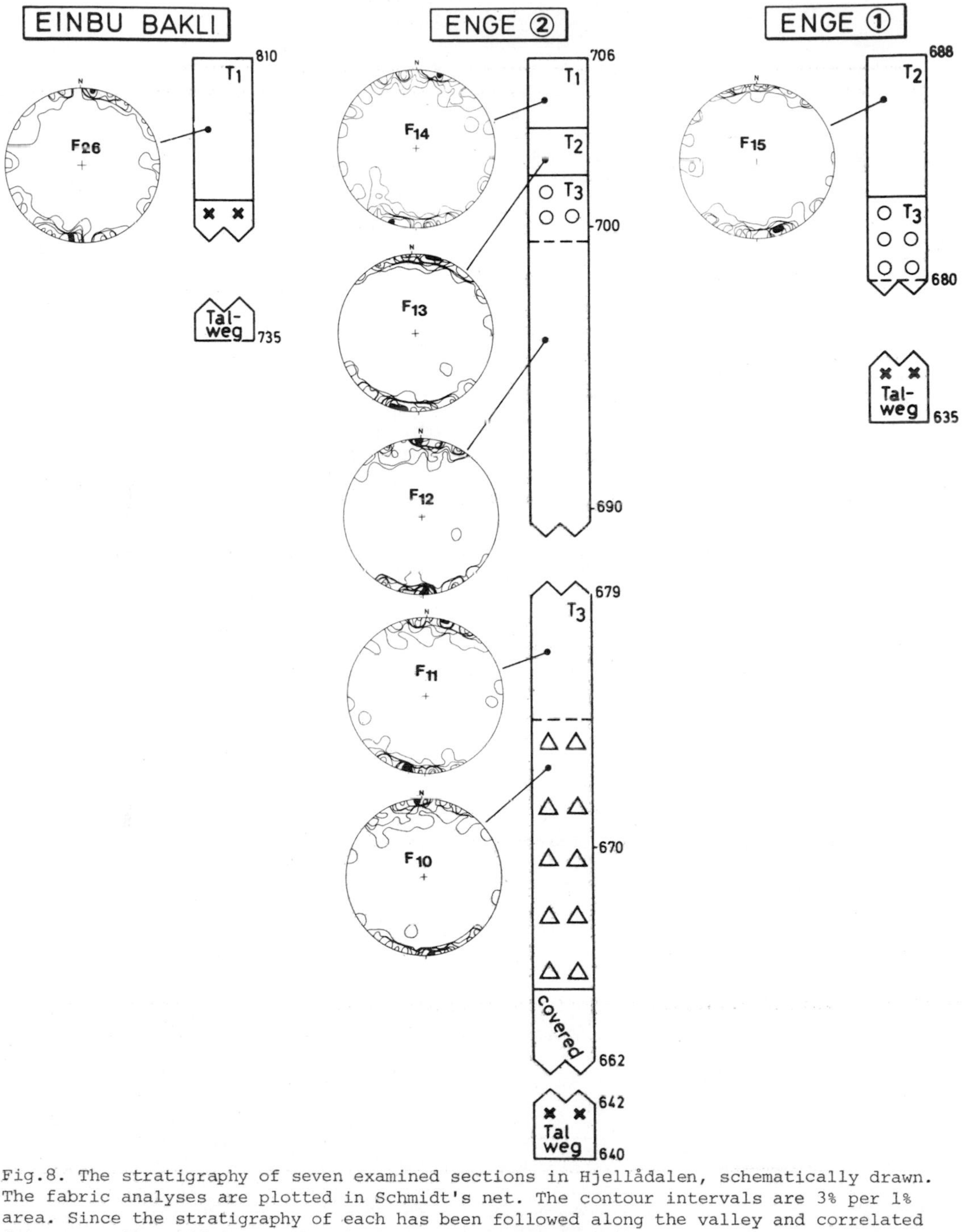

Fig.8. The stratigraphy of seven examined sections in Hjellådalen, schematically drawn. The fabric analyses are plotted in Schmidt's net. The contour intervals are 3% per 1% area. Since the stratigraphy of each has been followed along the valley and correlated from one valley side to another it is presumed that there has been a continuous accumulation of till with only a slight ice erosion during the last ice age. Pay attention to the lower block layer in till T_3. Location: see Fig.12.

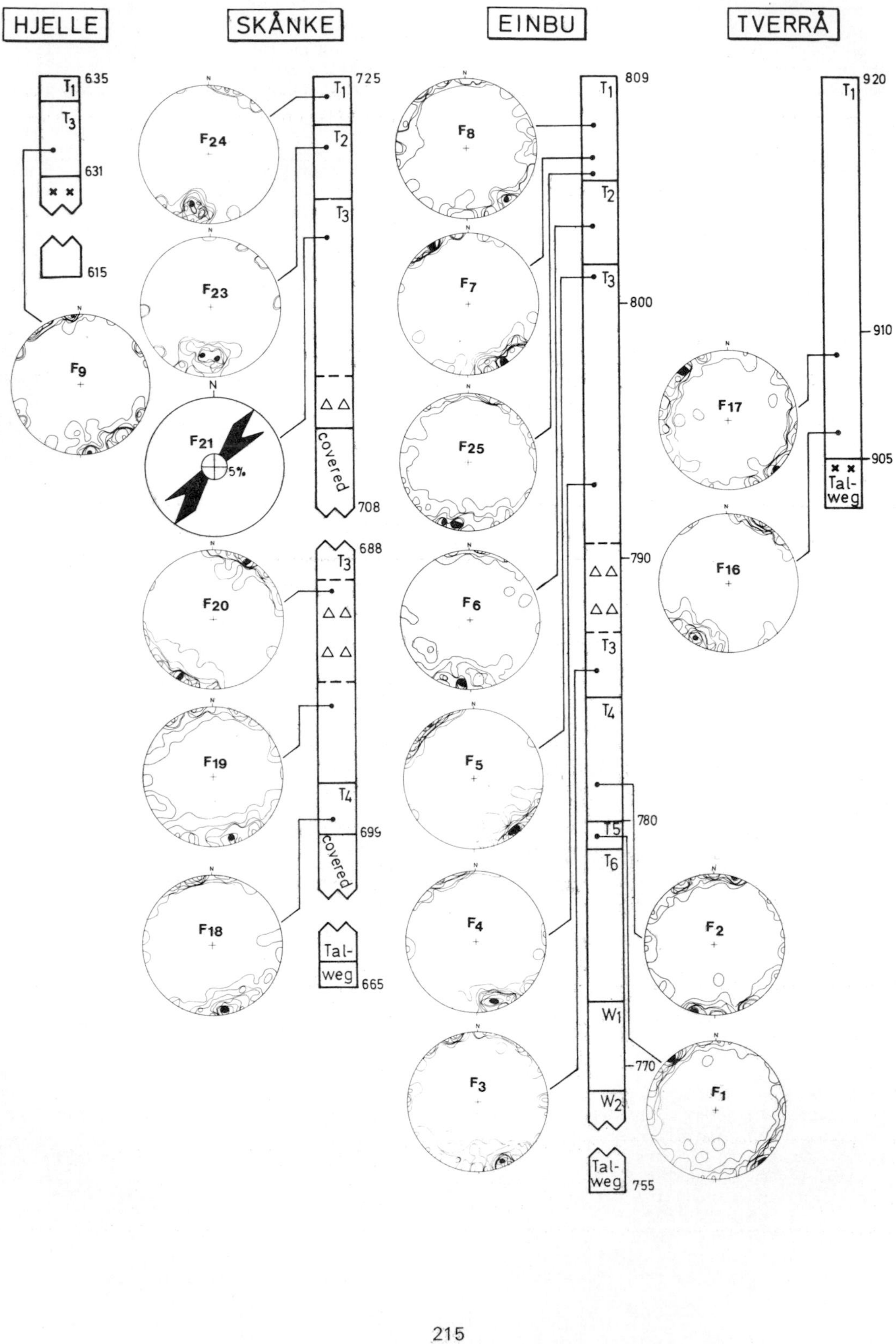
HJELLE
SKÅNKE
EINBU
TVERRÅ
T1
T3
635
631
615
F9
F24
F23
F21
5%
F20
F19
F18
725
T2
covered
708
688
T4
699
Tal-
weg
665
809
F8
F7
F25
F6
F5
F4
F3
800
790
780
T5
T6
W1
770
W2
755
920
910
905
F17
F16
F2
F1

consist of sand and gravel of at least eight metres (Fig.10). In the closest section west of the stream too, the same sediments (W_1), are observed, but not in any section further out into the Hjellådal.

The sediments are very likely to have been deposited in a small lake at a level at least 700 metres above sea level. The situation of the sediments (and correlation with younger tills further out the Hjellådal) shows that it hardly can be anything but ice which dammed the lake. The interpretation of the formation of this lateral lake is shown on Fig.12. Since the drainage which deposited the sediments, W_2, must have been considerably larger than the drainage today along the Einbudal, it is probable that the sediments are pro-glacial deposits in front of a glacier flowing out the Einbudal.

Above W_1, layers of six metres of silt and sand follow without any distinct border but generally the layers dip out from the valley side (T_6). The layers have little continuity and a typical trait is that a silt layer with lamina encloses blocks and bodies of gravel and sand. Especially in the upper two metres of the sediment the flow structures are distinct. As with W_1 the surface of T_6 is near horizontal. This sediment-type is merely observed in this one profile and the type has hardly had a great extension. The stone material contains a little more of the local gneiss and a little less of the longer-transported hornblendeschists than the sediment underneath, and also has a distinctly poorer roundness, cf. Fig.5.

The sediment is supposed to have been created by loose material slumped into water possibly in conjunction with advancing ice along the Einbudal. The sediment, T_6, interpreted thus as a pro-glacial deposit, is a flow till.

Above T_6 an approximately one metre thick till follows (T_5) with a distinct fissility and little stone and block content. The rock content is strongly dominated by the local gneiss, and the rounding was found to be poorer than in any other sample in the Hjellådal (see Fig.5, sample 34-76). The orientation is good, with a maximum direction of 135° and dip near nil or faintly towards the north-west. The till can be interpreted as a basal till of a valley glacier along the Einbudal after it had moved forward to the Hjellådal, cf.Fig.12.

T_5 is distinguished from T_4 by a layer of blocks and stones. The border is nearly horizontal. T_4 is brownish, while T_5 is grey. The tills in the Hjellådal change between grey and brown tones which are determined by the mass contents of clay and silt. More than 35-40 % clay and silt in material less than 20 mm gives a grey colour, a lesser content brown. Fabric analysis in T_4 shows that direction maximum is about 165° with dip near nil. The dip is important for the consideration of which way the basal till is deposited. On a horizontal level the accumulation has a faint dip against the current. Dip measurements in F_1 and F_2 also support the claim that the tills T_5 and T_4 are deposited against the south-east and south respectively.

A block-layer normally appears on the border between basal tills which have different fabrics. In the Einbu profile two such block layers are added to the foundation of the above-lying till and interpreted as the result of a change of direction in the ice movements. Thus the entire block-layer or part of it aquire the same orientation as the above-lying till. The petrography in the block-layer can on the other hand be in accordance with the petrography in the under-lying till. The author previously described a similar condition at Stenseng, (see Fig.1) but in that case the orientation of the till fabric corresponded exactly to that of the striae (Garnes & Bergersen 1977). A third type of block-layer in basal tills is in existence regionally in the Hjellådal, noteably in till T_3, but here it has no different orientation from the till above and underneath. The block-layer is here interpreted as belonging to the till-layer in which it appears (see Fig.8).

The petrography in T_4 differs from T_5 by substantially lesser content of local gneiss, which is exposed in the valley side underneath the deposits. On the other hand there is an increase of hornblende-schists of the same type as in W_2 and T_6, which may originate from the Einbudal, cf. Fig.7. The southerly orientation is the most important criteria to separate T_4 as a till in its own right. Since the same fabric trend is found in Skånke (see Fig. 8, F_{18} and F_{19}) it is possible that T_4 corresponds with ice movements towards the south which were topographically guided only to a small extent (see also Fig.12).

The border T_4/ T_3 is marked, too, by a block-layer which separates two different orientations but this block-layer is of little distinctness. T_3 includes the largest till bodies in the valley filling and is in several places observed to be at least from ten to fifteen metres thick.

The till has characteristic layers which show how it was accumulated. The layers often consist of stones in discontinuous layers and are, therefore, easily seen. It is conspicuous that these stones have a higher degree of roundness than in the rest of the till and are overrepresentative in long-transported rocks like basal gneiss and trondhjemite (see Fig.5, sample 38-76). These layers of "rolling-stones" are particularely typical for T_3 in the Hjellådal, but corresponding conditions are also observed in other northerly tributaries of Gudbrandsdalen. This phenomenon is not, on the other hand, observed in valley fillings in the Mid-Gudbrandsdal. The matrix in the layer of "rolling-stones" is found to have the same grain-size distribution as material in the external layers. The same is also valid for the till fabric orientation. This proves that the layers are not deposited in water but they are basal till. To what extent these layers represent an incorporation of material from sub-glacial rivers accidental or as a result of, for example, a period of increased melting and thus more melt-water will not be discussed here. But it is of some importance to mention that the largest of these layers is found in many sections as drawn in Fig.8. This layer, which can also be correlated throughout the valley, has a coarser material close to Talweg than close within the valley sides. The thickness of the layer is far greater in the vicinity of Talweg.

The fabric analyses F_3, F_4, and F_5 show a fairly strong agreement in direction and harmonize with ice-movements along the Einbudal. The dip measurements are, however, more difficult to interpret. While F_1 and F_2 were measured relatively far out from the valley side near the horizontal foundation which the surface of the sub-till sediments formed the other fabric analyses at Einbu were measured closer to the valley side. The tills in which these measurements have been made are deposited in near conformity with the dip of the valley side, consequently on a steep slope (circa 30°). If one takes this into consideration one will also note from these fabric-analyses something which is clear from observation, namely that the dip of the layers decreases when ascending the accumulation of T_3. This fact, along with the fact that the rock content shows that material has been transported from the north, tells that ice has moved southwards in the Einbudal. At Einbu the T_3 till has a strong grey colour caused by a high content of fine material. In sample 39-76 (Fig.5) the clay content is as high as 16%.

Measurements in T_3 tills in other localities (Skånke and Enge) show clearly that the ice movements that deposited the till were topographically guided, cf. Fig.8.

Withouth a clearly visual change-over from T_3 a new fine-grained basal till follows, its colour varying between grey and brown tones, but with brown dominating. The stone content (>20 mm) is very low, subjectively estimated to be not more than 5 %. The till is three to four metres thick and takes on a sandier (browner) tone as it ascends. The till lacks the distinct layers which are so typical of T_3, but it contains some minute injections of silt and clay. The texture is compact. The fabric in T_2 (F_6) is clearly different from the fabric in T_3. The dip from north to south is great despite the fact that the measurements have been carried out relatively far from the valley side. This indicates deposition towards the north, as a stoss-side moraine against the steep valley side. The rock content is distinctly different from T_3, especially with its high content of graphite-schist which presumably stem from the south-east part of the Hjellådal where such schist is exposed over a large area. It is unlikely that the rock can have been transported from the north. In addition a great part of the material in T_2 can have been reorientated from the T_3 till.

The T_2 till, which has also been identified at Skånke and Enge, has been hard to interpret. The common regional features in fabric / texture makes it likely that the till can be correlated to the inland ice phase after the ice divide had migrated to a position far south of the Hjellådal.

The most peculiar till but also the most extensive till in the Hjellådal is a very sandy, dark brown till (Fig.11). Contrary to the older tills which are almost entirely regarded as valley filling tills, T_1 is rather ubiquitous. The till has a varying thickness, increasing inwards along the valley (see Fig.8). The border to the underlying till is always very sharp. In one case, Enge 2, glacitectonic features have been observed in T_2 near the border with T_1 and these and also other indications show that the border between T_2 and T_1 could be an erosion border.

The T_1 till is poor in clay and silt content and is porous. The till is rich in blocks and the porosity causes the graphite-schists, which amounts to circa 50% of the blocks, to disintegrate easily and give the till a strong yellow-brownish colour. Parts of the till are sorted,

Fig.9. The gullied sections at Einbu, viewed towards the north-north-west. The section shown in Fig.5 is pointed out. The Einbudal valley is descending to the left.

Fig.10. Coarse sub-till sediments, W_2, exposed in the section shown in Figs.5 and 9, viewed towards the north. The layers dip to the right, evidently deposited from the Einbudal valley.

Fig.11. The basal part of the sandy till, T_1, at Einbu, viewed to the south-east.

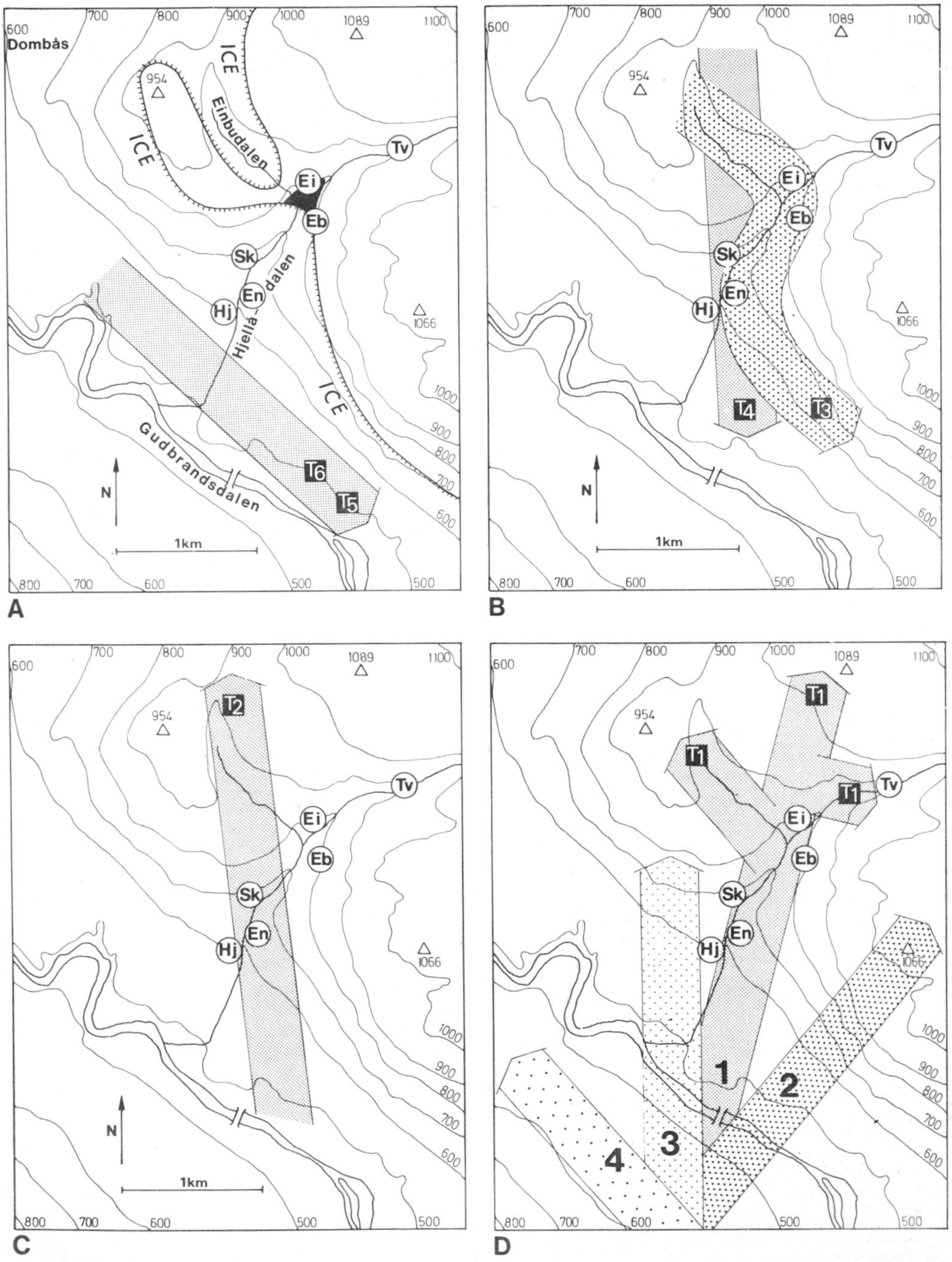

Fig.12. Reconstruction of four main ice movement phases and their corresponding tills T_{6-1} during the last ice age in the Hjellådal area. In phase A there occured an ice-dammed lake at Einbu (black). The ice movements during phases A,B, and C have been determined by till studies. The four successive ice movements, one (oldest) to four (youngest), in phase D are based upon the examination of numerous striae and till fabric analyses. The localities are: Hjelle (Hj), Enge (En), Skånke (Sk), Einbu (Ei), Einbu bakli (Eb), and Tverrå (Tv).

particularly the upper part, which often has floating structures, too. The till has a very good orientation to ca. 15° and a strong dip on the particles towards the south. This orientation is found in all localities from Einbu down the valley (see Fig.8). At Tverrå the orientation was found to be strongly topographically dependent. This can be interpreted to be a result of the fact that the valley filling furthermost into the valley is younger than further out. The direction $10\text{-}20^{o}$ agrees with a number of striae observations covering a large geographical area.

It is evident that orientation and texture/structure change upwards in T_1. The orientation shows a deflection in harmony with the topography in the upper layers. The uppermost one to two metres, for example at Einbu where a number of control measurements have been carried out, show a chaotic orientation. This seen in relation to the numerous flow structures and sorted parts in this section of T_1, makes it possible that the upper part represents ablation till or till material which has slid down from the last ice bodies during the ice-melting (i.e. flow till). T_1 can petrographically and mineralogically be distinguished from the underlying tills. The high content of basic minerals: amphibole, pyroxene, and epidote should be noted. This tendency is also valid in T_2. These minerals can probably be correlated to a type of graphite-schist which contains bands of amphibolite of which these tills have a high content. T_1 (and to some extent T_2) have a surprisingly high content of both angular and rounded stones compared with underlying tills. This may indicate that this ice phase has eroded strongly both in solid rock (angular particles) and in older deposits (rounded, polycyclic material).

If one compares the results of fabric analyses throughout T_1 with those of the deglaciation period in the investigated area by examined striations one finds a close relationship between the two. Fig.12 shows schematically the relationship between the till stratigraphy in T_1 and the deglaciation as it can be interpreted from acquired data. If this is correct, the T_1 till blankets belong chronostratigraphically to phase D in Fig.3. The T_1 tills, are, therefore, probably deposited by a melting ice sheet contrary to the underlying tills, which are all interpreted as being deposited by accumulating ice. The youngest parts of T_1 are, therefore, in this case deposited by decreasing ice closer to the front than older parts of T_1. This correlates well with the observations which show, for example, more sorted layers in the upper parts.

By looking at the petrography vertically in the tills there is a striking change in rock types. The lowest layers mainly contain underlying bedrock. Further up the vertical profile this bedrock diminishes in favour of longer-transported rocks, see Fig.13. As most of Norwegian tills, the Hjellådal tills are short-transported. Even in the upper part of the stratigraphy, no more than ten to twenty percent of the stoney material (20-60 mm) has been transported as far as three kilometers, the bulk of the material has been transported over much shorter distance. One exception is the characteristic layers of "rolling-stones" which contain circa twenty percent of "allochthonous" stone material. The average content of all analyses at Einbu is found to be nine percent of stoney material transported longer than three kilometres.

6 CONCLUSIONS

1. The water-laid sub-till sediments, W_2 and W_1, are presumably deposited at the end of the last interstadial period in central South-Norway, at the beginning of the last ice age. The most important arguments for this are:

a) The sub-till sediments are very ubiquitous in the investigated area (see Fig. 2). These have formerly chronostratigraphically been dated to "The Gudbrandsdalen Interstadial" (which after the most recent ^{14}C datings seems to have existed more than 40,000 years ago).

b) The established lithostratigraphy which can be followed along the Hjellådal discloses that large accumulation of till has taken place in the valley since the water-laid sediments were deposited. Deposits older than the sediments have not been found. It is, therefore, presumed that the valley was rather emptied of older deposits when the last ice-free period ended. During the course of the last 9,000 years intensive slope processes and river-erosion have removed most of what was accumulated during the last ice age. The previous ice-free period has consequently, existed for a long time, probably more than 9,000 years.

c) The stratigraphy in all tills is found to correlate with the reconstructed ice movement phases from the last ice age (see Fig.12).

2. The large filling of tills in Hjellådalen

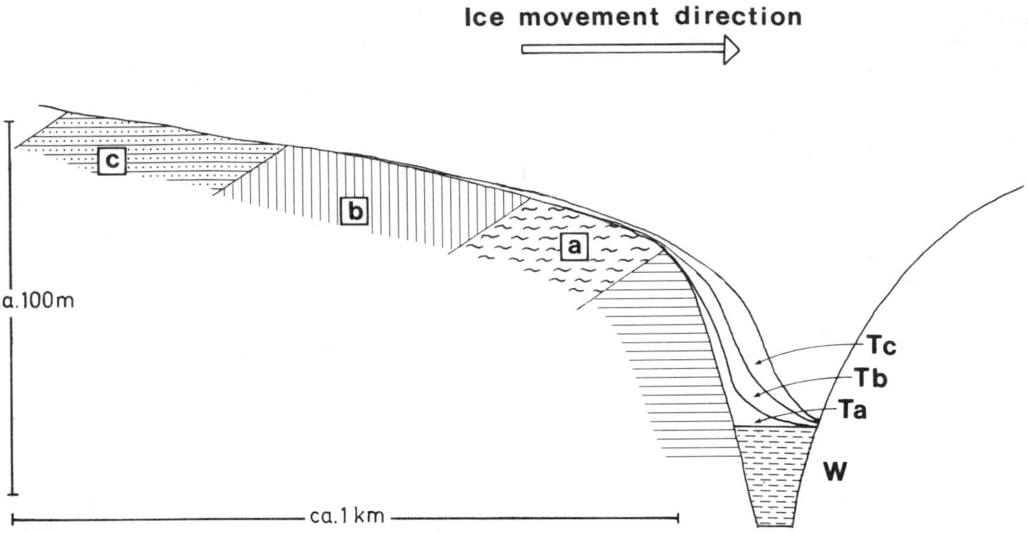

Fig.13. The content of different rock-types in a till section in relation to the distance from the exposures of the bedrock. Simplified from the Einbu section. The lowest till, T_a, resting on older sediments, W, consists of material derived mainly from the rock-type a. Further up the section the content of this rock is decreasing fast and type b and type c gradually increase.

is found to have been deposited mainly by ice movements flowing along the valleys. The tills T_{5-3} are from streams down the valleys and the tills T_2 and T_1 up the valleys. This mode of accumulation distinguishes the valleys from tributary valleys further south in Gudbrandsdalen where most ice movements are often found to be across valleys, cf. Fig.1.

3. The tentative classification of the filling in the Hjellådal till in six units is based on comprehensive analyses. Most of the basal tills in the valley, namely T_{5-3} are deposited while the ice divide was situated close to the main water divide. The above-lying till covers, T_2 and T_1, are interpreted as being deposited while the ice divide was located far south of the water divide. While the basal tills T_{5-2} are interpreted as being deposited by accumulating masses of ice, the till T_1 was deposited by decreasing inland ice. The most striking difference is that T_1 has a less fine grain content in matrix and a more complex petrography than the others. All the tills mainly consist of local material, only nine percent of the stone fraction being transported three kilometres or more. The percentage of "allochthonous" material, however, gradually increases when the till layers become thicker.

4. The accumulation of basal tills in an area has been found to be dependant on the area's position in relation to the ice divide. The different regional extension of tills from different ice movement phases in the central inland ice area can most easily be explained by regarding the inland ice divide as moving gradually and not sudden away from the main water divide towards the south.

7 ACKNOWLEDGEMENTS

This investigation is a part of the IGCP project "Quaternary Glaciations in the Northern Hemisphere" supported by The Norwegian Research Council for Science and the Humanities. The work was also supported by grants from "Nansenfondet og de dermed forbundne fond". I am very indepted to Ole Fredrik Bergersen for his helpful advice during the study.

8 REFERENCES

Bergersen, O.F. 1973, The roundness analysis of stones. A neglected aid in till studies,

Bull.geol.Instn Univ.Uppsala 5:69-79.
Bergersen, O.F. & K. Garnes 1972, Ice movements and till stratigraphy in the Gudbrandsdal area. Preliminary results, Norsk geogr.Tidsskr. 26:1-16.
Garnes, K. 1973, Till studies in the Gudbrandsdal area, eastern central Norway, Bull.geol.Instn Univ.Uppsala 5:81-92.
Garnes, K. 1975, Øst Jotunheimen. Beskrivelse og vurdering av de geomorfologiske og kvartærgeologiske forhold for "Landsplan for naturområder/forekomster". Report.Geol.inst.Univ.Bergen, 25 p.
Garnes, K. 1978, Zur Stratigraphie der Weichseleiszeit im Zentralen Südnorwegen, In H.Nagl (ed.),Beiträge zur Quartär-und Landschaftsforschung. Festschrift zum 60. Geburtstag von Julius Fink, p.195-220. Wien.Hirt.
Garnes, K. & O.F. Bergersen 1977, Distribution and genesis of tills in central south Norway, Boreas 6:135-147.
Garnes, K. & N. Heintz 1978, ^{14}C dateringer av noen norske og sovjetiske mammutfunn i kvartærgeologisk perspektiv, In Jubileumskr.Lab.Radiologisk Datering. Trondheim.(In print).
Galehouse, J.S. 1969, Counting grain mounts: Number percentage VS. Number frequency, J.sedim.Petrol. 39:812-815.
Guezou, J.C., M.-J. Poitout & N. Santarelli 1972, Le complexe de Trondheim et son soubassement dans la région de Lesja-Dombås (Oppland, Norvége centrale), Sciences Terre 17:273-287.
Lundqvist, J. 1969, Beskrivning til jordartskarta över Jämtlands län, Sver.geol. Unders.Ca. 45:1-418.
Mangerud, J. 1965, Dalfyllinger i noen sidedaler til Gudbrandsdalen, med bemerkninger om norske mammutfunn, Norsk geol.Tidsskr. 45:199-226.
Vorren, T.O. 1977, Weichselian ice movement in South Norway and adjacent areas, Boreas 6:247-257.

9 SUMMARY

Detailed investigations have been carried out in large till sections in Hjellådalen, a tributary mountain valley to Gudbrandsdalen near the northern main water divide in South-Norway. Hjellådalen have been a till trap during most of the last ice age (Weichsel), showing a till stratigraphy which can be traced from section to section. Comprehensive analyses of texture and structure of the sediments correlated with regional studies of striae and deglaciation phenomena have given the following picture of the till stratigraphy: The lowest stratigraphical unit consists of pro-glacial sediments presumed to belong to the very beginning of Weichsel or to the end of "The Gudbrandsdalen Interstadial" (^{14}C dated to have existed more than 40.000 years ago). Above the sediments several till layers have been distinguished. The lower and thickest of the Hjellådal valley infill of tills, T_{6-3} represent accumulations from an early stage of the last ice age when the ice divide still was situated close to the water divide. The upper tills, T_2 and T_1, correspond, respectively, to the maximum of the inland ice and the deglaciation period. In both cases the ice divide was situated far south of the water divide. The uppermost basal till, widespread all over the area, are easily distinguished from the underlying basal tills. The subdivisions of this till are correlated to different ice movement phases during the deglaciation of the inland ice. The youngest of these movements is closely controlled by topography. All the tills are composed of very short-transported material except for some characteristic boulder layers. A continuous migration and not a sudden shifting of the main ice divide during the whole of the ice age explains the accumulation and distribution of different basal tills and the distribution of erosion phenomena as well.

Moraines on the northern slopes and foothills of the Macgillycuddy's Reeks, south-west Ireland

W. P. WARREN
Geological Survey of Ireland, Dublin, Ireland

1 INTRODUCTION

1.1 Location

The MacGillycuddy's Reeks are a ridge of peaks forming the root of the peninsula usually called Iveragh after the barony of that name which occupies a large part of the peninsula. Iveragh lies between Dingle Bay and Kenmare River. The Reeks proper lie between the Gap of Dunloe to the east and the river Caragh to the west and are delineated by the valley of the river Laune to the north and Cummeenduff Glen to the south (Fig. 1). More precisely, they lie within the 10km grid square 077080. The mountains reach their highest altitudinal expression in Carrauntoohil (1041m), the highest mountain in Ireland, which rises 1000m above the valley of the river Laune in a distance of 6km.

1.2 Geological setting

The Reeks are composed of Old Red Sandstone rocks of Devonian age and are structurally a product of the Hercynian or Armorican orogeny. The sandstones belong largely to the Grey and Purple Sandstone Formations but are flanked to the south, east and north by rocks of the Green Sandstone Formation (Walsh 1968). A narrow corridor of Carboniferous limestone stretches from Lough Leane to the sea west of Killorglin. It is bounded on the south by the so called Armorican Front, which separates it from the Devonian sandstones and delineates geologically the northern frings of the Reeks, and on the north by the scarp of a superjacent Namurian shale and sandstone sequence.

1.3 Glacial setting

These mountains have endured glacierization both by autochthonous cirque and valley generated ice and by extraneous ice of southern provenance (Wright 1927). Ice from the south, which overrode and breached the major ridges of north-east Iveragh, pushed against, and split on, the southern slopes of the MacGillycuddy's Reeks. It formed two major outlet glaciers to the north, one of which moved eastwards down Cummeenduff Glen and northwards through the trough of Lough Leane, and the other moved north through the valley of the river Caragh (Glen Car). It reached its highest point at 700m O.D. on the southern flanks of Carrauntoohil and its level was maintained sufficiently in Cummeenduff Glen to breach the main ridge at Alohart (640m) and at Ballagh (595m) (Warren 1977). It cut a spectacular breach 250m deep, diffluent to the Cummeenduff Glen glacier, in its escape north at the Gap of Dunloe.

The two main glaciers and the ice that passed through the Gap formed a large piedmont glacier north of the Reeks. This glacier extended as far north as the Kilcummin Moraine (Warren 1977) which extends in a great arc from Ballahacommane (W 020905), five miles east of Killarney, through Kilcummin (V 992950), Ballyhar (V 927974), Castlemaine (Q 836030), Cromane (V 700980) and Glenbeigh (V 668901), where it emerges from the hills at its western extremity (Warren 1977). Till, bearing erratics of clear southern provenance, covers the ground north of this moraine. This indicates that ice, spreading from an accumulation area south or south-east of the Reeks, spread 18km north as far as the southern flanks of Sliabh Mis and 25km north-east to Sliabh Luchra where it achieved an altitude of at least 275m O.D.

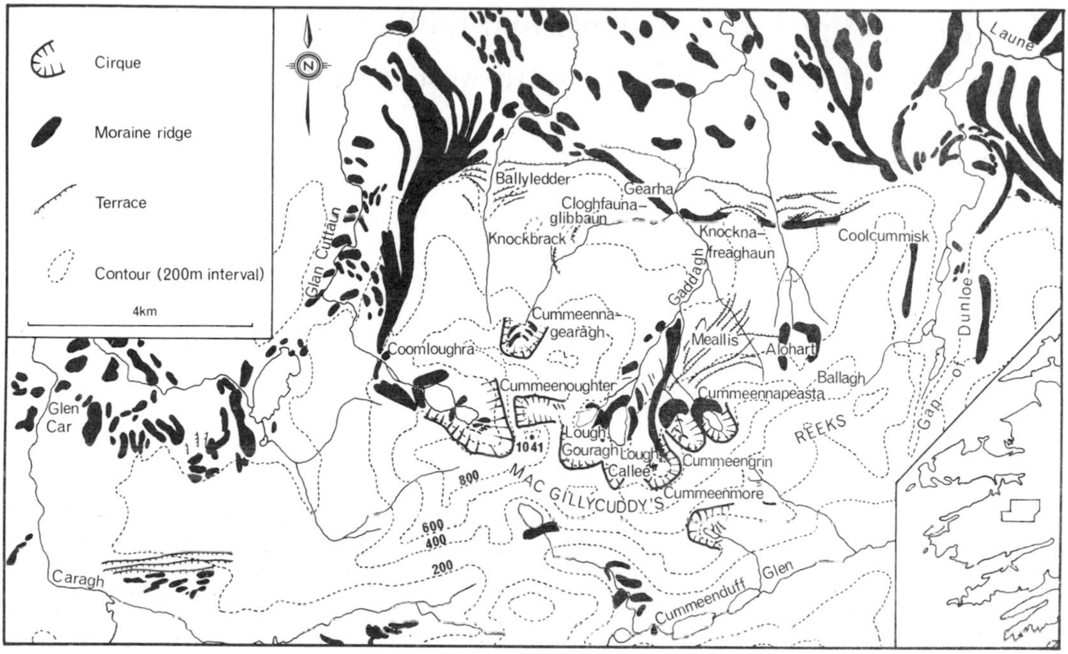

Figure 1. The MacGillycuddy's Reeks

The pattern of retreat of the general ice body from its maximum extent until it reached, or readvanced to, the Kilcummin Moraine is enigmatic owing to the paucity of morainic landforms north of that moraine. But south of the moraine, which is complex (it passes laterally from fluvio-glacial kame terraces in the east to gravel moraine in the north-east, to till moraine in the north and north-west), deglaciation was active and left a classical legacy of concentric arcuate moraines and linear kames as the ice front retreated to its two main outlet points (Fig. 2).

It is not proposed to describe the deglaciation of the foreland, rather, the intention is to illustrate the manner in which the general ice body impinged on the northern foothills of the Reeks and the relationship that existed between this ice and the local cirque and valley glaciers. To this end the Gearha Moraine and associated deposits and the moraines of Glen Gaddagh in particular are described.

2 THE GEARHA MORAINE

The Gearha Moraine, which runs along the northern foothills of the Reeks is the southern correlative of the Kilcummin Moraine. It emerges from the steep slopes of these foothills 1.5km west of the Gap of Dunloe as the uppermost and, at this point, most prominent constructional glacigenic feature. Eastwards for a distance of 400m the associated ice limit can be traced, in direct lateral continuity with the moraine, as a distinct line of contact between rounded and clearly glacial boulders and the flat angular periglacial hill slope material above. This limit can even be picked out in the walls of the fields where their constituent stones change abruptly from well rounded and striated glacial clasts to angular flat periglacial fragments. From here at 200m O.D., the moraine falls gradually westward with a gradient of 1:76.6 to cross the mouth of Alohart Glen at 147m O.D. and Glen Gaddagh at 135m O.D.

This is the uppermost clear glacigenic feature at the mouth of Alohart Glen. However, on the north-west slope of Coolcummisk hill there is a series of terraces above the level of this moraine. The lack of lateral continuity of these terraces and the tendency of the loose stones at the surface to a flat, angular, platey character indicates a periglacial origin, as does the very flat broad nature of their treads. At the centrepoint of its traverse across the mouth of the glen, where it is cut by the Owenacullen River, the moraine has an

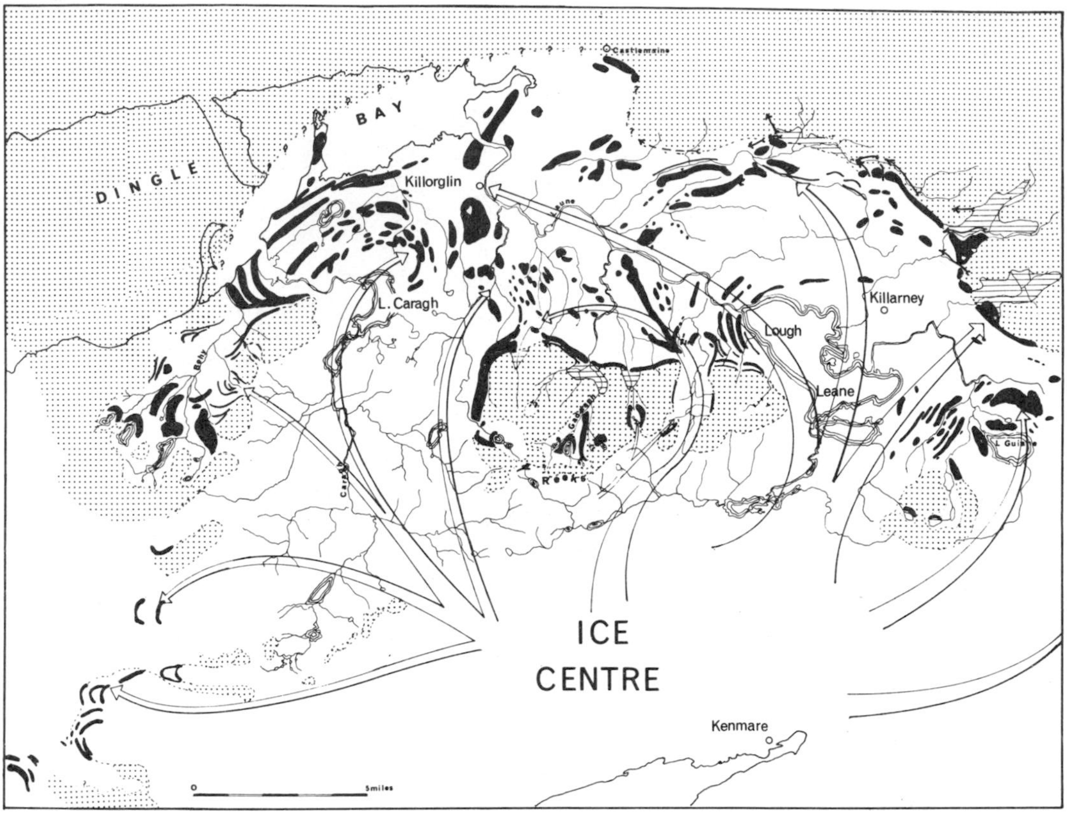

Figure 2. Ice limits associated with the Kilcummin Moraine. Black areas indicate moraines, horizontally shaded areas reconstructed proglacial lakes, black arrows meltwater channels, open arrows direction of ice movement and stippled areas are outside the limits of the Kilcummin event.

amplitude of 25m, and, although there is no clear section through the deposit, the loose well rounded rubble at the surface indicates that this is basically a gravel moraine at this point. This feature is easily traced across the north flank of Knocknafreaghaun, west of which it again lobes slightly into Glen Gaddagh as it crosses its mouth.

The river Gaddagh cuts a 30m deep section in this moraine as it dissects it from south to north and it is seen to be composed of a coarse, ill sorted, yet clearly bedded gravel dipping to the south (Fig. 3). These steeply dipping gravels, given their morphological configuration, would be best interpreted as fore-set beds of a sub-lacustrine moraine. This interpretation is confirmed in a set of rhythmic lake clays that occur distal to the moraine within the glen. This stream section is the type site of the moraine, which is here named the Gearha Moraine after the townland in which its type site occurs. The moraine continues to the west of this glen, but with an increased gradient of 1:32.8.

As the moraine approaches Ballyledder from the east it bifurcates so as to swing northward away from the hill slopes of Cloghfaunaglibbaun and Knockbrack as a disjoint ridge, and to continue a westward course as a distinct moraine terrace. The northward swinging ridge merges with a corresponding north-eastward trending lateral moraine ridge extending out from Glan Cuttaun. The terraced fork of the Gearha moraine merges with the higher terraces of the Glan Cuttaun lateral moraine; and there is an intermediate series of moraine terraces and ridge fragments indicating the pattern of deglaciation between the deposition of the upper moraine terrace and the lower ridge (Fig. 4).

The Glan Cuttaun lateral moraine complex is continuous along the eastern slopes of

Figure 3. The Gearha Moraine; steeply dipping beds seen in section.

Glan Cuttaun from its initiation under Coomloughra to its conjunction with the Gearha Moraine 6km to the north of Ballyledder. It is composed of a series of ridges and terraces each of which marks an accumulation of glacial or periglacial debris marginal to a former position of the glacier that occupied the glen. It is topped by a sharp ridge which falls 100m from its point of inception (396m O.D.) to the point at which it merges with the eastern slope of Skregbeg 4km to the north-east. The ridge re-emerges from the northern slope of Skrebeg and is a mirror image of the Gearha moraine as it swings north to join the lower ridge of that moraine at Ballyledder leaving a series of terraces above it which match the terraces of the Gearha moraine (Fig. 4). Stream sections in these terraces at Ballyledder indicate that they are composed of ill-sorted glacial gravels containing many striated stones.

3 THE UPPER GLACIAL LIMIT

There is further evidence of more extensive glaciation in this area of the foothills of the Reeks. This is seen to best advantage within the glens, particularly Glen Gaddagh and Coom Shannera (Ballyledder). Within Glen Gaddagh a series of terraces ascend the southern slopes of Knockbrack and Cloghfaunaglibbaun. These slope consistently upstream into Coolroe having swung around the eastern slope of Knockbrack and Cloghfaunaglibbaun from the north. At first sight they resemble solifluction terraces, but the indications are that they are lateral moraine terraces associated with ice that pushed into the glen from the north:-

1. Unlike the terraces in Alohart, but similar to those at Meallis further up Glen Gaddagh, they are individually continuous over a considerable distance (up to 1.9km).
2. They all possess a continuous declination up valley.
3. Two, preserving their terraced form, swing round the head of Coolroe Glen.
4. One of these (2T2 Fig. 4) occurs in a thick plug of till at the head of the glen. The till, 10m of which is seen in section, shows no disturbance or evidence of surface flow such as would be associated with solifluction terraces.

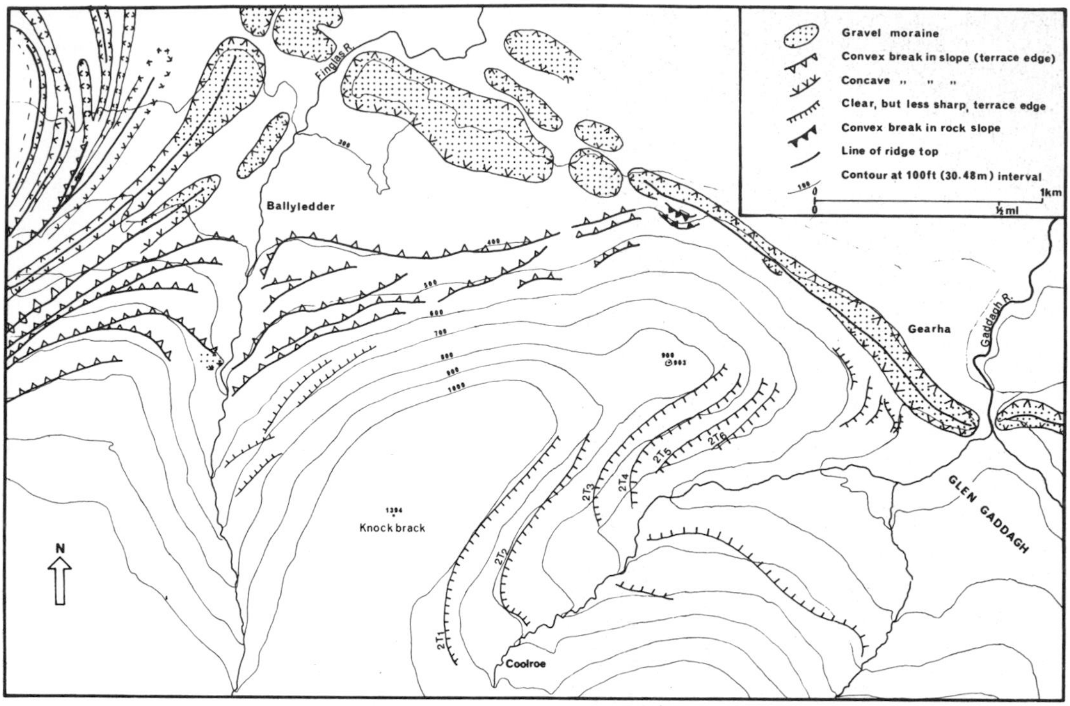

Figure 4. The Gearha Moraine at the mouth of Glen Gaddagh, associated moraines and terraces at Ballyledder and upper terraces at Coolroe.

5. The other (2T4) terminates in a blockfield as it swings across the glen. The blockfield is underlain by blocky boulderclay and is interpreted as a partly washed out moraine.

Till fabric analyses carried out at three points in a section through 2T2 indicate a preferred stone orientation 83° - 263°. Such a preferred orientation seems more consistent with ice pushing up valley than with a local emanation from Cummeennageragh, for the cirque is almost due south of the sample site. Furthermore, although the till here is composed entirely of sandstone and quartz, further down valley small quantities of chert were encountered in till outside the Gearha Moraine. The moraine itself contains both limestone and chert as does the till immediately proximal to it, and the nearest reasonable provenance of the chert, which is consistently present, is the limestone lowlands to the north.

On the western slope of Knocknafreaghaun, above, and almost paralleling, the terrace of the Gearha Moraine, another distinct terrace slopes down into Glen Gaddagh from the north-east. This terrace is problematic for, although it may represent the margin of a glacier that pushed up into the glen, it may, alternatively, be interpreted as a southward dipping bedding plane. Walsh (1968) recorded a dip of 70° to the south at about this point, but it is possible that he based this on the configuration of the terrace. If this is the case then the interpretation of the terrace is an open question; however, it should be noted that the regional dip, taken from other measurements, is to the south.

Further west, at Ballyledder, three further terraces, with declinations of 3° - 5° to the east, occur above the lateral moraine terraces that lobe into Coom Shannera. The uppermost of these reaches 305m O.D. Although there is no evidence that any one of them was deposited in association with glacier ice and they may simply be solifluction terraces, the possibility remains that they may be ice marginal features similar to those at Meallis in Glen Gaddagh.

GLACIO-LACUSTRINE DEPOSITS

4.1 Glen Gaddagh

Due north of the summit of

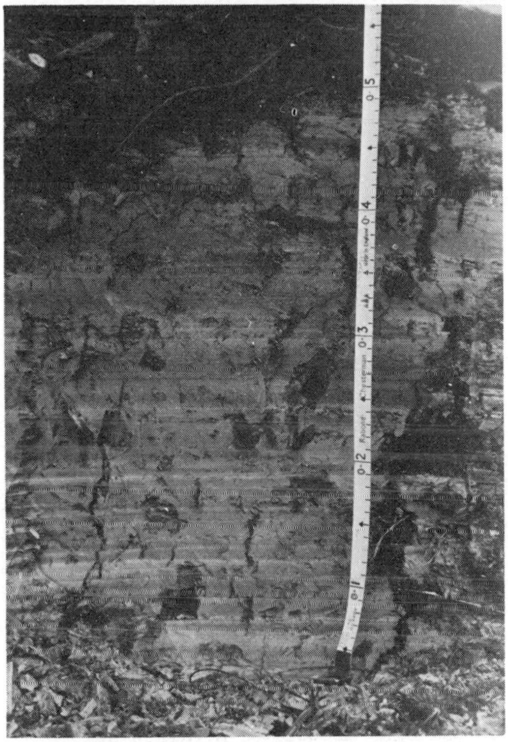

Figure 5. Sequence of rhythmic clays distal to the Gearha Moraine in Glen Gaddagh.

Cloghfaunaglibbaun (276m O.D.) an 8m deep rock-cut ravine takes the form of an in-and-out ice marginal drainage channel associated with the Gearha Moraine as it extends west of Glen Gaddagh. Its position, just 1 km west (downglacier) from Glen Gaddagh suggests the possible former existence of a proglacial lake, held up in Glen Gaddagh by the ice of the Gearha moraine. The existence of such a lake is confirmed in a set of rhythmites 0.5m thick in Glen Gaddagh just 130m outside the Gearha Moraine at an altitude of ca. 110m O.D. (Fig. 5). It is clear that the rhythmites resulted from a ponding of the lower part of Glen Gaddagh by the ice of the Gearha Moraine for the ice front could not have blocked the mouth of the glen at any of its discernable later positions. Further, an outlet channel at 144m O.D. occurs between the Gearha Moraine and the north-eastern slope of Cloghfaunaglibbaun.

Fifty nine couplets were counted (in the laboratory) in the rhythmite deposit, but, as it is overlain unconformably by a medium-coarse river gravel it is impossible to estimate the original thickness of the deposit. Assuming that these are varves, then the proglacial lake, supported by the Gearha Moraine, stood in Glen Gaddagh for at least 59 years. As there are no published varve curves for this part of Ireland, and as there are no dated varve curves for Ireland as a whole it is difficult to ascertain whether or not these are varves and there is no ready means of dating them if they are.

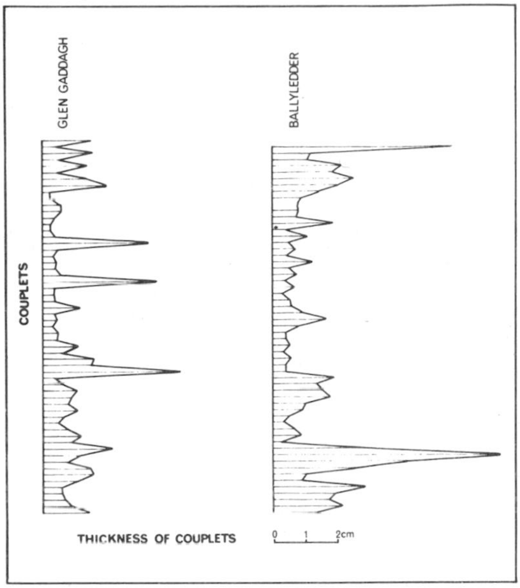

Figure 6. Curves for rhythmites in Glen Gaddagh and Coom Shannera.

Within the context of this study the most important conclusion stemming from the occurrence of rhythmites is that these are deposits of a still-water environment proximate to an ice front - conditions which favour varved sedimentation. But, whether or not they are annual couplets or varves, these rhythmites substantiate the former existence of a lake, which could only have been proglacial, in Glen Gaddagh.

4.2 Coom Shannera

Coom Shannera is a small river valley head between Skregbeg and Knockbrack. Here, in the townland of Ballyledder, is the area in which the two main distributary glaciers of the Kenmare ice sheet separated. The upper limit of glaciation is not clearly defined here, but moraine terraces of the Gearha Moraine and its contemporary, the Glan Cuttaun Moraine rise to 177m. Outside these moraines, within Coom Shannera, a stream-cut section reveals 0.77m of

rhythmic bands of silt/clay underlying 4.5m of local head. Fifty-eight couplets were counted at this site, and although there is no obvious correlation between the curve representing these deposits and that representing the rhythmites of Glen Gaddagh (Fig. 6) it is clear that lacustrine conditions could only have obtained in this open valley in response to a major blockage of its outlet such as could only reasonably have been caused by the ice which is clearly evinced by the Gearha and Glan Cuttaun moraines. Owing to the incomplete nature of both sets of rhythmites the lack of obvious correlation between their curves does not necessarily mean that they are not varves.

5 CIRQUE AND VALLEY MORAINES

At the head of Glen Gaddagh four distinct cirques, Coomcallee, Coomgouragh, Cummeenmore and Cummeenoughter, go to form the Coomcallee compound cirque (cf Evans and Cox 1974). Technically, Cummeennapeasta and Cummeengrin (Fig. 1) are separate from the compound cirque and may be regarded as elements of a cirque complex which also includes the Coomcallee compound cirque. The Hag's Teeth Moraine is the largest and most distinctive moraine rampart associated with the Coomcallee cirque complex. It is 2.2 km long from its point of emergence at the north corrie wall of Cummeenmore to the point of its dissection by the Gaddagh river. In this distance it falls 305m. Its distal front rises 45m above the valley floor where its greatest angle of slope is 22°. Its proximal slope is variable along its length, being 30° above Lough Callee, but ranging between this and 18° elsewhere. It is composed, as revealed in the section cut by the Gaddagh river, of a stony Old Red Sandstone till with a sand dominated matrix.

5.1 Moraines outside the Hag's Teeth Moraine

Outside the Hag's Teeth Moraine the floor of the glen is covered with till and fluvio-glacial outwash; and till extends up to the 305m (1000 ft) contour. In the north eastern part of the glen, in Meallis, the surface expression of the surficial deposits takes the form of a series of hillside terraces (Fig. 7). On first sight these seem to be solifluction terraces but detailed morphometric and morphographic mapping revealed that they are composed in part of loose angular soliflucted debris and in part of compact local till, and that they run as continuous features from the lateral portion of the Hag's Teeth Moraine along the hill slope down valley to the north-east. The most continuous of these is traceable without a break for 4 km. In order to map these features accurately, and as objectively as possible in an area of very limited control (the hillside is uncultivated bogland) a series of points at 100m intervals was surveyed along each terrace front. This was done to a base line A-B (Fig. 7) 1.11 km long with a variable third control as a check on the accuracy of the intersection of the sightings on the end points of the line. The hand held prismatic compass was used for the purpose and found to be remarkably accurate when each reading was checked after initial recordings. All slope measurements were made here using an abney level. A hand-levelled traverse was made across all the major terraces using an abney level and thirty metre tape; compensation was later made for the hill slope inclination (Fig. 8).

All of the major terraces emerge from the hillside that rises on the distal side of the Hag's Teeth Moraine under Cummeennapeasta Lough. Initially five terraces can be recognised, but, progressing eastwards, secondary terraces emerge from these and usually run sub-parallel with them producing a total of 12 terraces. The lower eight terraces slope consistently down valley to the north east. The slope of the upper four is not consistent; initially they fall down valley, but approximately half way along their lengths they begin to rise up the hillside. In this way they resemble, though in much larger dimensions, the solifluction lobes that are common above them on the steeper hillside. In plan the upper terraces are more irregular than the lower ones. They resemble a series of conjoined lobes, whereas the lower features form regular lines following the sweep of the glen.

All of the lower eight terraces incline laterally (slope) down-valley. The lateral tread inclination varies from terrace to terrace and along each terrace, but it is never greater than 7° nor less than 1°. In general the gradient of each terrace is initially steep (4° - 6°) it then decreases to 1° - 3° as the terrace swings round to the north where the glen is widest, steepening once more, in the case of those that continue to the mouth of the glen, to 7°.

The lowermost terraces T1 and T2 (Fig. 7) form a single feature - a block moraine that extends north eastwards as a pair of terraces, but swings north and

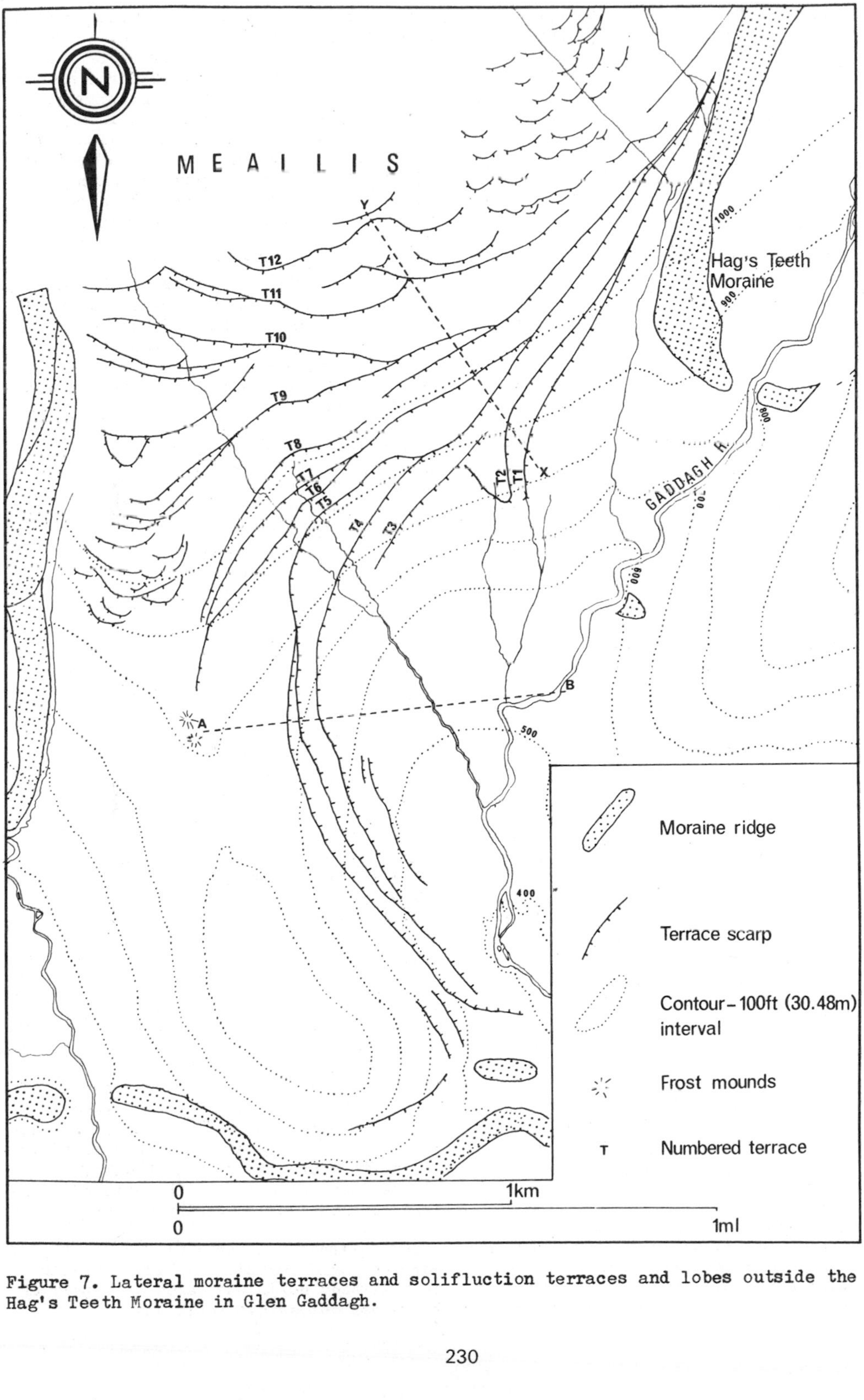

Figure 7. Lateral moraine terraces and solifluction terraces and lobes outside the Hag's Teeth Moraine in Glen Gaddagh.

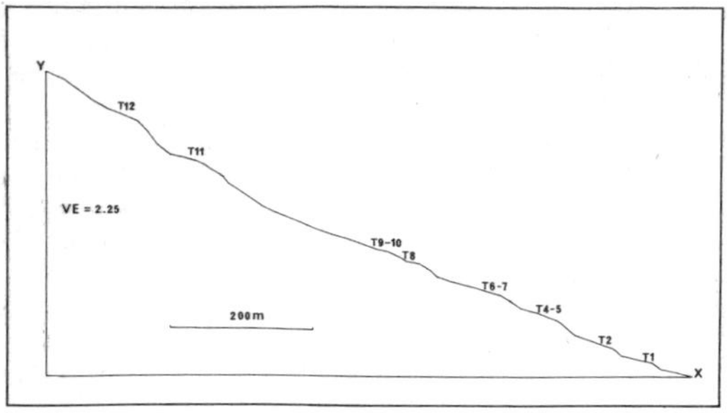

Figure 8. Profile across the terraces at Meallis

north-north-west away from the hillside, 500m north-east of the frontal part of the Hag's Teeth Moraine, to display a distal slope of 2°. The projected arc of this moraine meets a mound 8m high (composed of a coarse stony till) on the west bank of the river Gaddagh. Together these features are taken as a partly over-washed precursor of the Hag's Teeth Moraine.

The five terraces above these, up to and including T7 are composed of till. Above T7 stream sections show that the terraces are composed of very angular local debris. Terraces 3, 4, 5, 6 and 7 are composed totally of till, divisible into two distinct facies - a lower, matrix dominated till with well rounded clasts frequently overlaid by or passing up into stony till also with well rounded stones but less frequently striated. The upper 0.5m of these deposits is frequently composed of more angular, and on occasions, clearly fragmented till clasts. Evidence of down slope movement of this facies is usually clear from the down slope alignment of the a axis of the a-b plane of contained clasts (see Lundqvist 1949).

T8 is the uppermost terrace that includes identifiable till in its composition. A stream section cut in this terrace shows the following succession.

3. Angular soliflual material. 1.4m
2. Stony till showing some evidence of frost disturbance. 1.0m
1. Matrix dominated till with no evidence of disturbance. 0.7m

Fabric analysis of the debris forming T8 revealed that while the clasts of the upper facies followed strongly the direction of the hill slope in orientation and dip, the fabric of the underlying, undisturbed till revealed a strong west - east trend in the orientation of the a axis of the clasts. The vector mean and the mode lie very close to one another (Fig. 9) with respect to each deposit and show that the fabric of the till is 60° out of line with that of the upper deposit and the hill slope, and approaches the line of the valley.

Above the level of T8 along Meallis stream cutting (the only open section in the deposits above this level) there is no evidence of glaciation. The upper terraces and lobes are composed entirely of angular fragments of local rock which show a preferred fabric, and are imbricate, down-slope. In short they are solifluction terraces (see Watson and Watson 1967). Where the stream intersects T10, very clear evidence of down slope movement of frost shattered rock was noted. Here, well cleaved shales with cleavage dip to the south are frost shattered and overturned so that the resultant angular stones, with very little relative displacement in a vertical sense, lie dipping north to form a very coarse angular head. This is clearly illustrated where a band of quartz between cleavage planes could be used to trace its fragments down slope.

The amplitude of the terrace rises is variable. However two categories are clear: the lower nine terrace fronts vary in amplitude from 4.5m to 14.2m while the upper three and the large lobe above them possess amplitudes within the range, 21.3m - 30.2m. Also, the slope of the terrace fronts differs as between the upper three and the lower nine. The slopes of the lower terrace fronts vary between 12.5° and 15.5° while the upper terrace fronts and the large lobe above them range between 17.5° and 21° in slope (Fig. 7). The slope of the treads does not show any preference with regard to the upper and lower terraces, it ranges from 6.5° - 10°. As has been noted above, the treads of all of the lower eight terraces are inclined laterally to the north-east the longest (T5) being in excess of 4km. The upper terraces tend to lobe, falling initially from the south-west but on reaching their lowest point they begin to rise once more. In addition, the upper terraces tend to be composed of a number of smaller lobes. T9 is initially inclined down valley, but from the point at which it meets T10 it runs sub-horizontally, rising no more than $\frac{1}{2}^{\circ}$, to the east. T10 is a steep fronted terrace rising steeply to the east.

It is concluded that Glen Gaddagh was once filled with ice to at least the point at which Meallis stream intersects T8, as this is the highest known occurrence of glacial till. This being the case and bearing in mind that T1 and T2 form a block, or block-covered, moraine and that the intervening terraces between these and T8

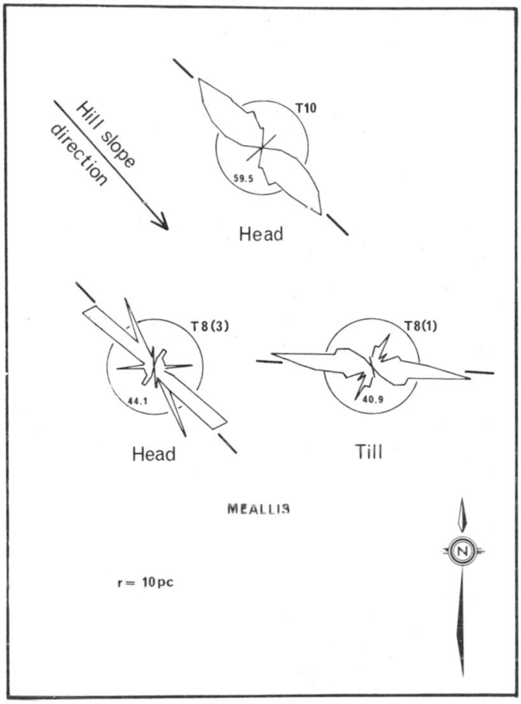

Figure 9. Fabric diagrams from lower till in T8 and head in T8 and T10. Straight lines outside circles indicate vector means and figures inside vector magnitudes.

are largely composed of glacial till, it is suggested that the lower eight terraces are ice marginal deposits or essentially lateral moraine terraces. The fact that they are partly composed of soliflual material need not alter this conclusion, for it is reasonable to assume an accumulation of hill slope debris along an ice margin particularly above the line of equilibrium of the glacier. The extreme length of some of the terraces (4km) and the fact that one (T1 - T2) swings onto the floor of the glen, where it displays a reverse or distal slope, re-enforce this conclusion. Further, it would be expected that true solifluction terraces would yield clear down-slope fabrics. Above T8 ice may or may not have been in contact with the solifluction terraces; there is no evidence one way or another. It is, however, clear that some of the features, particularly the lobes north of the eastern extremity of T9, are typical both in plan and cross-section of stone-banked solifluction lobes as described by Galloway (1961) and, being fronted by large blocks (which enable one to pick them out easily in the field), they are thought to be the non ice-contact solifluction lobes pene-contemporaneous with the ice-contact terraces. Frost mounds immediately north of these lobes, on the flat ground just above the 800 foot (244m) contour further testify to periglacial conditions. It is concluded that ice, prior to the deposition of the Hag's Teeth Moraine, spread out of Coom Callee to fill Glen Gaddagh to 1.75km north of the Hag's Teeth Moraine.

5.2 Moraines inside the Hag's Teeth Moraine

Within the Hag's Teeth Moraine is a series of moraine mounds and ramparts and lateral moraine terraces and many fragments of such. These can be traced back into Cummeenmore the innermost lying within this cirque hollow. Loughs Callee and Gouragh are each enclosed by a large arcuate moraine. The proximal slope of each of these is extremely variable but in each case falls within the range 11.5° - 18.0° except to the east of Crompaun na Callee Island in Lough Callee where the moraine banked up against the hillside attains a slope of 30°. The Lough Callee moraine has a maximum amplitude of 21m and that of Lough Gouragh 12.3m. The glaciers which deposited the moraines at Lough Callee and Lough Gouragh were fed, in part at least, by ice from the higher cirques of Cummeenmore and Cummeenoughter. Within each of the moraines damming the loughs is a further complex of moraines, one in Cummeenmore and the other in Cummeenoughter. That in Cummeenmore is best developed and most complex.

In Cummeenmore nine distinct moraines are recognised. These can be sub-divided into four distinct groups. Each group represents a period of stabilisation of the ice-front, its individual ridges representing minor fluctuations within these periods. These groups are referred to as moraines in the text and are numbered 1-4 in Figure 10. Moraines 1 and 2 seem to mark halt positions of the ice-front during its retreat. Moraine 3, however, is a remarkable feature clearly representing a significant re-advance of the ice-front. South of the stream that drains Lough Cummeenmore the moraine may be traced southward, initially as a ridge running parallel with, and inside moraine 2, but 100m south of the stream it rises up on the proximal side of this moraine, falls down its distal slope and swings in a wide arc to follow the line of moraine 2 once more, but now it is 70m outside this moraine. Moraine 3 gradually peters out to form a discrete line of boulders running up the south wall of the cirque. North of the point at which this moraine crosses moraine 2 it widens from a

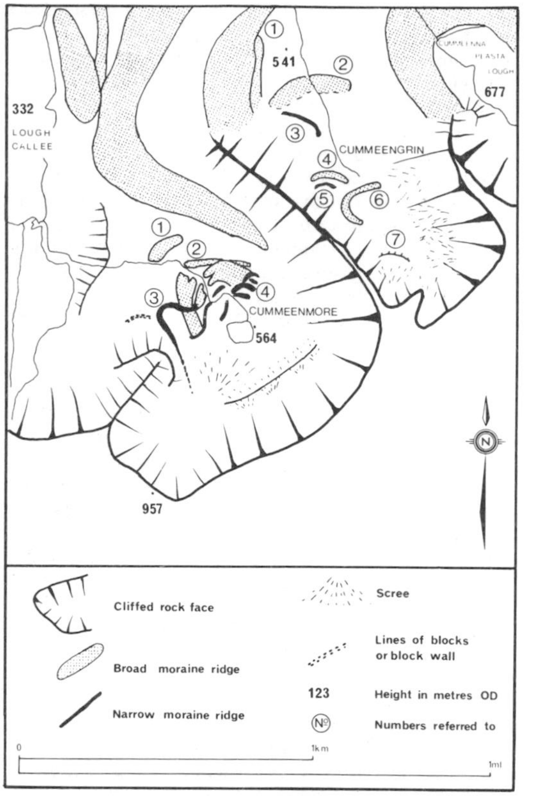

Figure 10. Cummeenmore and Cummeengrin.

narrow, steep-sided, blocky moraine to form a wide, double ridged feature with a much more subdued outline. North of the stream this moraine assumes proportions similar to those of the moraines damming Loughs Callee and Gouragh. Surprisingly, the ice-lobe that deposited moraine 3 does not seem to have significantly altered the morphology of moraine 2 where it overrode it south of the stream. However, north of the stream, moraine 3 also occludes part of moraine 2, but on this occasion the portion overridden is completely masked by debris of moraine 3. The final group is composed of 4-5 narrow, steep sided (20°) moraine ridges which show considerable evidence of ice-front fluctuation in their juxtaposition. Perhaps the most striking aspect of the moraines of this cirque is the fact that, although there is a general trend towards 'fresher' moraine morphology from the outer to the inner moraines, moraine 3 changes its morphological characteristics along its length from those of the inner group to those of the outer.

A line of large angular blocks, piled loosely to form a rampart extends almost due west from the outer lobe of moraine 3. It does not fit the pattern of the moraines, for it is very close to, and slightly arcuate towards, the north-facing side-wall of the cirque (Fig. 10). It differs also from the moraines in that it is poorly vegetated and the boulder interstices are unfilled. Furthermore, there is no evidence of a glacial movement behind it, and, bearing in mind the vertical thickness of firn (30-150m) required to produce plastic ice (Sugden and John 1976, Patterson 1969), it is unlikely that a dynamic glacier could have developed there, yet remanie ice could have persisted for some time. Such ramparts have been described by, inter alios, Kinahan (1894), Russell (1933), Bryan (1934), Watson (1966), and attributed to the occurrence of a perennial or recurring snow-patch which provided a canopy for frost riven stones to override and collect at the bottom to form a rampart. They were originally termed cloghsnatty (correctly clocha snachta) in Co. Galway where Kinahan (1894) reported he observed them forming during a particularly cold winter and where the local term, clocha snachta (snow stones), suggests their mode of deposition was well known. They were subsequently termed protalus ramparts by Bryan (1934) and Watson (1966). This process was also observed in the corries of Lugnuquilla, County Wicklow (Kinahan 1894) and can be readily observed in areas of colder climate. But it can be difficult and at times impossible to distinguish between a protalus rampart and a small cirque block moraine. Quite clearly blocks can and do accumulate in this way at the snouts of steep glaciers producing features intermediate between moraine and protalus rampart.

The other cirques on the northern side of the the MacGillycuddy's Reeks contain similar groups of moraines. Coomloughra contains within its basin three large moraines of the type that impound Loughs Callee and Gouragh, a group of narrow block moraines and two inner protalus ramparts. This cirque is flanked on its western side by a large lateral moraine rampart which indicates that ice extended beyond its basin. Cummeenoughter contains only four small narrow block moraines similar to the inner moraines in Cummeenmore, but their position (Fig. 1) suggests that they cannot be purely nivation deposits. Cummeennageragh contains four small moraine ridges of the inner type at Cummeenmore. Cummeennapeasta contains only one visible moraine which dams the lough which drowns the whole cirque basin, and Cummeengrin contains a varied series of morainic landforms which bears

further description.

Cummeengrin lies outside the Coomcallee complex. The cirque does not contain a basin in that it slopes consistently in one direction. Its back and side walls form a large amphitheatre and it is fronted by a large (breached) moraine reaching 20m in amplitude on the proximal side. This moraine is a composite one with a narrower ridge on its crest and a moraine terrace at a lower level along its proximal side. 200m inside this moraine is a distinct, but subdued, moraine ridge which follows an arc concentric with that of the outer moraine. A further 300m inside this moraine an exposed rock surface is polished and striated along the line of the cirque floor. Within these moraines there are five other ramparts which are problematic with regard to genetic classification. They are all essentially block features, and although the cirque was clearly occupied by a large cirque glacier during the deposition of the outer moraines, all evidence diagnostic of glacial action now lies outside the block moraines. The block moraines from outer to inner are orientated; 31°, 6°, 6°, 346° and 5°.

It is a feature of each of these moraines that the proximal side bears a very slight amplitude (<1.5m). While this may be attributed in part to the overall slope of the cirque floor, it also illustrates that the medium of deposition did not over-deepen the area behind the "moraines". Furthermore, all but "moraine" 6 (Fig. 10) are oriented so squarely towards the north-facing side wall of the cirque, and are sited so close to it, that it is improbable that a sufficient thickness of snow could have developed here to produce plastic ice. Moraine 6 on the other hand, is oriented towards the back wall and is so positioned that it is difficult to conceive how a firn or snow patch sufficiently large to produce this feature could retain a non-ice crystalline structure, bearing in mind that the block moraine, if the agent of deposition was ice, would be largely composed of protalus which has rolled or slid down a steeply inclining snow bank. The situation is enigmatic. However, it is reasonable to conclude that the medium of deposition, whether snow or ice was perennial over some period because the ramparts are clear, well-formed features that would have necessitated the recurrence of ephemeral ice or snow bodies in exactly the same position from year to year during the period of deposition. As to whether these are glacial moraines or nivation, protalus ramparts is, perhaps, impossible to conclude, but it seems clear that they owe their genesis to the occurrence of a solid canopy over which frost riven blocks could slide to accumulate in a pile at the bottom. Watson (1966) with reference to the nivation cirques and deposits near Aberystwyth in Wales, was unwilling to conclude as to the crystalline state of the medium responsible for their formation. The point he stressed is that there is no evidence of glacial erosion associated with them. This is as much as can be concluded regarding the inner deposits of Cummeengrin.

In broad terms the pattern, recorded elsewhere (Farrington 1934, Seddon 1957, Manley 1959, Lewis 1970), of an inner group of narrow, steep sided blocky moraines and an outer group, much more subdued and, surficially at least, less blocky and broader, can be seen within the Coomcallee cirque complex and Coomloughra. Such patterns, recorded with the aid of palynological evidence (Seddon 1957, Lewis 1970) have been used in an attempt to typify moraines belonging to two separate glacial events. This approach is rejected here as it embodies a denial of uniformitarianism. Clearly, without supportive bio-stratigraphic evidence, correlation cannot be made from cirque to cirque, much less mountain range to mountain range, on the basis of morphological similarity of moraines. The lateral transition from one morphological "type" to another along a single moraine ridge in Cummeenmore (moraine 3) lends weight to this argument. It is significant that in this case the part of the moraine that corresponds to the "inner type" is nearest the north-facing side wall, and in the most shaded portion, of the cirque, while the "outer type" lies in the most exposed part (cf Derbyshire and Evans 1976). This sort of locational difference is likely to promote two processes that will lead to differences in morphology.

1. Closeness to the side-wall will probably result in greater accumulation of supra-glacial blocky debris.

2. Lower shade availability away from the wall will produce greater ablation and a higher input of englacial and sub glacial debris, while shade from insolation will probably make for a relatively slow rate of ablation and consequent low rate input of such debris.

The bimodality evident in cirque moraine morphology probably has its roots more in these factors rather than in specific age differences.

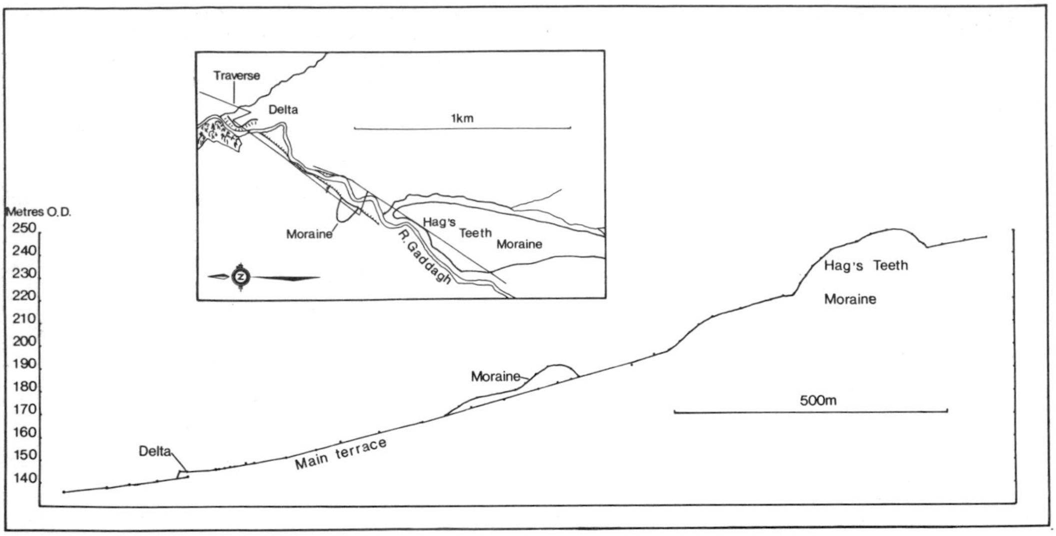

Figure 11. Levelled profile of the outwash terrace from the Hag's Teeth Moraine.

6 MORAINES ASSOCIATED WITH GLACIAL BREACHING

A massive moraine which reaches 80m in height impounds an ice scoured hollow at Alohart. Wright (1927) described this as a cirque moraine, but an examination of the back wall of the hollow and the top of the ridge above it revealed striae to within 30m of the col surface, which is ice scoured and flanked by a lateral block moraine on its eastern side. It was therefore concluded that the large moraine was deposited by a diffluent offshoot of the north-east moving glacier in Cummeenduff Glen, which had formed an expanded foot glacier at the base of a steep ice fall. The radiating pattern of striae focusing, as it does, on the apex of the back wall under the col tends to confirm a glacial input from a single point. A true cirque would be expected to have a broader zone of accumulation.

This glacier, having breached the sharp ridge of the Reeks at 640m O.D. was starved of ice once its surface level in Cummeenduff Glen fell below that point. Its marginal nature is exemplified in its single large moraine delineating an area of clear ice scoured rock peppered only with the occasional large block, a legacy of the sudden and rapid disintegration of what was in its final stages a remarkably clean ice lobe once it had lost its external source of supply.

The ridge is further glacially breached just 1km east of Alohart at 595m O.D. Ice fed the Gap of Dunloe glacier through this breach at Ballagh. It retreated actively, gradually depositing a series of fine recessional moraines before it was finally cut off from its source of supply.

7 MORAINE CORRELATION

It was attempted to ascertain the relationship between locally generated ice in Glen Gaddagh and the more general ice sheet that blocked the mouth of the glen. Levelling along the outwash terrace associated with the local glacier produced a thalweg which indicated that glacial outwash from the Hag's Teeth Moraine grades from this moraine to an almost flat surface at 145.6m O.D. (Fig. 11). Although the terrace at this point is composed of a medium grained well rounded gravel there is no identifiable delta bedding. The outlet notch of Glacial Lake Glen Gaddagh was accurately levelled to 144m which is very slightly (1.6m) below the level of the present lowest point of the outwash terrace. The terrace, however, has been partly eroded by the modern stream and its lowest point was, obviously, somewhat lower. Thus, the evidence points to the contemporaneous accumulation of the Gearha Moraine of the general ice sheet and the Hag's Teeth Moraine of the local glacier.

8 CONCLUSION

No attempt is made to date the major depositional features, as this would

entail a detailed discussion of stratigraphic sequences outside the area in question. Clearly, however, the Gearha Moraine relates to a major stabilisation of the ice front, a readvance or a completely separate glacial advance. The Hag's Teeth Moraine represents a contemporaneous event in Glen Gaddagh. Although this too was preceded by a more extensive valley glacier, there cannot have been precise contemporaneity between this and the early expansion of the general ice, for the lateral moraines of the valley glacier extend well below the uppermost deposits of the general ice.

Nor can precise dates be put on the cirque moraines, however, cirque glaciers are known to have developed in the latter part of the Late Glacial Period in the Lough Nahanagan cirque in County Wicklow (Colhoun, Synge and Watts, in press) and may have existed in other cirques in the Wicklow Mountains. In south-west Ireland Pollen Zone III, or the Nahanagan Stadial (Mitchell 1976) of the Late Glacial Phase is represented in polleniferous deposits by a sparse herbaceous vegetation and an indication of unstable soils, indicating extremely cold conditions at Long Range, between Lough Leane and Upper Lake, near Killarney (Watts 1963). Accordingly, it is probable that the MacGillycuddy's Reeks harboured some glaciers at this time, and it is probable that some, at least, of the cirque moraines refer to glaciers of this period. Such moraines, if they exist, may only be distinguished from others deposited under similar conditions by means of bio-stratigraphic evidence. Such evidence was not available within this area.

It is conceivable that some of the protalus features may be of considerably younger age than Pollen Zone III, for G.H. Kinahan (1894) observed such features in the process of formation in Connemara and in County Wicklow as recently as the Winter of 1876-77. Thus the colder conditions that are known to have obtained between 1600 and 1850 and termed The Little Ice Age (Brooks 1949) may well have produced some of the nivation features.

9 REFERENCES

Brooks, C.E.P. 1949, Climate through the ages. (2nd ed.) London, Benn.

Bryan, K. 1934, Geomorphic processes at high altitudes, Geogrl Rev. 24:655-656

Colhoun, E.A., F.M. Synge & W.A. Watts, in press, The cirque moraines at Lough Nahanagan, Co. Wicklow, Proc.R.Ir.Acad.B.

Derbyshire, E. & I.S. Evans 1976, The Climatic Factor in Cirque Variation. In E. Darbyshire (ed.), Geomorphology and Climate, p. 447-494. London, Wiley.

Evans, I.S. & N. Cox 1974, Geomorphometry and the operational definition of cirques, Area 6:150-153.

Farrington, A. 1934, The Glaciation of the Wicklow Mountains, Proc.R.Ir.Acad. 42B:173-209.

Galloway, R.W. 1961, Periglacial phenomena in Scotland, Geogr.Annlr. 43:348-363.

Kinahan, G.H. 1894, The recent Irish Glaciers, Ir.Nat. 3:236-240.

Lewis, C.A. 1970, The Upper Wye and Usk Regions. In C.A. Lewis (ed.). The Glaciations of Wales, p. 147-173. London, Longman.

Lundqvist, G. 1949, The orientation of the block material in certain species of flow earth, Geogr. Annlr. 31:335.

Manley, G. 1959, The late-glacial climate of North-West England, Lpool Manchr geol.J. 2:188-215.

Mitchell, G.F. 1976, The Irish Landscape, Glasgow, Collins.

Patterson, W.S.B. 1969, The physics of glaciers. Oxford University Press, Oxford.

Russell, R.J. 1933, Alpine landforms of western United States, Bull.geol.Soc.Am. 44:927-950.

Seddon, B. 1957, Late-glacial cwm glaciers in Wales, J. Glaciol. 3:94-99.

Walsh, P.T. 1968, The Old Red Sandstone west of Killarney, Co. Kerry, Ireland, Proc.R.Ir.Acad. 66B:9-26.

Warren, W.P. 1977, North East Iveragh. In C.A. Lewis (ed.), South and South West Ireland, INQUA X Congress 1977 Guide-book for excursion A15, p. 37-45. Norwich, Geo Abstracts.

Watson, E. 1966, Two nivation cirques near Aberystwyth Wales, Biul. Peryglac. 18:95-113.

Watson, E. & S. Watson 1967, The Periglacial origin of the drifts at Morfabychan, near Aberystwyth, Geol.J. 5:419-440.

Watts, W.A. 1963, Late glacial pollen zones in western Ireland, Ir.Geogr. 4:367-376.

Wright, W.B. 1927, The Geology of Killarney and Kenmare, Mem.geol.Surv.Ire.

The patterns of deglaciation and associated depositional environments of till

EUGENIUSZ DROZDOWSKI
Polish Academy of Sciences, Toruń, Poland

Deglaciation is closely related to the negative net balance of a glacier, that is to the excess of ablation over accumulation. The mass loss greater than the mass gain is resulted in wastage of the glacier and the release of land and water surfaces from ice masses covering. However, the pattern of deglaciation and associated depositional environments of till may vary greatly in space and time, reflecting the effects of different environmental factors, above all the climatic factor. Therefore, the down-wasting glacier and associated formation of till seem to be more readely elucidated if examined from the glaciological point of view, in terms of the relationship between the climatic regime of a glacier and the types of ice wastage.

TYPES OF ICE WASTAGE

Appreciation of the climatic regime of a glacier is good expressed by the vertical gradient of net annual specific balance at the border of accumulation area introduced by Shumskiy /1946/ as "energy of glaciarization", and restated by Meir /1962/ as "activity index". It reflects the degree of intensity of the accumulation-ablation processes which in turn control the rate of mass exchange and the dynamics of the glaciers.

The value of the activity index decreases in general with growth of climatic continentality; for the glaciers in south-central Alaska, for instance, it amounts to 22 mm/m, in Tybet to 2-3 mm/m, and on the tops of Tian-Shan to 0.5 mm/m /Kotlyakov 1968/. That means a decrease of the amount of accumulation and ablation and lowering of mass exchange with growing continentality of climate. In dry continental climate, under conditions of dominance of the anticyclonic type of weather not only increment of the mass is low but also the mass loss is great because of intense sublimation. This affects the type of ice wastage and associated depositio-

nal sedimentary environments.

Taking into account the above described regularities, it seems reasonable to distinguish two main types of the wastage of glacier ice: ice wastage taking place in humid maritime climate, and ice wastage proceeding in dry continental climate. Each of these types takes place under specific geometrical and dynamical conditions of a glacier, depicting different value of the activity index.

The first type of ice wastage is going on in considerable narrow marginal zone of the ablation area. Since the snowfall is very high and comes in winter season which is most suitable for accumulation, the equilibrium line altitude is very low, so that the glacier is still active near its snout. Under these conditions the most important processes involved in ice wastage are surface melting due to conduction of heat from air and codensation of water vapour due to moist air advection /cf. Ahlmann 1948, Schytt 1967, Paterson 1969/. Very efficient are also the mechanical and thermal action of meltwater and meteoric water which penetrate the glacier. Such conditions favour subglacial deposition and, when the base of a glacier is frozen to its bed /cold glaciers of polar or subpolar types/, large production of till of melt-out type. Present-day examples of this type of ice wastage provide glaciers in Iceland /e.g. Thorarinsson 1943, Okko 1955, Kozarski and Szupryczyński 1973/, south-central Alaska /Tarr and Martin 1914, Clayton 1964/, and Svalbard /e.g. Gripp 1929, Klimaszewski 1960, Boulton 1972a, Drozdowski 1977/.

The second type of ice wastage, taking place in continental climate, differs considerably from the previous one both in its areal extent and character of processes involved. As a result of small quantities of atmospheric precipitation, falling mostly on the transition seasons or summer season which are not suitable for accumulation, the activity index is very low and, in consequence, the equilibrium line is placed relatively high on the glacier surface, commonly above the line of maximum snowfall /Kotlyakov 1968/. The ice wastage encompasses a broad zone of the ablation area and is going on mainly by melting due to solar radiation and sublimation under conditions of dry anticyclonic weather. Such conditions tend to cause the formation of englacial and ablation tills of the sublimation type, like those encountered in the Antarctic by Shaw /1977/.

THE PLEISTOCENE GLACIAL CYCLES AND CHANGES OF PATTERN OF DEGLACIATION

The processes of ice wastage related to the Pleistocene ice sheets differ from the present-

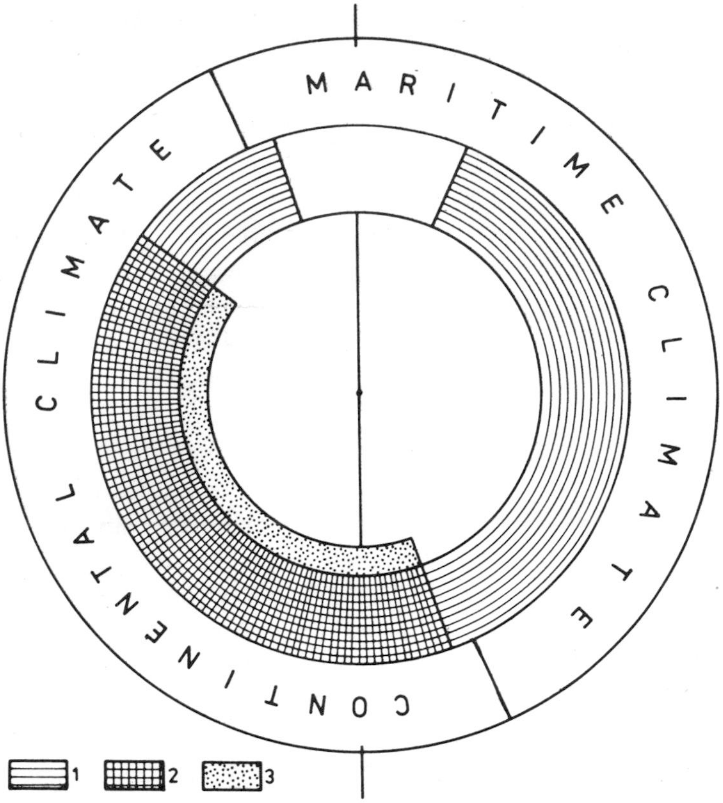

Fig. 1. Diagram of changes of the climate type and related glacial regime during the glacial cycle
1 - maritime glacial regime, 2 - continental glacial regime, 3 - accumulation of loess and loess-like sediments

- day equivalents not only by scale but above all by rythmic changes of the type of a climate and related glacial regime during the glacial cycles /Fig.1/. In the first half of the glacial cycle prevailed cold and moist maritime climate because only this type of climate makes it possible for the initiation and a rapid growth of an ice sheet /Jahn 1950, Barry et al. 1975/. An ice sheet of maritime regime was then formed, characterized by high value of activity index and low equilibrium line altitude. It persisted till the formation of the anticyclonic system above the surface of the ice sheet stimulated by high snow albedo /Brooks 1949/ and cooling of North Atlantic Ocean /Donn and Eving 1966, Adam 1976/. From this time begun a phase of prevelance of continental type of arctic cli-

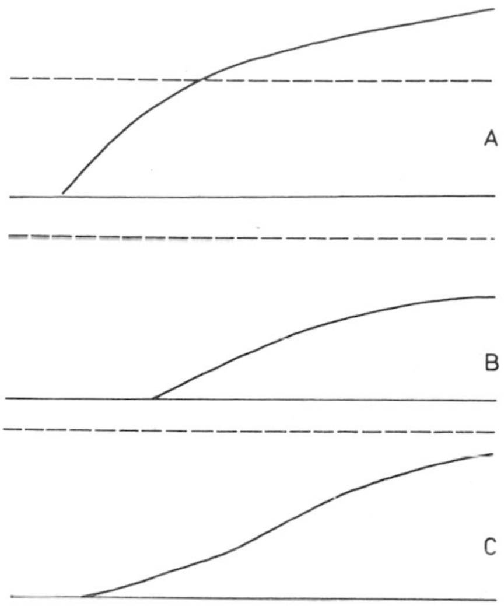

Fig.2. Scheme of change of the equilibrium line altitude
A - situation in maritime climate
B - situation in continental climate
C - situation after change from maritime to continental type of climate

mate. In consequence, superimposition of different glacial regimes resulted. The continental glacial regime of low activity index and high equilibrium line altitude was superimposed on the maritime glacial regime of opposite characteristics /Fig. 2/, subduing thus the dynamics of the "maritime" ice sheet and stagnation of a broad zone of its ablation area. That pattern of deglaciation correspond to, what in the literature is called regional stagnation /Flint 1930, Black 1977/, areal deglaciation /Niewiarowski 1963, Galon 1968, Bartkowski 1969, Asyeyev 1973/, or large-scale stagnation /Clayton 1967/. The width of the stagnation zone and at the same time the effectiveness of ice wastage were presumably controlled by the amount and rate of the climate changes. The more abruptly and deep was the change of the arctic climate type from maritime to continental type, i.e. the more large was the extent of the ice sheet, the more destructive was the press of continental regime upon the maritime regime of the ice sheet.

This pattern of deglaciation took place most probably during the culmination phase of each large glacial cycle, after reaching an extreme extension by the ice sheet. Therefore, it is suggested here to term it "culmination-phase deglaciation" or simply "areal deglaciation", but defined only in relation to the described above climatic-glacial event.

During the second half of the glacial cycles /cf. Fig.1/, in conditions of renewed growth of oceanisation of the arctic climate, as was the case in the centre of glaciation in Fennoscandia /Hoppe and Liljequist 1956, Lamb and Woodroffe 1970/, the ice sheet adjusted relatively quick to changing conditions of mass exchange, approaching to those in initial phase of the glacial cycle. Very important was the direction of climate changes because, in contradistinction to the culmination-phase deglaciation, the superimposition of maritime glacial regime on continental regime led to

Fig. 3. Spitsbergen, 1975. Marginal zone of Aavatsmarkbreen showing thick cover of ablation till and related supraglacial ice-moraine ridges resulted from downwastng of the glacier surface

lowering of the equilibrium line altitude, which may have even drived temporaryly the dynamics of the ice sheet in spite of a general amelioration of climate. Thus at the end of glacial cycles the processes of ice wastage approximated in their manner and range the first distinguished type of ice wastage, proper to maritime climate. The relative contribution of the various processes to the vanishing ice sheet was of course changed from place to place, depending on such local or regional factors, as relief and geology of the ice bedrock, its tectonics, contact of ice margin with water bodies etc. /i.g. Hoppe 1959, Bryson et al. 1969, Lundqvist 1973/.

In any rate, this pattern of deglaciation differs from the previously described culmination-phase deglaciation mainly in narrower zone of ice stagnation, an increase of melting due to influence of heat from air and condensation of water vapour, as well as grater activity of meltwater and meteoric water on the surface of the ice sheet. Present-day equivalents of this deglaciation pattern can be observed at the margins of the contemporaneous cold glaciers but of corse on much smaller scale /Fig.3/. In contrast to the previously distinguished deglaciation pattern, for this pattern the term "down-phase deglaciation" is suggested.

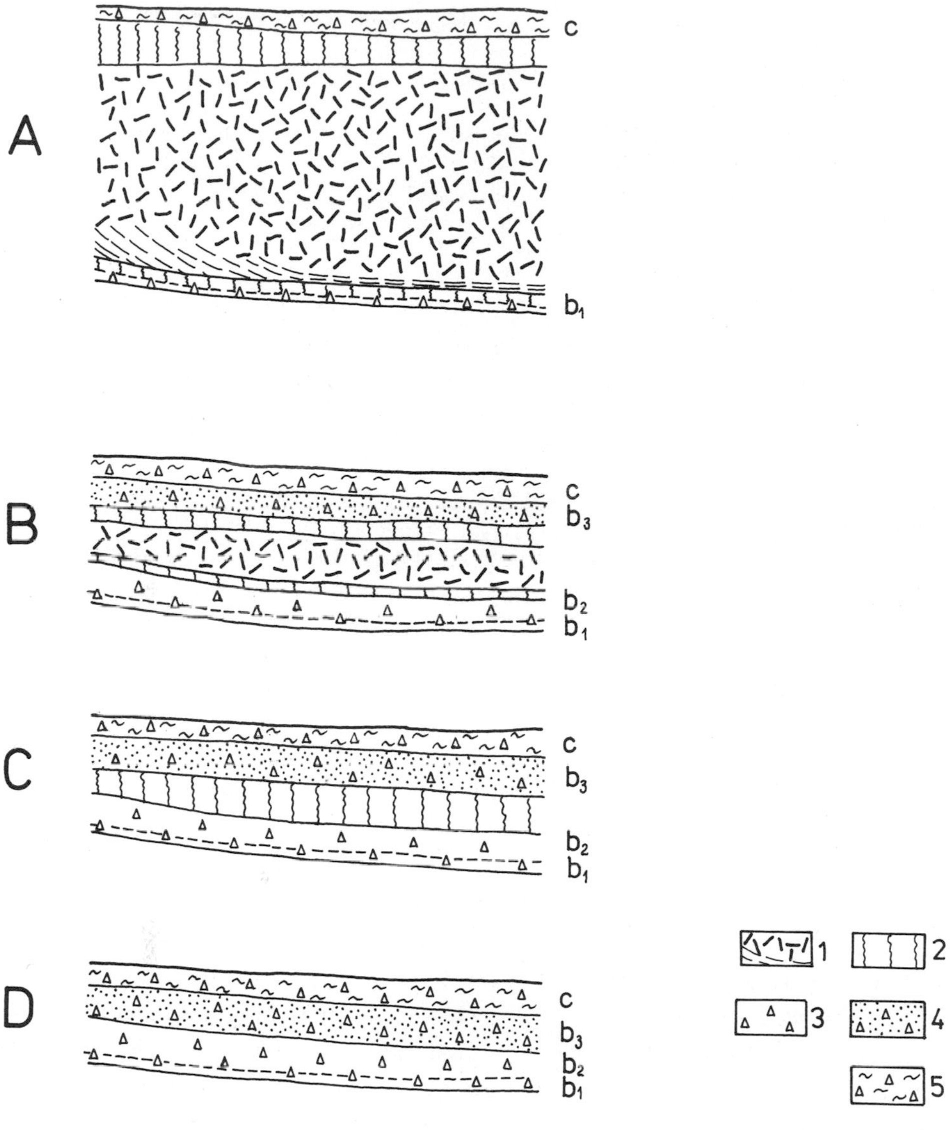

Fig. 4. Formation of till in the downwasting ice sheet far away from its snout

1 - glacier ice and frozen to the bed regelation ice being below pressure melting point, 2 - glacier ice at the pressure melting point, 3 - subglacial till, 4 - englacial till, 5 - different kind of ablation till, c - ablation till lying in situ and flow till, b_1 - subglacial till released from basal ice at the beginning of stagnation, b_2 - subglacial till of melt-out type, b_3 - englacial till resulted from sublimation and melting of ice

FORMATION OF TILL ASSOCIATED WITH DEGLACIATION

A great deal of information concerning the formation of till in stagnant ice has already been collected, referring both to modern glaciers and ancient ice sheets /cf. Asyeyev 1974, Embleton and King 1975, Sugden and John 1976/. However, a number of remarks related to the distinguished patterns of deglaciation, supported by field evidence, can be added. It is worth noting that these remarks contribute only to the deposition of till on land. Deposition in water bodies or from active ice is not considered here.

Commonly, within cold ice sheet wasting down three main depositional environments of till depended on the position within the ice sheet are developing, namely: supraglacial, subglacial, and englacial. The spatial distribution of these environments at some distance from the glacier snout is presented in Fig.4. The main types of deposited till are signed by letter symbols used in the earlier author´s publications /Drozdowski 1974,1976/. From the scheme it may be concluded that a continuous stratum of till is formed as a result of gradual accumulation of till, going on simultaneously from the top and base of the stagnant ice. First of all

Fig. 5. Loess-like sediments accumulated in an ice crevasse during areal deglaciation of the lower Vistula region in the Middle Würm

Fig. 6. Supraglacial loess-like lacustrine deposits resting on ablation till. Note the collapse structure which resulted from mélting of the underlying ice

ablation till /c/ and subglacial till of melt-out type /b_1 and b_2/ are areated. The depositional environments shift gradually downwards and upwards at the rate controlled by provision of heat derived from the surface and from the base /geothermal heat flux/. The formation of till, i.e. the release of glacial debris from ice proceed under conditions of overburden. Ablation till is accumulated on the top surface of buried ice, whereas subglacial till - on the glacier bed or previously deposited lodgment till associated with the advance of the ice sheet.

The culmination-phase deglaciation is evidenced mainly by englacial depositional environment of till, in which the release of englacial debris is dominated by sublimation rather than melting.

Sublimation as a dominant process of ice wastage is postulated for Mindel and Riss ice sheets on lowland areas of East Europe by many Russian and Polish authors /cf. Asyeyev 1974, Szponar 1975/. It took place also in the lower Vistula region in nothern Poland during the wastage of the Middle-Würmian Scandinavian ice sheet /Drozdowski 1975/. The continental type of climate is here evidenced by eolian loess-like sediments.Two genetical groups of these sediments are recognized /Drozdowski 1979/: niveo-eolian, accumulated within the crevasses of wasting ice in winter season /Fig.5/ and loess--like lacustrine, deposited in shallow lakes developed on the surface of downwasting ice sheet /Fig.6/. From the climatological point of view the occurence of

loess-like deposits in situation pointing to their accumulation during the wastage of the ice sheet indicated the existence of dry anticyclonic winds, just as today at the margins of Greenland and Antarctic ice sheets. This confirms the suggestion already expressed by Jahn /1950/ who shifted the loess sedimentation period beyond the culmination phase of the glacial cycles /cf. Fig.1/.

After termination of deposition of englacial till /of both melt-out or sublimation type/, tripartite sequences resulted, containing three different types of till. They tend to occur in a horizontally persistent stratum which can be followed for considerable distances, if they were formed within large patches of stagnant ice. A good example provides the sections on the valley-sides of the mentioned lower Vistula River, where the tripartite structure of the till stratum can be followed for a distance of more than 100 km /Olszewski 1974/. This confirms the principal attribute of the Pleistocene areal deglaciation, i.e. the downwasting of a broad marginal zone of the ice sheet. Such a pattern of deglaciation embodied, first and foremost, the fast expanding ice lobes, in which the ice was much thinner than in the main portion of the ice sheet, and therefore much more susceptible to stagnation.

In places where the ice sheet was disintegrated and meltwater penetrated it, till sequences interbedded frequently by glaciofluvial sediments and redeposited ablation till /flow till/ resulted. Such a type of sequence is more common in the areas where down-phase deglaciation took place.

However, the significance of thermal and dynamical regimes of the glacier for the formation of ablation and englacial tills should be stressed here. These types of till can originate in cold glaciers and only then, when the ice before stagnation underwent compressive stress which caused development of the upward component of the ice flow and abundant supply of debris from the substratum and basal debris-rich portions of the ice sheet /cf. Goldthwait 1951, Weertman 1961, Boulton 1972b, Schytt 1974/.

DISCUSSION

The described areal patterns of deglaciation resemble to certain degree the depositional conditions of the so-called "supraglacial sediment association" distinguished by Boulton /1976/ and considered by him as equivalent to areal deglaciation. This view seems to be correct as far as the down-phase deglaciation is concerned, in which stagnation of ice encompasses a narrow marginal zone shifting progressively up-glacier, like in modern cold glaciers.

However, the culmination-phase deglaciation differs considerably from this pattern. The differences concern the climatic environments above all, which can substantially change the regime of the glacier, its geometry and dynamics. Of particular significance is the rapid change of the type of arctic climate from maritime to continental. This circumstance, which does not have its analogues example in modern glaciers, gave rise to superimposition of two different glacial regimes, considered here as a basic cause of stagnation of large marginal areas of the ice sheet. Such an explanation is supported mainly by regionally persistent multi-till sequences containing distinctly developed englacial till, and by the occurence of loess-like sediment accumulated on down-wasting ice sheet.

Nevertheless, it is plausible to distinguish, after Boulton /1976/, still another pattern of deglaciation, connected with, what is termed by him "subglacial sediment association". That pattern is associated with temperate glaciers and is presumably equivalent to the "normal glacier retreat" /Flint 1929/ or "frontal deglaciation" /Galon 1968/. It occured probably on some areas glaciated in the Pleistocene, especially in the final phases of the glacial cycles.

REFERENCES

Adam, D.P. 1975, Ice ages and the thermal equilibrium of the Earth, Quat.Res. 5: 161-172.

Ahlmann, H.W:son 1948, Glaciological research on the North Atlantic coasts, R.geogr.Soc.Res. Ser.1.

Asyeyev, A.A. 1974, Drevniye materikoviye oledeneniya Evropi. Moskva, Izd.Nauka.

Avsyuk, G.A. 1950, Ledniki ploskikh vershin. Trudi Inst.Geogr. AN SSSR 45.

Barry, R.G., J.T.Andrews and M.A. Mahaffy 1975, Continental ice sheets: conditions for growth, Science 190: 979-981.

Bartkowski, T. 1969, Deglacjacja strefowa deglacjacją normalną na obszarach niżowych /na wybranych przykładach z Polski zachodniej i północnej/, Bad.Fizjogr.Pol.Zach. 23: 7-33.

Black, R.F. 1977, Regional stagnation of ice in northeastern Connecticut: an alternative model of deglaciation for part of New England, Am.J.Sci. 256: 715-728.

Boulton, G.S. 1972a, Modern Arctic glaciers as depositional models for former ice sheets, Q.J.geol. Soc.Lond. 128 /4/: 361-393.

Boulton, G.S. 1972b, The role of thermal regime in glacial sedimentation. In R.J.Price and D.E. /eds/, Polar geomorphology, Inst. Br.Geogr.Pub. 4: 1-19.

Boulton, G.S. 1976, A genetic classification of tills and criteria for distinguishing tills

of different origin. In W.Stankowski /ed./, Till its genesis and diagenesis, UAM Geografia 12: 65-80.

Brooks, C.E.P. 1949, Climate through the ages. London, Benn.

Bryson, R.A., W.M.Wendland and J.T. Andrews 1969, Radiocarbon isochrones on the disintegration of the Laurentide ice sheet, Arctic and Alpine Res. 1 /1/: 1-13.

Bülow, K. 1927, Die Rolle der Toteisbildung beim letzten Eisrückzug in Norddeutschland, Z. Dtsch.Geol.Ges. 79, 314: 273-283.

Clayton, L. 1964, Karst topography on stagnant glaciers, J.Glaciol. 5 /37/: 107-112.

Clayton, L. 1967, Stagnant-glacier features of the Missouri Coteau in North Dakota. In L.Clayton and T.F.Freers /eds/, Glacial geology of the Missouri Coteau, Grand Forks, North Dakota, North Dakota Geol.Surv., Misc.Ser.30: 25-46.

Donn, W.L. and M.Ewing 1966, A theory of ice ages III, Science 152: 1706-1712.

Drozdowski, E. 1974, Geneza Basenu Grudziądzkiego w świetle osadów i form glacjalnych. Prace Geogr. IG PAN 104.

Drozdowski, E. 1975, Penultimate period of deglaciation in the Grudziądz Basin, lower Vistula River valley: an interstadial--like interval of the Middle Würm, Geogr.Polonica 31: 213-235.

Drozdowski, E. 1976, Stratigraphy and genesis of till in the section at Sartowice Dolne, the lower Vistula valley. In W.Stankowski /ed./ Till its genesis and diagenesis, UAM Geografia 12: 253-257.

Drozdowski, E. 1979, Chronology and glacial sedimentary environments of the Baltic /Würm/ glaciation in the lower Vistula region, Trans.Inst.Br.Geogr. /in print/.

Embleton, C. and C.A.M.King 1975, Glacial geomorphology. London, Edward Arnold.

Flint, R.F. 1929, The stagnation and dissipation of the last ice sheet, Geog.Rev. 19: 256-289.

Flint, R.F. 1930, The glacial geology of Connecticut: Connecticut Geol. and Nat.History Survey Bull. 47.

Galon, R. 1968, On types of deglaciation of the Scandinavian inland ice, Actorum Geographicorum Debrecen, Ser.7, 14: 87-91.

Goldthwait, R.P. 1951, Development of end-moraines in east-central Baffin Island, J.Geol.59: 567--577.

Gripp, K. 1929, Glaziologische und geologische Ergebnisse der Hamburger Spitzbergen Expedition 1927. Abh.Nat.Ver.Hamburg, 22.

Hoppe, G. 1959, Glacial morphology and inland ice recession in northern Sweden, Geogr.Annlr 41: 193-212.

Hoppe, G. and G.H.Liljequist 1956, Det sista nedisningsförloppet i Nordeuropa och dess meteorologiska backgrund, Ymer 76: 43-74.

Jahn, A. 1950, Less, jego pochodzenie i związek z klimatem epoki lodowej, Acta Geol.Pol.1 /3/: 257-302.

Klimaszewski, M. 1960, Studia geomorfologiczne w zachodniej części Spitsbergenu między Kongs--Fjordem a Eidembukta. Zeszyty Naukowe UJ 32, Prace Geogr.1.

Kotlyakov, W.M. 1968, Snezhniy pokrov semli i ledniki. Leningrad, Gidrometeoisdat.

Kozarski, S. and J.Szupryczyński 1973, Glacial forms and deposits in the Sidujökull deglaciation areas, Iceland, Geogr.Polonica 26: 255-311.

Lamb, H.H. and A.Woodroffe 1970, Atmospheric circulation during the Last Ice Age, Quat.Res. 1 /1/: 29-58.

Lundqvist, J. 1973, Isvsmältningens förlopp i Jämtlands län. Sv.geol.Unders. C 681.

Meir, M.F. 1962, Proposed definitions for glacier mass budget terms, J.Glaciol. 8: 252-261.

Niewiarowski, W. 1963, Some problems concerning deglaciation by stagnation and wastage of large portions of the ice sheet within the area of the last glaciation in Poland, Rept 6th Conf. INQUA /Warsaw, 1961/, Łódź 3: 245-256.

Okko, V. 1955, Glacial drift in Iceland, its origin and morphology. Bull.Comm.géol.Finl.170: 1-133.

Olszewski, A. 1974, Jednostki litofacjalne glin subglacjalnych nad dolną Wisłą w świetle analizy ich makrostruktur i makrotekstur. Stud.Soc.Sci.Tor., Sect.C, 8,2.

Paterson, W.S.B. 1969, The physics of glaciers. Oxford, Pergamon.

Shaw, J. 1977, Tills deposited in arid polar environments, Can.J. Earth Sci.16, 6: 1239-1245.

Schytt, V. 1967, A study of "ablation gradient", Geogr.Annlr 46 /3/: 327-332.

Schytt, V. 1974, Inland ice sheets - recent and Pleistocene, Geol. För.Stockh.Förh. 96: 299-309.

Sugden, D.E. and B.S.John 1976, Glaciers and landscape. London, Arnold Publ.

Szponar, A. 1974, Etapy deglacjacji w strefie przedgórskiej na przykładzie przedpola Sudetów Środkowych. Acta Univ.Wratislav. 220, Studia Geogr.21.

Shumskiy, P.A. 1947, Energiya oledeneniya i zhizhn lednikov. Moskva, Geografgis.

Tarr, R.S. and L.Martin 1914, Alaskan glacier studies. Nat. Geogr.Soc., Washington.

Thorarinsson, S. 1943, Oscillations of the Icelandic glaciers in the last 250 years, Geogr. Annlr 25: 1-54.

Weertman, J. 1961, Mechanism for the formation of inner moraines found near the edge of cold ice caps and ice sheets, J.Glaciol. 3:965-978.

Changing conditions of glacial erosion and deposition reflected by differentiation of glacial deposits at Rozwady (Swietokrzyskie Mountains)

L. LINDNER & H. RUSZCZYNSKA-SZENAJCH
Warsaw University, Warsaw, Poland

1 INTRODUCTION

The Rozwady site (Fig 1) is situated in north-western margin of Swietokrzyskie Mountains, very close to the Polish Lowland area. The exposure is situated almost in the highest part of an elevation built of Lower-Jurassic yellowish platy sandstones, which are covered here by a several metrethick glacial series corresponding to the maximum stadial of the Middle Polish (Riss) Glaciation (Lindner 1971). The glacial series shows considerable lithological differentiation in comparison with lowland glacial deposits of the same age.

The exposure at Rozwady was demonstrated by the authors during the "Field Symposium - The Quaternary of the Western Part of Swietokrzyskie Region"(June 1977), and a preliminary report concerning genetic differentiation of the glacial deposits was published in Polish (Lindner+Ruszczynska-Szenajch 1977).

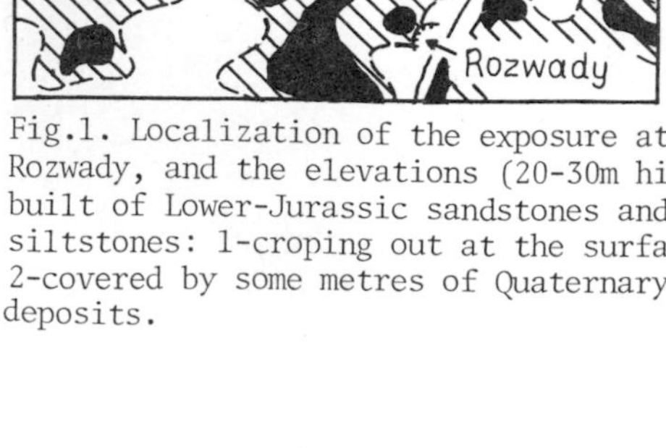

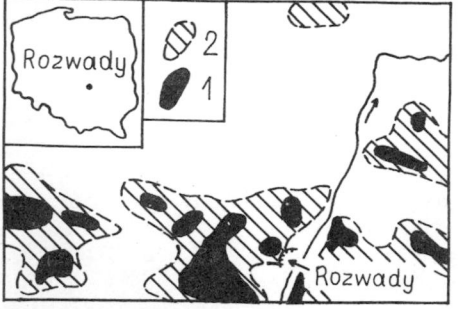

Fig.1. Localization of the exposure at Rozwady, and the elevations (20-30m high) built of Lower-Jurassic sandstones and siltstones: 1-croping out at the surface, 2-covered by some metres of Quaternary deposits.

2 GENETIC HORIZONS WITHIN THE GLACIAL SERIES AT ROZWADY, AND THE PROCESSES RESPONSIBLE FOR THEIR DEPOSITION

At a first glance the observer may distinguish two main units within the glacial series, each of them showing a distinct twofold character (Fig.2). The lower unit is represented by a yellowish till, covered by brown till, and the upper unit is composed of coarse-clast material - tightly packed in the lower horizon, and loosely arranged in the upper horizon.

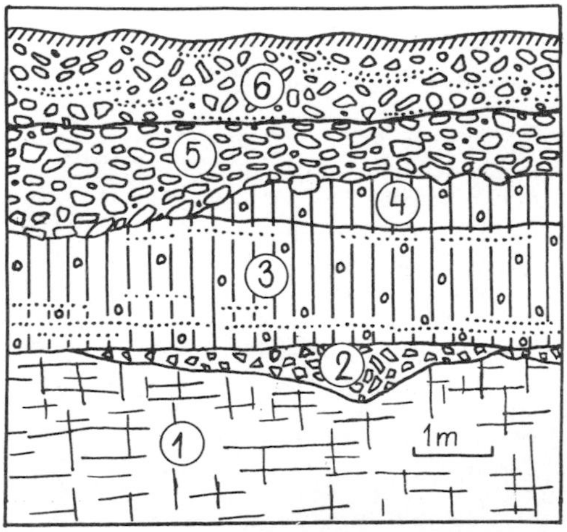

Fig.2. Generalized profile of the sequence of glacial deposits at Rozwady; 1-bedrock (sandstones), 2-redeposited weathered bedrock material, 3-yellowish lodgement till, 4-brown lodgement till, 5-lower (lodgement) horizon of coarse-clast series, 6-upper ("flow") horizon of coarse-clast series.

2.1. Bedrock surface

In the northern part of the exposure the bedrock surface underlying the glacial series shows a groove about 1,5 m deep and about 3 m wide (Fig.3). Some of sandstone joint blocks occurring at the bottom of the groove are disturbed from their primary horizontal position and dip to the NW (Fig.cit.). Subsequent works exposed the north-western continuation of the groove, which was becoming shallower (to about 0,5 m) and narrower (to about 1 m) in that direction.

These features as well as the deposits infilling the groove (described beneath) point to its origin as a subglacial scar, cut by boulders carried in the lowermost part of the ice advancing from NW direction. In some places (to NW) the boulders probably ploughed through the weathered sandstone cover and only their most protruding parts cut through the fresh rock carving within the sandstone a comparatively shallow and narrow groove, while in other places (Fig.3) they affected mainly fresh sandstone, resulting in formation of the wide and deep groove, which shows also the consequences of strong dynamic pressure exerted on sandstone plates.

In another part of the exposure the surface of sandstones shows rather "natural" character - without any features of cutting or smoothing by ice.

2.2 Redeposited weathered bedrock material

The described groove is filled with weathered sandstone material containing scarce Scandinavian pebbles. All this material is squeezed into the groove. This looks like an overturned fold with its axial plane dipping to the NE and closing to the north-east (Fig.3).

Thin patches of weathered sandstone material occur also in other parts of the exposure. The material is either coarse, composed mainly of small sandstone blocks, or comparatively fine - in some places almost clayey. It contains sometimes sporadic particles of Scandinavian origin. The weathered material lies on the sandstones, but it does not change gradually into a fresh rock as in a normal weathering profile, but shows more or less clear traces of mechanical redeposition.

This evidence suggests that during the phase of ice-advance into the sandstone culmination subglacial conditions did not favour complete removal of the weathered sandstone material, but that this deformed beneath the ice, although local erosion of the underlying bedrock did occur . The pure lithological character of the redeposited weathered material suggests that its redeposition has not been accompanied by effective melting-down of debris transported by the overriding ice i.e. by glacial deposition sensu stricto.

Fig.3. Groove at the top of sandstones, filled with redeposited weathered sandstone material, covered by bedded yellowish till.

2.3 Yellowish lodgement till

The weathered sandstone material is covered everywhere by the yellowish till. The till is comparatively sandy, composed mainly of local sandstone material. It indicates a phase of glacial deposition, though it was neither a continuous nor uniform process.

The most complete profile of the till may be examined in the northern part of the exposure, where the till lies on a somewhat lower bedrock surface and attains its maximum thickness of about 3 m. In the lower part the till is of clearly bedded character (Fig.3) and individual till beds are often separated by thin layers of non-stratified pure ground sandstone material. The till is tightly packed, and its stratification is conformable with the underlying bedrock surface. Many large clasts within the till show analogical arrangement and also show a tendency to occur more frequently within the lowermost parts of individual beds.

These observations point to a lodgement

sensu stricto origin for the till. The process of till deposition was interrupted by episodes of non-deposition accompanied by movement of finely-ground sandstone material, which was spread thinly by the moving ice over already accumulated till bed (s). The thin "smears" of more clayey weathered material, occurring in some places within the lowermost parts of the till, suggest that local bedrock material has been redeposited in unfrozen state. It seems most probable that these breaks in glacial deposition were controlled by the temperature changes underneath the ice (cf. Weertman 1962).

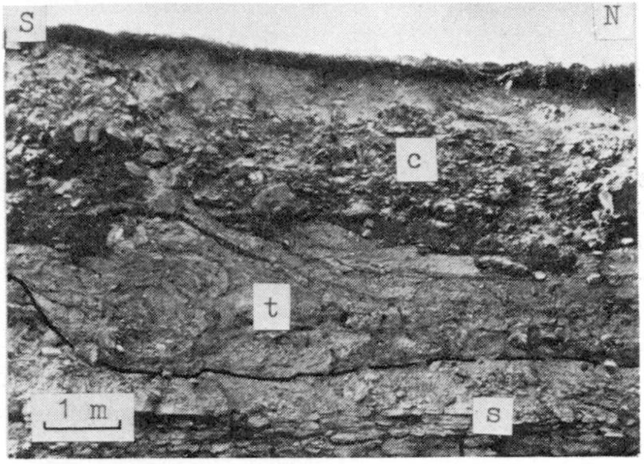

Fig.4. Yellowish bedded till (t) disturbed against the step of underlying sandstones (s), truncated by overlying coarse-clast series (c).

The more abundant occurrence of larger clasts at the lowermost parts of till beds can be explained by G.S. Boulton's conclusions (1975), that large particles may be lodged first (due to strong friction drag), when the conditions favouring lodgement begin to exist. The upper part of the yellowish till in the northern part of the exposure is represented by a non-bedded massive till. A boulder about 0,5 m diameter is "drowned" within the till. It lies near the contact with the underlying stratified till, and clearly marked structural lines beside and above the boulder show traces of the process of its sinking down within non-consolidated and water-saturated till.

In the southern part of the exposure the sandstone bedrock forms a conspicuous step inclined to the north, beds of yellowish till are much disturbed and pressed against the step (Fig.4). It is possible that the deformations formed when the material

Fig. 5. Yellowish till with ground sandstone material at the top (g), covered by brown till which is truncated by coarse-clast series.

was frozen to the moving ice and formed the lowermost debris layers in transport. The overladen ice could not override the sandstone step, it was strongly compressed and disturbed, and left here through decollement as a debris-laden dead icemass (analogical examples from recent glacial environments see in: Boulton 1970, Lawrushin 1976). The ice-sheet moved over this detached disturbed mass, and covered it with normal lodgement till. When the dead ice which cemented the deformed morainic mass melted out, the debris turned into melt-out till showing well preserved glacidynamic structures.

In some places at the southern end of the exposure the yellowish till is almost entirely massive, similar to the upper horizon of that till in the northern part of the exposure. Irregular sandy rafts occur within that massive till. The sands show completely obliterated primary stratification and they contain thin smeared layers of till. In some places larger rafts appear to have disintegrated into smaller irregular lumps, suggesting redeposition of that sandy material in an unfrozen state (Ruszczynska-Szenajch 1976a).

Beds of clearly stratified sands showing no traces of redeposition occur in some places at the top of the massive yellowish till. Their thickness does not exceed 30-40 cm. As the yellow till grades up into the brown till which is considered to be a lodgement till (see next section) it seems likely that these sediments have been deposited subglacially. They probably indicate intense melting of the ice and the existence underneath the ice-sheet

of running water which transported the melted-down debris and deposited it as thin fluvioglacial beds. These sands are covered by some decimetres of yellowish till beds representing again the "cold" facies of that till. So, the temperature regime underneath the ice changed abruptly from comparatively warm to considerably cold. This trend was even more stressed at the final stage of deposition of the yellowish till, what is marked by a continuous occurrence, in the southern and northern parts of the exposure, of a bed composed of almost pure ground sandstone material (Fig.5) indicating mechanical redeposition of locally-derived material.

The pure ground sandstone material forming regular layers within the yellowish till indicates a local source and "glacitectonic transport" underneath the moving ice. The local character of the till also points to local incorporation of that material into the ice. Thus, the areas of ice-erosion and ice-deposition were not more than several kilometers from each other and were probably separated by a zone characterised by no-erosion and no-deposition, where mechanical processes of glacial tectonics worked alone.

2.4 Brown lodgement till

The yellowish till is overlain by the brown till, which forms a discontinuous layer - occurring mainly in the southern part of the exposure - and whose thickness doesn't exceed 40 cm (Fig. 5). The intense brown colour of the till is due to the presence of Scandinavian material.

The boundary of the brown till with the underlying yellowish one is usually clearly marked (Fig.cit.) but it never cuts the yellowish till beds, showing good conformity with them. In places where the brown till covers immediately the massive part of the yellowish till, their contact doesn't form a sharply marked line. These features indicate a rather continuous process of deposition of both tills, without any obvious break.

The structure of the brown till is usually massive and compact, and the orientation of flattened pebbles and boulders is often horizontal. The conspicuous feature of that till is also the occurrence of larger clasts within its lower part or exactly at its bottom-surface (Fig.cit.). In some places thin beds of yellowish

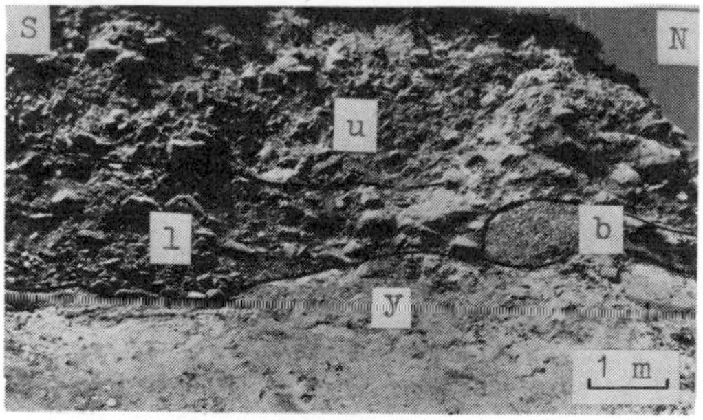

Fig. 6. Yellowish till (y), covered by brown till (b) containing large "rotten" Scandinavian granite, truncated by lower (lodgement) horizon of coarse-clast series (l), overlain by upper ("flow") horizon of coarse-clast series (u).

ground sandstone material occur in the upper parts of this typical "Scandinavian" till. All those features suggest the lodgement origin of the brown till. It was deposited in almost one continuous process, which had begun with horizontal lodgement of the largest transported clasts, and then changed into steady deposition of till-mass, which in final stages was being interrupted by episodes of no-deposition and mechanical redeposition of bedrock material from the nearest vicinity.

The process of deposition of the brown till was very similar then to that of accumulation of the yellowish till. Though, these two tills differ in composition (i.e. almost purely local material in the yellowish till and far-travelled Scandinavian material in the brown till). It seems that after the lower, local material had melted out from the base of the ice, melting penetrated up into the overlying Scandinavian material, which was then also deposited subglacially. Thus, the depositional profile reflects here a reversal of the erosional sequence: last eroded - first deposited, or : first eroded - last deposited (compare also observations from recently glaciated areas - Boulton 1972). Such regular sedimentation may be expected mainly within lodgement till sensu stricto i.e. deposited underneath the moving ice (and also in some basal melt-out tills), which gives another argument for the interpretation of lodgement origin of both tills.

2.5. Surface of ice-erosion

The top of the brown till and, in places the top of yellowish till, are cut by a clearly marked erosional surface, which is the base of the overlying coarse-clast series. The erosional-and-mechanical processes are marked here by small grooves and larger depressions cutting both tills (Fig. 5 and 6) and directed from W to E. This direction is confirmed also by the arrangement of the lowermost sandstone plates of the coarse-clast series, which lie immediately on the erosion surface and dip to the west. In some places comparatively large sandstone boulders occur within the grooves (Fig.5). We suggest that they represent the tools carried by the ice which have carved that grooves (similar to observations from recently glaciated areas - Boulton 1974). Sandstone boulders cut also some clearly marked beds of yellowish till blanketing the above desribed glacidynamic deformations (Fig.4).

The above evidence indicates, that the erosion surface was formed underneath an ice-sheet whose lowermost part has been laden with coarse sandstone debris. Some larger and more protruding sandstone boulders served as carving tools which cut the underlying till and formed shallow depressions and grooves. The discordant nature of the surface suggests that a considerable amount of material has been removed. This has probably occurred through freezing-on to the glacier sole. The direction of ice movement at that time was different from the direction prevailing during the deposition of the underlying tills. The question remains, whether the erosion represents only changed thermic conditions and change of ice-flow direction, or an interval when the area was ice-free.

2.6. Coarse-clast series, the genetic equivalent of lodgement- and flow-till

The 2 m thick coarse-clast series overlying the erosion surface consists mainly of large sandstone particles ranging from boulder to gravel size with abundant sand (Fig. 6). The material is usually sharp-edged and predominantly local with a small amount of Scandinavian finer particles. At a first glance the series resembles coarse slope-deposits such as "head" from the British Islands. But in this place there was no slope which might give rise to such deposits. There are two distinct coarse-clast horizons.

The lower horizon of the coarse-clast series is characterised by comparatively tight packing of the material and by regular arrangement of sandstone plates which lie conformably with the surface of the substratum although it sometimes shows an imbricate structure with westerly dip (Fig.6). In its lowermost parts the horizon shows in some places (especially in shallow depressional forms) a concentration of larger particles (Fig.cit.). All these features, together with the fact that it overlies a clearly pronounced ice-erosion surface, suggest this horizon to be lodgement till, deposited underneath moving ice. The question arises whether the lack of fines is a result of specific sedimentation processes or it is due to supply of coarse material only. The impoverishment in fine particles during the process of deposition of lodgement till is known from areas characterised by the possibility of free escape of melt-water (into e.g. karst underground system - Ruszczynska-Szenajch 1976 b), but such a situation did not exist at Rozwady. Thus, the later situation seems the most probable, i.e. the ice supplied coarse material only. The local character of that material points to its short transport within the ice, preventing far-advanced grinding processes, but abundance of coarse and sharp-edged particles in comparison with (also local) yellowish till may point to very effective erosional processes in the source area, quickly creating new regelation layers one beneath another. In such conditions most of the frozen-up debris has almost immediately gone into suspension, where - according to G.S. Boulton (1974) - grinding is much less effective than in the zone of traction. So the material deposited at Rozwady was nearly identical - according its grain size and shape - with the material taken from the near vicinity. It is also noteworthy that the zone of ice-erosion which had prevailed here during the preceding phase was shifted, and at Rozwady gave way to deposition. The first stages of this deposition were marked by some concentration of coarser fragments at the base of the horizon, which were lodged first when the thermic regime just began to favour deposition.

The characteristic feature of the upper horizon of the coarse-clast series is the loose arrangement of particles - in contrast to the tight packing within the lower horizon. This feature, which results in greater permeability, caused the post-depositional manganese concentration which occurs consistently at the boundary of these two horizons. The setting of coarse particles within the upper horizon shows in some places traces of gravitational sliding during the process of deposition. Another characteristic feature of the upper horizon of coarse-clast series is the occurrence in situ of stratified fluvioglacial sand which occur at the base, in the middle, and sometimes at the top part of the horizon. This points to the deposition of this horizon within the ablation zone. The surface melting of debris was accompanied by formation of small streams carrying sandy material, as in modern processes (Boulton 1972) where sands deposited by these streams are covered by coarse particles which slide down the surface of melting debris-laden ice, and which may be regarded then as an equivalent of flow-till. The upper horizon of the coarse-clast series is usually thicker than the lower one, and in some places nearly the whole series is represented by this deposit indicating final deglaciation of that area. The lack of fine particles within this horizon is, most probably, caused by the same reasons as that of the lower horizon, though the ablation zone might also create some better conditions for the impoverishment in silt and clay during the process of deposition.

An interesting characteristic of the whole coarse-clast series is its low content of Scandinavian material, although the whole series has been undoubtedly deposited by a Scandinavian ice-sheet. It may point to decollement of the basal part of the ice-sheet at the foot of the bedrock elevations surrounding the locality from the west, and to comparatively short-lasting glacial deposition at Rozwady underneath the moving ice. This short time didn't allow the successive ice-masses laden with Scandinavian rocks to override the detached dead ice and the elevations and to reach the area. However, one cannot exclude the existence of very effective lodgement conditions in more northerly lowland areas, which might result in considerable impoverishment of the ice-sheet in Scandinavian debris.

3 SOME COMPARISONS WITH RECENTLY GLACIATED AREAS AND WITH THE LOWLANDS GLACIATED DURING THE PLEISTOCENE

The questions discussed in the previous chapter illustrate the fact that some areas of middle latitudes characterised by elevated relief and hard bedrock glaciated during the Pleistocene may supply close analogies in glacial processes and their geological products to recently glaciated areas in polar and subpolar zones, especially to the very clear evidence and interpretations of processes by G.S.Boulton, whose works concern a wide range of glacial geological problems and prove very helpful in the interpretation of many Pleistocene events.

The actualism method cannot be, as yet, so helpfully applied in the studies of thick Pleistocene glacial series occurring on vast lowland areas characterised usually by a soft substratum with low relief. These series are very often considerably deformed by syngenetic glacitectonic processes.

The difference in the geological record and most probably in the character of processes, between recently accumulated glacial deposits and Pleistocene lowland glacial series is hardly a result of differentiated climates according to latitude (e.g. Rozwady is situated in a very close neighborhood to the lowlands), and it is rather much more a consequence of the bedrock-and-relief conditions and the corresponding water conditions.

4 ACKNOWLEDGEMENT

The authors are very much indebted to Dr. G.S. Boulton, University of East Anglia, Norwich, for his kind improvements of the manuscript.

5 REFERENCES

Boulton, G.S. 1970, On the deposition of subglacial and melt-out tills at the margins of certain Svalbard glaciers. J.Glaciol 9:231-245.

Boulton, G.S. 1972, Modern Arctic glaciers as depositional models for former ice-sheets. J. Geol. Soc.Lond. 128:361-393.

Boulton, G.S. 1974, Processes and patterns of glacial erosion. In D.R. Coates (ed.), Glacial geomorphology, p.41-87. New York.

Boulton, G.S. 1975, Processes and patterns of subglacial sedimentation: a theoretical approach. In A.E. Wright and F. Moseley (ed.), Ice ages: ancient and modern, p. 7-42, Liverpool.

Lavrushin, J.A. 1976, Strojenie i formirowanie osnownych moren materikowych oledenienij (in Russian), Izd. "Nauka", Moskwa.

Lindner, L. 1971, Stratygrafia plejstocenu i paleogeomorfologia polnocno-zachodniego obrzezenia Gor Swietokrzyskich(Pleistocene stratigraphy and paleogeomorphology of the north-western margin of the Holy Cross Mountains, Poland.)Studia Geol. Polon. 35:1-113 (Engl.summ.).

Lindner, L. + Ruszczynska-Szenajch, H. 1977, Zagadnienie genetycznego zroznicowania glin zwalowych i osadow pokrewnych (On the genetic differentiation of tills and akin deposits). Przeglad Geol. 25: 432-438.

Ruszczynska-Szenajch, H. 1976a, Depresje glacitektoniczne i kry lodowcowe na tle budowy geologicznej poludniowo-wschodniego Mazowsza i poludniowego Podlasia (Glacitectonic depressions and glacial rafts in mid-eastern Poland). Studia Geol. Polon. 50:1-106 (Engl. summ.).

Ruszczynska-Szenajch, H. 1976b. Examples of differentiation of tills due to bedrock conditions prevailing in the place of deposition. In: W. Stankowski (ed.) Till - its genesis and diagenesis, p. 81-89, Poznan.

Weertman, J. 1962, Stability of ice-age ice caps. US Army Cold Reg.Res.Engng. Lab. Res. Rept. 97 (12 pp).

C. Progress reports

Die quartären Ablagerungen im übertieften Wolfratshausener Zungenbecken (Oberbayern)

H. JERZ
Bayerisches Geologisches Landesamt, München, Bundesrepublik Deutschland

Das Wolfratshausener Becken im Süden von München repräsentiert ein Zungenbecken des Isarvorlandgletschers -- ebenso wie das benachbarte Becken mit dem Starnberger See.

Im Norden wird das Wolfratshausener Becken von Endmoränen begrenzt, im Süden durch den Molasseriegel bei Penzberg vom Stammbecken des Gletschers, dem Kochelsee-Becken, getrennt.
Das heute von Isar und Loisach durchflossene Wolfratshausener Becken enthielt nach den drei letzten Vorlandvergletscherungen ausgedehnte Seen. Der letzte der sog. Wolfratshausener Seen ist noch im Würm-Spätglazial ausgelaufen (vgl. Jerz 1969).

Ergebnisse von neueren Untersuchungen wie geologische Kartierungen, Aufschlußbohrungen, geophysikalische Messungen führen zu der Vorstellung, daß das Wolfratshausener Becken in seinen wesentlichen Umrissen in der 3. letzten Eiszeit, also in der Mindel-Eiszeit im klassischen Sinne, angelegt worden ist. Die Tiefenerosion des damaligen Gletschers reichte bis etwa 20-30 m unter die heutige Geländeoberfläche (= 560 - 550 m ü.NN bei Wolfratshausen).
Während der 2. letzten Eiszeit, der Riß-Eiszeit, war im Wolfratshausener Becken insbesondere eine Tiefenerosion wirksam. Die Beckensohle reicht noch mehr als 130 (150) m unter den heutigen Talboden (= ca. 450 m ü.NN).
Eine deutlich geringere Glazialerosion ist für die letzte Eiszeit, die Würm-Eiszeit, festzustellen. Überraschenderweise sind sogar in zentralen Bereichen des Wolfratshausener Beckens noch Moränen, Schotter und sogar Feinsedimente (Seetone) der vorletzten Eiszeit erhalten (vgl. Bohrung Achmühle in Fig.1). Möglicherweise bildet die kürzere Zeitdauer der letzten Vorlandvergletscherung den hierfür maßgeblichen Grund (vgl. auch Jerz 1978, im Druck).

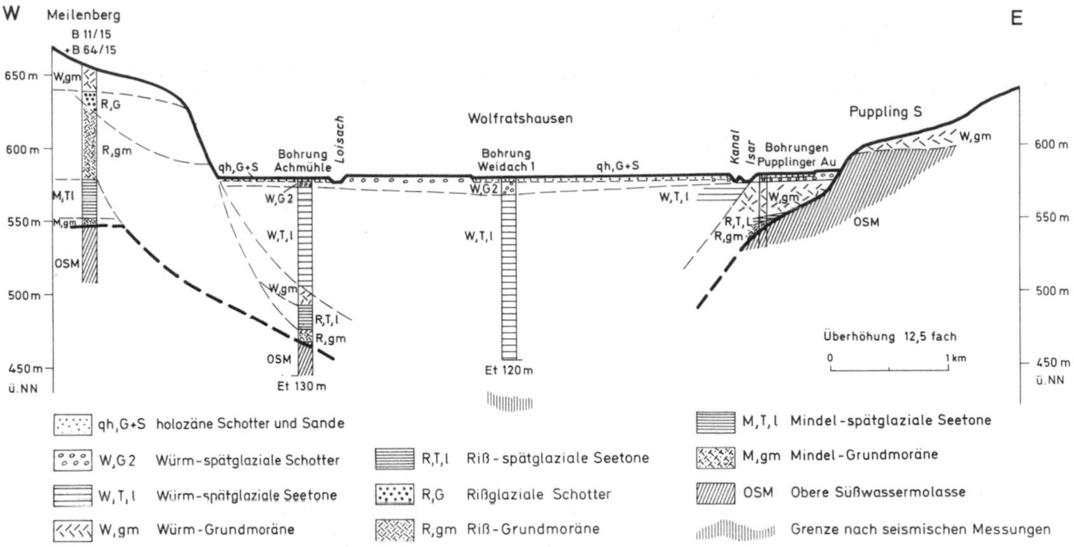

Fig.1. West - Ost - Profil durch das nördliche Wolfratshausener Becken (Überhöhung 12,5 fach).

Im folgenden werden die einzelnen Substrate aus der quartären Füllung im Wolfratshausener Becken anhand der Aufschlußbohrungen Pupplinger Au, nahe dem Ostrand des Beckens (s.Fig.1), kurz beschrieben und miteinander verglichen.

1. Die pleistozänen Sedimente werden fast im gesamten Auenbereich der Isar und Loisach von durchschnittlich 2 - 3 m (max. 10 m) mächtigen postglazialen Schottern und jungholozänen Flußmergeln abgedeckt.
Lagerung: locker.

2. Darunter folgen Feinsedimente des spätglazialen Wolfratshausener Sees. In den Bohrungen Pupplinger Au sind sie 10 - 15 m mächtig, in zentralen Bereichen des Beckens erreichen sie über 100 (120) m. Es handelt sich vorwiegend um graue bis blaugraue Tone und Schluffe mit Feinschichtung (Bänderung). Darin eingetiefte etwa Süd-Nord verlaufende Erosionsrinnen sind mit überwiegend sandigen Sedimenten aufgefüllt. Es handelt sich vermutlich um Strömungsrinnen im ehemaligen Wolfratshausener See.
Zustandsform der Seetone: steif (bis weich plastisch);
Lagerung der Seesande: locker, in wassergesättigtem Zustand zu Grundbruch neigend.

3. Das Liegende der spätglazialen Seetone bildet eine würmzeitliche Grundmoräne. Ihre Mächtigkeit schwankt zwischen wenigen Metern und über 25 m; ihr Relief ist sehr unregelmäßig. In randlichen Bereichen des Beckens mit hochanstehender Grundmoräne sind bis zu 20 m tiefe Glazialfurchen nachgewiesen, die mit jungen Seetonen ausgefüllt sind.
Für den Nordteil des Beckens lassen sich in verschiedenen Bohrungen Oszillations- bzw. Ablationsmoränen von der Grundmoräne s.str. unter-

scheiden. Sie verzahnen sich dort mit feingeschichteten lakustrisch-glazialen Ablagerungen, welche als Stausedimente zwischen dem Eisrand und dem zerfallenden Eis (Toteis) abgesetzt worden sind.
Die Würm-Grundmoräne ist im allgemeinen sehr schluffreich. Es konnte darin reichlich umgelagertes Seetonmaterial, vielfach in Form von Glazialschuppen, nachgewiesen werden.
Lagerung bzw. Zustandsform der Grundmoräne: dicht, halbfest; Lagerung der Ablations- bzw. Oszillationsmoräne: weniger fest (steif).

4. Unter der Würm-Grundmoräne ist im Beckenbereich häufig ein älterer, bis zu 10 m mächtiger Seeton erhalten, der als Riß-spätglaziale Bildung angesehen wird. Sein präwürmzeitliches Alter erscheint durch weiter im Süden bei Herrnhausen erhaltene Schieferkohlen (bis zu 2 m mächtig) gesichert.
Zustandsform: halbfest.

5. Als älteste quartäre Ablagerung im eigentlichen Wolfratshausener Becken gilt eine vermutlich rißzeitliche Grundmoräne, welche die Beckensohle auskleidet. Sie ist meist zwischen 5 und 10 m mächtig. Darin eingelagert sein können Lagen mit mehr oder minder kiesiger Moräne, die wassererfüllt sind. Ihr Grundwasser steht unter hydrostatischem Druck, der Auftrieb kann bis über die Geländeoberfläche reichen.
Lagerung bzw. Zustandsform: dicht, halbfest bis fest.

6. An der steilen Westflanke des Beckens treten zwischen Wolfratshausen und Schäftlarn noch Ablagerungen des ehemals breiteren (und weniger tiefen) mindelzeitlichen Gletscherbeckens zum Vorschein: stark komprimierte Seetone, an der Oberfläche ± stark aufgeweicht; sie werden dem drittletzten Spätglazial zugerechnet und sind hier von rißzeitlichen Schottern und Moränen überdeckt. Die Grenzschicht zur Molasse bildet eine vermutlich mindelzeitliche Grundmoräne (vgl. Fig.1: E - W - Profil).
Lagerung bzw. Zustandsform: dicht, halbfest bis fest.

7. Den Untergrund bilden Schichten der Oberen Süßwassermolasse mit Mergel, Sandmergel und Sandstein (Tertiär).
Zustandsform: fest, sehr fest bis hart.

Zusammenfassend lassen sich aus dem Vergleich der quartären Ablagerungen im Wolfratshausener Becken folgende Angaben machen:
Die Würm-Grundmoräne ist im allgemeinen schluff- und tonreich (mit bis zu 40 % Ton); sie weist vielfach Material aus älteren Seeablagerungen auf. Die Rißmoräne -- wie auch die Mindelmoräne -- enthält reichlich vom Molasseuntergrund abgeschürftes und aufgenommenes Material, kenntlich an den graugelben bis graugrünen Gesteinsfarben und am höheren Sandanteil der Moränen. Bei den Zustandswerten bestehen keine gravierenden Unterschiede; dies gilt auch bei mehr-

maliger Vorbelastung der älteren Grundmoränen durch das hier 250 - 300 m mächtige Gletschereis. Deutliche Unterschiede ergeben sich dagegen im Vergleich der nicht vorbelasteten und der vorbelasteten glazigenen Feinsedimente. Die Konsistenzzahlen für die jungen, Würm-spätglazialen Seetone liegen im Mittel im steifplastischen, die entsprechenden Kennziffern für die vorbelasteten älteren Seetone im halbfesten Zustandsbereich. Die Wassergehalte vermindern sich bei den vorbelasteten Feinsedimenten auf mehr als die Hälfte.

Summary: The quaternary deposits in the overdeeponed glacial basin of Wolfratshausen.
The basin of Wolfratshausen represents one of the wellknown glacial formed basins in the alpine foreland. It was determinated in its essential lines by the glacial erosion during the Mindel glaciation, then it was strongly overdeepened during the Riß glaciation and finally (only) overformed during the Würm glaciation.
In the central parts of the basin there are various glacial deposits preserved: moraines and varves of the two last glaciations. The Riß ground moraine contains a lot of sand and marl material eroded from the Molasse bedrocks, the Würm ground moraine more silt material from older glacial varves.
The consistency limits for the different ground moraines (basal tills) are similar. They differ evidently for glaciolacustrine sediments which are formerly compacted by the glacier or not.

Literatur:

Deutscher Normenausschuß, 1953, DIN 4020, Bautechnische Boden-Untersuchungen, Richtlinien. Berlin und Köln.

Jerz, H. 1969, Erläuterungen zur Geologischen Karte von Bayern 1:25 000, Blatt Nr. 8134 Königsdorf. München, Bayerisches Geologisches Landesamt.

Jerz, H. 1978 (im Druck), Das Wolfratshausener Becken, seine glaziale Anlage und Übertiefung. Eiszeitalter und Gegenwart 29 (Edith-Ebers-Symposium 1977).

Jerz, H. 1979 (in Druckvorbereitung), Geologische Karte von Bayern 1:25 000, Blatt Nr. 8034 Starnberg Süd. München, Bayerisches Geologisches Landesamt.

Penck, A. & E. Brückner 1901, Die Alpen im Eiszeitalter, Band 1. Leipzig.

The effect of glacial-lacustrine relationships on the deposits of the Huron Lobe during the Late Wisconsinan, south eastern Michigan

DONALD F. ESCHMAN
University of Michigan, Ann Arbor, Mich., USA

My recent work has been in the southern part of the Black River basin, located in Sanilac and St. Clair counties, a short distance northwest of Port Huron, Michigan. The river flows south along the distal edge of the Port Huron Moraine for about 65 kilometers before turning east to cross the moraine and enter St. Clair River a short distance downstream from its head at the southern end of Lake Huron. In the area of study the Black River and its only significant tributary, Mill Creek, are incised up to 30 meters below the lake plain related to Lake Whittlesey.

There are at least four tills of Late Wisconsinan age exposed in the stream cuts along Black River. In order to facilitate the discussion in this paper I use local, as yet unofficial, names for each of the tills. Their sequence is as follows :

(youngest)	Jeddo Till
	Fisher Road Till
	Black River Till
(oldest)	Mill Creek Till

The Jeddo Till, the surface till of the Port Huron Moraine, dates from about 13,000 years ago. The Mill Creek Till overlays a unit of sand and fine gravel which in turn is on top of an organic zone dated at 48,300 ± 800 BP (QL - 1215). Thus, all four tills are probably of Late Wisconsinan age, although there is a possibility that the Mill Creek Till, which in places seem to be very slightly weathered, may be of Middle Wisconsinan age.

Within the area, each of the tills has characteristics indicative of deposition in a body of standing water; some of the clearly lacustrine units which are found between tills are characterized by structures that, in turn, suggest their deposition near a glacial margin. The sedimentary features which indicate such an ice-lake relationship vary greatly in scale, and some of them are much more definitive than others. In this paper I describe each line of evidence suggesting such a relationship of glacier ice and lake to demonstrate both the variety in kind of evidence and in scale. I begin with evidence found within the tills that suggests their deposition in a lacustrine environment and then describe some large scale sedimentary structures from one of the inter-till lacustrine deposits that, to me, seem best explained as having been the result of deposition in a lake near an ice margin.

Before going on I must point out that the location, one which prior to Port Huron time, at least, lay well within the drainage area tributary to the glacially deepened basin of Lake Huron, is one where a glacier would be expected to terminate in a proglacial lake whenever the ice front was in the vicinity. The sediments which separate the several tills in section are nearly everywhere thinly laminated lacustrine silts and sands. In at least one place beach sands related to Glacial Lake Arkona are found directly beneath Jeddo Till along the east bank of Black River. As an aside, it seems obvious to me that in a locale such as this there must have been a sequence of proglacial lakes, like the classic one related to the last ice retreat first described by Leverett and Taylor (1915) and subsequently elaborated on by Hough (1958, 1963) and many others, each time an ice front passed across the area - during the Mid-Wisconsinan Interstadial, and even some of its subdivisions, as well as during the several interglacials.

Beginning with the largest scale sedimentary evidence suggesting that glacial ice terminated in a lake within the area, the Fisher Road Till nearly everywhere consists of two till members separated by a lacustrine

unit averaging less than one meter in thickness. The two till members of the Fisher Road Till are nearly indistinguishable except for the fact that the upper member generally has a somewhat higher content of silt and clay. Upon closer look one finds that the basal portion of the upper member of this till contains thin silty laminae and, in many places, small clasts (up to a centimeter or so in diameter) of fine silt and clay "ripped up" from the underlying lacustrine member. On the other hand, the lacustrine unit within the Fisher Road Till in many places contains thin (up to five centimeters thick) lenses of till. In one or two places lenses of brecciated lacustrine clays are found in otherwise undisturbed thinly bedded lacustrine silts and clays; such isolated brecciated zones, while by no means diagnostic of a near-ice lacustrine environment, certainly may well be the result of fluctuations of meltwater discharge, of lake level, or of sediment load such as characterize that environment. Taking these various kinds of evidence together I suggest that the Fisher Road Till in the area in which it has thus far been recognized was deposited by ice which terminated in a proglacial lake, and that its three members result from a retreat and readvance of the ice front.

When we look carefully at the other till units within the area we find much evidence supporting the idea of a subaqueousenvironment for their deposition as well - or at least evidence that fits such an environment. Within these till units, and in particular near the base of each, there is small-scale interbedding of till and stratified sediments. In some cases stratified sediments are perhaps fluvial in origin but frequently they are well sorted thinly laminated fine sands and silts more likely indicating a lacustrine environment of deposition. Again, clay and silt clasts are common in the basal meter or so of most of the tills in the area; sometimes these clasts, clearly of lacustrine sediments, are the only real evidence of such an environment within the particular part of the section at that locality.

One of the till units in the area, the Black River Till, is a very compact sandy till that is probably the equivalent of the Catfish Creek Till of southwestern Ontario. Evenson et al. (1977) demonstrated the subaqueous environment of deposition for the Catfish Creek Till in Ontario and many of the features they recognize as indicative of the lacusturine environment also occur in the Black River Till. In addition to these the Black River Till typically displays another feature which I suggest might well indicate a lacustrine environment of deposition for the unit. About a meter below the top of the till there is commonly a zone of boulders and cobbles, typically only one clast thick. The boulders do not form a true boulder pavement, for they rarely are in juxtaposition. When seen in section the boulder zone is horizontal. The till above the boulder zone is indistinguishable from that found below, so one must conclude that whatever caused the concentration of large clasts was a short-lived phenomenon. The localization of large clasts near the top of this unit seen in several exposures, and all falling at nearly the same elevation, suggests to me that it is the result of wave-winnowing of the till during its deposition approximately at the level of the surface of a proglacial lake.

I must admit that this explanation stems in part from an inability to explain the concentration of large clasts in any other way. There is no erosional irregularity or evidence of weathering at the level such as one would expect would result during winnowing by a subaerial process to develop a lag deposit. On the other hand, there is no other evidence of a lacustrine environment at this horizon, either.

The last line of evidence which suggests to me that glacial ice terminated in standing water during deposition of the drift in this area is from one of the interbedded, or intertill, lacustrine units itself. In one high stream cut the Black River till directly overlays a thick lacustrine sand and silt unit which, in its upper 4 or 5 meters, contains some of the largest and best developed ball and pillow structures I know of. The term "ball and pillow structure" is here used for elliptical masses of fine sand completely encased in silt or clay; such structures are formed by the foundering of a sand bed into a water-saturated mud (Potter and Pettijohn 1963: 150). Some of the individual sand masses at this locality are nearly 50 cm long and over 30 cm thick. The fine sand within a pillow is typically very well laminated, and the laminae in the outer part of most pillows are bent until they parallel the base and ends of the mass. The bending is the result of drag created by the movement of the sand as it sank through the mud.

As Potter and Pettijohn (1963) suggest it is clear that the structures form after deposition of the sand layer and that their formation is abrupt. The foundering seems

clearly the result of the liquefaction of the underlying muds. Kuenen (1958) created similar structures in the laboratory by shaking the vessel containing the sand and mud layers; as a result he suggested that the sudden liquefaction was the result of earthquake shocks. Other things which might well cause liquefaction of such water saturated muds include changes in pore water pressure either through changes in hydraulic head or through flow of water, and overloading due to rapid deposition.

Whatever the precise cause for the foundering of the sand bed that underlies the Black River Till at this site it seems to me that a proglacial lake with a nearby (advancing?) ice front is the perfect environment for nearly all of the potential causes of liquefaction of the mud. The nearby ice would be a ready source for much sand, and fluctuations in the level of the lake and in the influx of water into the bottom sediments from the glacier are to be expected. Perhaps even the calving of a large mass of ice from the glacier front, would create a shock wave within the water sufficient to cause liquefaction. In any case, I suggest that the ball and pillow structures within the unit underneath the Black River Till at this one site are also evidence for a depositional environment best described as one of a glacier terminating in a proglacial lake.

REFERENCES

Evenson, E.B., Aleksis Dreimanis & J.W. Newsome 1977, Subaquatic flow tills : a new interpretation for the genesis of some laminated till deposits, Boreas, 6 : 115-133.

Hough, J.L. 1958, The Geology of the Great Lakes. Univ. of Illinois Press, Urbana.

Hough, J.L. 1963, The prehistoric Great Lakes of North America, Amer. Scientist 51 : 84-109.

Kuenen, P.H. 1958. Experiments in geology, Trans. Geol. Soc. Glasgow, 23 : 1-28.

Leverett, Frank & F.B. Taylor 1915, Pleistocene of Indiana and Michigan and the history of the Great Lakes. U.S. Geol. Survey, Mon.53.

Potter, P.E. & F.J. Pettijohn 1963, Paleocurrents and basin analysis. Springer Verlag, Berlin.

Complex till sections in the western Swiss Plain

J. J. M. VAN DER MEER
Universiteit van Amsterdam, Amsterdam, Netherlands

The study area is situated in the western part of the Swiss Plateau (van der Meer, 1976), and was covered by the Rhône glacier during the Würm (Hantke, 1978). In this area a number of till exposures have been found. Of these, most show homogeneous tills, while some show structures. The thickness of the exposed till sections varies between 1 and 10 metres. By homogeneous is meant, that the tills do not show any structures other than fissility and that the texture and the stone content are the same. Not only is there no variation in one exposure, but there is also hardly any variation between exposures. Almost all of the textural analyses done on the fraction smaller than 2 mm show loams (USDA classification). All of the pits visited up until this summer (1978) showed these homogeneous tills, as did most of those visited during this summer.
Of the pits, showing complex till sections, four will be briefly described. For the location of these pits see fig. 1; no analytical data can as yet be given.

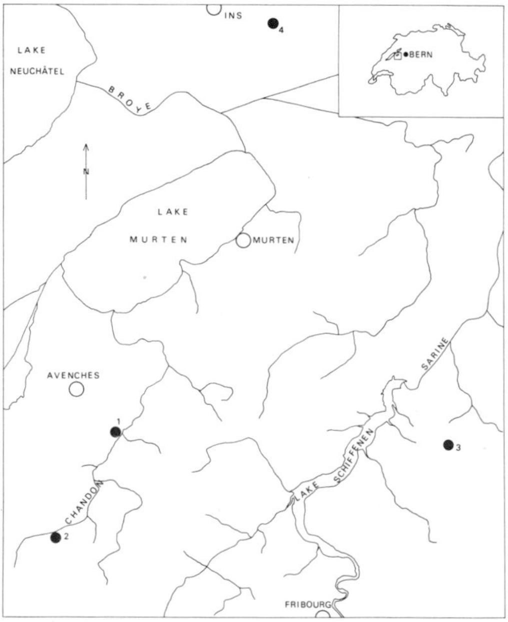

Fig.1. Location of the four sections

Pit 1: Donatyre

In this large gravel pit till is exposed only in the highest part. The pit is located on a fluvioglacial terrace in a meltwater valley. From bottom to top it shows: 1. molasse sandstone; 2. horizontally deposited gravel; 3. more deltalike deposited gravel and sand; some of the calcareous gravels show striae; 4. thin (up to 1 m thick) till-layers with intercalated sand, gravel and varvelike deposits (fig. 2). The till layers are all very compact and show faint internal layering. The till layers are not continuous but they do make contact with one another. In between they are separated by sand, gravel and varvelike deposits.

The stone content in the different till layers is the same, and the stones are uniformely spread in the layers. In one place a fold is to be seen in the till which is quite distinct, because of the varvelike deposit, which is folded with it. The top of this fold is cut by another till layer.
On the other side of the valley, in a gravel pit that shows the same stratigraphy, some pieces of wood found in the upper sands and gravels have been dated at 55.100 (+ 4500, -2900) BP (GrN-8105).

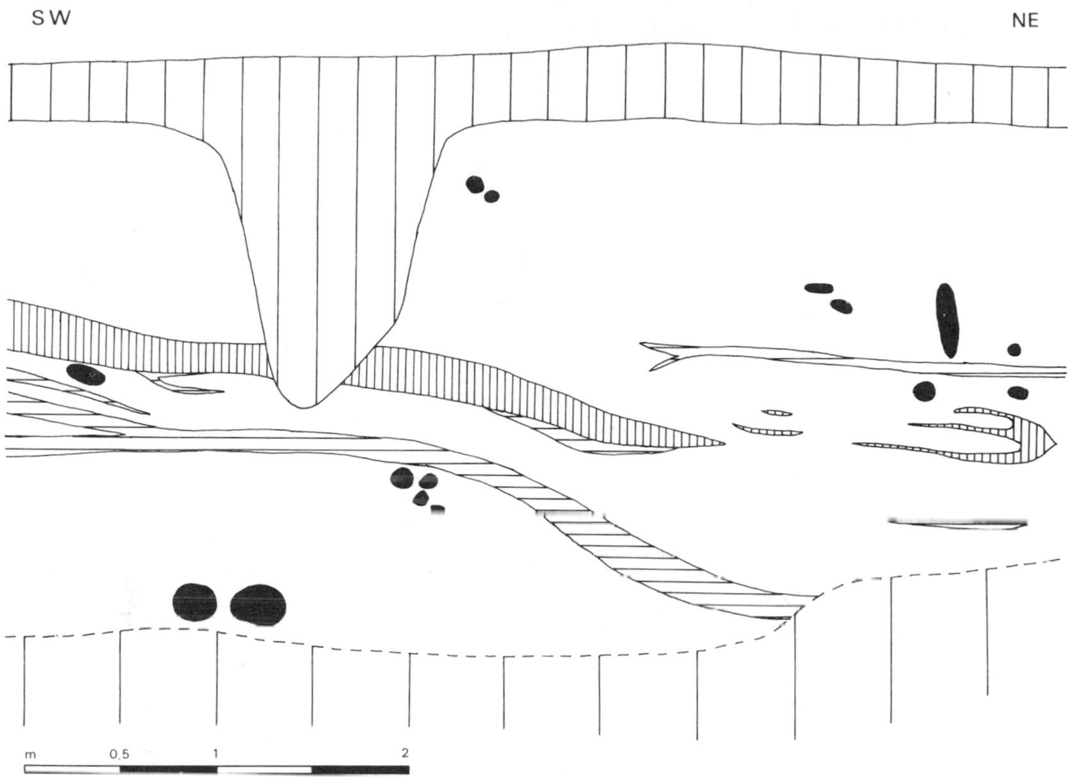

Fig. 2 Sketch of the wall in pit 1, near Donatyre. Horizontal and vertical scale the same. For the legend see fig. 3

Pit 2: Léchelles

This is an abandoned gravel pit not far south of the pit described above; it is situated at the beginning of the same meltwater valley. Though the till here is also underlain by gravels, there is not enough exposed to enable correlation with the two gravel deposits in the first pit with which they are comparable. In this pit there are two walls, which, although they are only some tens of metres

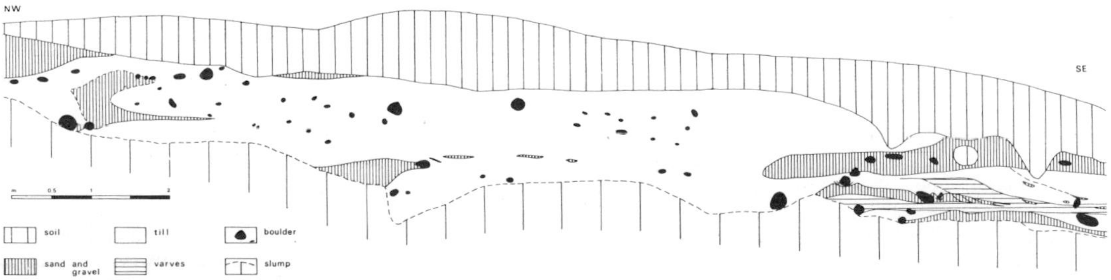

Fig. 3 Sketch of the first wall in pit 2, near Léchelles

apart, cannot easily be related to one another. The first wall is only exposed for a height of one and a half metre; it has a length of some thirty metres. The left half consists of fluvioglacial sands and gravels, which overlie the tills exposed in the right half of the wall (fig. 3). This till is compact, but shows also more sandy, loose parts. Beside this there are intercalations of sand, gravel and varvelike deposits. The boundaries between these different materials are not always clear. The second wall - which consists of two parts perpendicular to each other - is much higher and shows some four metres of till on top of gravel. The boundary between these two is gradual. The till is very compact and shows bands of silty sand which are internally layered (fig. 4).

Fig. 4. Part of wall 2 in pit 2, showing till with laminated bands, lying on top of gravel

These bands - up to 10 cm thick - are not horizontal, but they are sometimes bow-shaped and even seem to cut each other. There also occur very small sandy layers of some centimetres to some decimetres in length, either horizontal or running diagonally across the wall. These are especially to be seen in that part of the wall that is perpendicular to the part that shows the bands. The bands are not very clear on this part of the wall.

Pit 3: Galmis.

This gravel pit is situated on the W side of a hill, indicated by Crausaz (1959) as a drumlin. The wall in this pit is at least 150 m long and up to 12 m high (fig. 5). The stratigraphy is from top to bottom:
1. sand and gravel, up to 3 m thick;
2. till, up to 9 m thick; 3. sand and gravel, thickness at least 6 m. In the NNE part of the wall the till consists of three distinct types: a lower very gravelly till, a layered till and an upper "normal" till (fig. 6).
The upper till does show sand layers which are sometimes folded. The layered till consists of thin layers of till, with a normal distribution of gravel, separated by very thin clay layers, that do not contain any gravel (fig. 7). This layered till resembles a flow till carrying evidence of surface washing, described by Boulton (1971) from Svalbard.
The lower gravelly till shows some faint structures.
Going SSW the distinction between these three types of till disappears completely and only the upper till with some sand layers can be seen.
All of the till is very compact, as are the clay laminae in the layered till. Some of the sand layers in the upper till are quite loose though.
The boundary between the till and the lower gravel is diffuse, while that between the till and the upper gravel is sharper.

Pit 4: Müntschemier

This gravel pit is situated NE of Lake Neuchâtel. The largest wall is at least 150 m long and some 20 m high. It shows about 5 m of till on top of gravels, which can, in the W part of the pit, be divided into two deposits, separated by a coarser gravel layer (fig. 8).
At the boundary between the till and the gravel structures occur, that mostly resemble frost wedges. These structures consist of finely laminated sand and clay with some small gravel in strings. The wedges appear in the lower part of the till and at, or a little below the boundary with the gravel they turn to a more horizontal position, thinning to the east. Only one of these structures also has an extension to the west. Sometimes it seems as if these structures can be followed higher up in the till. It is as yet not certain whether they are really frost wedges or that they are shear planes. In three other pits in the

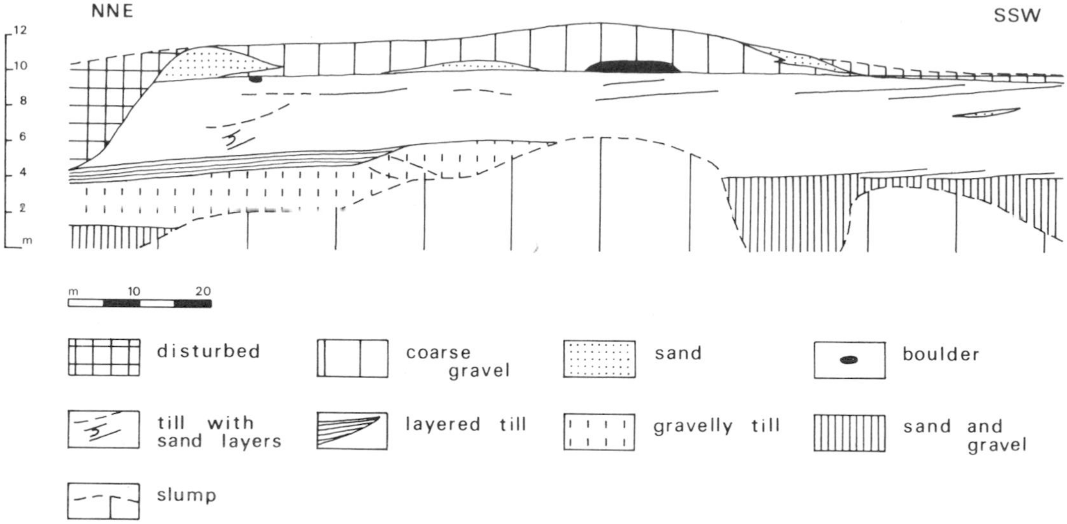

Fig. 5. Sketch of the wall of pit 3, near Galmis

Fig. 6. Part of the NNE side of fig. 5, showing from top to bottom: "normal" till-laminated till-gravelly till-gravel (scale in dm)

Fig. 7. Detail of the layered till in pit 3. The knife is 20 cm long

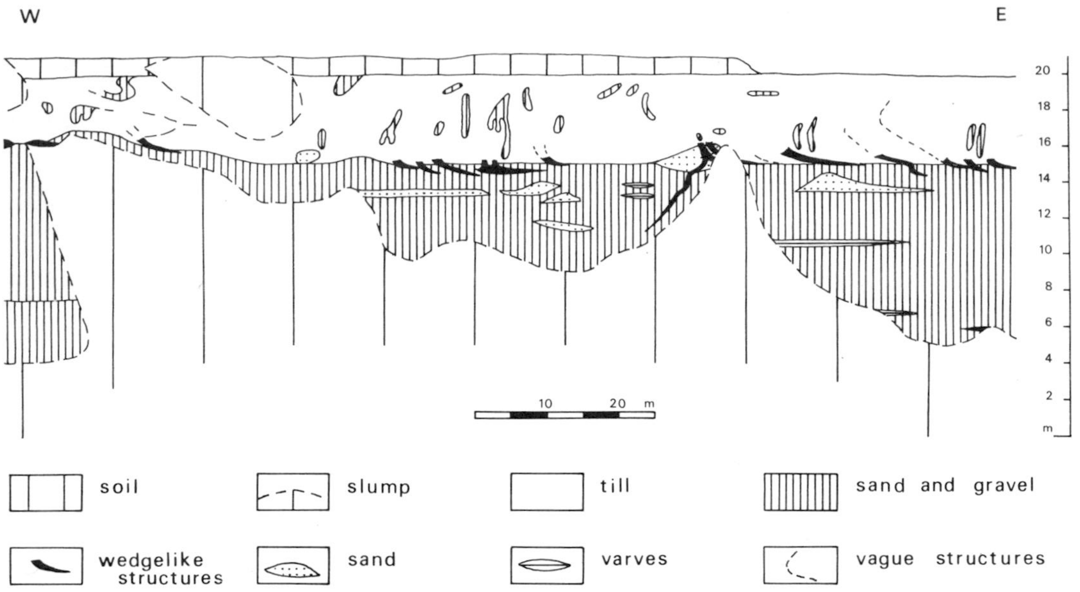

Fig. 8. Sketch of the wall in pit 4, near Müntschemier

study area the same kind of structures have been found, but there they had no visible connections with the till. Similar wedges have been described by Mangerud & Skreden (1972) and Macar (1969).
The till in this pit also shows concentrations of stones, mostly in more or less vertical zones. In one place a small fold is to be seen. In a gravel pit some 300 m south of this pit a 20 m high fold has been described by Portmann (1956).

Interpretation

The tills in the first three pits are tentatively interpreted as flow tills. In the fourth pit a more thorough study of the wedgelike structures is needed, before interpretation of the till is possible.

Acknowledgement

Mr's. W. den Besten and K. van Gijsel are thanked for their help in the field.
Mr. A.Eikeboom and Mrs. M.C.G.Keijzer-van der Lubbe are thanked for the reproductions and the typing of the manuscript respectively. Mr. B. de Leeuw is thanked for his help when the time ran short. Dr. A.C.Imeson kindly revised the english text.
I am thankful also to the Swiss colleagues, Prof. Dr. J.-P. Portmann and Dr. Ch. Schlüchter, for discussing the problems in the field.

REFERENCES

Boulton, G.S., 1971, Till genesis and fabric in Svalbard, Spitsbergen. In: R.P.Goldthwait (Ed.): Till, a symposium. Ohio State University Press, pp. 27-72.
Crausaz, Ch.U., 1959, Géologie de la région de Fribourg. Thèse, Fribourg, 119 pp.
Hantke, R., 1978, Eiszeitalter Band I. Ott Verlag, Thun, 468 pp.
Macar, P., 1969, A peculiar type of fossil ice fissure. In: T.L.Péwé (Ed.): The periglacial environment. McGill Queen's University Press, Montreal, pp.337-346.
Mangerud, J. & S.A.Skreden, 1972, Fossil ice wedges and ground wedges in sediments below till at Voss, Western Norway. Norsk Geologisk Tidskrift, Vol.52, pp. 73-96.
Portmann, J.P., 1954-55, Pétrographie des moraines du glacier Würmien du Rhône dans la région des lacs subjurassiens (Suisse). Bull.Soc.Neuch.Géographie, Tome LI, pp. 13-55.
Van der Meer, J.J.M., 1976, Cartographie des sols de la région de Morat (Moyen-Pays Suisse). Bull.Soc.Neuch.Géographie, Tome LIV, pp. 5-52.

Analysis of Pre-Pleistocene glacigenic rocks: Aims and problems

M.J.HAMBREY & W.B.HARLAND
University of Cambridge, Cambridge, England

1 INTRODUCTION

The study of ancient glacigenic rocks has a long, controversial history, the criteria for distinguishing such rocks being less obvious than for contemporary glacial deposits. It has long been clear that a standard scheme for describing alleged tillites (i.e. all rocks of mixed grain size having some glacial component) is needed, particularly if wide-ranging correlations are to be made. The Pre-Pleistocene Tillite Project was finally established by a working party of the International Geological Correlation Programme led by one of the authors (WBH) in 1974 with a view to gathering critical data on alleged tillite formations throughout the world, and to publish the data in compact, standardised form to facilitate comparison and analysis. Because of the difficulty of obtaining funds, the initiation of the project was delayed until May 1977, but it is anticipated that the work will be completed in 1980. The purpose of this paper is to discuss the aims of the Project, the importance of glacigenic deposits in the geological record, the problems that have arisen concerning the recognition of tillites and the difficulties with regard to terminology.

2 AIMS OF THE PROJECT

In order to analyse glacial events throughout geological history and to enable global correlations to be made, a reliable data base concerning tillites has long been needed. The world-wide survey, which is the aim of the Tillite Project, is intended to serve as a data base for studies of the following:

(a) sedimentology of glacial environments and establishment of criteria for their recognition;

(b) palaeoclimatic patterns and palaeogeological reconstruction;

(c) the processes that relate to climatic change and the origin of ice ages;

(d) time-correlation, especially of Precambrian rocks in which normal stratigraphic indicators are lacking.

The original intention for obtaining the necessary information was to issue a questionnaire to specialists throughout the world. This has now been reorganised into a set of "Instructions for Contributors" which will give scientists the opportunity to produce short papers in their own style, yet following a standard scheme based on checklists of important characteristics; this approach will also reduce the amount of editing that would otherwise have been necessary. The information to be requested for individual alleged tillites will be both descriptive and interpretative.

Descriptive information will include:

(1) names and synonyms;
(2) geographical distribution;
(3) stratigraphic and structural setting;
(4) boundary relations above and below the tillite unit;
(5) description of strata within the unit(s), including sedimentary structures;
(6) description and abundance of stones.

Interpretative information requested will consist of the following:

(1) chronostratigraphic age;
(2) geochronometric age;
(3) palaeolatitude;
(4) depositional environment and palaeogeography;
(5) palaeoclimatic sequence;
(6) palaeotectonic position.

The final publication will be fully referenced and indexed, and it is hoped that an initial analysis of the data so obtained will be made.

3 IMPORTANCE OF GLACIGENIC DEPOSITS IN THE GEOLOGICAL RECORD

If we did not live so near to an ice age with large areas of the world still under ice, it would have been difficult to attribute tills and tillites to glacial action. In the geological record glacial deposits are relatively unimportant in terms of volume, but they are crucial indicators of palaeoclimate, particularly in rocks lacking fossil flora and fauna. Continental glacial deposits tend to be thin when considered in relation to the deposits of marine sedimentation which make up much of the geological record. Thus examples of the characteristic features of a terrestrial glacial deposit, such as underlying striated pavements, are generally few and far between, and we must look to marine deposits for evidence of former glaciations. Even so, glaciomarine deposits generally attain a considerable thickness only in restricted areas. At the edge of a continental ice sheet most of the till is laid down in a narrow belt where the ice enters the sea. Thus, even if ice ages are frequent and prolonged, the total volume of glacially deposited material may be small in comparison with other types of sediment. However, it is important to note that the effects of glaciation may be transmitted over a much greater area as a result of advances and retreats, and by ice-rafting which, even if resulting in a small volume of sediment, nevertheless gives rise to some very distinctive facies. A modern analogy may be drawn here. Even today some 80 000 000 km^2 of ocean floor are receiving material that has melted out from floating ice (Heezen & Hollister 1964). For example, in the north Atlantic icebergs commonly reach as far south as Newfoundland (47°N), while around Antarctica bergs commonly float north of 50° latitude (Crowell 1964). Another marine facies that may well accompany extensive continental glaciation results from slumping and turbidity currents involving till deposited at the continental margin. Such turbidites may also enclose rafted stones, and distally may indeed simulate varvites.

Tillites have the best chance of being preserved in the geological record if they are deposited during continuous subsidence and sedimentation, as will occur in a geosyncline. In such situations deformation by folding may result in the rocks being taken deep into the crust, safe from immediate denudation, and in any case the deposits will be thickest in such environments.

The importance of ice ages throughout geological time was once fully anticipated and now, after a time of discussion and doubt (roughly between 1935 and 1965) is again increasingly being recognised as more glacigenic rocks are identified in many parts of the world.

From a review of the literature it is clear that ice ages are by no means unusual events. Table 1, which is based on reported occurrences of alleged tillites, indicates the frequency and widespread nature of pre-Pleistocene ice ages. Claims for glaciation have been made for all geological periods, although some of these have not been substantiated. When completed the Tillite Project should enable us to assess more reliably and thoroughly the areal extent, duration and importance of all known ice ages.

At least two major ice ages are known to have occurred during Precambrian times. Middle Precambrian tillites occur in several continents, but are exceptionally well-preserved in Canada. Tillites of Late Precambrian age have been found in all continents except Antarctica. This was either a period of world-wide refrigeration in which areas subjected to glaciation were situated near the equator or a period of rapid polar wandering, leaving a trail of diachronous tillites. A late Ordovician (- Silurian) glaciation, although probably less widespread, has left very fine indicators of ice action on land, notably in Africa. Late Palaeozoic times are represented by the well-known and extensive glacial deposits of Gondwanaland. Glacigenic sedimentation has also been claimed for the Mesozoic Era in South America. Finally, we have the late Cenozoic ice age, formerly conceived as being restricted to Quaternary time, but increasingly being traced back into late Tertiary times on the basis of observations on marine sediments. An outline of and references to tillites through space and time are given by Harland and Herod (1975), who pointed to the need for a systematic study of this kind.

4 CRITERIA FOR THE RECOGNITION OF TILLITES

The criteria used for deciding whether a particular rock is of glacigenic origin have been discussed frequently in the literature (e.g. Harland et al 1966; Schermerhorn 1966; Spencer 1971), and are summarised in a different form below, classed according to evidence for alternative environments. It is unusual for many of these characteristics to be preserved in the pre-Pleistocene record; some probably have rarely been observed, and alternative explanations might exist for features which in a Pleistocene deposit would clearly be glacigenic.

m.y.	Geological Era	AFRICA	ANTARCTICA	ASIA	AUSTRALASIA	EUROPE	NORTH AMERICA	SOUTH AMERICA	Name(s) of glacial age
2	Quaternary	X	X	X	X	X	X	X	Late Cenozoic
65	Tertiary		X	X		X	X		
136	Cretaceous						X?	X	(Mesozoic)
190	Jurassic		X				X?	X	
225	Triassic							X?	
280	Permian	X	X	X	X		X?	X	Late Palaeozoic
345	Carboniferous	X		X				X	
395	Devonian	X?		X?				X	
430	Silurian	X				X	X	X	Early (mid) Palaeozoic
500	Ordovician	X		X		X	X	X	
570	Cambrian	X		X	X	X	X	X	Late Precambrian, Eocambrian, Varangian
1500	Late Precambrian	X		X	X	X	X	X	
2500	Mid. Precambrian	X		X	X		X		Early Proterozoic

Table 1. Distribution of alleged tillites in space and time, based on a literature search. Some of these alleged tillites may eventually be proven to be of non-glacial origin. A question mark indicates a high degree of uncertainty.

4.1 Evidence for terrestrial glaciation

(a) Abraded bedrock surfaces (striations, grooves, pavements, roches moutonnées).

(b) Stone-rich beds

(i) with a great range of grain sizes (unsorted);

(ii) in which the stones have a preferred orientation (although such a fabric could arise from other sedimentary and tectonic processes);

(iii) with structures resulting from the imposition of external forces before consolidation (e.g. by ice push or by an overriding glacier);

(iv) with boulder pavements, i.e. levelled or striated upper surface;

(v) of irregular thickness, but tending to smooth out the underlying palaeo-relief.

(c) Depositional fossil landforms of distinctive form, preserved in the stratigraphic record (e.g. drumlins, moraines, eskers).

4.2 Evidence of glaciomarine (and glacio-lacustrine) deposition

(a) Great thickness (up to hundreds of metres) and lateral extent (up to hundreds of kilometres) or unsorted deposits of wide-ranging grain size.

(b) Fine-graded stratification, e.g. varved clay or siltstones (although these may be confused with the more common distal marine turbidites).

(c) Ice-rafted (drop) stones in finer sediments with disrupted laminae.

4.3 Evidence common to both terrestrial and aqueous deposition

(a) Lithologies of stones and undecomposed minerals variable (i.e. far-travelled).

(b) Striated and facetted stones.

(c) Presence of rock flour (individual grains indistinguishable under the petrographic microscope).

(d) Angular quartz grains with razor-sharp edges (if unaltered by subsequent chemical and mechanical processes), and characteristic conchoidal breakage patterns (as observed under the electron microscope).

4.4 Other evidence of a cold climate (periglacial phenomena) in terrestrial deposits

(a) Patterned ground.

(b) Involutions

(c) Ice-wedge pseudomorphs.

(d) Lobate flows on palaeoslopes (solifluction deposits).

Few of the above criteria in themselves demonstrate a glacial origin of a particular sediment. Some are strongly suggestive of a glacial origin, but normally, to be certain, a combination of these factors on a sufficient scale is necessary. Often such distinctive characteristics have been destroyed by subsequent deformation and metamorphism. Good evidence of a continental glaciation is the occurrence of a tillite lying on a striated pavement. However, care must be exercised in using even this criterion as proof of glaciation, as indicated by the general acceptance that the well-known subtillite striated pavement at Bigganjargga, northern Norway, has also been argued to have resulted from slumping and flowage of allochthonous till over soft sediment within the sedimentary sequence (Harland 1964). As previously mentioned, glaciomarine tills have the best chance of preservation in the geological record, and the criteria normally used for recognising and classifying Quaternary glacial deposits are of little value when applied to more ancient deposits. Nevertheless, the presence of drop stones of sufficient size and quantity in marine sediments is one characteristic that can normally be used on its own as proof of glacial origin. But, as the distribution of icebergs today indicates, the source area could well have been many hundreds of kilometres distant. Adjacent evidence, however, may indicate changes of sea level and a glacial environment.

The Tillite Project is designed to encourage scientists to analyse objectively alleged tillites, and to decide, on the basis of the criteria listed above, whether they are in fact of glacial origin. We also hope that the Project will stimulate further research, particularly where definitive criteria have not been reported.

5 PROBLEMS OF TERMINOLOGY

Over the years a varied vocabulary has built up in connection with tillites, largely independently of Quaternary studies. This unsatisfactory state of affairs is further complicated by the different usage to which certain terms are put. It is obvious that the most serious problems arise in connection with rocks that look like tillites but are of uncertain origin. "Tilloid" seems to have been the term first applied to a "till-like deposit of doubtful origin" (Blackwelder 1931) and semantically the "-oid" ending (Gk. - like) would support such a non-genetic meaning (e.g. Harland et al 1966). On the other hand, "tilloid" was later applied to "non-glacially deposited till-like rocks generally laid down by gravity transport" (Schermerhorn and Stanton 1963; Schermerhorn 1974), and this genetic non-glacial definition was adopted by Pettijohn (1975). The term is widely used in both senses today, and no clear preference has emerged on this issue for a particular definition, either in recent publications or from correspondents of the Tillite Project. For the purposes of the Project, therefore, "tilloid" is taken to include tillite-like rocks of uncertain origin and tillite-like rocks of non-glacial origin. However, in the case of till-like indurated sediments, "tillitoid" might be more appropriate and this could leave the tilloid issue unresolved, with tilloid signifying either of the above two alternatives.

Other non-genetic terms have been used to describe some tillite-like rocks, although not exactly to correspond to the original definition of tilloid. Of these "diamictite" and "mixtite" are the most favoured, and each has its strong supporters. Diamictite has been defined as "any non-sorted terrigenous sediment that consists of sand and/or larger particles in a muddy matrix" (Flint et al 1960), while Schermerhorn (1966) introduced the term "mixtite" to include "rocks with a wide range of grain sizes and characterised by a sparse to subordinate fraction (phenoclasts of all shapes and sizes) englobed in a ground mass composed of varying proportions of sands, silts and clay". Both these terms are wider in concept than tilloid/tillitoid, containing rocks that do not look like tillites, such as agglomerates, pebbly mudstones, talus, turbidites etc.

The Tillite Project has been encouraging scientists to make known their preferences regarding terminology in the hope that some consensus might emerge and that certain recommendations can be made. We need to distinguish petrographic and petrogenetic classifications of tillite-like rocks, and

consultations between Quaternary and pre-Pleistocene geologists are in progress. It is hoped that as a result classifications can be devised that are applicable to both unconsolidated and lithified deposits. Of course, it is unlikely when dealing with ancient glaciations that one can postulate precisely the mode of deposition, hence a genetic classification of the type proposed by Dreimanis (1977) will have only limited application to most older rocks, especially in the case of water-laid deposits. Nevertheless, such a classification is useful to encourage students of pre-Pleistocene rocks to think in terms of glacial processes, and equally to challenge Quaternary geologists with the four-dimensional evidence of ancient glaciations.

6 CONCLUSIONS

The study of ancient ice ages poses special problems which do not generally apply to Quaternary deposits, notably imperfect exposure, frequent deformation and metamorphism, lack of external evidence of climate and topography. Characteristics which enable a glacial deposit to be recognised are generally less easily observed in pre-Pleistocene rocks, particularly as these more often than not are marine. Close contact is necessary between Quaternary and pre-Pleistocene geologists, particularly to standardise terminology and classifications. It is anticipated, now that formal contact has been established between INQUA and the Tillite Project, that the present unsatisfactory situation will be rectified. Such contact should also encourage research along lines that would be of benefit to both disciplines. As an example, we think more emphasis should be given to the study of present-day glaciomarine sedimentation, a process which is of such great importance in the geological record. Conversely, Quaternary geologists may find much of help in interpreting the importance and characteristics of present-day glaciations by studying the better exposed glaciomarine tillites.

When the Pre-Pleistocene Tillite Project is completed it is hoped that the evidence for glaciations throughout geological time will be clarified, using as a firm data base the contributions of many scientists.

7 ACKNOWLEDGEMENTS

The Pre-Pleistocene Tillite Project is financed largely by the Royal Society of London, with additional support from UNESCO, which we gratefully acknowledge.

8 REFERENCES

Blackwelder, E. 1931, Pleistocene glaciation of the Sierra Nevada and Basin Ranges. Bull. Geol. Soc. Am. 42: 865-922.

Crowell, J. C. 1964, Climatic significance of sedimentary deposits containing dispersed megaclasts. In A. E. M. Nairn (ed.), Problems of Palaeoclimatology, London, Interscience.

Dreimanis, A. 1977, Terminology and development of genetic classification of materials transported and deposited by glaciers. Programme of Abstracts, INQUA 10th Int. Congr., Birmingham, Aug. 16, 1977: 7-10.

Flint, R. F., J. E. Sanders & J. Rodgers 1960, Diamictite, a substitute term for symmictite. Bull. Geol. Soc. Am. 71:1809.

Harland, W. B. 1964, Critical evidence for a great Infra-Cambrian glaciation. Geol. Rundschau 54: 45-61.

Harland, W. B., K. N. Herod & D. H. Krinsley 1966. The definition and identification of tills and tillites. Earth Science Reviews 2: 225-256.

Harland, W. B. & K. N. Herod 1975, Glaciations through time. In A. E. Wright & F. Moseley (eds.), Ice Ages: Ancient and Modern. Geol. J. Spec. Issue No. 6, Liverpool.

Heezen, B. C. & C. Hollister 1964, Turbidity currents and glaciations. In A. E. M. Nairn (ed.), Problems in Palaeoclimatology, London, Interscience.

Pettijohn, F. J. 1975, Sedimentary Rocks. Third Edn., New York, Harper & Row.

Schermerhorn, L. J. G. 1966, Terminology of mixed coarse-fine sediments. J. Sediment. Petrol. 36: 831-835.

Schermerhorn, L. J. G. 1974, Late Precambrian mixtites: glacial and/or non-glacial? Am. J. Sci. 27: 673-824.

Schermerhorn, L. J. G. & W. I. Stanton 1963, Tilloids in the West Congo Geosyncline. Quart. J. Geol. Soc. Lond. 119: 201-241.

Spencer, A. M. 1971. Late Precambrian glaciation in Scotland. Mem. Geol. Soc. Lond. No. 6.

Varves and glaciolacustrine sedimentation

Introductory remarks

M. STURM
EAWAG-ETH, Dübendorf, Switzerland

According to the "Glossary of Geology" (Gary, McAfee & Wolf 1972) the term varve is defined as: "A sedimentary bed or lamina or sequence of laminae deposited in a body of still water within one year's time; specif. a thin pair of graded glacio-lacustrine layers seasonally deposited (usually by meltwater streams) in a glacial lake or other body of still water in front of a glacier. A glacial varve normally includes a lower "summer" layer consisting of relatively coarse-grained, light-colored sediment (usually sand or silt) produced by rapid melting of ice in the warmer months, which grades upward into a thinner "winter" layer, consisting of very fine-grained (clayey), often organic, dark sediment slowly deposited from suspension in quiet water while the streams were ice-bound. Counting and correlation of varves have been used to measure the ages of Pleistocene glacial deposits. Etymol: Swedish varv, "layer" or "periodical iteration of layers" (DeGeer 1912)".

The "Colloquium on Varves" held during an INQUA "Symposium on Moraines and Varves" at Zurich in September 1978 has demonstrated that varves are complex and diversified sedimentary features and represent a great variety of depositional environments. From the presented data it became evident that definitions of varves have to be reconsidered: varves are not formed exclusively in glacio-lacustrine conditions, varves are not exclusively caused by meltwater streams, and varves are not simply thin light/coarse and dark/fine couplets. The knowledge of quaternary geologists as well as of sedimentologists and limnologists has to be considered, if an appropriate definition of varves is envisaged. Hopefully, a revised definition will be prepared within the near future by Working Group (4) "Quaternary lacustrine sediments" of the INQUA Commission on Genesis and Lithology of Quaternary Deposits.

The following articles represent the majority of the contributions to the meeting at Zurich: The first article deals with the hypothetical hydrological set-up for the formation of clastic varves in oligotrophic lakes (M. Sturm). The 3 following papers describe varve-like structures and true varves from recent lakes; sediments are reported from European and North-American lakes with differing trophic states (A. Lambert; St. D. Ludlam; M. Boyko-Diakonow). Glaciolacustrine, lacustrine and marine varves are the basis for interpretation and dating of varve series from Asia, Europe and N-America as presented in the next two articles (St. Kempe & E.T. Degens; D.J. Schove). Sedimentological investigations of deposits from a Canadian tidal lake and from glacial lakes in Canada and Ireland respectively, present data on "inverse" varves and on proximal depositional environments of clastic varves (G.M. Ashley; J. Shaw & J. Archer; J.M. Cohen). The last article deals with the paleo-environment of an ancient lake in Japan (S. Horie).

REFERENCE

Gary, M., McAfee, R. & Wolf, C.L. 1972, Glossary of Geology - American Geological Institute, Washington D.C., 823 p.

Origin and composition of clastic varves

M. STURM
EAWAG-ETH, Dübendorf, Switzerland

INTRODUCTION

Since many decades geologists were impressed by the occurrence of regularly laminated sediments, showing alternations of light and dark coloured laminae. When representing a time-span of one year a pair of a light and a dark lamina was called v a r v e (DeGeer 1912). Varves were restricted by definition to the glaciolacustrine environment and were primarily used for dating and correlation of glacial sediments in Scandinavia and N-America (DeGeer 1912, Antevs 1922, Sauramo 1923). At a time when other aging methods such as dating with radio-isotopes were not yet available varve-counting and correlation of varve series over large distances proved to be an unique tool for stratigraphic analysis of Pleistocene deposits.

Ever since, scientists were speculating on the depositional mechanism of varve-formation (Antevs 1951, Kuenen 1951); varves were thought to be exclusively of glaciolacustrine origin. It was only within the last 10 years that an increasing number of articles on varves were published, presenting sedimentological data of varve series not only from the glacial environment (Mathews 1956, Agterberg & Banerjee 1969, Ashley 1975) but also from proglacial (Østrem 1975, Gustavson 1975), lacustrine (Ludlam 1969, Kempe 1977, Sturm & Matter 1978) and marine environments (Gorsline 1977, Kelts & Hsü 1978, Degens et al. 1978). Varves s.l. were described in these studies within a striking variety of structures, thickness, composition and colour, reflecting different environmental conditions during their formation.

Thus, varves may be used to interpret the former nature of depositional basins in terms of climate, mineralogy of drainage area, changes of water-level, temperature, trophic state (Geyh et al. 1971, Kelts 1978). However, a correct interpretation of the depositional environment of varved sediments can be carried out only, if adequate sedimentological information exists, especially on the formation of varves in recent environments.

In order to facilitate such an interpretation this paper will describe distinct sedimentary structures such as graded beds and rhytmical lamination and their appropriate hydromilieu. In a theoretical approach it will be shown that given hydrological parameters will produce characteristic sedimentary structures as depositional equivalent.

The considerations presented in this paper are based on a theory of varve formation in an oligotrophic alpine lake (Sturm & Matter 1978).

SUSPENDED MATTER

It is assumed that the formation of varves depends on the existence of suspended matter (SM) within the water column as proposed by DeGeer 1912, Sauramo 1923, Kuenen 1951, Gustavson 1975 and others.

Five different sources are considered to be responsible for the supply of SM to a depositional basin:

i. RIVER: SM deriving from all kind of subaerial runoff like rivers, brooks, torrents; including material from subaerial landslides, avalanches, calving glaciers etc.

ii. AIR: SM transported to a basin as dustparticles by wind, rain, snow, volcanic erruptions

iii. AUTOCHTHONUOUS: SM formed within the water column of a basin by chemical precipitation, bioproduction(algal growth, fecal pellets, residues of mollusces, vertebrates, etc.)

iv. RESUSPENSION: SM derived from the activities of bottom currents, bioturbation, degasing of sediment

v. UPDWELLING: SM imported to a basin by groundwater flow, subaqueous springs, volcanic exhalations.

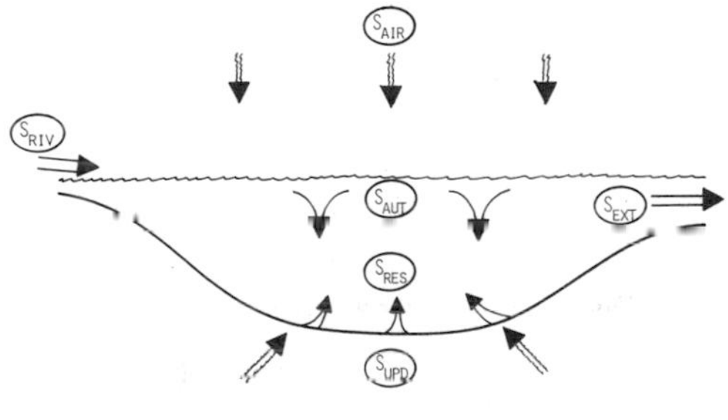

Figure 1: Sketch of hypothetical sources of suspended matter in a depositional basin (not to scale)

According to the sources of SM(see Fig. 1) the finally deposited sediment will be the sum of this amounts of material minus the amount of SM exported from the basin (e.g. by river outflow):

$$S_{DEP} = (S_{RIV} + S_{AIR} + S_{AUT} + S_{RES} + S_{UPD}) - S_{EXT}$$

This scheme of different sources of suspended matter will be valid for any kind of basin wether lacustrine or marine, wether glacial or non-glacial.

However, suspended matter of an oligotrophic alpine lake basin with predominantely clastic sedimentation will have its origin nearly exclusively from river tributaries (Fig. 2).

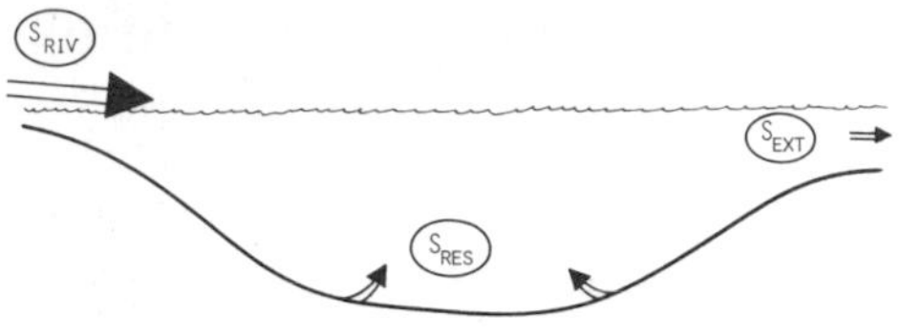

Figure 2: Sketch of dominant sources of suspended matter in an oligotrophic basin with large sediment-ladden tributaries (not to scale)

In such a lake the finally deposited sediment will equal the sum of suspended material by river import and resuspended material minus the export of SM by outflow:

$$S_{DEP} = (S_{RIV} + S_{RES}) - S_{EXT}$$

The river import of suspended particles outweighs by far the amount of material from other sources. Occasional occurrence of subaqueous slides, bottom-currents, and (less important) degasing of methane will add certain amounts of resuspended particles (S_{RES}) to the total of suspended matter. Particles from other sources such as dust (S_{AIR}), skeletons of diatoms and calcite crystals (S_{AUT}), and particles suspended in groundwater (S_{UPD}) can be neglected because of their insignificant quantities.

In a hypothetical basin as described above, the river input of SM may be either continuous, or it may be discontinuous introducing SM to the lake at certain times (e.g. springtime) of the year only.

STRATIFICATION OF THE WATER BODY

Stratification of a lake (or sea) is assumed to be the second important parameter for the formation of varves, at least in clastic environments (Gustavson 1975, Sturm 1975, Sturm & Matter 1978). Density differences are believed to be exclusively responsible for stratification of the water in a basin. Such density differences are caused mainly by gradients of temperature, salt concentration, and concentration of suspended matter.

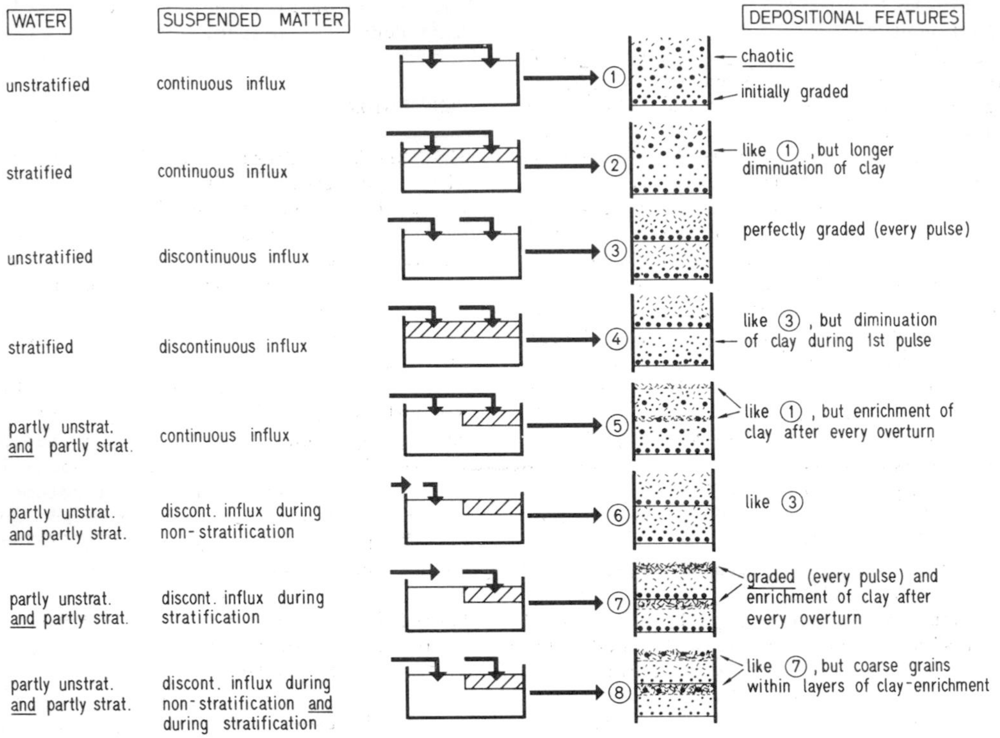

Figure 3: Idealized sedimentary features as a result of two hydrological parameter (stratification of water and influx of suspended matter) in an oligotrophic lake with clastic deposition. Case 7 represents the structure of an ideal clastic varve (heavy dots stand for sand, light dots for silt and light dashes for clay)

The water column of a hypothetical lake may show one of the following modes of stratification:

i. unstratified water all the year through
ii. stratified water all the year through
iii. water is alternatingly stratified at certain times of the year (e.g. summer) and unstratified at the rest of the year (e.g. winter).

HYDROLOGY AND SEDIMENTARY STRUCTURES

Suspended matter a n d stratification of the water column are considered to be the govering agents of varve formation in an oligotrophic lake with predominantly clastic sedimentation. It has to be stated for the following considerations that the term "suspended matter" is assumed to involve equivalent parts of sand, silt and clay particles. Moreover, it is assumed that bottom-currents will not interfere the depositional conditions in the hypothetical basin. On the basis of these assumptions the formation of different sedimentary structures will be discussed in the following (see also Fig. 3).

Cases ①, ②, ⑤ in Fig. 3 demonstrate that continuous transport of suspended matter into a lake will cause the formation of 'chaotic' deposits, i.e. the deposition of an unorganized sediment, showing no kind of grading or regular lamination. Chaotic deposition will occur at any mixing state of the lake. However, in annually (i.e. partly) stratified lakes the clay particles may partially be trapped within the epilimnion during stratification, causing the formation of a faintly visible clay-enriched layer within the chaotic sediments, when finally deposited during times of non-stratification (case ⑤ in Fig.3). Uniform, continuous influx of SM into a lake - wether stratified or not - can thus not be responsible for the forma-

tion of laminated and varved sediments respectively.

Different deposits will be caused by discontinuous pulses of suspended matter (cases ③, ④ and ⑥ to ⑧ in Fig. 3). Within non-stratified lake water and within permanentely stratified water sand, silt, and clay particles of a suspension pulse will settle exclusively in accordance to STOKES' law; due to the absence of counteracting forces, such as turbulence induced by density gradients, graded beds will be deposited on the lake bottom, showing complete fining from sand to clay (③, ④, ⑥ in Fig. 3). Although sediments formed under such circumstances will exhibit often a succession of more or less regular layers (i.e. graded beds) which represent each a pulse of SM, they do not fit the observed, clearly bimodal sedimentary structure of classical varves; because varves show basal coarse layers, toped by non-gradational distinct clay layers.

However, case ⑦ in Fig. 3 is believed to represent conditions, which induce the distinct bimodal depositional mechanism of a classical varve couplet: Suspended matter is introduced to a lake by discontinuous influx during the period of stratification. The coarser particles start settling immediatly, depositing a basal graded bed of sand and silt. The bulk of clay particles, being trapped during stratification within the epilimnion, will start settling after overturn only and cause the deposition of a distinct, non-gradational clay lamina on top of the previously formed coarser layer.

A final hydrological situation is presented in case ⑧ in Fig. 3: Pulses of suspended matter are introduced to a stratified lake during the period of stratification as well as during non-stratification. Deposition of SM from pulses during stratified water will be equal to case ⑦. Additional suspension pulses during the period of non-stratification, however, will cause the coarse particles of this pulse to settle time-equivalent to the clay particles still under way of the preceding pulse (which was introduced during stratification). Thus the coarses will deposit a layer of sand and silt within the clayey lamina on top of a varve couplet.

This kind of depositional mechanism describes best the unusual appearance of coarse sand layers within clayey winter layers of varve couplets, as it was observed by the author during a field-trip to glacial Lake Hitchcock (U.S.A.), and as it has been described recently by John Shaw from a succession of Pleistocene varves of glacial Lake Penticton (Canada, Shaw 1978).

CONCLUSIONS

The foregoing considerations show, that given hydrological parameters of a depositional basin such as the influx of suspended matter and the stratification of the water column cause predictable sedimentary features. In other words, the sedimentary record of a basin can be used, to interpret former hydrological situations of the basin properly.

Moreover, it is demonstrated that the bimodal sedimentary structures of clastic varves may occur only, if discontinuous (seasonal) influx of suspended matter is matched to the stratified water column of a partly (annually) stratified lake. During other lake conditions (non-stratification, continuous influx of SM, etc.) quite different sedimentary structures would be formed in the sediments of a lake(chaotic sediments, complete graded beds, etc.).

Finally it has to be stated that the hypothetical cases discussed fit the model of an oligotrophic lake with predominantly clastic sedimentation. Thus, different considerations on the deposition of varves are needed for such lakes, where the trophic state or the sources of suspended matter are different or for the marine environment.

ACKNOWLEDGMENTS

Financial support by a grant from US-NSF (OCE 75-13844 to A.Lerman) is gratefully acknowledged. The author appreciates helpful comments by A.Lerman, G.Ashley, J.Hartshorn, A.Dreimanis, J.Westgate and S.Ludlam. An earlier draft of this paper was critically reviewed by Chr.Schlüchter.

REFERENCES

Agterberg, F.P. & I.Banerjee 1969, Stochastic model for the deposition of varves in glacial Lake Barlow-Ojibway,Ontario, Canada. - Can.J.Earth Sci. 6 : 625 - 652.

Antevs, E. 1922, The recession of the last ice sheet in New England. - Amer.Geogr. Soc.Res.Ser. 11, 120 p.

Antevs, E. 1951, Glacial clays in Steep Rock Lake, Ontario, Canada. - Bull.Geol.

Soc.Amer. 62 : 1223 - 1262.

Ashley, G.M. 1975, Rhytmic sedimentation in glacial Lake Hitchcock, Massachusetts-Connecticut. - In: Glaciofluvial and Glaciolacustrine Sedimentation (ed. by A.V.Jopling and B.C.McDonald) Spec.Publ. SEPM 23 : 304 - 320.

DeGeer, G. 1912, A geochronology of the last 12,000 years. - 11th Int.Geol.Congr.1910 Stockholm, 1 : 241 - 253.

Degens, E.T., P. Stoffers, S. Golubic & M. D.Dickman 1978, Varve chronology: estimated rates of sedimentation in the Black Sea deep basin. - Initial Rep.DSDP 42/2 : 499 - 508.

Geyh, M.A., J.Merkt & H.Müller 1971, Sediment-, Pollen-, und Isotopenanalysen an jahreszeitlich geschichteten Ablagerungen im zentralen Teil des Schleinsees. - Arch. Hydrobiol. 69 : 366 - 399.

Gorsline, D.S. 1977, Changes in the depth of the oxygen minimum over a glacial cycle. - 10th INQUA Congr.1977 Birmingham.

Gustavson, T.C. 1975, Bathymetry and sediment distribution in proglacial Malaspina Lake, Alaska. - J.sed.Petrol. 45 : 738 - 744.

Kelts, K. 1978, Geological and sedimentary evolution of Lakes Zurich and Zug, Switzerland. - Thesis unpubl.no 6146, ETH-Zurich, 224 p.

Kelts, K. & Hsü, K.J. 1978, Freshwater carbonate sedimentation. - In: Lakes, Chemistry Geology Physics (ed. by A.Lerman) Springer, 295 - 323.

Kempe, S. 1977, Hydrographie, Warvenchronologie und Organische Geochemie des Van Sees, Ost-Türkei. - Mitt.Geol.Pal.Inst. Univ.Hamburg 47 : 125 - 228.

Kuenen, P.H. 1951, Mechanics of varve formation and the action of turbidity currents. - Geol.För.Förh.Stockholm 73 : 69 - 84.

Østrem, G. 1975, Sediment transport in glacial meltwater streams. - In: Glaciofluvial and Glaciolacustrine Sedimentation (ed. by A.V.Jopling and B.C.McDonald) Spec. Publ.SEPM 23 : 101 - 122.

Sauramo, M. 1923, Studies on the Quaternary varve sediments in Southern Finland. - Comm.Géol.Finlande Bull. 60 : 164 p.

Shaw, J. 1978, Winter turbidity current deposits in late Pleistocene glaciolacustrine varves, Okanagan Valley, British Columbia, Canada. - Boreas 7 : 123 - 130.

Sturm, M. 1975, Depositional and erosional features in a turbidity current controlled basin (Lake Brienz). - 9th Int.Congr.Sedimentology 1975 Nice, 5 : 385 -390.

Sturm, M. & A. Matter 1978, Turbidites and varves in Lake Brienz (Switzerland): deposition of clastic detritus by density currents . - In: Modern and Ancient Lake Sediments (ed. by A.Matter and M.E.Tucker) Spec.Publ. Int.Assoc.Sedimentology 2 : 147 - 168.

Varve-like sediments of the Walensee, Switzerland

ANDRÉ M. LAMBERT & KENNETH J. HSÜ
ETH-Zentrum, Zurich, Switzerland

Introduction

The Swedish word "varv", according to De Geer (1912, p.253) means "as well a circle as a periodical iteration of layers". The original Swedish varves are glacial marine clays distinguished by their periodical laminae of different colour and grain. The term is, therefore, essentially descriptive. De Geer (1912), however, presented arguments for his interpretations that the coarse laminae of varves are deposited by bottom currents in proglacial lakes, and those currents owe their origin to annual melting of glaciers. These conclusions led him to assume rhythms of the coarse laminae, or of the varve-couplets, and thus permitted him to construct a geochronology for the last 12,000 years.

Both of De Geer's interpretations have been controversial. Antevs (1951), for example, thought that glacial melt-water should be less dense than lake water, and that the sediment-laden meltwater entering glacial lakes should spread out as plumes, or turbidity-overflows. He apparently had no quarrel with the interpretation of annual rhythms, as he himself assumed annual settling of silt-sized clasts from the overflows during the summer, and of clay-sized particles during the winter.

Kuenen (1951) and most modern workers (Mathews, 1956; Ludlam,1967; Ashley, 1972; Gustavson, 1975) supported De Geer and qualified his "bottom-currents" with the modern term "turbidity underflows". They are suspension currents and not wind- or wave-generated bottom currents. The authors also inclined to assume an annual rhythm for the varves (e.g. Gustavson, 1975, p. 262).

The purpose of our investigation is to use bottom-current meters to survey turbidity underflows in a Swiss lake, the Walensee, to relate the strength of those underflows to flood stages of its tributaries, to determine by coring and by sediment-harvesting if the flood cycles are annual, and to assess the reliability of varves as a tool of geochronology. Our investigations supported De Geer's first contention that sediment-laden floodwaters could generate turbidity underflows to deposit varves, but threw doubt on his second interpretation that varves or varve-like sediments are necessarily annual! Our results also confirmed the long-overlooked findings of W. Stumpf, an engineer of the Swiss Federal Hydrological Agency who found non-annual "varves" in sediment traps.

The field work for our investigation was carried out during 1969-1977 as a part of a limnogeological research program of the Swiss Federal Institute of Technology (ETH). We have been supported financially by grants from the Swiss National Science Foundation and from the ETH. We are indebted to the cooperation of many of our colleagues

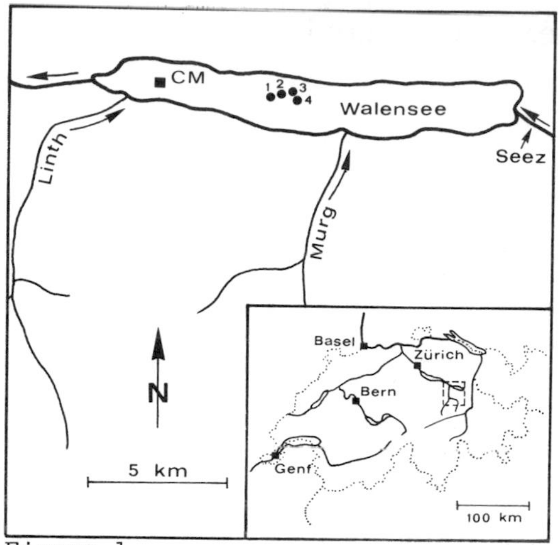

Figure 1.
Index map of the Walensee and its main tributaries. CM = Current meter station from April to October 1977. 1,2,3,4 = position of gravity cores shown in Figure 2.

particularly Kerry Kelts, André Bärfuss and Kurt Ghilardi of our Institute, Neil Marshall from the Scripps Institution of Oceanography, La Jolla, California, and Dr. Zimmermann from the Water Supply Agency of Zurich.

Geographic Framework of the Walensee

The Walensee is a long, narrow and deep lake, with a maximum depth of 145 m (Fig. 1). The lake is bounded by steep slopes on both sides, reaching 25° or more locally. The drainage basin is 1061 km^2 in extent, and the bed rocks are mainly sedimentary. The terrane is rugged; high peaks over 3000 m in elevation are partly covered by glaciers. The climate is temperate and humid. Altogether 35 rivers flow into the lake, of which the Linth is by far the largest, contributing 25 m^3/s of inflow (average for the last 12 years), or 60% of the total, to the lake. Except for the second largest (Seez, 8%), all the rest contribute less than 1% of the inflow. The annual average concentration of suspended materials from the Linth is 84 mg/l, and the outflow of the Linth Canal has only 4 mg/l suspensions. The annual sedimentation rate is about 87,000 tons per year, or an average of 1.7 mm/yr (for sediments with a density of 2.5 g/cm^3), or of 2.5 mm/yr (density of 1.5 g/cm^3). However, much of the sediments are accumulating on the subaqueous fan of the Linth, where the sedimentation rate reaches 90 mm/yr, and in the basin center where the rate is about 10 mm/yr. The steep flanks are mostly by-passed by sediments. In addition to the suspension-load, the bed-load of rivers accumulates as deltas at the mouth of rivers.

The Linth River was diverted into the lake through the construction of a canal. This course-correction in 1811 led to a doubling of the drainage area of the lake and to a drastic change of the sedimentary regime. The addition of the Linth resulted in increases of sedimentation rate, and changes of sedimentary facies: Above fine grained reddish brown lacustrine muds a sequence follows characterized by grey coarse graded sands (Lambert, 1978). The datum 1811 can thus be easily identified in most cores from the western part of the basin by a marked increase of mean grain-size and a distinct change of color.

"Varves" and Associated Sediments

The sediments of the lake have been studied by means of 33 piston cores of up to 7 m length and 40 gravity cores of up to 1 m length (Lambert, 1978). In this article we shall deal with the genesis of laminated varve-like sediments. Laminated sediments are commonly found in the deeper part of the lake (below 80 m). In areas proximal to the deltas the stratification is often destroyed by intense burrowing by organisms, which are mostly Tubifex (Florin, 1972). Where the laminations are preserved, the sediments consist of a succession of graded detrital layers ranging from 1 to 6 mm in thickness (Fig. 2, 4). Coarse silt or sandy layers form the bases and clay the top of those graded

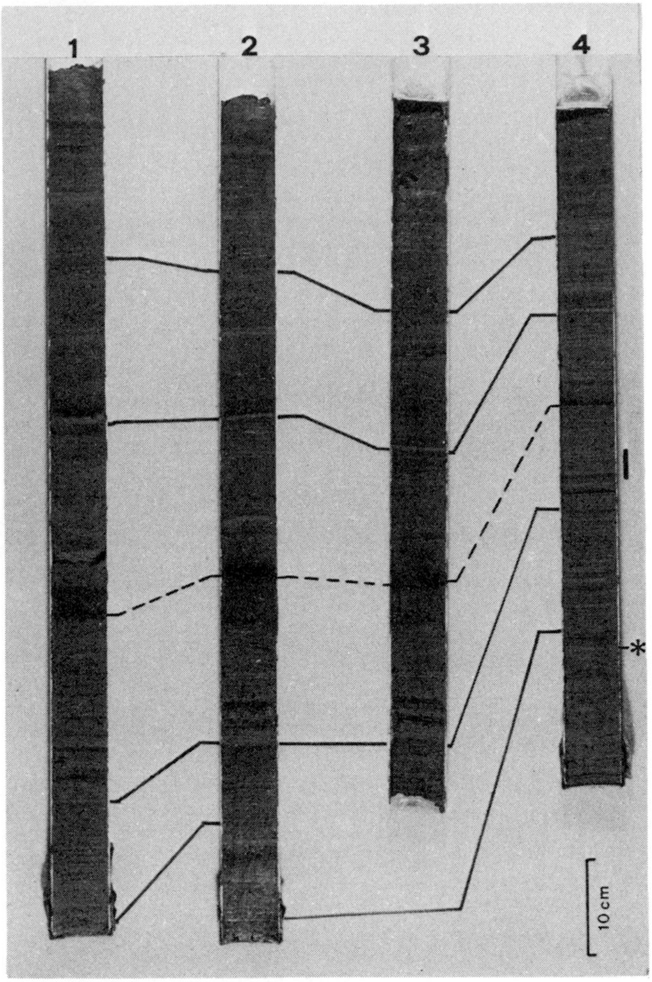

Figure 2. Gravity cores with sediments from the central part of the lake (Positions see Fig.1). A detail of core No. 4 (vertical bar) is enlarged in Fig. 4. The dotted line correlates a graded sandy layer, decreasing in thickness with increasing distance from the Linth mouth. It was most probably deposited by the catastrophic Linth-flood of 1944. The deposits of 1911/1912, harvested in a sediment trap by the Swiss Federal Hydrology Agency, could be recognized in core No. 4 (asterisk).

laminae. A conspicuous feature of these deposits is the variety of colors characterizing different layers. It ranges from deep reddish brown to reddish gray, locally also yellow and pink. There is, however, no regular reiteration of a certain succession of colours. Individual laminae could be correlated for a distance of at least several hundred meters. We have observed significant variations of lamination-thickness.

The constituents of detrital silts and sands are mainly quartz, feldspar, carbonate. The clay minerals are mainly micas (75-80%) with subordinate chlorite. The mineralogy reflects the composition of bedrock in the drainage basin.

The "varves" are associated with graded beds, slump beds and burrowed muds. The graded beds are mostly sandy. Evaluation of the average sedimentation rate based on the 1811-facies change permitted to identify the deposits of the catastrophic flood of August 25th, 1944, as a particularly notable sandy layer, several centimeters thick (Fig. 2). A few gravelly layers, up to 75 cm thick, and with clasts up to 25 mm in size, are present near the base of the steep slope in front of small deltas; these coarse clastics were apparently deposited by turbidity underflows of unusually high density or collapsing delta front parts.

Cyclicity of "Varves"

Sixteen of our 33 long piston-cores penetrated the 1811 datum, and we counted the number of varve-like silt laminae or "varves" above the marker. The "varves" deposited in the years between 1811-1971 is not 160 as predicted by the assumption of annual cyclicity, but ranges 300 to 360 in number (Lambert, 1978). We also counted some 70 coarse silt laminae for the thirty-year interval since the deposition of the 1944 flood-deposit. The sedimentary record thus indicates that more than two graded silt layers, on the average, were deposited annually.

A more direct method is to sink a sediment-trap on the lake bottom to harvest the annual sediment-accumulation. Our attempt was frustrated because the ropes anchoring our sediment-trap were cut and stolen. Fortunately our colleagues in the Swiss Federal Hydrological Agency called our attention to a previous successful investigation by W. Stumpf (1916).

Stumpf installed a sediment-trap in the Walensee at a depth of 120 m on May 11, 1911, and recovered the trap 364 days later on May 9th, 1912. He found a layer of sediments of about 9 mm thick in the trap and he identified five distinctly coloured "varves" of silts and clays; the colours in stratigraphically descending order are grey, yellow, red, grey and yellowish grey. He repeated this experiment for a period of 159 days during the summer to winter 1912. This time he found a 6-mm deposit of grey, fine-grained silt without distinctive layering. The precise description of these sediments and especially of their color-succession permitted us to recognize the 1912 datum in our gravity cores (No. 4 in Fig. 2). We were then able to count 171 "varves" for the past 66 years, again an average of more than 2 laminae per year.

Turbidity Underflows

We have used bottom-anchored current meters to study occurrences and physical properties of turbidity underflows in the Walensee, and of their relation to flood stages of the tributaries. Our first campaign in 1973 discovered that the underflows up to a few meters thick, and with a maximum of 30 cm/s speed, are active in springtime during the snow-melting season in the Alps, when the lake does not yet have a density-stratification (Lambert, Kelts and Marshall, 1976). A second series of surveys from April through September 1977, with current-meter position shown by Figure 1, indicated that hyperpycnal inflow also occurred when the thermocline was established: during this period the main tributary Linth induced 6 underflows of more than 4 cm/s speed (threshold speed of our current-meters). One event lasted for 13 hours and reached velocity maxima of 50 cm/s several times (Fig. 3). The underflow arrived at the site of current meter 1/2 km downslope in the lake, 1 1/2 hours after the high-water surged past the Linth measuring-station, 4 km upstream (average speed of 80 cm/s). There is little doubt that the underflow was an underwater

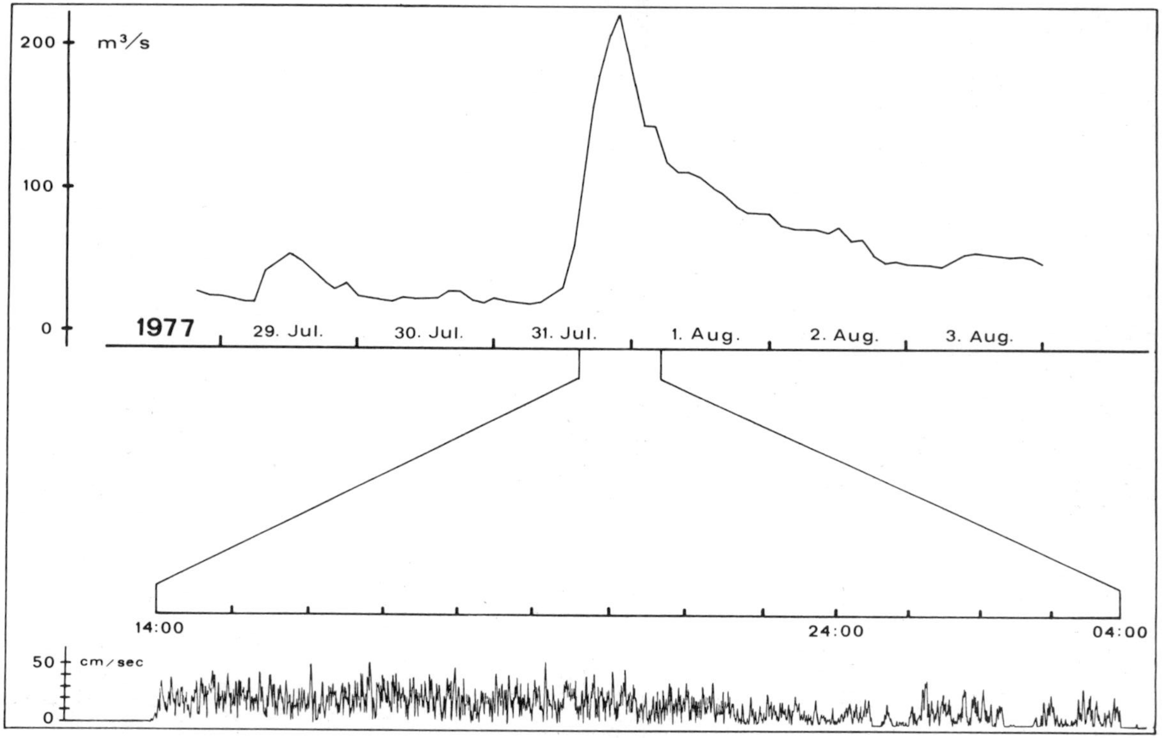

Figure 3. Upper part: Discharge of the Linth river with the annual peak due to heavy rains during July 31st, 1977. Lower part: Turbidity underflow recorded by a bottom-anchored current meter in 85 m depth, 3 m above the lake floor (see Fig. 1 for position). The underflow arrived 1 1/2 hours after travelling 4 1/2 km from the Linth measuring-station.

continuation of the Linth flood. Our previous measurements provided the same evidence (Fig. 10 in Lambert et al., 1976). However, these investigations have also shown that turbidity underflow is not present in every Swiss lake; our meters have failed so far to detect any underflows in Lake Zurich, for example. We also fail to register any notable underflow-activities in the Walensee except after major rainstorms in the Spring or Summer when the flood-water has a higher density than the lake water. During the 1977-flood, for example, the density of the inflowing Linth-water was about 1.008 g/cm^3 when the underflow was active and thus definitely denser than the bottom water of the lake that had a density of 0.9996 g/cm^3. Echo sounding revealed the presence of a channel on the upper reaches of the subaqueous fan of the Linth. The channel is 6 m deep and 70 m wide, at a location where the water-depth is 75 m (Lambert, 1978). The channel attests to the power of the underflows as an erosive agent. It is also reasonable to assume that the underflows have been channelized like an underwater stream, and did not spread out like a sheetflood.

Flow rate records by the Swiss Federal Hydrological Agency since 1908 indicate that the Linth floods are not annual occurrences. There may be a rainstorm during the snow-melting season in the late Spring or early Summer which leads to a high-water stage, with the flood-water dense enough to continue on down into the lake as silt-laden underflows. Not uncommonly, there have been one or more late Summer floods after torrential rains which might feed one or more additional underflows. It is, therefore, not surprising that Stumpf was able to register as many as five "varves" from May 1911 to May 1912. In fact, the flow-rate record of the Linth (Naef, 1911, 1912) covering the same period shows at least three major peaks (May 19th, June 26th, October 6th). In addition other streams, such as the Murg (Fig. 1), also had high-water stages which may have continued as underflows to deposit the red laminae on the lake bottom that year. On the other hand, there have been dry years when neither the snow-melt nor the storm flood was dense enough to form silt-bearing underflows. Then the flood water would be spread out as overflows, or proceeded as interflows. On those occasions only the clay and finest silt particles were carried out as suspensions and deposited as mud in the lake. Stumpf had a "lean" year after the rich harvest of the preceding period, as he found no "varves", but only gray mud after May 1912. Again this result is compatible with the hydrometric records of the Linth: the 1912 curve is unusually "smooth" and indicates low flow rates.

Over a long period of observations, there seems to be a more regular pattern. The records during the last few decades show that there are on the average two major impulses of suspension-load to the Walensee. This observation may explain our sedimentary record of about two "varves" per year since 1811.

Discussion

The studies of lake sediments and of underflows have presented enough circumstantial evidence that the "varves" of the Walensee were deposited by continuously fed turbidity-underflows, which represent underwater continuation of the Linth flood. We are planning for direct observations, either by an underwater camera, or in small submersible. Meanwhile, the turbidity measurements by the Water Supply Agency of Zurich (Zimmermann, pers. communication) suggests that solid matter transported by underflows into the deeper parts of the lake forms a nepheloid layer near the lake bottom. The graded bedding may have a complex origin. The fluctuations of the current strength may have prevented the deposition of finer suspensions; only sand-size particles could settle out until maximum velocity is reached. Differential settling of suspended silts and clays during the flow of ebbing strength may

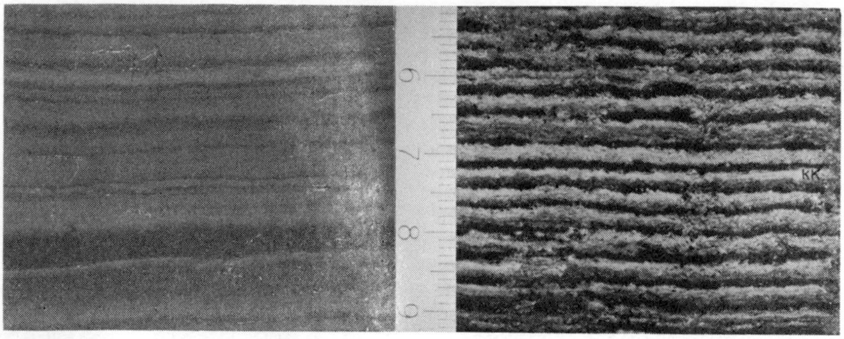

Figure 4. Left: Sediments from Walensee consisting of differently coloured, graded laminae of silt and clay (Core No. 4, Fig. 2). Scale in cm. Right: Varves from Lake Zurich (30 cm below the lake bottom). Light layers are the summer chemical precipitates; dark laminae are detritus deposited during winter. (Photography by K.Kelts).

have been responsible for the coarse to fine gradation observed in the "varves". This postulated mechanism of deposition may explain why coarse turbidite sands may have oscillating grain-size variation, or even reverse grading locally; only finer-grained turbidites and laminated silts and clays are characterized by more regular finer-upwards size-grading.

Both the sedimentological and hydrological records of the Walensee are unmistakable that "varves", or graded laminae of silt and clay are not necessarily deposits of annual cycles. The frequency of "varve"-deposition in the path of the Linth underflows depends very much upon the magnitude and frequency of the Linth high-water stage. We might further point out that laminae of different colors indicate that detritus has been supplied at different times from different tributaries which drain the petrographically inhomogeneous catchment area (red Permian sandstones, grey Mesozoic limestones and buff-brown Flysch deposits, yellow Globigerina shales, etc.). Each tributary delivers sediments to its own subaqueous fan which interfingers more or less with neighboring fans. As a consequence a vertical succession of laminae may have been deposited from underflows of different origin. Thus the number of laminae deposited annually at any given spot depends not only upon the floodstage of any particular stream, but also upon that of some adjacent streams, which may issue turbidity underflows of long travel-distances.

We do not intend to make an unwarranted generalization that all varves are not deposits of annual cycles. Figure 4 shows varves from the mesotrophic Lake Zurich (Kelts, 1978), where the light laminae represent chemical sedimentation prevailing during summers and the darker laminae detrital sedimentation during winters. A comparison of those varves with the non-annual "varves" of the oligotrophic Walensee shows a difference: the annual rhythm of Lake Zurich varves are more regular, while the irregularity of the Walensee "varves" reflects the fickle unpredictability of weather. Some of the detrital varves in the geological record resemble superficially the Lake Zurich varves more than those of the Walensee (see, for example, Pettijohn, 1975, fig. 8-6). Perhaps the laminated sediments of Walensee should not be called varves. Perhaps the varves should be more precisely defined, with a quantitative specification of the regularity of the rhythm. We might choose such parameters like

the ratio of standard deviation of thickness to the average thickness of the varve interval, the percentage of laminated sediments per unit of sedimentary-thickness. Such a definition must wait for a systematic, descriptive study of the various measurable parameters of many varve deposits, and is beyond the scope of this work. In conclusion, we wish only to reiterate that the "varves" or varve-like laminated silts and clays of the Walensee are deposits of turbidity underflows, but are not deposits of annual cycles.

References

Antevs, E. 1951, Glacial clays in Steep Rock Lake, Ontario, Canada, Bull. Geol. Soc. Amer. 62:1223-1262.

Ashley, G.M. 1972, Rhythmic sedimentation in glacial Lake Hitchcock, Massachusetts, Connecticut, Univ. Mass., Amherst, Dept. Geology Publ. 10: 148 p.

Florin, J. 1972, Ergebnisse aus Benthos-Untersuchungen am Walensee, Schweiz. Verh. int. Ver. Limnol. 18:461-466.

Geer, G. de 1912, A Geochronology of the last 12'000 years, Int. Geol. Congr. 1910, Compte Rendu 11:241-253.

Gustavson, T.C. 1975, Sedimentation and physical limnology in proglacial Malaspina Lake, Southern Alaska, Spec. Publ. Soc. Econ. Paleont. Mineral. 23:249-263.

Kelts, K.R. 1978, Geological and Sedimentological Evolution of Lake Zurich and Lake Zug,Ph.D. Thesis, ETH, 256 p.

Kuenen, P.H. 1951, Mechanics of varve formation and the action of turbidity currents, Geol. Fören. Förh. Stockholm 73:69-84.

Lambert, A. 1978, Eintrag, Transport und Ablagerung von Feststoffen im Walensee, Eclogae geol. Helv. 71/1:35-52.

Lambert, A.M., K.R. Kelts & N.F. Marshall 1976, Measurements of density underflows from Walensee, Switzerland, Sedimentology 23:87-105.

Ludlam, S.D. 1967, Sedimentation in Cayuga Lake, New York, Limnology and Oceanography 12:618-632.

Mathews, W.H. 1956, Physical limnology and sedimentation in a glacial lake, Bull. Geol. Soc. Amer. 67:537-552.

Naef, J. 1912, Graphische Darstellungen der schweizerischen hydrometrischen Beobachtungen 1911 und 1912, Abt. f. Landeshydrographie des Schweiz. Departements des Innern.

Pettijohn, F.J.(ed.) 1975, Sedimentary Rocks, 3rd Ed. Harper & Row, Publ., 628 p.

Stumpf, W. 1916, Methode der Deltavermessungen der Abteilung für Wasserwirtschaft. Ann. Schweiz. Landeshydrographie, Bd. II

Rhythmite deposition in lakes of the northeastern United States

STUART D. LUDLAM
University of Massachusetts, Amherst, Mass., USA

1 INTRODUCTION

Rhythmically layered lake sediments commonly have formed under a variety of climates and in many geologic periods. For example, within 2 hours drive of Amherst, Massachusetts, the location of the University of Massachusetts, laminated lake sediments of Triassic, Pleistocene and Post-glacial age can be found. These sediments were deposited in tropical, pro-glacial and temperate climates, respectively (Hubert et al. 1976; Schafer and Hartshorn 1965; Ludlam 1976). At the present time, rhythmically layered sediment or sediment containing occassional laminae are being deposited within 12 Massachusetts and New York lakes. Seven additional lakes are known in the northeastern United States in which rhythmites have been or are currently being deposited. Some of the rhythmites consist of annually deposited couplets, that is, true varves. The present paper attempts to review what is known about these layered sediments, particularly those currently being deposited in lakes of the northeastern United States. The research was supported in part by funds provided by the Water Resources Research Act of 1964, as amended.

2 GEOGRAPHIC DISTRIBUTION

In the eastern United States, 19 extant lakes are known to have deposited laminated sediments at some time since the last glaciation. These lakes are listed in Table 1. The apparent clustering of lakes depositing rhythmites in New York and Massachusetts is at least partly artificial in that the greatest effort has been expended in locating them in these areas. Their apparent confinement to northern states also reflects the distribution of lakes in general. In the eastern United States, lakes are uncommon south of glaciated areas. Lakes of eastern Canada also produce rhythmically layered sediment. A prominent example is Little Round Lake, Ontario (Tippett 1964), a 7.4 ha meromictic lake with a maximum depth of 16.2 m (Daley et al. 1977).

3 TYPES OF RHYTHMITES

In our region, the variety of lake types in which varves and other rhythmites are being deposited is impressive (Table 1). Lakes in which these sedimentary structures are well developed vary in depth from about 14 m to 406 m, and in area from 0.08 to 82,400 km^2. The lakes also vary greatly in productivity, from oligotrophic to eutrophic, and are subject to different degrees of mixing. Five of the lakes are monomictic, that is, the water column mixes from autumn through winter and spring. During the summer a thermocline separates an upper, warmer water layer (epilimnion) from a lower, colder layer (hypolimnion). Other lakes are dimictic, their water columns mixing only in spring and fall. In winter, mixing is prevented by ice cover, in summer, by thermal stratification. In a third group of lakes, the meromictic lakes, the water column does not mix completely at any season since the deepest water layer, the monimolimnion, is stabilized by chemical stratification. Above the monimolimnion these lakes show a seasonal pattern of stratification similar to dimictic lakes.

As would be expected from this variety of lakes, the composition and mechanism of formation of the varves and rhythmites they contain are highly variable. To a degree, the type of layering found in a lake sediment in our region reflects the phytoplankton productivity of the lake. In oligotrophic situations physical and chemical mechanisms are often responsible for the formation of the couplets. Periodic influx

Table 1. Some characteristics of lake-sediment systems in which post-glacial laminated sediments are found. If each couplet of a dark and pale lamina is an annual deposit this is indicated in the fourth column. The degree to which the laminae are preserved and the stratification pattern of the lake are given in the fifth and sixth columns, respectively.

Geographic Area and Lake	Surface Area (ha)	Maximum Depth (m)	True Varves	Degree of Preservation	Lake Type	Reference
CONNECTICUT						
Lower Linsley	9	15	no	good	dimictic	Vallentyne & Swabey 1955
GREAT LAKES						
Erie	2,570,000	64	unknown	poor	monomictic	Kemp 1969
Ontario	1,950,000	244	unknown	good	monomictic	Kemp 1969
Superior	8,240,000	406	unknown	good	monomictic	Dell 1974
INDIANA						
Myers	38	18	unknown	good	dimictic	Stahl 1959
MASSACHUSETTS						
Larkum	8	14	unknown	moderate	dimictic	Ludlam 1976
Laurel	67	16	probably	moderate	dimictic	Soukup 1975
Lower Goose	91	14	no	very poor	dimictic	Ludlam 1976
Onota	250	20	unknown	poor	dimictic	Ludlam 1976
Richmond	88	16	no	moderate	dimictic	Ludlam 1976
Stockbridge Bowl	155	15	probably	moderate	dimictic	Ludlam 1976
Upper Goose	17	9	no	very poor	dimictic	Ludlam 1976
MICHIGAN						
Douglas	1,510	27	no	good	dimictic	Wilson 1944
Third Sister	4	18	yes	excellent	dimictic	Eggleton 1931
NEW YORK						
Cayuga	17,200	132	yes	excellent	monomictic	Ludlam 1967
Fayetteville Green	26	52	yes	excellent	meromictic	Ludlam 1969
Fayetteville Round	13	51	probably	excellent	meromictic	
Jamesville Green	3	16	unknown	excellent	meromictic	Kellogg 1967
Seneca	17,500	188	unknown	excellent	monomictic	Woodrow et al. 1969

of sediment laden streams, littoral erosion and thermally controlled precipitation of calcite are among the dominant mechanisms of couplet formation. In eutrophic lakes biological mechanisms also become important and couplets may form in response to periodic blooms of diatoms or other phytoplankton as well as to physical and chemical mechanisms.

Cayuga Lake is an example of a large, monomictic, oligotrophic lake. The varves in this lake are composed almost entirely of allochthonous material, that is, material formed outside the lake. Most of the sediment, particularly organic detritus, enters the lake via streams during late winter and early spring thaws and is carried into the lake basin by turbidity currents. The varves that form nearshore contain organic rich layers which are deposited in late winter and early spring and organic poor layers deposited during the remainder of the year. Each varve is about 1 to 3 cm thick after compaction by 30 years accumulated sediment. The characteristics of these varves vary with sediment source and distance from the source (Ludlam 1967). A somewhat similar situation exists in Third Sister Lake, Michigan. In this small, dimictic, eutrophic lake, allochthonous clay layers alternate with dark layers containing a high percentage of autochthonous organic material (Eggleton 1931), that is, organic material formed within the lake.

In meromictic, oligotrophic, Fayetteville Green Lake, the varves consist almost entirely of autochthonous calcite. Most of this calcite is precipitated from the epilimnion during late spring and early summer because of increasing temperature (Brunskill 1969). As it is deposited the calcite dilutes the organic material which is produced all year. This mechanism has formed varves that are a few millimeters to a fraction of a millimeter thick in fresh sediment (Ludlam 1969). In the small, shallow, dimictic lakes of western Massachusetts uncompacted varves or other laminae are usually about a cm or less in thickness. In eutrophic Stockbridge Bowl, intense spring diatom blooms produce sedimentary layers rich in diatom frustules which alternate with layers in which organic detritus is more abundant. In other moderately productive western Massachusetts lakes, the occurrence of varves seems to be related to seasonal changes in algal productivity or seasonal breakdown of stratification (Ludlam 1976).

4 DISTRIBUTION OF RHYTHMITES IN LAKE SEDIMENTS

Commonly, when rhythmites are found in cores of lake sediment, they occur at one or more levels intercalated in sediment which is more or less uniform in consistency. This occurs in Lake Erie and Linsley Pond, and occassional layers, separated by massive sediment, are found in Richmond Pond. A particularly interesting observation is that in all the small lakes in western Massachusetts where sediment laminae have been found, the laminations are confined to the most surficial sediments. Several theories have been advanced to explain this. Ludlam (1976) suggested recent increases in productivity may have caused the formation or preservation of the laminae. In discussions with the author, Robert Wetzel and Gregg Brunskill advanced the idea that older laminae might have been destroyed by water turbulence generated by great storms, particularly during periods of mixing. The latter theory predicts that currently observed varves may be destroyed at some time in the future, and a new series of varves would then begin to be deposited over the storm mixed sediments. The storms themselves might form the sedimentary layers by suspending and winnowing the sediment into layers according to the settling rates of the particles. A single storm could produce numerous layers if wind driven turbulence abated slowly.

The storm hypothesis is not supported by sedimentary analyses of pigments and planktonic microfossils in Stockbridge Bowl and neighboring Richmond Pond. In each lake the two types of profiles are in agreement. They indicate that productivity in Stockbridge Bowl has been increasing since the 1930's while in Richmond Pond it has increased very suddenly and only in the last few years. It seems very unlikely that within each lake the two lines of evidence would give the same results while between neighboring lakes of similar depth they would give such dissimilar results if the storm hypothesis applied to these lakes.

The storm hypothesis has validity, however. In lakes where the lateral distribution of rhythmites has been studied, varved or otherwise laminated sediments are usually confined to relatively deep water. In some lakes this distribution is related to the mechanism of varve formation (Anthony 1977). However, in two lakes in our own region, Fayetteville Green Lake and Stockbridge Bowl, the mechanism controlling varve deposition operates throughout the area of

the lake. In these cases the lack of varves in shallow water sediments indicates that mechanisms must exist which destroy varves that might have formed there. Three major agents capable of destroying varved lake deposits as they form have been identified in our area. First, Storr (1962) presented evidence that in large lakes, seiches could erode sediment at a considerable water depth. Seiches are often caused by wind action. Second, Margaret Davis (1973) demonstrated the erosive power of wind generated water turbulence. In Frain's Lake, a small (500 m long, 10 m deep), dimictic lake in Michigan, the upper mm of sediment was resuspended during periods of mixing at 10 m water depth. Stronger winds during mixing would undoubtedly have caused reworking of a greater thickness of sediment. Third, Ronald Davis (1974) showed that the activities of burrowing organisms could rapidly obliterate any minor sedimentary structure.

Evidently, for laminated lake sediment to exist, two conditions must be met. First, there must be a mechanism whereby laminae are formed, second, there must be another mechanism or mechanisms that allow the laminae to be preserved by moderating or eliminating the action of destructive agents. In our temperate lakes, the latter mechanisms include a number of chemical and physical phenomena which may best be understood by examining the distribution of varves within and among lakes.

5 MECHANISMS CONTROLLING THE OCCURRENCE OF RHYTHMITES

Certain relationships are immediately evident. All lakes in our region that are currently depositing laminated sediments are thermally stratified, and some are meromictic, that is, permanently chemically stratified. Although sediments in Sandusky Bay, a large, shallow (<4 m) arm of Lake Erie, are finely laminated (Wilson 1944) and seem to be an exception to this general rule, it is known that they are not currently forming but represent deposits of an earlier, higher lake stage (Antevs 1953). Varved glacial clays are also exposed in Lake Superior (Dell 1974). The one exception to the rule is a reservoir backed up by the Arkport flood control dam near Hornell, New York. This reservoir is usually dry and only fills during flooding. Wood (1947) demonstrated that during one 2 week period of spring flooding 2.5 to 17.8 cm of sediment were deposited and in many areas showed three distinct layers similar in general appearance to glacial varves.

Although the effects of thermal stratification are far reaching, it has one obvious effect. Stratification protects the sediment beneath the thermocline from wind driven water turbulence for as long as the stratification exists.

Anaerobic conditions in the bottom waters of lakes also seem to be extremely important in determining whether or not rhythmically layered sediments will be preserved. Both productive and unproductive lakes have been observed to deposit varved or otherwise laminated sediments as long as oxygen is absent from the bottom waters for an appreciable fraction of the year (Ludlam 1976). To date, among our smaller, shallower lakes, none has been found in which laminae are preserved and in which oxygen is abundant in the bottom water for all or almost all of the year. Sedimentation rates in the lakes depositing laminae range between 100 and 1400 g m^{-2} yr^{-1} and a year's deposit may measure up to a cm in thickness before burial.

In Stockbridge Bowl and Fayetteville Green Lake the lateral extent of varves is correlated to oxygen content of the water immediately overlying the sediment. In Fayetteville Green Lake well developed sequences are confined to those areas in which the overlying water lacks oxygen throughout the year. In Stockbridge Bowl, varves are found where oxygen is absent from the overlying water for between 1 month (8 m water depth) and 5.5 months (14 m water depth) of the year. In this lake oxygen profiles were made at the deepest point of the lake. They do not necessarily reflect the oxygen concentrations near the sediment-water interface at shallower depths. It is likely that at the sediment-water interface the period of anoxia was greater than 1 month at 8 m depth, the greatest depth at which a well preserved sequence of varves was found.

The relationship between oxygen and the occurence of varved or otherwise laminated sediments in smaller lakes with low sedimentation rates seems well established in our area. It is doubtful if anaerobic conditions alone have a direct effect on the deposition or preservation of most of these sedimentary structures, but rather anoxia limits the activities of benthic organisms (Jónnasson 1969) that would otherwise disturb the sediment. Other factors also control these organisms, including sediment type (Kinney 1972) and the availability of food (Jónnasson 1969). It must be pointed out that anaerobic conditions normally do not occur in our lakes except in water layers protected from turbulent mixing and atmospheric contact by thermal or chemical stratification. Thus, the effects of protection from bioturbation and turbulent mixing are linked.

The relationship between hypolimnetic anoxia and preservation of rhythmites does not apply to our largest lakes. In four of these lakes, Cayuga and Seneca Lakes in New York and Lakes Ontario and Superior among the Laurentian Great Lakes, rhythmites have been discovered although oxygen is perennially abundant in the bottom water Because these lakes are very large, water turbulence will be greater at a given depth than in a smaller lake in the same region (Hutchinson 1957). Evidence of erosion of sediments at a considerable water depth exists in both Seneca Lake (Storr 1962) and Lake Superior (Dell 1974). However, the velocity of wind driven water currents decreases with water depth (Davis 1973) and under ideal conditions the decrease should be logarithmic (Hutchinson 1957). Since the rhythmically layered sediments in all these lakes have been described from water of considerable depth, it is likely that the thickness of the overlying water column protects the sediment to some degree from wind driven currents. The consistency of the sediment may also play a role. In Lake Ontario, the layers are 0.1 to 10 mm thick in silty clays (Kemp 1969). Despite an abundance of oxygen in this lake, clay substrates in deep water did not support a large or diverse benthic fauna (Kinney 1972), and are probably more resistant to disruption than the softer, less cohesive sediment common to many of our smaller lakes. In Cayuga Lake, the sediments are also quite often of a clayey consistency but here the varves are from 1 to 3 cm in thickness after 30 years compaction. Millimeter thick laminae are found within the varves. In this situation it is likely that the ability of the benthos and water movements to stir the sediment and obliterate the layered structure is exceeded by the sedimentation rate. Interestingly, in 18 m deep Third Sister Lake, the only small lake in our region with varved sediment that is known to have both hypolimnetic anoxia and an abundant bottom fauna (Eggleton 1931), the varves are not only relatively thick (1.5 - 3.0 cm) but also contain a great deal of clay. It is possible that in some of the larger lakes, rhythmites of considerable age have been exposed or brought near the sediment-water interface by erosion as in Lake Superior (Dell 1974). In these situations it is not necessarily known under what circumstances the layers formed.

The last relationship, already noted, is that among the lakes of western Massachusetts with varved or otherwise rhythmically laminated sequences these structures have been deposited only in the last 5 to 35 years. To discover why this was so, cores of surficial sediment from lakes with and without laminated sequences were studied for sedimentary pigment concentrations. The concentration of these pigments theoretically is related directly to phytoplankton productivity in the lake at the time of sediment deposition. In all those lakes where varved or otherwise laminated sediments were found, an associated rise in sedimentary pigment concentration was also found. This rise begins shortly before or at the same point that laminated sediments first occur in a profile and usually continues into the most recent sediment. In one case where a significant, recent rise in pigment concentration occurred, laminations were absent. In that lake, oxygen was found in the hypolimnion for almost the entire period of stratification, and burrows of insect larvae were abundant in the sediment. The increase in pigment concentration in the youngest sediments probably signals a trend toward rising productivity. This argument is strengthened by the fact that in 2 of the lakes, profiles of sedimentary pigments and planktonic microfossils were in agreement. In addition, the longest and most clearly marked varve sequences were found in the two most productive lakes.

A rise in productivity would increase the probability of having varved sediment deposited and preserved in a lake. This is because it increases the probability of large periodic blooms of diatoms and other algae which are often a cause of rhythmite deposition. In addition, a rise in productivity in lakes is often paralleled by a decline in hypolimnetic oxygen concentration. It is interesting to speculate that in some situations, particularly in the earlier stages of eutrophication in deep lakes, the growth of the profundal benthic fauna will be increased (Wetzel 1975) before the limiting aspects of low oxygen concentration occur. In these cases increased productivity may at first decrease the probability that sedimentary laminae might survive.

From the foregoing it is evident that the interrelationships which control the formation and preservation of rhythmites in lakes are of considerable complexity. The major mechanisms operating or presumed to operate in our region are summarized in Figure 1. It is evident that thermal stratification has particularly far reaching consequences. It is seasonal warming of the epilimnion of Fayetteville Green Lake that is responsible for the seasonal precipitation of calcite which defines the spring-summer layer of each varve (Brunskill 1969). Thermal stratification controls the periodic occurrence of hypolimnetic anoxia in our lakes which in turn

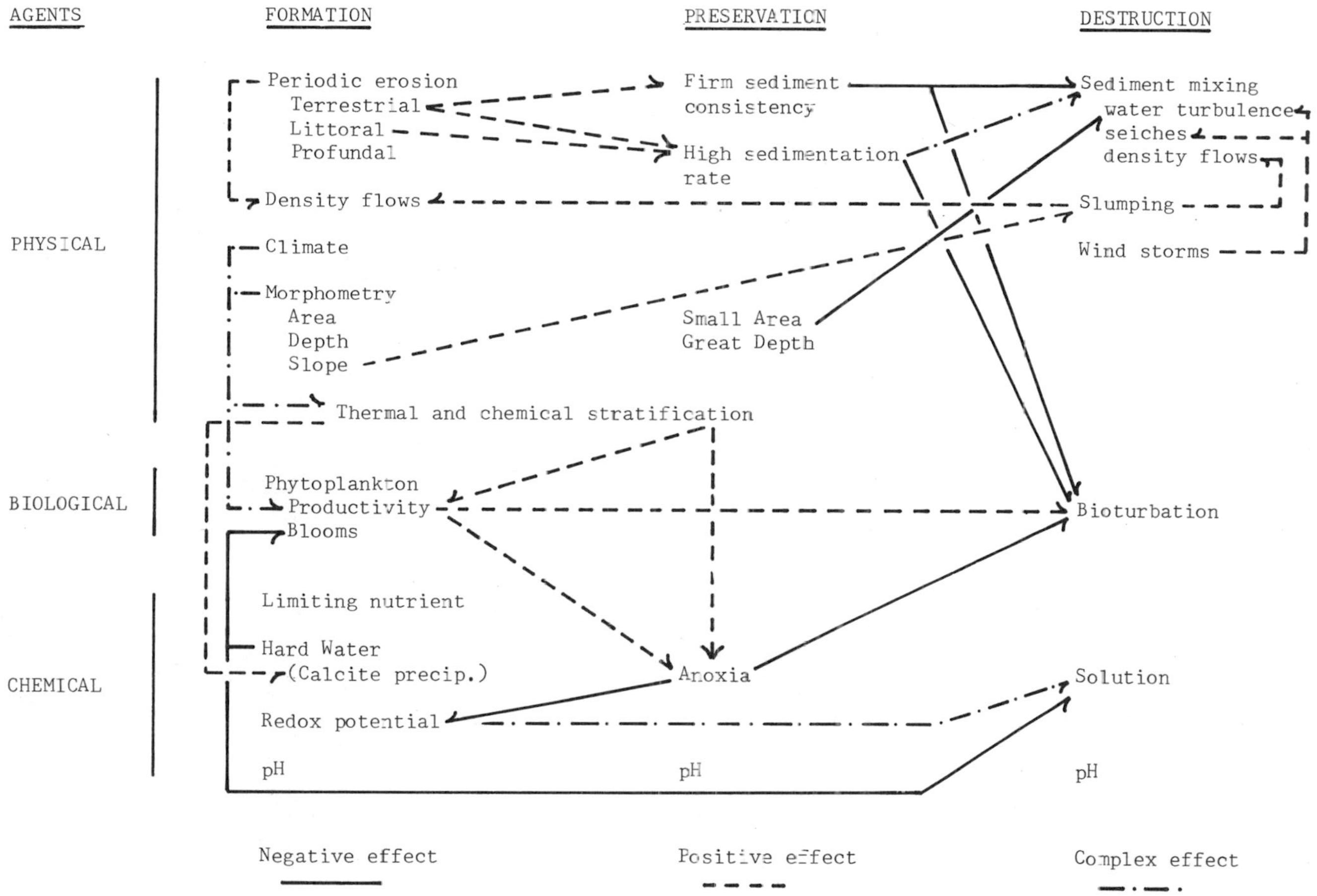

Figure 1. A simplified diagram illustrating some major agents known or suggested to influence the occurrence of rhythmites in lakes of the northeastern United States. Relationships between agents are indicated by lines. The effect of pH is highly complex so no attempt has been made to indicate its relationship to other agents.

may force the formation of laminae in iron rich lakes (Anthony 1977) and possibly in several lakes in western Massachusetts. Thermal stratification contributes to the periodicity and productivity of phytoplankton which in death may form both sedimentary laminae and the food supply for the benthos which may destroy those same laminae. And, of course, stratification protects the sediment beneath the thermocline from a great deal of water turbulence.

The depth of the thermocline is determined in large measure by the intensity of wind action across the lake surface. The amount of kinetic energy transferred from the wind to the lake is directly proportional to the area and fetch of the lake and inversely proportional to the amount of protection offered to the lake by surrounding hills and vegetation (Hutchinson 1957). Thus, the depth of the thermocline can be used as an indicator of the relative amount of wind induced water turbulence in a lake.

6 DISCUSSION AND CONCLUSIONS

In the eastern United States, varved or otherwise laminated lake sediments are relatively common. They occur in a wide variety of temperate lakes, including monomictic, dimictic and meromictic lakes. The deposition of the varves and laminae is due to a wide variety of phenomena operating both within and without the lake. Water movements and bioturbation may destroy sediment structures as they are deposited in many lakes. When these structures are found in modern temperate lakes a mechanism must exist which moderates or eliminates the effect of these destructive agents.

Of the three major lake types in our area, the monomictic lake provides minimum protection from sediment disturbances. Cayuga Lake, for example, mixes for 5 to 6 months of the year (Henson et al. 1961) and when the sediments were studied the bottom water contained abundant oxygen and an abundant bottom fauna. Preservation of sediment structure seems to depend on high sedimentation rate and in some areas a clayey, relatively firm sediment. Other large lakes support only a small benthic fauna in sediment types where rhythmites are found.

Meromictic lakes offer maximum protection from sediment disturbances. The deepest water lies beneath a chemocline and may not be subjected to direct wind mixing for centuries or longer. This water layer is protected from all but minimum turbulence. Because the bottom waters of meromictic lakes mix so slowly with the overlying waters, microbial activity can maintain anaerobic conditions and produce hydrogen sulfide, effectively preventing the growth or survival of large populations of macroscopic organisms. It is not suprising that some of the best, recent, varved lake deposits in the northeastern United States are found in meromictic lakes. Meromixis does not ensure the occurence of varved deposits, however. Although the three meromictic lakes in New York State discussed herein all have varved or rhythmically layered sediments, meromictic Hall Lake, Washington, does not (Culver 1977).

The highly stable stratification of the water column which defines the meromictic condition is chemical in nature, but ordinary dimictic conditions in conjunction with hypolimnetic anoxia for a significant portion of the year seem sufficient to preserve laminations less than 0.1 mm thick in Douglas Lake, Michigan (Eggleton 1931; Wilson 1944). In the dimictic lakes of western Massachusetts, which are only 13 to 16 meters deep, winter stratification with ice cover may last 3 to 4 months and summer stratification 5 to 6 months. Mixing of the water column lasts only 2 to 4 months of the year. This pattern of stratfication offers considerable protection to the sediment beneath the thermocline, particularly in protected lakes where periods of stratification are longer and periods of mixing shorter. Some of these lakes may also skip a period of mixing from time to time. Because none of them is particularly deep, the volume of the hypolimnion, or bottom water is small. It is in contact with a relatively large area of organic sediment and receives a considerable amount of organic detritus from overlying waters, particularly in the more productive lakes. As a result, in most of these lakes oxygen is lost rapidly from the hypolimnion during stratification so that anaerobic conditions prevail in the hypolimnion for 4 months or more of the stratification period. In some cases, hydrogen sulfide is also produced. This combination of circumstances favors the preservation of sedimentary structure so that in 7 of the 11 lakes studied in western Massachusetts some sort of layering was evident in the sediment (Ludlam 1976). On Cape Cod, Massachusetts, similar lakes are subjected to considerably more wind action with shorter periods of stratification and anoxia. None has as yet been shown to be depositing laminated sediments (Soukup, personal communication).

Varve deposition or preservation becomes more likely as lake productivity rises in the dimictic lakes of western Massachusetts. This is probably due to increasingly intense algal blooms which contribute to varve formation and to the occurrence of increasingly lenghty anaerobic periods in the

hypolimnion which limit the activity of the benthos.

7 REFERENCES

Antevs, E. 1953, Geochronology of the Deglacial and Neothermal ages, J.Geol. 61:195-230.

Anthony, R.S. 1977, Iron-rich rhythmically laminated sediments in Lake of the Clouds, northeastern Minnesota, Limnol. Oceanogr. 22:45-54.

Brunskill, G.J. 1969, Fayetteville Green Lake, New York. II. Precipitation and sedimentation of calcite in a meromictic lake with laminated sediments, Limnol. Oceanogr. 14;830-847.

Culver, D.A. 1977, Biogenic meromixis and stability in a soft-water lake, Limnol. Oceanogr. 22:667-686.

Daley, R.J., S.R. Brown & R.N. McNeely 1977, Chromatographic and SCDP measurements of fossil phorbins and the postglacial history of Little Round Lake, Ontario, Limnol.Oceanogr. 22:349-360.

Davis, M.B. 1973, Redeposition of pollen grains in lake sediment, Limnol.Oceanogr. 18:44-52.

Davis, R.B. 1974, Stratigraphic effects of tubificids in profundal lake sediments, Limnol.Oceanogr. 19:466-488.

Dell, C.I. 1974, The stratigraphy of northern Lake Superior late-glacial and post glacial sediments, Proc.17th Conf.Great Lakes Res. 179-192.

Eggleton, F.E. 1931, A limnological study of the profundal bottom fauna of certain fresh-water lakes, Ecol.Monogr. 1:231-332.

Henson, E.B., A.S.Bradshaw & D.C. Chandler 1961, The physical limnology of Cayuga Lake, New York, Cornell Univ.Agr.Expt. Sta.Mem. 378:1-63.

Hubert, J.F., A.A. Reed & P.J. Cary 1976, Paleogeography of the East Berlin Formation, Newark Group, Connecticut Valley, Amer.J.Sci. 276:1183-1207.

Hutchinson, G.E. 1957, A Treatise on Limnology. I. Geography, Physics, and Chemistry. New York, John Wiley & Sons, Inc.

Jónasson, P.M. 1969, Bottom fauna and eutrophication. In Eutrophication: Causes Consequences, Correctives, p. 274-305. Washington, Natl.Acad.Sci.Publ. 1700.

Kellogg, C.W. 1967, Plankton and sediment studies in Jamesville Green Lake. Unpublished ms.

Kemp, A.L.W. 1969, Organic matter in the sediments of Lakes Ontario and Erie, Proc.12th Conf.Great Lakes Res. 237-249.

Kinney, W.L. 1972, The macrobenthos of Lake Ontario, Proc.15th Conf.Great Lakes Res. 53-79.

Ludlam, S.D. 1967, Sedimentation in Cayuga Lake, New York, Limnol.Oceanogr. 12:618-632.

Ludlam, S.D. 1969, Fayetteville Green Lake. III. The laminated sediment. Limnol. Oceanogr. 14:848-857.

Ludlam, S.D. 1976, Laminated sediments in holomictic Berkshire lakes, Limnol.Oceanogr. 21:743-746.

Schafer, J.P. & J.H. Hartshorn 1965, The Quaternary of New England. In H.E. Wright, Jr. & D.G.Frey (Eds.), The Quaternary of the United States, p. 113-128. Princeton, Princeton.

Soukup, M.A. 1975, The limnology of a eutrophic, hardwater New England lake, with major emphasis on the biogeochemistry of dissolved silica, Ph.D.thesis, Univ. Mass.,Amherst.

Stahl, J.B. 1959, The developmental history of the chironomid and Chaoborus faunas of Myers Lake, Invest.Indiana Lakes Streams 5:47-102.

Storr, J.F. 1962, Delta structures in the New York Finger Lakes and their relation to the effects of currents on sediment distribution and aquatic organisms, Great Lakes Res.Div.Inst.Sci.Technol., Univ.Michigan 9:129-138.

Tippett, R. 1964, An investigation into the nature of the layering of deep-water sediments in two eastern Ontario lakes, Can. J. Botany 42:1693-1709.

Vallentyne, J.R. & Y.S. Swabey 1955, A reinvestigation of the history of Lower Linsley Pond, Connecticut, Amer.J.Sci. 253:313-340.

Wetzel, R.G. 1975, Limnology. Saunders, Philadelphia.

Wood, E. 1947, Multiple banding of sediments deposited during a single season, Am.J. Sci. 245:304-312.

Woodrow, D.L., T.R. Blackburn & E.C. Monahan 1969, Geological, chemical and physical attributes of sediments in Seneca Lake, New York, Proc.12th Conf. Great Lakes Res. 380-396.

The laminated sediments of Crawford Lake, southern Ontario, Canada

MARIA BOYKO-DIAKONOW
Penticton, B. C., Canada

1. CRAWFORD LAKE VARVES

Crawford Lake ($43^{\circ}28.1$'N, $79^{\circ}56.9$'W) is situated in Silurian Guelph-Amabel dolomite, one kilometer west of the Niagara Escarpment and 65 kilometers southwest of Toronto (Fig. 1). The lake probably originated as a cavern formed in the dolomite bedrock (Fig. 2). Crawford Lake has an area of 2.5 ha, a maximum depth of 24 m and a sediment thickness of at least 4.5 m (Fig. 2 , 3).

Testing of the deep water sediment in 1970 showed it to be banded with alternating light and dark laminae (Fig. 4,5). Each light lamina is followed gradually by the overlying dark lamina, but the latter formed a sharp boundary with the overlying light lamina. This repetition indicates that the lower light layer and overlying dark layer are related and form a "couplet". A light carbonate-rich layer was observed at the sediment-water interface in summer (June) and a dark organic-rich lamina overlay it the following winter (March, Fig. 5). This indicates that a complete couplet is probably deposited each year.

The geological setting, limnological features and characteristics of the couplets are similar in Fayetteville Green Lake, New York (Ludlam 1969) and Crawford Lake. Since the Fayetteville Green Lake couplets are known to be true varves the argument that the Crawford Lake couplets are annual is probably correct.

2. CARBONATE-ORGANIC VARVE FORMATION

The couplets in Crawford Lake are identical to those described from the deep water sediment of Fayetteville Green Lake (Ludlam 1969). Both lakes are meromictic. The water in the deep basin is stratified permanently. Incomplete mixing preserves a lower, more dense layer (monimolimnion) below a circulating or mixing layer (mixolimnion). The density gradient between these layers is the chemocline which contains a characteristic dense bacterial plate. Laminated sediments in both lakes were collected only from that part of the basin below the chemocline. The constantly cold and quiet anoxic water of the monimolimnion permits undisturbed accumulation of seasonally deposited sediments. Meromixis insures against mixing of the sediment by bottom feeding fauna and spring and fall turnover (Davis 1967; Davis 1968). The deposition is uniform and simultaneous, as indicated by the lateral continuity of "marker" bands (Ludlam 1969).

Annual deposition provides an explanation for the abundance of calcite in the light laminae and its paucity in the dark laminae. According to Brunskill (1969) the seasonal input of calcite, regulated by temperature, is the variable responsible for couplet formation in Fayetteville Green Lake. In May or early June, calcite crystals reach the sediment, initiating couplet formation. This sudden calcite precipitation would explain the

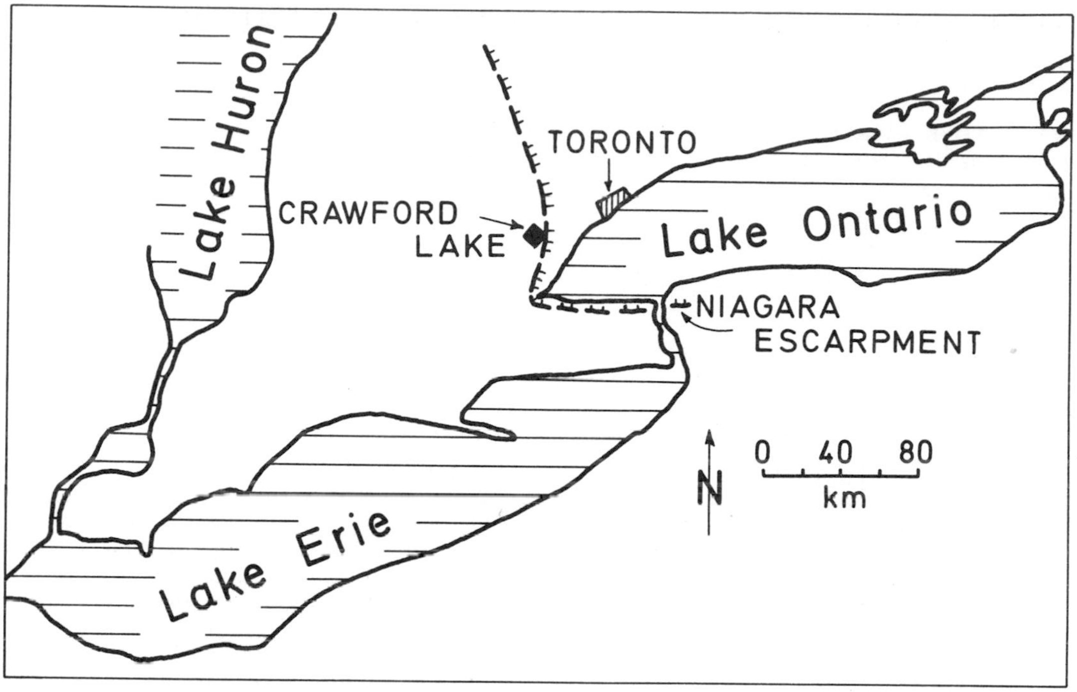

Figure 1. Schematic map of southern Ontario with location of Crawford Lake

sharp contact with the underlying dark lamina in Crawford Lake. Calcite rich materials are deposited until midsummer when the rate of calcite supply decreases slowly in relation to the organic matter and clastic material that are deposited throughout the year. This could explain the gradual contact between a pale lamina and its overlying dark lamina.

In Fayetteville Green Lake photosynthesis played only a secondary role in couplet formation (Brunskill 1969), and after Ludlam (personal communication) photosynthesis is not responsible for varve formation. This is not necessarily true in Crawford Lake. Tippett (1964), who has worked with varved sediments from McKay Lake and Little Round Lake in eastern Ontario, felt that biological agents such as Chara and phytoplankton were the principle cause of calcite precipitation. The role of photosynthesis in calcite precipitation should be examined more closely in Crawford Lake.

There is one major difference between the sediments of Crawford Lake and Fayetteville Green Lake. In the latter lake, there is a marked frequency of "massive" layers between sections of laminated sediment. These are attributed to turbidity currents and slope of the basin (Ludlam 1969). They are rare or absent in Crawford Lake.

3. VARVE-DATED POLLEN RECORD

Direct observations and analogies to known varved sediments indicate that the sequential couplets in Crawford Lake each represent one year's accumulation of sediment. The annual pollen rain is locked into the couplets which provide an absolute chronology.

The layered sediments from Crawford Lake were collected by freezing the sediments in situ with a frigid finger sampler (Swain 1973). The frozen sediment was embedded in Carbowax 1540 and 200 couplets were counted backwards from the sediment-water interface of 1971 (Fig. 4,5). A comparative study was made of this 200 year interval using the pollen record of the sediment and the hi-

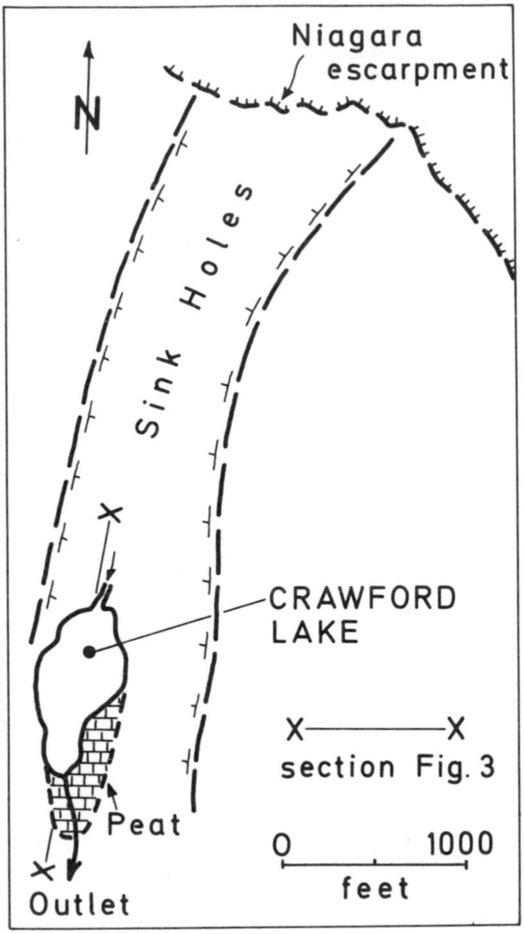

Figure 2. Crawford Lake - situated in a sinkhole area close to the Niagara Escarpment

story of European settlement derived from land patent records and written records of forest clearance (Boyko 1973). Pollen analyses were performed by two individuals (author and J.H. McAndrews) each working separately on alternate five-couplet segments and independently of the historical records. The latter records complemented or aided in explaining the interpretation derived from the varve dated pollen record.

It is generally accepted that European weeds accompanied the spread of settlement from the western shores of Lake Ontario northward at the beginning of the nineteenth century. One of these weeds, Rumex acetosella (sorrel) appears in the Crawford Lake sediment in the early 1820's. Settlement started in the 1790's. The approximate 30 year difference is explained by the type of settlement practices, that is, land grants with absentee landlords as opposed to actual settlement, and by the rate of spread of sorrel. Sorrel was possibly introduced as a crop seed contaminant, as were many other introduced weeds, and spread as fields were cleared.

The distinct rise of native Ambrosia (ragweed) and its subsequent increase is identified in all recent sediments in the lower Great Lakes region and it is attributed to European settlement. The Ambrosia rise has never been dated. At Crawford Lake it begins in the mid-nineteenth century (1846-1851).

In southern Ontario the distribution of native ragweed depends largely on soil disturbance. The appearance of Ambrosia after Rumex acetosella in the pollen record suggests that the latter weed had a greater chance of success in the initially cleared fields, probably because it was sown with the crop. Native Ambrosia had to spread from its sparse, naturally disturbed habitats such as river banks and bluff slopes. Historical records indicate that about 35-40 % of the surrounding study area of 3885 sq km was cleared of forest when Ambrosia began to appear consistently in the pollen record.

A sharp decline in total pollen influx after 1871 corresponds to near maximum forest clearance in the surrounding area. The woody plants provided the majority of the pollen rain (over 90 %) before European settlement and once these were removed the succeeding herbs, including crops and weeds, were poorer pollen producers, mainly because the majority of them are insect-pollinated and/or they reproduce vegetatively.

After World War II there was a general movement of people from the farms to the cities. Consequently the area of farmland was reduced and in the pollen record this shift is evidenced by an increase in woody plant pollen from the early 1950's.

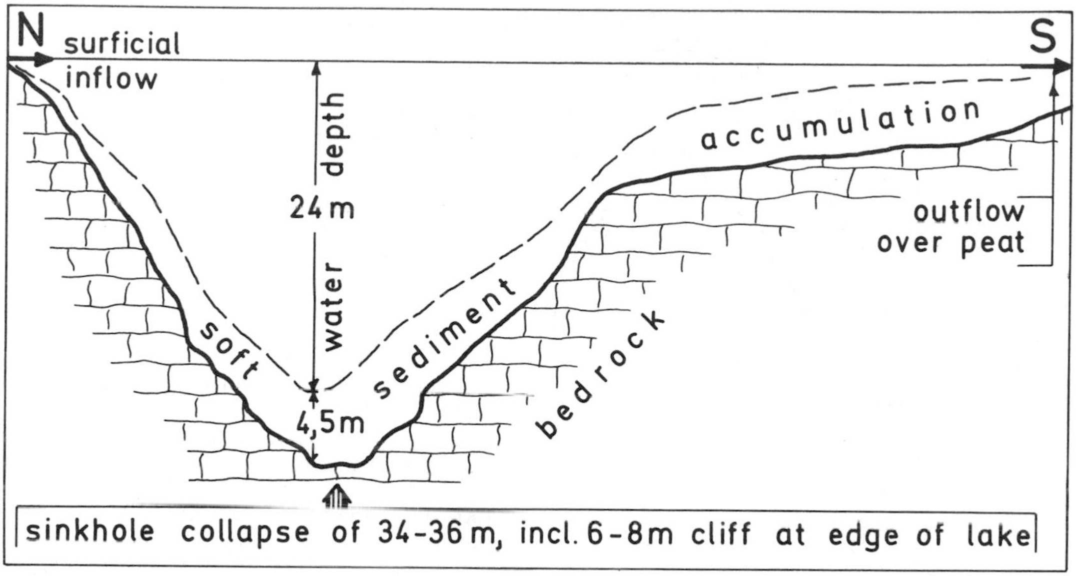

Figure 3. Crawford Lake - longitudinal section and bathymetry

The increase is caused by succession of shrubs and trees which replaced the herbs of the abandoned fields.

4. DISCUSSION AND CONCLUSIONS

The Crawford Lake sediments are laminated. Each couplet of a dark and underlying pale lamina appears to be an annual deposit similar to the varves of Fayetteville Green Lake. Since both these lakes are meromictic a brief search was made for potentially meromictic lakes. Three features were used to identify these lakes:

1. Great water depth compared to surface area.
2. Wind protection by steep slopes and forest cover of the slopes.
3. Oxygen concentration approaching zero with depth.

Two lakes, Found Lake and Greenleaf Lake, in Algonquin Park, Ontario, had these characteristics. Both had laminated sediments. Evidently there are more lakes with laminated sediments in Ontario than previously believed.

The varves of Crawford Lake were used to establish an absolute pollen chronology for the region extending 200 years into the past. Rumex acetosella pollen first appears in the early 1820's, about 30 years after settlement. A rise in native Ambrosia pollen began about 25 years later, between 1846 and 1851. Maximum forest clearance occurred after 1871 while the abandonment of farmland after World War II is clearly seen in a rise in woody plant pollen in the early 1950's.
Crawford Lake offers an excellent site for the study of deposition of varved lake sediments. The sediments offer an opportunity of paleoclimatological studies based on an absolute chronology.

5. ACKNOWLEDGEMENTS

This study was part of a Masters of Science thesis program under the supervision of Dr. J.H. McAndrews, Botany Department, University of Toronto and Department of Geology and Mineralogy, Royal Ontario Museum, Toronto. Dr. McAndrew's participation in this study is gratefully acknowledged. - I express my sincerest thanks to Dr. S.D. Ludlam, Amherst, for his review of my article and for his most valuable improvements of this paper.

Figure 4. Frigid finger sampler illustrating water-sediment interface of Crawford Lake laminated sediments in March

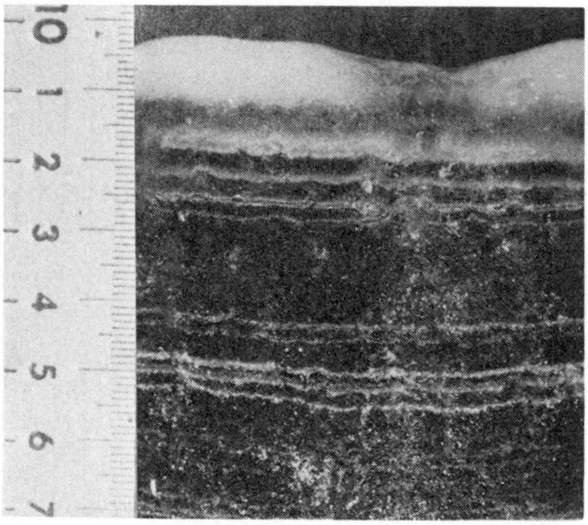

Figure 5. Crawford Lake laminated sediments, including water-sediment interface (dark winter lamina of 1971-72). Scale is in centimeters.

6. REFERENCES

Boyko, M. 1973, European impact on the vegetation around Crawford Lake in southern Ontario, M. Sc. Thesis Dept. Botany, University of Toronto.

Brunskill, G.J. 1969, Fayetteville Green Lake, New York. II. Precipitation and sedimentation of calcite in a meromictic lake with laminated sediments, Limnol. Oceanogr. 14:830-847.

Davis, M.B. 1968, Pollen grains in lake sediments: redeposition caused by seasonal water circulation, Science. 162:796-799.

Davis, R.B. 1967, Pollen studies of nearsurface sediments in Maine Lakes. In E.J. Cushing and H.E. Wright (eds.), Quaternary Paleoecology, p. 143-173. New Haven, Conn., Yale Univ. Press.

Ludlam, S.D. 1969, Fayetteville Green Lake, New York. III. The laminated sediment, Limnol. Oceanogr . 14:848-857.

Swain, A.M. 1973, A history of fire and vegetation in northeastern Minnesota as recorded in lake sediment, Quat. Res. 3:383-396.

Tippett, R. 1964, An investigation into the nature of the layering of deep water sediments in two eastern Ontario lakes, Can. J. Bot. 42:1693-1709.

7. APPENDIX

The figures 1, 2 and 3 have been redrawn after handouts prepared by the author for ROM-members field-trip to Crawford Lake on May 3, 1972 (Ed.).

Varves in the Black Sea and in Lake Van (Turkey)

STEPHAN KEMPE & EGON T. DEGENS
Universität Hamburg, Hamburg, Germany

1 INTRODUCTION

In sedimentology, rhythmites are well recognized phenomena. Environmental perturbations occurring at regular intervals produce sediment layers of alternating physical and chemical characteristics. The frequency of these alternations may be counted in days, months, years and time intervals of much longer duration. Once the clock-controlled sequence is assessed, rhythmites prove to be a powerful tool in dating, for example, trees, otholits, shells, corals, sinter deposits or lake and ocean sediments.

In this presentation, we will restrict ourselves to the "classic" rhythmites which we recognized in Black Sea and Lake Van deposits. In addition of dating the sediments, insight was gained on the nature of environmental perturbations which left distinct "fingerprints" in the fabric of the sediment sequence.

To form annually stratified sediments in a body of water, two conditions are to be met: (i) the sediment material must yield discernable deposits in response to seasonal variations, and (ii) the freshly deposited sediment must remain intact and not altered by bottom currents or burrowing organisms.

Both conditions are met in the present Black Sea and Lake Van (Fig. 1, Table 1): the tributary rivers show a distinct hydrological cycle and the biological activity varies considerably during the year. Furthermore, the Black Sea is stratified at a water depth of about 200 m (Degens and Ross 1974): and the resulting anoxic conditions at depth exclude the presence of higher organisms. The lack of benthos in the deep waters of Lake Van is caused by its unusual water chemistry, i.e., the high soda content (Degens and Kurtman 1978).

2 VARVES IN THE BLACK SEA AND THEIR ENVIRONMENTAL INTERPRETATION

The first meter of sediment of the euxine abyssal plain is finely laminated. Within this sediment section, three distinct units (Fig. 2) are discernable (Degens et al., 1978a):

- The topmost unit is a coccolith ooze, measuring 30 to 50 cm. At certain places it can be interrupted by turbidites. Of significance is the systematic alternation between white and dark layers with up to 50 double laminae per cm of sediment.
- Unit two is a microlaminated dark-brown sapropel whose average thickness is 40 cm. The lamination can be recognized in the form of black and brown microlayers which alternate in a regular fashion; up to 100 double layers were counted per cm of sediment.
- Unit three consists of a series of alternating light and dark lutites with no obvious micro-layering.

The microlaminations in the coccolith and sapropel units were interpreted as varves, i.e., as annual rhythmites, and a varvecount was conducted. The result of this count showed unit one to be about 1000 years old, while unit two yielded different ages in different cores. The maximum varve-age for this unit was 4000 years. The age appears to be controlled by bathymetry. In the central part of the Black Sea basin ($\sim$ 2200 m water depth), the anoxic water body started to form 5000 years ago. With the continuous rise of the H_2S-O_2 interface, more and higher parts of the basin floor became progressively incor-

porated into the anoxic water layer, thus, at a water depth of 500 m, sapropel formation started almost 3000 years after its first appearance on the euxine abyssal plain.

It is interesting to note that previous ^{14}C datings have assigned ages of 3000 and 7000 years for the start of unit one and two (Degens and Ross 1974), whereas ^{14}C and varve ages are in excellent agreement for the sapropel unit, i.e., 4000 years duration. A 2000 years discrepancy exists for the coccolith unit (1000 versus 3000 years). This is attributed to the dilution of recent carbon by ^{14}C older carbon eroded from the surrounding tributary area.

Noteworthy is the fluctuation in rates of sedimentation across the profile (Fig. 3). The plot illustrates (i) an increasing sedimentation velocity over the past 1500 years and (ii) a very noisy rate pattern. In contrast, rates observed in the sapropel unit are lower by a factor of three and rather constant. The onset of the drastic increase in rate of deposition appears to be connected to the advance of man and agriculture in the tributary regions of the Black Sea. Deforestation caused a gradual change in the direction of a "Kultursteppe" (Degens et al. 1976).

The alternating dark and light layers are interpreted as a reflection of the annual biological cycle. The light layer forms during the spring bloom of the coccolithophoridae, while the dark laminae correspond to the sedimentation of terrigenous material and organic debris which slowly settle throughout the year.

Cores taken by the GLOMAR CHALLENGER in the course of the Deep Sea Drilling Project reveal a multitude of differently varved sections: they are described and interpreted elsewhere (Degens et al. 1978b).

3 VARVES IN LAKE VAN AND THEIR SIGNIFICANCE FOR PALEOCLIMATOLOGY

Lake Van is situated in East Anatolia/Turkey. Besides being the world's fourth largest closed lake (i.e., it has no outlet other than evaporation) by volume (see Table 1) it is the largest soda lake on earth with a salt content of 21°/oo, of which 8.3°/oo are either Na_2CO_3 or $NaHCO_3$ (Kempe 1977, Kempe et al. 1977).

In 1974, nine sediment cores were taken from various places throughout the Lake Van basin (Fig. 1). These cores proved to be valuable records in assessing the lake's past (Kempe 1977, Kempe and Degens 1978).

The recovered sediment section (Fig. 4) is quite colorful and displays, after oxidizing, a variety of shades between brown, redbrown, green, grey and white. Intercorrelation between the cores is rapidly established by marker layers, i.e., 19 major ash-layers and series of redbrown markers in the top and bottom sections of the longest cores (St 2 and 3). In most cores, the sediment is finely layered, with the exception of the lower parts in cores from somewhat more shallower waters (St 1, 10, 14). These unlaminated series must have formed under the influence of the epilimnion diurnal convection, which appears to be strong enough to destroy the sediment structure. The beginning of the varved section above these series mark a stage when the lake level has reached a stand of about 50 m above the bottom, which at present is the depth of the epilimnion-thermocline in the lake in late summer.

The tedious work of counting the varves produced not only varve model ages for a number of environmental incidents, but also a detailed record of the variation in the rate of sedimentation (Fig. 5). The longest core (core St 3, Fig. 4) is a complete section of the Holocene, just reaching late glacial times (8450 B.C.). The sedimentation rate curve shows marked alterations during the past 10 000 years: again there is a pronounced increase in rate of sedimentation in recent times possibly due to the advance of agriculture. Over the past 800 years, sedimentation rate rose to 18 cm/100 years, while before the rate stayed at 5 cm/100 years for several thousand years. Around 5000 and around 8000 B.C. rates were in the order of 5 to 15 cm/100 years. From this data and the complete documentation of Lake Van expedition (Degens and Kurtman 1978), the following scenario emerges:

During glacial times, lake level stood 70 m higher than today, as evidenced by coastal terraces. Onset of a warmer and dry climate at the beginning of the Holocene caused the lake level to drop by as much as 340m below its present stand. With the advent of a more moist climate it started rising again to its present mark around 6000 to 6500 B.P. (Fig. 6).

The varves contain more information than just a reconstruction of the overall climatic history of that region. They reveal general weather conditions of every single year and that for the past 10 000 years. With longer cores, this record might even be extended far into glacial times. Gener-

ally, Lake Van varves consist of two layers, a lighter and a darker one. This pattern often changes to a cycle having up to five layers per year. Two of these five may be rather thin and dark; possibly, they are residues of intensive plankton blooms in spring and autumn. Apart from these biologically induced patterns, the layering itself seems to be principally influenced by the hydrological cycle in the Van area. From December through February, air temperatures stay below 0° C, and rivers are fed during this time only by Ca-rich discharges of groundwater. These waters, upon reaching the lake, precipitate pure aragonite. In addition, in February and March the whole water column of the lake is completely turned over, thus spreading a white aragonite winter stratum very evenly across the lake basin. Melting of the snow cover in the surrounding high mountains, and with the advent of spring precipitation in April and May, rivers increase their erosional activity and supply terrigenous detritus which forms a darker band on top of the white layer. It is this darker band which is responsible for the year to year variation in varve thickness. This suggests a very close relationship between runoff volume and varve thickness.

It was found that the lake level fluctuates in close correlation with sunspot activity (Kempe 1977) (Fig. 7). Increase in sunspot activity will cause a change in the overall water balance of the lake, that is, more rain fall and consequently more river discharge. This process lags about one year behind the sunspot cycle (Fig. 7).

In Fig. 8 the thicknesses of 100 varves from around 7000 B.C. are plotted, and a 10-year periodicity comes to light. This is supported by Fourier analysis which reveals that the dominant frequence has a period length of ten years (for comparison: the modern cycle has 10.7 years). These findings prove the interdependence of the system: sedimentation-erosion-water balance-sunspot activity.

4 SUMMARY

Varves are very useful tools for age assignments and reconstructions of past climatic conditions. In the Black Sea varves have helped to better understand the Holocene history of this unique body of water; and in Lake Van the climatic history of this region over the past 10 000 years could be elucidated. Especially striking is the recording of the early Holocene sun activity by means of varve thickness variation in Lake Van. The introduction of widespread agriculture seems in both study areas to result in an increase in erosion several times its former average.

5 REFERENCES

Degens, E.T. & D.A. Ross 1974, The Black Sea - Geology, Chemistry, and Biology. - Amer. Ass. Petrol. Geol. Mem. 20: 1-633.

Degens, E.T. & F. Kurtman 1978, The Geology of Lake Van. - Maden Tetkik ve Arama 169: 1-158. Ankara.

Degens, E.T., A. Paluska & E. Eriksson 1976, Rates of soil erosion. - In B.H. Svensson & R. Söderlund (eds.), Nitrogen, phosphorus and sulphur-global cycles. SCOPE Report 7, Ecol. Bull. 22: 185-191. Stockholm.

Degens, E.T., W. Michaelis, K. Mopper & S. Kempe 1978a, Warven-Chronologie holozäner Sedimente des Schwarzen Meeres. - Neues Jb. Geol. Paläont. Mb. (in preparation).

Degens, E.T., P. Stoffers, S. Golubić & M.D. Dickman 1978b, Varve chronology: estimated rates of sedimentation in the Black Sea deep basin. - Initial Reports of the Deep Sea Drilling Project, Vol. 42, Part 2: 499-508.

Fonselius, S.H. 1974, Phosphorus in Black Sea. - In: E.T. Degens & D.A. Ross (eds.), The Black Sea-Geology, Chemistry and Biology. - Amer. Ass. Petrol. Geol. Mem. 20: 144-150.

Kempe, S. 1977, Hydrographie, Warven-Chronologie und organische Geochemie des Van Sees, Ost-Türkei. - Mitt. Geol. Paläont. Inst. Univ. Hamburg 47: 125-228.

Kempe, S. & E.T. Degens 1978, Lake Van varve record: The past 10420 years. - In E.T. Degens & F. Kurtman (eds.), The Geology of Lake Van, Maden Tetkik ve Arama 169: 56-63. Ankara.

Kempe, S., F. Khoo & Y. Gürleyik 1978, Hydrography of Lake Van and its drainage area. - In E.T. Degens & F. Kurtman (eds.), The Geology of Lake Van, Maden Tetkik ve Arama 169: 30-44. Ankara.

Ross, D.A., Uchupi, E., Prada, K.E. & McIlvaine, J.C. 1974, Bathymetry and microtopography of Black Sea. - In: E.T. Degens & D.A. Ross (eds.), The Black Sea - Geology, Chemistry and Biology. Amer. Ass. Petrol. Geol. Mem. 20: 1-10.

Zeist, W. van & H. Woldring 1978, A pollen profile from Lake Van; a preliminary report. - In E.T. Degens & F. Kurtman (eds.), The Geology of Lake Van, Maden Tetkik ve Arama 169: 115-123. Ankara.

Table 1. Physical-chemical parameters of the Black Sea and Lake Van

	Black Sea	Lake Van****
Surface (km^2)	423 000*	3 574
Volume (km^3)	534 000*	607
Maximum depth measured (m)	2 206*	457
River inflow (km^3)	320**	2.5
Inflow from Mediterranian Sea (km^3)	200**	---
Precipitation (km^3)	230**	1.72
Evaporation (km^3)	350**	4.2
Outflow to Mediterranian Sea (km^3)	400**	---
Tributary area (km^2)	190 700***	12 520

*Ross et al. (1974), **Fonselius (1974), ***Degens et al. (1976), ****Kempe (1977)

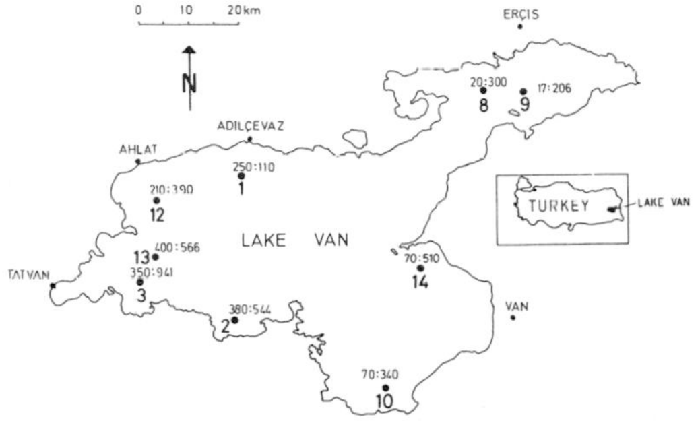

Figure 1: Sketch map of Lake Van, Turkey. Bold figures: station numbers; small figures: water depth (m), core length (cm); (after Kempe 1977)

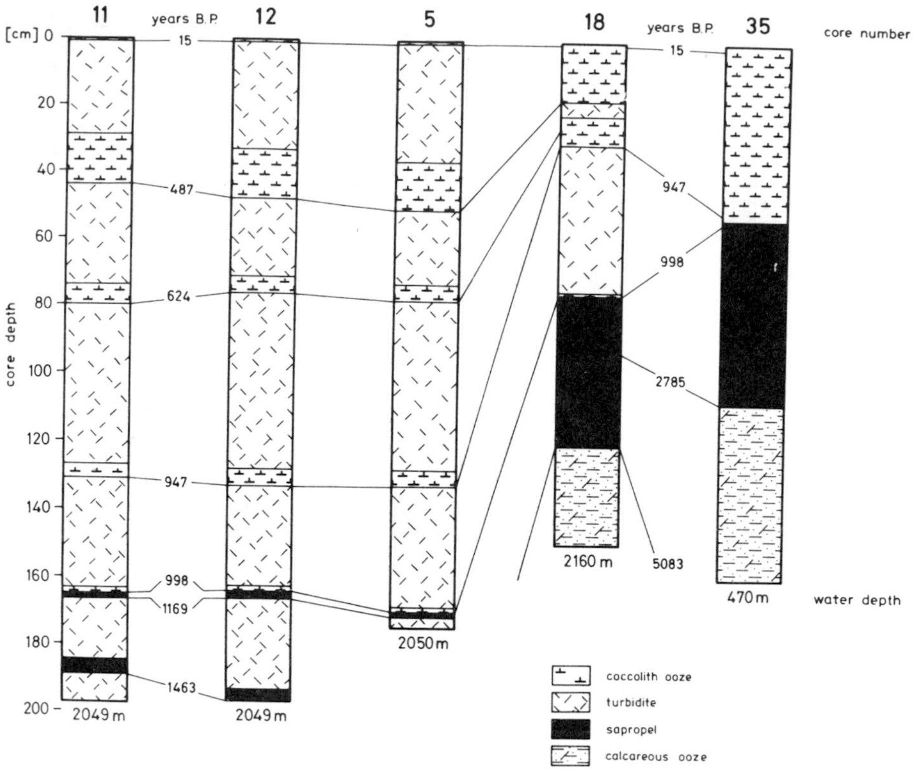

Figure 2: Stratigraphy of the upper meters in various cores from the Black Sea (from Degens & Ross 1978)

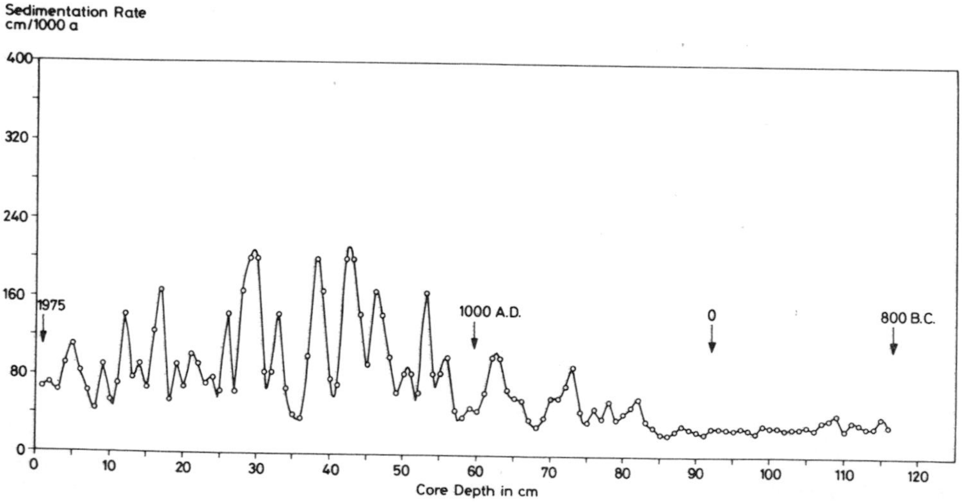

Figure 3: Sedimentation rate curve, core 35, Black Sea (from Degens & Ross 1978)

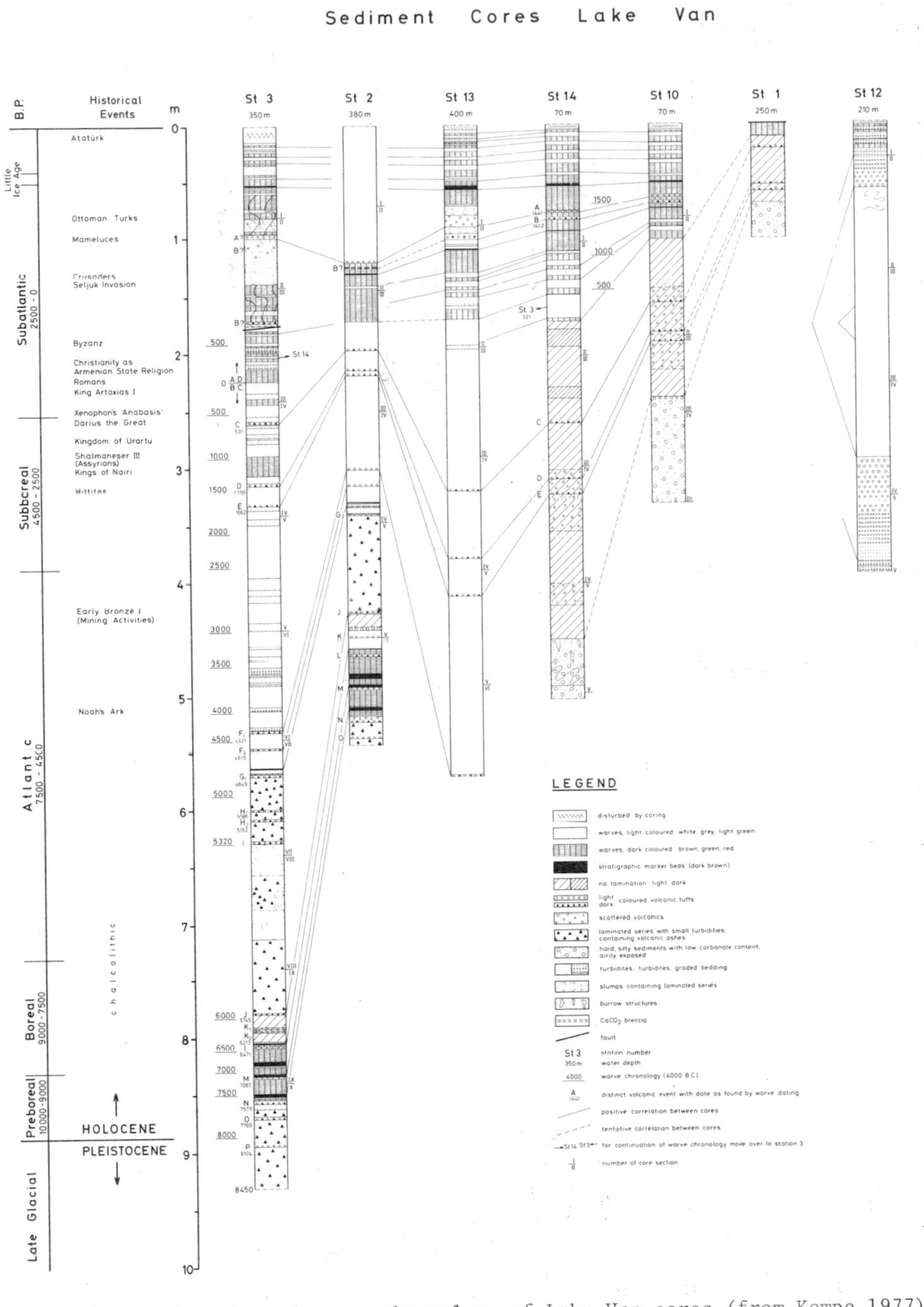

Figure 4: Stratigraphy and varve-chronology of Lake Van cores (from Kempe 1977)

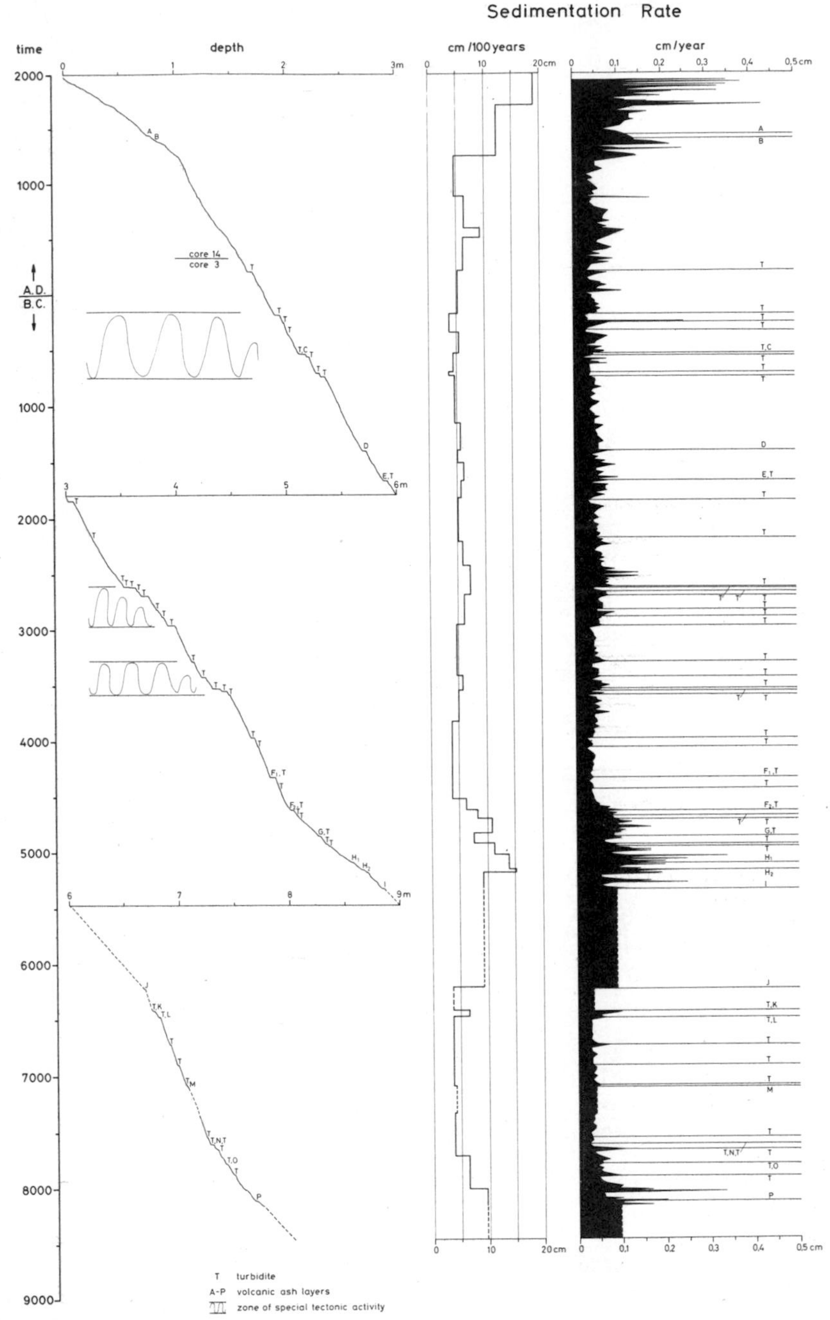

Figure 5: Sedimentation rate curve for Lake Van cores 14 (top) and 3 (bottom) (from Kempe 1977)

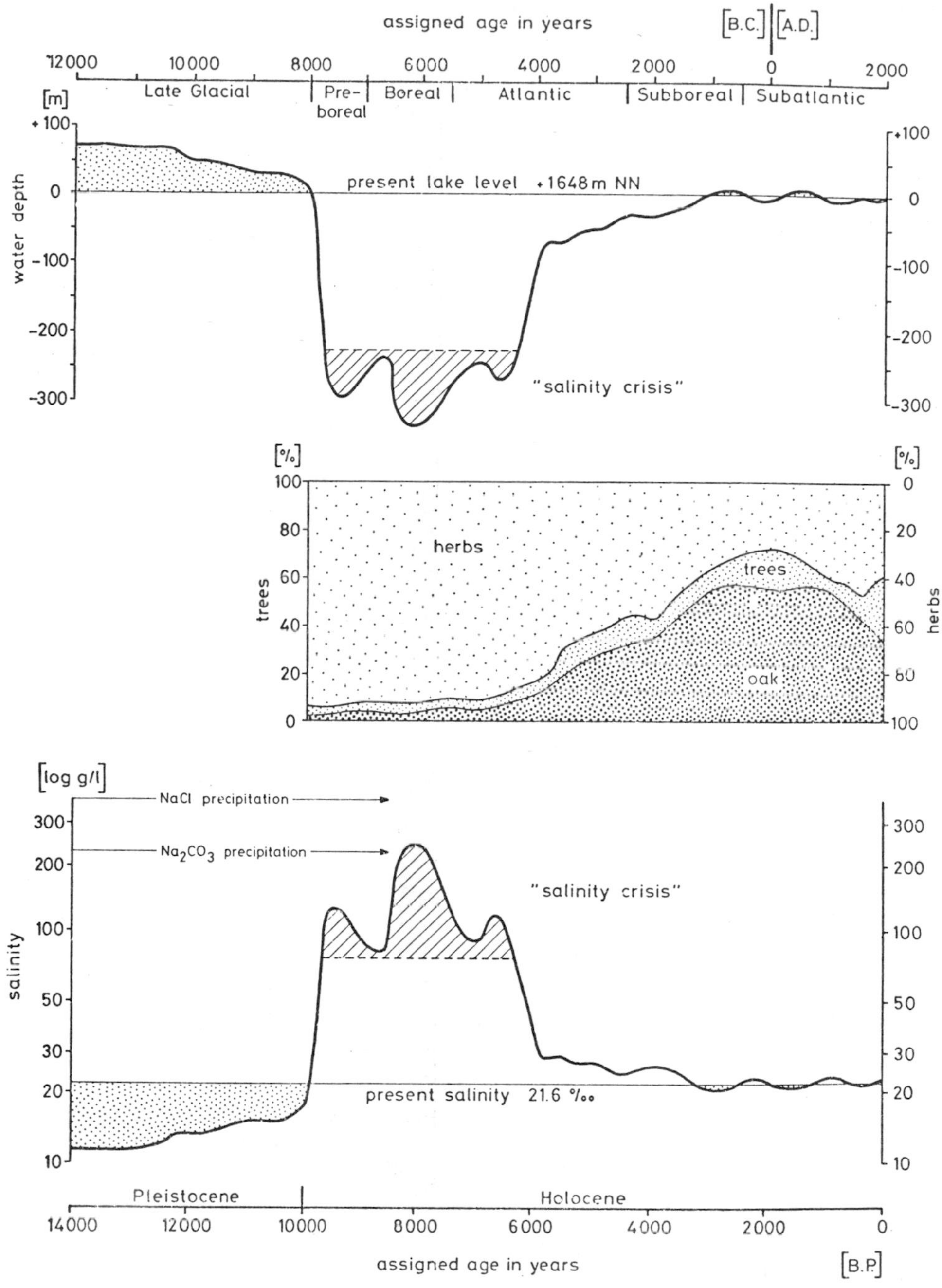

Figure 6: Lake Van level fluctuations in the Holocene (top graph). The comparison with the pollen record clearly shows (middle graph) the late change to a more moist climate in the middle Holocene. Despite the enormous regression in lake volume, the salts were not concentrated enough to form salt deposits (bottom graph) (from Degens & Kurtman 1978)

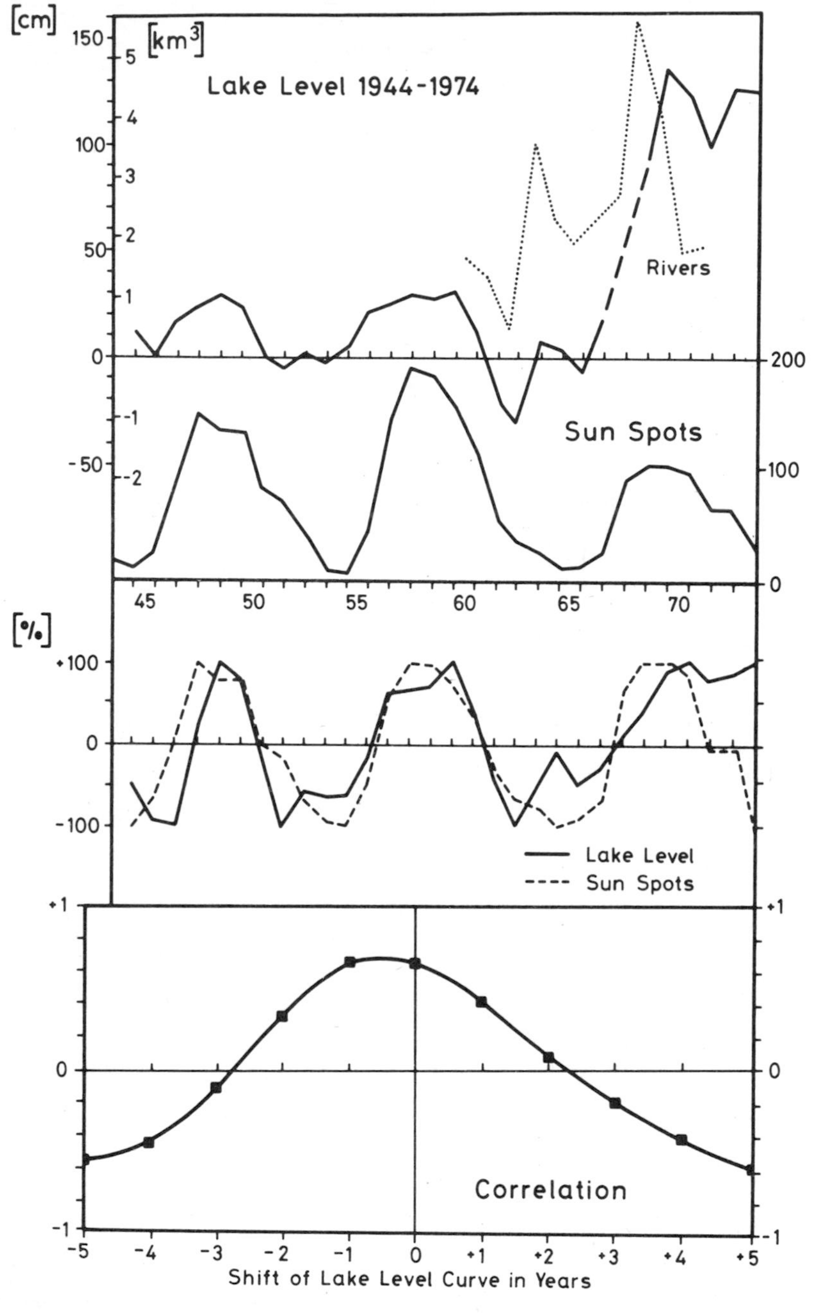

Figure 7: Correlation of recent lake level fluctuations with sunspot activity (from Kempe 1977)

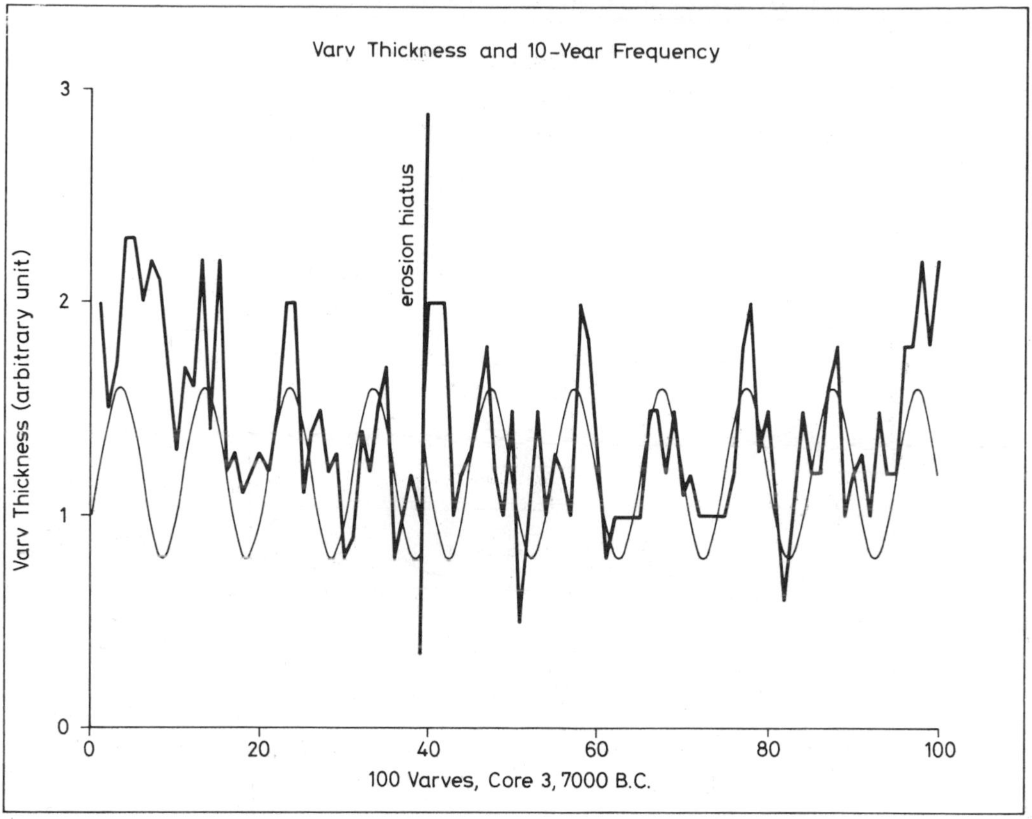

Figure 8: Varve thickness of 100 varves from Lake Van. Note the 10-year frequency; as an optical aid a sinuscurve with a 10-year period is given

Varve-chronologies and their teleconnections, 14000-750 B.C.

D.J.SCHOVE
St David's College, Beckenham, Kent, England

1 PREPARATION OF DATA

Varve-chronologies can be accurately dated and it will be possible to bridge the gap between Holocene varves and tree-rings c 6000 B.C. and glacial varves c 7500 B.C. The recommended procedure forms the topic of this paper.

Charts of varves or tree-rings cannot be matched by skeleton plots or visual comparison except in particular localities, although this method is useful for glacial varves in the west-east direction parallel to the ice-front. For long-distance comparisons we must search for and isolate those non-random elements found in weather to-day. Cross-correlation with other varves or tree-rings can then be applied.

Late Glacial varves in North America and Europe through a 3900-year period have thus (Schove 1969,1971,1978) been fitted into a coherent transatlantic chronology, but this is still a floating one fixed only in Radiocarbon or b.p. time (c 12450/8250 b.p.).

Holocene varves can be counted back from the present-day, but the series available at present are not accurately counted or dated; the Lake Saki varves (USSR) have been redated here and placed in the absolute B.C. time-scale provided by the bristlecone pines (USA).

Collaboration is needed so that long series can be constructed, replicated and cross-dated with tree-rings especially in the period of the missing millenium, 6300-8250 tree-ring years B.C.

1.1 Primary Series

Time series of measurements should, where-ever possible, be made of the thickness not only of each complete varve but also of each part of the couplet (e.g. winter clay and summer silt) separately.

Peculiar colours or structural features of unusual varves should be noted. X-ray densitometer plots may prove essential for correct numbering.

Replicated series should then be cross-dated against one another by using skeleton plots or the standard tree-ring programs as developed in Belfast and Hamburg. The measurements can be normalized so that the standard deviation is $\pm$ 1 and the (running) mean is zero. However, proximal (summer) and distal (mainly winter) varves are best made into separate series, and for some parameters (e.g. turning points of thirty-year running means) the original data may be preferable.

The standardized series are amalgamated into regional means and plotted on the usual scales of·2 years per cm and 10 years per inch. The author would be grateful for any charts of long series in these forms.

1.2 Secondary Series

From a primary series a, b, c ...... secondary series can be obtained to highlight non-random elements. An 'Oddness Index' useful for regional teleconnections is derived from first differences of ring widths with alternate signs reversed (e.g. b - a, b - c, d - c, d - e) and then using running means in groups of two (Schove 1974: 167). This series is A, B, C ...... where

$$B = \frac{2c - b - d}{2}$$

A Biennial Index for inter-continental teleconnections is obtained by using

three-year moving means of the squares of this index:

$$\beta = \frac{A^2 + B^2 + C^2}{3}$$

This Biennial Series β, γ, σ reflects the amplitude (linked with sunspot activity) but not the phase of the Biennial Oscillation.

Alternatively, it may suffice to note the central dates of biennial cycles which are characteristic of specific decades (e.g. the 1530s A.D.) or centuries. Triennial cycles should also be noted, as these are often simultaneous in different continents, as in the period A.D. 1880/1910 when sunspot cycles were weak.

1.3 Other Non-random features

Other useful parameters (Shove, 1971a) include the turning-point years of ten- and five-year means. The moving five-year means reflect what I term the pressure parameter effect (Shove 1976); this is the tendency for air to flow from regions where (in the season in question) the atmosphere bulges (e.g. the Upper High over the Indian Ocean) to regions of the Upper Troughs (e.g. the Davis Strait). A clue to the phase of the eleven-year sunspot cycles may sometimes be derived indirectly in this way, but statisticians have shown that running means give rise to spurious cycles and five-year running means yield pseudo-cycles of 10-15 years. Years of very wide and very narrow rings are noted in the 'Decile' test. Years of sudden changes up or down with phases of abnormal variability or complacancy are useful especially in lower latitudes.

The central years of the abnormal features should be noted and each abnormality weighted on a scale - 4 to + 4, ± 1 being regarded as loosely corresponding to the standard deviation. They are conveniently plotted on scales of 10 years and of 100 years to 1 cm as skeleton plots. Cross-correlation with all possible lags usually yields the correct position of any extensive overlap between two series.

Curves can be drawn freehand, and in particular curves for the biennial versus triennial tendencies may make it possible to obtain approximate synchronization visually without further arithmetic. The Swedish and North American glacial varves were cross-dated first in this way. Some abnormalities are nevertheless time-transgressive: two-year cycles thus occur (in the period A.D. 800 - 1978) about 11 years earlier in German summer-sensitive rings and in the Nile Flood levels they occur several years earlier still.

2 LATE GLACIAL

Code-letters used for the algebra of well-known glacial varves are as follows:

D N. Sweden (Döviken zero)
V Central Sweden (Swedish dates f.Kr)
N South Sweden (Nilsson's f.Kr. dates)
F Finland (Relative to Sauramo's zero)
C Canada (Timiskaming)
Q Quebec
Br Bracebridge-Huntsville
A S. New England
B N. New England

2.1 Trans-Baltic Solutions

Swedish varves have usually been assigned a year 'before Christ' or 'f.Kr.' on a Swedish time-scale (N or V). Finnish dates (F) differ from such negative years by a number of years which varies from 8226 in the Late Glacial (F - N) to 8140 in the Early Holocene (F - V). The errors appear to lie in the Swedish rather than the Finnish counts, and no corrections to Sauramo's numeration have yet been reported by Niemala.

2.2 Trans-Atlantic Solutions

Wide and narrow proximal varves often correspond to warm and cold summers, but warm seasons in eastern Canada more often correspond to cold seasons in north-west Europe so that the year-to-year tendency (especially in distal varves where the winter portion is more important) may be in opposition on the two sides of the Atlantic.

The Finnish and Canadian series in the Early Holocene (Pollen-zones IV and V) were thus found to be related by the formula (varying according to century):

F - C = 52 to 63

The writer's main Canadian series is built up especially from the Timiskaming curves of Antevs (1925). In relation to the zero of the Timiskaming series, Antevs' Quebec series (from Maniwaki, north of Ottawa 47°N 76°W) was found to be 83 years earlier and the Espanola series (from north-east of Lake Huron 47°N 82°W) was found to be first 78 (for Espanola 3/4) and later (for Espanola 5, separated by an unconformity) 59 years earlier.

The main Canadian series thus extends from - 83 to about + 2000 and was shown to correspond with the period c 8175 to c 6063 'B.C.' (f.Kr.) in the N. Swedish varve-chronology (V). This is known to correspond to radiocarbon dates c 10300/ 8250 b.p. or c 8350/6300 b.c. It is expected to correspond with tree-ring dates 9350/7250 $\pm$ 200 B.C. on the extrapolated bristlecone scale.

Current dates in N. Sweden are linked to the Döviken zero (6923 f.Kr. on the V-scale or about 9050 b.p. on the C14 scale) and this year seems to correspond to year 1166 $\pm$ 1 in the Canadian series of Antevs (1925) and Hughes (1965).

No satisfactory series of varves for the cold period of European pollen-zone III could be identified for north America, but the zero of Antevs' 'Burke Falls' series ($45\frac{1}{2}^{0}$ N $79\frac{2}{3}^{0}$ W) is about 251 years older than the first set of the Espanola series (46°N 82°W) which together constitute a chronology that is not yet linked with Europe. An approximate C14 date is needed.

The Huntsville and Bracebridge series from east of Lake Huron (45°N 79°W) were on the other hand teleconnected first with one another (the Huntsville zero is 187/ 186 years later than that of Bracebridge), and the combined series was cross-dated with the U.S.A. (B) series for northern New England, the postulated zero of this is c 6939 $\pm$ 5 years earlier than that of Bracebridge. Cross-dating with the Baltic was made for each of these series, which were shown to commence in the Boelling period and to continue in the Alleroed, thus corresponding to the Two Creeks period in the wide sense. These years were connected with the conventional Swedish varve-years (N in pollen-zones Ib and II) by the formula:

$$Br - N = 10,030 \pm 4$$

The United States varves were discussed separately (Schove 1969) but no link with Europe has yet been found for the A-series. New series from Poland, The USSR and Esthonia are needed for this.

3. HOLOCENE

Glacial varve series are now being counted backwards from the present day, notably at Wisconsin and in Finland, using new techniques of obtaining cores from modern lakes which freeze every winter. Varve-structure is preserved by refreezing after coring.

Other types of varves are being studied and are found even at the sea bottom off the California coast and in the Red Sea. The ice-layers found in Greenland and Antartica have annual layers which are being dated by tephrochronology in Denmark back at least to A.D. 1783.

The well-known varves of Faulensee, Switzerland and of Yugoslavia need to be re-investigated as the dates originally assigned to them were incorrect.

Some varves have been counted but not yet measured. The Lake of the Clouds varves are thus too thin for measurements and in any case are not weather-sensitive. German varves studied at Schleinsee, S.W. Germany by Geyh and his collaborators (1971) will be of special interest as they cover the period 3250/ 7200 B.C. and almost span the missing millenium. Measurements would be invaluable.

The chronology of Holocene varves has nowhere been firmly established. This applies to the best-known series from Lake Saki in the Crimea, USSR; the methods adopted for glacial teleconnections were used to find the correction necessary at c 1500/1000 B.C.

3.1 Lake Saki varves since 2000 B.C. (d)

The silt deposits of the western shore of Lake Saki, Crimea, USSR were measured and counted by Perfilev in 1929. The mineral portion depended especially on precipitation and the micro-organisms in the hot season affected the silt deposited so that one varve could be distinguished from its neighbours. Annual measurements were published by Schostakovitch (1934).

One possible dating check is the saw-tooth or zigzag signature found in the 1530s in tree-rings in Germany and the USSR as far east as Siberia (and also in vintages in France). Certainly it occurs in the Saki varves at the right decade and Dr S. Eddy tells me that the counting may have been correct back to Roman times. In our algebra this series is termed d.

3.2 Tree-ring series c 2000 - 750 B.C.

Four tree-ring series are available for comparison through the first third of the Saki period:

a. USA: Bristlecone Pine 5142 B.C.+..
b. Ireland: Oak c 4000/1000 B.C.
c. Asia Minor: Juniper c 1600/750 B.C.
e. Central Europe: Oak c 2000/1000 B.C.

The first of these series was dated in absolute chronology (Ferguson 1969) and the last series (Becker) is not yet published.

The Irish and Asia Minor series have meanwhile been cross-dated with the Bristlecone series using the author's methods described above.

3.3 Asia Minor Tree-Rings 1562 - 767 B.C.(c)

Archaeological timber from the Hittite and Archaic period at Gordion, Turkey, has been used (Kuniholm 1977) for a coherent floating chronology by Kuniholm. These ring-widths reflect especially winter and spring rainfall, and were therefore expected to cross-date the Lake Saki series from the opposite side of the Black Sea. The correct dates were found by tests which included direct comparison with the Bristlecone (α) Series, as the curve of biennial and triennial cycles showed good agreement.

In the Decile Test a lag of 5470 gave 51 points with the same signs in extreme years; a lag of 5469 also gave a non-random result with 19 similar and 51 opposite signs. This is because the preceding year's weather also affects the Asia Minor trees. The year numbered as 1000 (estimated archaeologically by Kuniholm as c 1540 B.C.) was therefore dated (Schove in progress) 1531 B.C. and the series was dated 1562 - 767 B.C.

3.4 Irish Tree-Rings c 3980 - 1070 B.C.(b)

This important series has been published (Pilcher et al. 1978) as a floating chronology estimated as c 3998 - 1088 $\pm$ 10 B.C. (dated by comparison with Bristlecone C14 dates). This series had a higher serial correlation than the bristlecone and did not readily yield a more precise direct cross-dating. Cross-dating was nevertheless eventually possible with the Saki varves (d) using e.g. dates of maxima and minima of five-year means.

3.5 Central Europe

Professor Becker's tree-ring series are not yet published but are mentioned because of their potential value for dating European varves at Faulensee, Lake Saki and elsewhere.

3.6 Interrelationships

Lake Saki curves (d) were at first assumed to be within a century of the correct dates but this failed to lead to any cross-dating. Wide-aperture tests (Biennial, Triennial etc.) with the other series on the assumption that the error was less than 700 years led to the conclusion that the error lay between 100 and 200 years and narrow-aperture tests eventually led to solutions 145/153 years in the period c 1600/1000 B.C. Through most of this period the assumption of a 145-year error made sense of visual matching with the 30-year and 5-year running means of Asia Minor curves. If the Saki 'B.C.' date is regarded as 'minus d' the following formulae applied:

$$b - d = 2676/2684$$
$$a - d = 8149 \pm 5$$
$$c - d = 8130 \pm 5$$

The tree-ring series were related as follows:

$$a - b = 5470$$
$$c - b = c5448 \pm 2$$
$$a - c = 22 \pm 2 \text{ (Partly inferred)}$$

3.7 Tephrochronological prospects

The Lake Saki varves would repay further investigation in this period, and detection of the year with Santorini dust would yield the date of the Thera eruption in the Aegean sea. Babylonian observations of Venus were missing in a year regarded as 1635 B.C. and the hypothesis of Weir that there was an eruption in that year would be tested.

3.8 Climatic implications c 1500/1000 B.C.

The several series examined above suggest:

1. There was a warm moist period c 1550-1300 B.C. but occasionally prolonged droughts as in 1429/24, 1416/11 and c 1352 B.C. affected both sides of the Black Sea.
2. Dryness prevailed in the Black Sea area as the global temperature fell in the 13th century B.C., and severe droughts occasionally affected both the Saki varves and the Asia Minor tree-rings (e.g. c 1295, c 1265).
3. Moist (but globally colder) conditions returned in the period c 1220/1140 B.C., notably in the period c 1200/1190 conventionally associated with the end of the Bronze Age and the Trojan Wars.
4. Further periods of drought in N.W. Asia Minor occurred c 1150, c 1130, 960s, 924/1, 888, 845/839 B.C. The climate was globally cold to c 950 B.C. and then became warmer, and the last of these droughts synchronized with the period of Ahab in Palestine.

3.9 Future work

Collaboration in the collection, replication and dating of long series of varves and tree-rings, especially through the missing millenium c 8250/9200 B.P. or c 6300/7250 B.C. is requested. The approximate substade (Schove 1978a) or radiocarbon century and palaeomagnetism of any floating chronology provides the first clues to the dating and Oddness and Biennial indices will normally lead to an approximate dating. For precise dates it is helpful to know the meteorological factors and seasons involved and if modern varves are available for comparison 'Response Functions' as used in tree-ring analysis should be calculated.

REFERENCES

Antevs, E. 1922, The recession of the Last Ice Sheet in New England, Amer. Geogr. Research Series No. 11 (and other papers).

Ferguson, C.W. 1969, A 7104-year Annual Tree-Ring Chronology, Tree-ring Bulletin 29 (3-4): 3 - 29.

Geyh, M.A., J. Merkt & H. Muller 1971, Sedimentological pollen-analytical and isotopic studies of annually laminated sediments in the central part of the Schleinsee, Germany, Arch. Hydrobiol. 69: 355-399.

Hughes, O.L. 1965, Surficial geology of part of the Cochrane District, Ontario, Canadian Geol. Soc. of Amer., Special Paper 84: 535-565.

Kuniholm, P.I. 1977, Dendrochronology at Gordion on the Anatolian Plateau, Pennsylvania Univ. Doctoral Dissertation.

Pilcher, J.R., J. Hillam, M.G.L. Baillie & G.W. Pearson 1977, A long sub-fossil oak tree-ring chronology from the north of Ireland, New Phytologist, 79: 713-729.

Schostakowitsch, W.B., Bodenablagerungen der Seen und periodische Schwankungen der Naturerscheinungen, Memoires de l'Institut Hydrologique, Leningrad, 13: 95-140.

Schove, D.J. 1969, A varve teleconnection project. In M. Ters (ed.), Etudes sur le Quaternaire dans le Monde, pp. 927-935, Paris.

Schove, D.J. 1971, Varve teleconnection across the Baltic, Geogr. Annaler, 53A: 214-234.

Schove, D.J. 1974, Dendrochronological dating of oak ... A.D. 650-906 (recte A.D. 490-746), Medieval Archaeol. 18: 165-172.

Schove, D.J. 1976, In S. Bezzaz, D. Dalby (eds.) Drought in Africa, 2. London, International African Institute.

Schove, D.J. 1978a, Tree-ring and Varve Scales combined, Palaeogeography, Pal. Pal., 25 (1978): 209 - 233.

Schove, D.J. 1978b, Tree-ring and Varve Teleconnections, c 4000 - 750 B.C. (Submitted to Nature, London).

APPENDIX

The figures below have been kindly drawn for me by H. Bolliger (EAWAG-ETH, Switzerland) and help to illustrate this paper and that which has now appeared in Pal. Pal. Pal., 25 (1978): 209-233. Progress in measuring and dating (back to the Middle Ages) ice-varves in Greenland is meanwhile reflected in papers by C.U. Hammer (cf. J. Glaciol., 20, 1978, 3-26) and varves back to 8500 B.C. have been studied by Degens and Kempe (cf. this volume).

The period for which varve or tree-ring measurements are most needed is illustrated in Figure 2. The way in which 2- and 3-year cycles can be used for first approximate synchronization is illustrated in Figure 3. Decile Tests use weighted outstanding maxima and minima from each series (cf. Schove, 1971, p. 224 and 231-2) and help in exact synchronization. A sample of very wide tree-rings in Turkey, when 5570 years were added to the Asia Minor Scale (as in Table 1), was found to synchronize with 15 wide rings (over 125) but only 5 narrow (less than 75) in the U.S.A.

Thirty year tests show sometimes good agreement as shown in Figure 5 Annual curves occasionally do, as in Figure 6.

In general, no single test has high significance but once many independent tests lead to the same correlation the results can be trusted. The correlations shown in Figure 4 are approximations covering a wider time-range than those of Table 1.

Table 1. The four scales c 1500/1000 B.C. with equivalent dates on the Bristlecone scale + 65007000.

- 1500	1400	1300	1100	- 1000	
1501 B.C.				1001 B.C.	
+6500	U.S.A. Scale			+7000	Bristlecone
c6479	Irish Scale			c6979	add 21
+1030	Asia Minor Scale			+1530	add 5570
-1645	U.S.S.R. Scale			-1145	add 8145

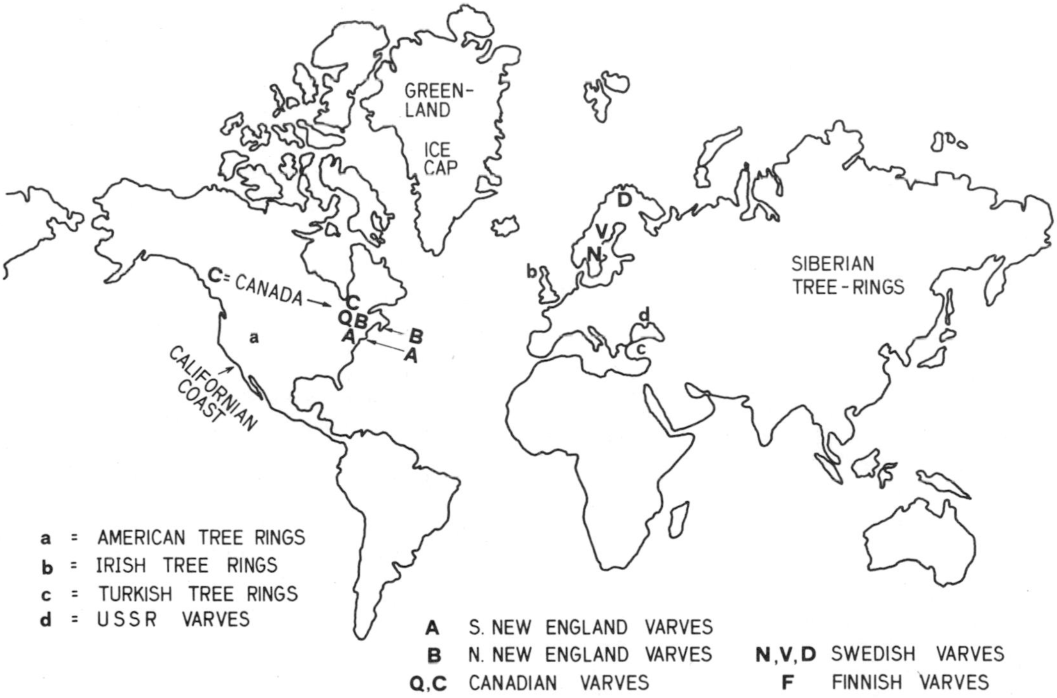

Figure 1. Map showing sites of the four scales (a to d) and of the principal Late Glacial varve series.

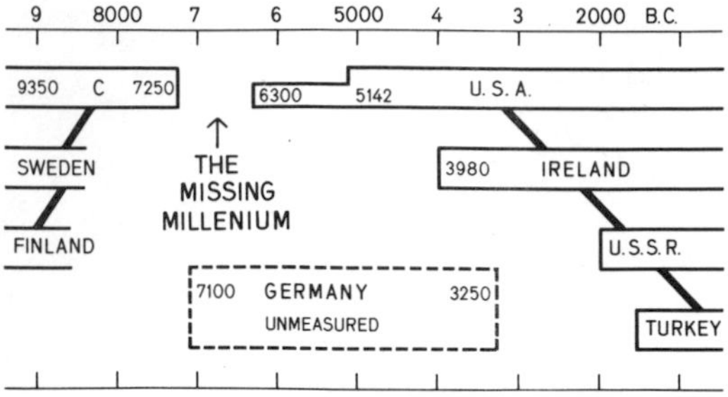

Figure 2. The gap c 7250/6300 B.C. and the S.W. German varves that may provide an overlap.

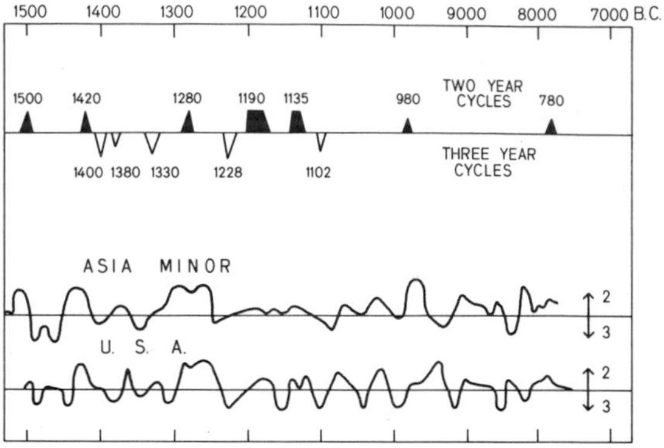

Figure 3. Two and three-year cycles common to most tree-ring and varve series, with curves for Asia Minor and the U.S.A.

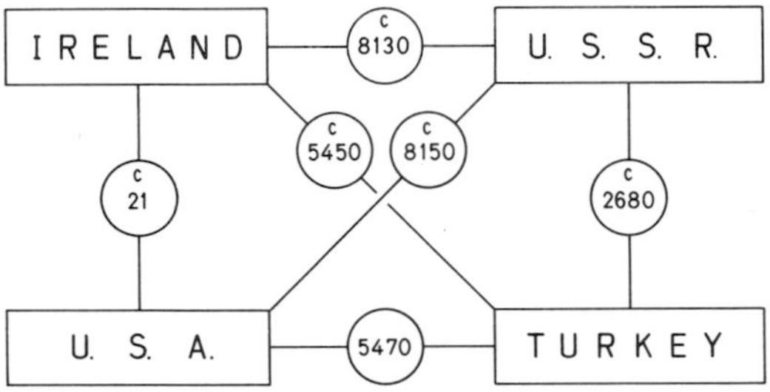

Figure 4. The approximate interrelations of the four chronologies with the differences in years between each pair.

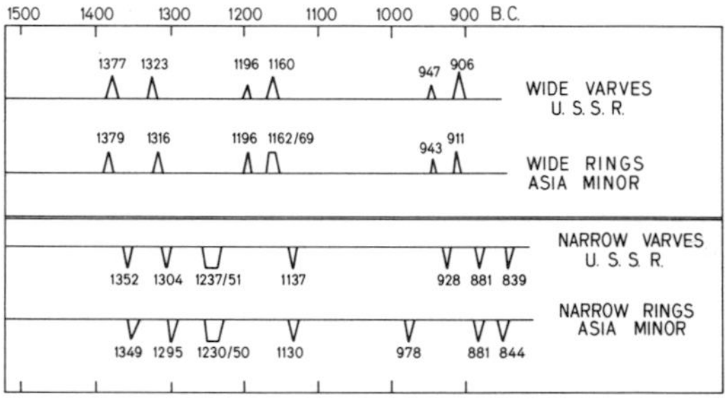

Figure 5. Central dates of thirty-year WIDE (wet) and NARROW (dry) periods common to Lake Saki varves and Asia Minor Tree-Rings.

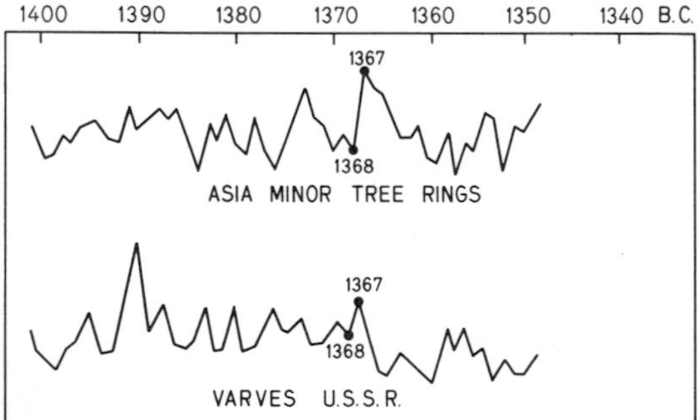

Figure 6. Tree-rings in Asia Minor and Varves in the USSR, 1400/1340 b.C. It is assumed that the error in Lake Saki varves is 145 years (Note the thirty-year wet period 1391/62 B.C. in the tree-rings, cf. Fig. 4).

Sedimentology of a tidal lake, Pitt Lake, British Columbia, Canada

GAIL M. ASHLEY
Rutgers — The State University of New Jersey, New Brunswick, N.J., USA

ABSTRACT

Pitt Lake is situated in a relict fjord within the Coast Mountains of British Columbia, 30 km inland from the port of Vancouver. The lake is freshwater, but is directly connected to the ocean via Pitt River and Fraser estuary. Although the Fraser estuary and the entire Pitt system is tidal, salt water seldom reaches to within 10 km of the Fraser-Pitt confluence. Flow reverses in response to tides, causing Pitt River to fluctuate 2 m and Pitt Lake as much as 1.2 m per tidal cycle. There is an upstream movement of sediment in Pitt River from Fraser River, evidenced by identical mineralogy of Pitt River and Fraser River sediments, a decrease in grain size from the Fraser to Pitt Lake, and a predominance of flood-oriented bedforms in the river channel. A delta of 12 km^2 area has accumulated at the lower (draining) end of the lake. Morphology of the delta is considered to be an excellent example of sediment diffusion and deposition from a simple jet into a low energy lacustrine environment. Cores in the delta topsets and lake bottom sediments reveal silt and clay rhythmites, interpreted as varves. The coarse layers are deposited during winter when discharge of Fraser River is low and tidally induced discharge in Pitt system is high. The fine layers are deposited during spring run-off when additional fines are added to the lake from the Pitt basin and the tidal effect is significantly reduced because of increased run-off. ^{137}Cs dating of sediments shows that as much as 1.8 cm/yr are accumulating in the active portions of the delta with an estimated $150 \pm 20 \times 10^3$ tonnes deposited annually.

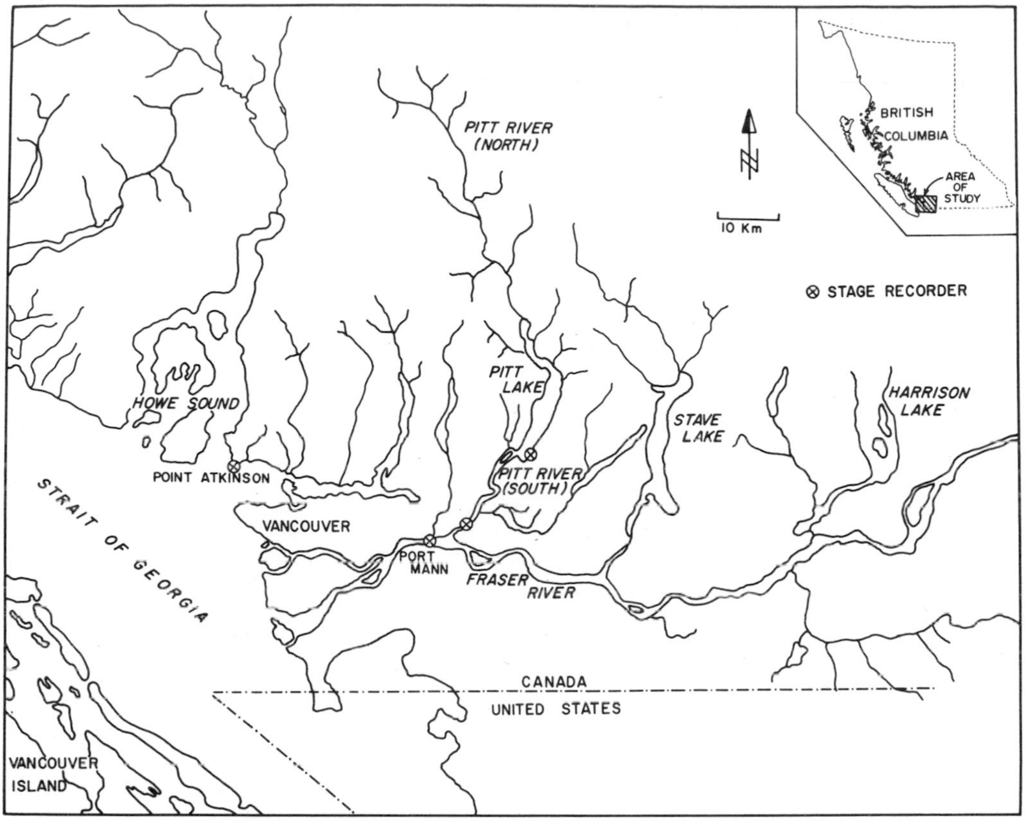

Figure 1. Location map of Pitt tidal system

INTRODUCTION

Pitt River (North) - Pitt Lake - Pitt River (South) system is situated in a glacially scoured valley within the Coast Mountains of British Columbia, Canada, approximately 30 km inland from the port of Vancouver (Fig. 1). The valley of the Pitt, 70 km in length opens abruptly into the Fraser lowland. Pitt River (North) drains 816 km^2 including several mountain glaciers and provides a mean discharge of 80 $m^3 \cdot sec^{-1}$ to the lake. A prominent sill, 5 km from northern end of lake, allows only clay-sized sediment to be carried to the lower end of the lake.

Pitt River (South) and Pitt Lake are tidal, being connected to the Pacific Ocean (Strait of Georgia) by the lower Fraser River. Although water levels in the Pitt system respond to the tides, salt water seldom extends closer than 10 km downstream of the Fraser - Pitt confluence. Rising water (flood tide) in the Strait retards flow of the Fraser and raises its elevation progressively eastward until the water level at the Fraser-Pitt confluence is higher than in Pitt River (South). Flow in the Pitt then reverses and water diverted from the Fraser flows northward up Pitt River (South) into Pitt Lake. As the water elevation falls (ebb tide) in the Strait, Fraser River flow is accelerated. The surface elevation is lowered progressively eastward until the level at the Fraser-Pitt confluence is less than that of Pitt River (South). Flow then reverses in the Pitt system and drains toward the sea. The elevation and magnitude of water level oscillations in the Pitt system are a function of the complex interaction of Pitt basin drainage, Fraser River discharge, and the tidal prism. Pitt River (North) is not included in this study and the term Pitt River will hereafter refer to Pitt River (South).

Upstream movement of sediment in Pitt River from Fraser River toward Pitt Lake is indicated by: (1) a predominance of bedforms related to flood tide in the river

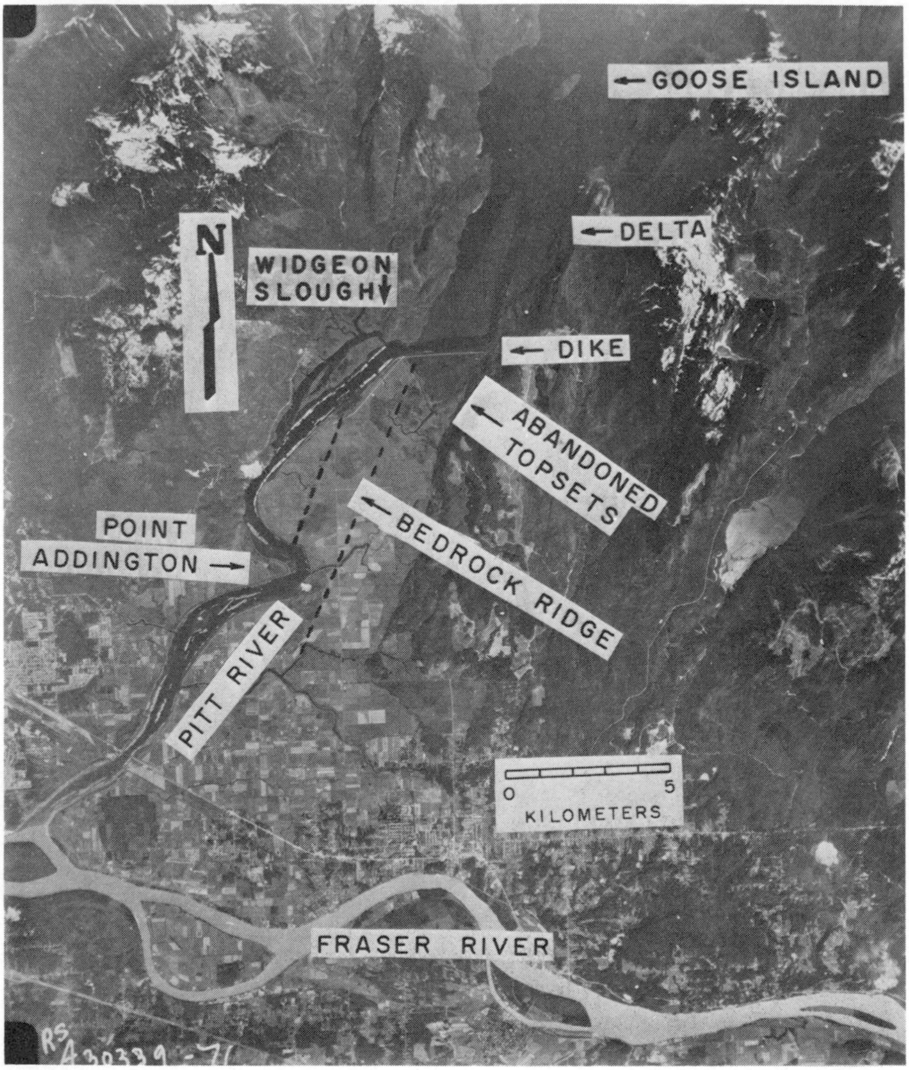

Figure 2. Aerial photo showing the main physiographic features of the lower Pitt system. The bedrock high is outlined with dashed lines. Its extent is based on isolated bedrock knobs that protrude through the flood plain.

channel; and, (2) a decrease in grain size of channel sediment from the Fraser to the lake. In addition, velocity and stage measurements demonstrate that flood flows have higher peak velocities and that flood flows persist for a shorter time period than ebb flows.

A large tidal delta with a surface area of 12 km^2, has accumulated at the southern (draining) end of the lake (Fig. 2). Because of the unusual position of the delta there has been speculation on whether it is actively growing or a relict feature from earlier post-glacial time. The purposes of the present study were threefold: first, to determine if the delta is active and to estimate the present deltaic sedimentation rate; second, to examine the hydraulics of the lake channel and to evaluate the effect of bidirectional flow on sediment dispersal and delta morphology; and third, determine processes of sediment dispersal in the lake and estimate lacustrine sedimentation rates.

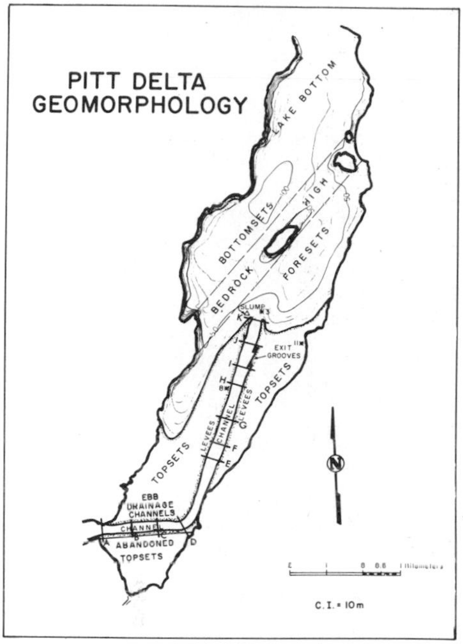

Figure 3. Geomorphology of Pitt tidal delta with lake bathymetry. Depth contour interval is 10 m. A topographic "high" on the lake bottom connects islands and the bedrock ridge bordering the southwest side of lake. This "high" is coincident with the 70 m depth contour and 6 mean grain size contour (Fig. 11) and is used as an arbitrary division between delta foresets and bottomsets - lake bottom. Cross section (A-K) locations of Figure 5 are illustrated. Cores used for ^{137}Cs dating shown by *.

GEOLOGIC HISTORY

Following the most recent deglaciation (15,000 - 11,000 B.P.) of British Columbia the melting ice left numerous elongate lakes in interior valleys along a coastline dominated by fjords. However, in early postglacial time the exact location of the shore fluctuated as a result of a complex interaction of eustatic sea level changes and crustal rebound (Mathews et al., 1970). During the period of instability, ocean waters flooded past the mouth of Pitt Valley, as is evidenced by marine shells (12,690 ± 190 B.P.; I-5959, Mathews, 1973) collected at an elevation of 107 m on the east side of Pitt valley. Isostatic uplift of the Fraser lowland began around 13,000 B.P. and was essentially complete by 8,000 B.P. (Mathews et al., 1970). Fraser River, supplied with abundant glacial sediment, rapidly constructed a delta westward and by 8,290 ± 140 B.P. (G.S.C. 229, Dyck et al., 1965) "Pitt Fjord" was isolated from the sea at its southern end by this delta. It is likely that a short tidal channel maintained a connection between the fjord and the Fraser estuary. Tidal currents flowing through this channel must have carried sediment from Fraser River into the fjord, building a flood tidal delta which continued to grow northward as Fraser delta progressed westward. By 4,645 ± 95 B.P. (I-7047; Mathews, 1972 pers. comm.) the leading edge of Pitt delta stood at least 20 km north of Fraser River near the present outlet of Pitt Lake (Fig. 1). The dated material was a log found in delta topsets 10 m north of the channel and buried under 60 cm of sediment. At some time during this period "Pitt Fjord" was flushed of saline water and became Pitt Lake; at the present time salt water is not found anywhere in lake. As the sea-land relationship has been much the same as at present since 5,500 B.P. (Mathews et al., 1970), it is possible that Pitt Lake has been in existence for approximately 6000 years.

The boundary between Pitt River flood plain and Pitt Lake tidal delta has been a transitional one throughout their development. At present, dikes and ditches have created two entities, but the division is artificial. Historically, water flow and sedimentation have been a continuum from river to lake.

During the last 4,700 years the delta front has advanced from the present lake outlet approximately 6 km north into Pitt Lake at the average rate of 1.28 m/yr. However, with the change from paraglacial to nonglacial conditions sediment supply would decrease (Ryder and Church, 1972). In addition, containment of the Fraser River within the last century may also have been important in altering sediment supply to the Pitt system. Thus this progradation rate most likely has decreased exponentially starting at meters per year and decreasing to the probable present rate of centimeters per year.

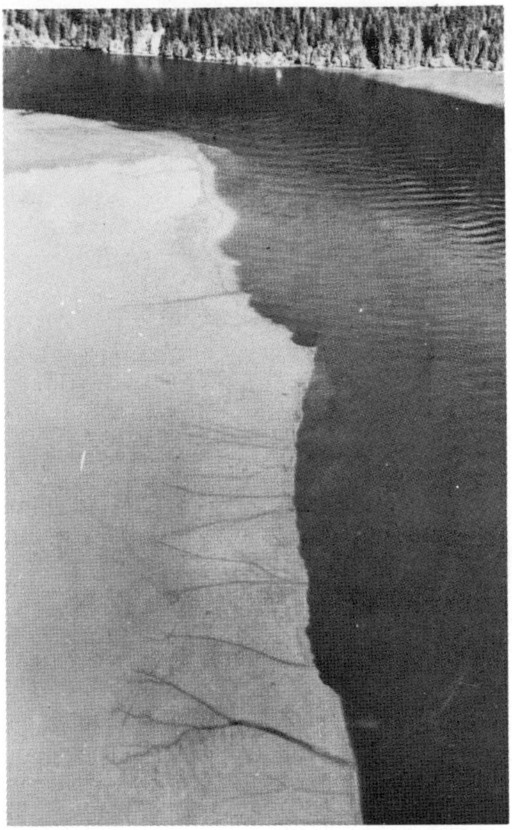

Figure 4 A. Oblique aerial photo looking east at right-angle bend. Note ebb drainage channels and scalloped margin of topsets. Wind-generated waves cover water surface.

GEOMORPHOLOGY

Delta

The present delta surface covers 12 sq. km. (5.8 km long and 2.2 km wide) and contains a single distributary channel with a right-angle bend (Fig. 3). Minor erosion in the form of a scalloped channel margin occurs near the bend (Fig. 4A) but a study of aerial photos dating back to 1940 indicates little change in 35 years. The channel is incised in the reach between the lake entrance and the bend, with nearly vertical channel banks in some places (Fig. 5). However, the channel banks gradually change from steep to gentle slopes toward the end of the delta (Fig. 5) in conjunction with a gradual shallowing of the channel. The delta surface is highest at its southern margin and slopes down (6.0×10^{-4}o) toward the topset/foreset slope break. During low water in Pitt Lake the southernmost kilometer of delta is exposed and minor southerly draining channels in a dendritic pattern are eroded into the topsets (Fig. 3; Fig. 4A) by water draining from the delta surface into the channel during ebb flow. The drainage channels are 2 - 3 m wide and 1 m deep at their widest cross section. Levees border both sides of the major delta channel (Fig. 4B) and a few minor flood exit grooves are eroded diagonally through the levees (about 1 km from end) marking crevasse splays where sediment is carried up out of the channel onto the delta surface (Fig. 3).

A longitudinal profile along the thalweg (Fig. 6) shows the lake entrance and right-angle bend to be extremely deep (30 m) whereas the reach between the deeps is more shallow (10 m) and thus probably is an area of (temporary) deposition. The section of channel from bend to delta front is a ramp which shallows from 30 m to 4 m in a gentle slope of 0.3^{o} to 0.05^{o}. The thalweg appears to have one partial meander which is similar in length (λ_M = 600 m) to meanders of Pitt River (South) (Ashley, 1977). The channel bottom projects as a wedge-shaped tongue (flanked by levees) into the lake (Fig. 3) pointing northeast in the direction of delta-front progradation (east side of island).

The delta topset surface is flat and devoid of any major topographic features with the exception of the ebb drainage channels and the levees. Occasional scour holes (0.5 m deep) on the surface reveal a horizontally stratified, highly cohesive sediment. The occurrence of nearly vertical banks bordering the delta channel supports this conclusion. The binding agent is thought to be organic in nature as little clay is present in the sediment. Isoetes echinospora dur. (quillwort), is ubiquitous on the delta surface. Roots of this plant are thin, white, threadlike filaments which do not decompose readily: they were found in all surface samples. It is interpreted that this macrophyte is an important agent in binding the sediment. Other macrophytes (Scirpus validus Vah., Myriophyllum hippuroides Nutt., and Potamogeton sp.) also populate the flats. Other possibilities for cementation are micro-organisms. Blue-green algae were found to be rare whereas diatoms are relatively abundant. Mucus produced by the diatoms may act as a temporary binding

Figure 4 B. West side delta topsets. Levees can be seen bordering topset margin in the foreground. Boat wake (30 m) for scale.

agent for the delta sediments. Unio clams, spaced 1 m apart, occur on almost the entire surface and appear to feed at the sediment/water interface, causing little noticeable bioturbation in the underlying sediments.

Lake bottom

The foreset slope (Fig. 3) was examined by depth sounding. The contact between topsets and foresets is a sharp break in slope and is outlined in the figure. The foreset/bottomset contact is a gradational change in slope from $4^{\circ} - 1^{\circ}$ to an essentially horizontal surface and is positioned arbitrarily at the bedrock ridge (Fig. 3). This ridge line approximates the 70 m depth contour and is also coincident with the mean grain size contour of 6 ϕ. Delta foresets range in slope from $1^{\circ} - 2^{\circ}$ near the east shore to $4^{\circ} - 6^{\circ}$ at the end of the main channel and the "side-sets" bordering the western embayment have a slope of $10^{\circ} - 20^{\circ}$. In general, the entire foreset-bottomset slope is gentler to the east side of Goose Island than to the west and the general shape of the delta indicates that sedimentation has occurred consistently on the east side of the lake. The foresets are generally smooth with a gentle concave upward profile. Only one slump feature (Fig.3) was noted on the entire foreset apron, indicating a relatively stable slope. The slump has a relief of about 3 meters and occurs just west of the fan created at the end of the distributary channel. Seismic data from the lake (Mathews, unpublished data, 1976) indicates that a topographic high (Fig. 3) exists between the bedrock ridge (bordering the southwest side of lake) and Goose Island essentially splitting the lower end of the lake into two basins.

Based on its geomorphology, the Pitt Delta can best be described as a single talon of a birdfoot delta which has been welded to the eastern lake shore. A depositional model for the birdfoot delta includes progradation of the distriburary channel into relatively deep water (Scruton, 1960). Sediment is conveyed along the delta channel and is brought out periodically and deposited on the delta surface as levees. Along each side of the major channel are interdistributary troughs which slowly fill with fine-grained sediment. As the distributary channel extends into standing water, it broadens, becomes more shallow and gradually loses its identity (Reineck and Singh, 1975). The Pitt appears to fit this general model. It is clearly dominated by fluvial processes. The depositional environment is one of very low energy and little reworking of the

fluviatile sediments occurs by waves or tidal currents.

Bed configurations in the channel

In conjunction with the study of Pitt Delta geomorphology an examination was made of the bed configuration of the delta channel bottom. Soundings were made over an 18-month period, under both ebb and flood flows. Using depth sounding records (Raytheon, model ≠DE-119), side-scan sonar records (Klein, model ≠2000), and visual observation by divers, two bedform types were found; ripples/spacing ratio = 1:10; spacing 60 cm) and sand waves (spacing ratio = 1:30, spacing 5 m). Ripples are ubiquitous on the sandy substrate of the channel bottom, as well as on the sandy delta topsets. Small sand waves (10 - 15 m spacing and 0.15 - 0.3 m high) are found in the area between the outlet and the right-angle bend and on a portion of the ramp north of this bend (Fig. 6A, B). Larger sand waves (25 m spacing and 0.7 m high) occur only in reaches of rapidly shallowing depth (Fig. 6A). All bedforms were found to be related to flood tide, which is interpreted as reflecting the dominant flow conditions and direction of net sediment transport (see Ashley, 1977).

HYDRAULICS

Tides

The main driving force behind the hydrodynamics of Pitt Lake is the tide. The mixed, mainly diurnal tide in the Strait of Georgia produces one or two tidal cycles a day in the lake, depending upon the nature of the tidal curve. Water level (stage) data used (Fig. 7) are unpublished records of the Water Survey of Canada. Minor features such as small stage fluctuations and quick short changes in flow direction are damped between

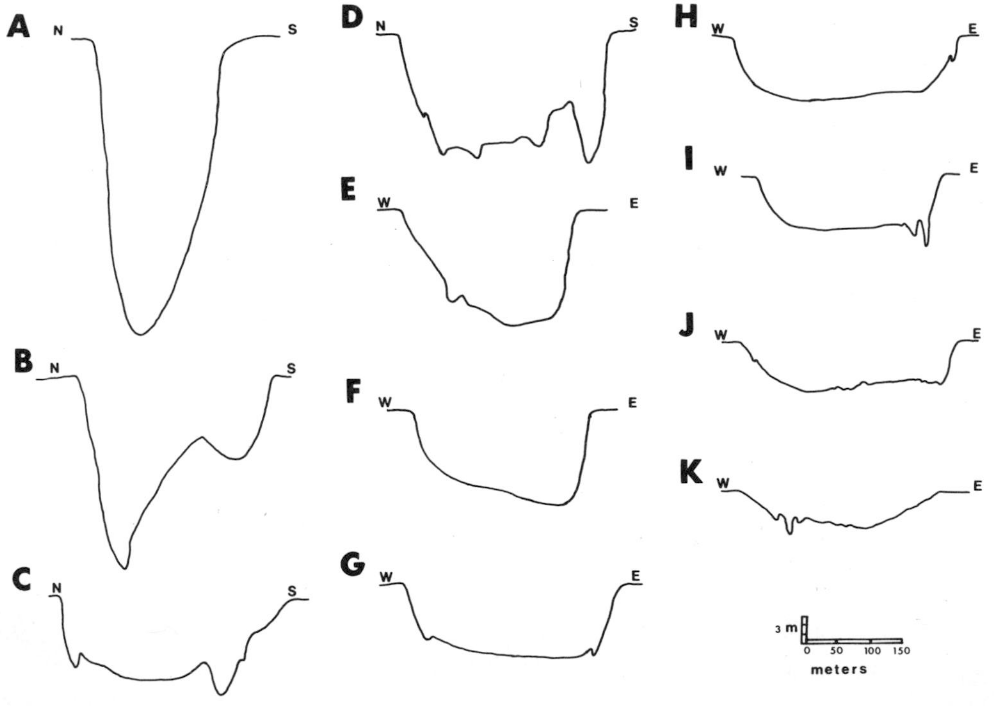

Figure 5. Cross sections drawn from depth sounding profiles (8X vertical exaggeration). Profile locations shown on Figure 3. A general shallowing of the channel and change of channel bank slopes from almost vertical to gently dipping occurs along the delta from outlet to the end.

the ocean and lake and are not expressed in lake stage. When flow conditions persist for several hours (symmetric diurnal tides in the Strait), diurnal lake stage curves are produced (Fig. 7A). On the other hand, highly asymmetric tidal curves, such as shown in Figure 7B, produce only one complete cycle a day. During winter, a delay of 5 hr. 15 min. occurs between high tide in the Strait and high tide in the lake, while a 6 hr. 20 min. delay occurs for passage of low tide from the Strait to the lake. During the freshet when the contribution of Pitt basin drainage is high, it takes 15 hr. 30 min. for either high or low tide to pass from the Strait to the lake. Lake stage level fluctuations varied from 0.27 m to 1.16 m within a tidal cycle, during the year (1973) of stage data examined in detail. Table I gives maximum, minimum, and mean ranges for four representative months.

In addition to tidally induced oscillations in water level in Pitt Lake, the absolute level of these oscillations changes seasonally with a maximum during freshet run off (May - July) and a minimum during winter (Dec. - Feb.). Discharge (Q) contributed to Pitt system from Pitt River (North) and small streams surrounding the lake varies from 210 m^3/sec. (freshet) to 30 m^3/sec. (winter) (Water Survey of Canada, 1966). The result is that during the freshet more than 50 % of water moving through the Pitt Lake - Pitt River (South) system is contributed by basin drainage contrasting with only 5 % during the winter.

The magnitude of discharge flowing through Pitt River (South) (Ashley, 1977) and into Pitt Lake is directly related to the magnitude of the tidal range in the Strait, if Fraser and Pitt basin discharge are constant. With high discharges in the Fraser and Pitt during the freshet the tidal effect in Pitt Lake is small; however, when discharges of Fraser and Pitt systems are low (winter), the tidal effect is great.

A.

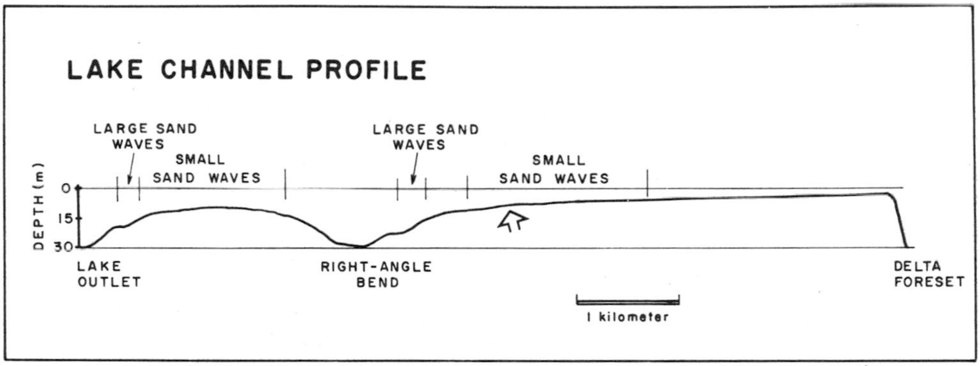

B.

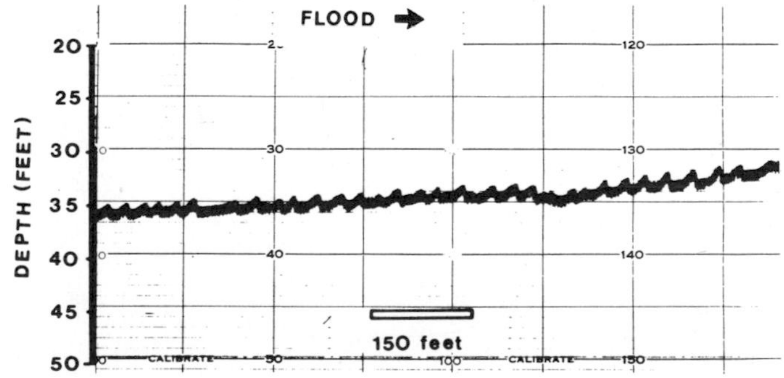

Figure 6 A. Profile of thalweg of delta channel. Deep areas occur at outlet and right-angle bend. Large sand waves (L.S.W.) are found in areas of shallowing channel. Small sand waves are found to within 3 km of the end of the delta. The location of depth soundings taken in B is indicated by arrow.

Figure 6 B. Depth sounding of small sand waves in delta channel: note flood orientation. Spacing is 8 m; height is 30 cm.

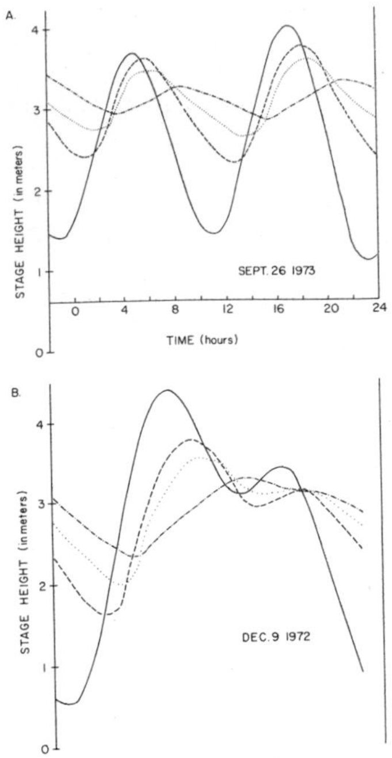

Figure 7. Time stage curves for Strait of Georgia ———, Fraser River – – – –, Pitt River ...., and Pitt Lake –·–·–·–,
-A. Semi-diurnal tide in the Strait creates semi-diurnal fluctuations in Pitt Lake.
-B. The effect of mixed, mainly diurnal tide in the Strait is damped by the time it reaches the lake causing only one fluctuation in lake level.

Peak flows estimated for both seasons and both flow directions are compared in Table II. Estimates were made using lake area, lake stage curves, velocity measurements at lake outlet, and cross sectional area of lake outlet. Thus, there are not only pronounced seasonal differences, with winter having the highest discharge ($2400\ m^3 \cdot sec^{-1}$) but also during either season the largest flood discharge is greater than the largest ebb discharge. The greater difference in discharge between flood and ebb of freshet (Table II) compared to flood and ebb discharge of winter, is due to the increase in volume of water moving through the Fraser-Pitt systems (during the freshet). Water added by streams draining into Pitt Lake raises the elevation of the lake surface as much as 3 m. The elevation of Fraser River is also increased and the net effect is a decrease in the water slope of the upstream flow (flood) into the Pitt. The raised lake elevation also accentuates the time-stage asymmetry of the tidal cycle and thus changes the proportion of time devoted to flood (lesser) and ebb flow (greater). The ebb current flows for a longer period of time (65 % - 75 % of total) at a lower discharge which produces a lower ebb water slope compared to the winter.

Flow pattern over the delta is significantly different on flood and ebb. Flow entering the lake from the river is mainly confined within the deep distributary channel (Fig. 8A). The flow is then deflected northward at the east side of lake causing considerable scour (35 m deep). Continuing along the delta, flood currents generally remain confined in the channel to within 1.5 - 1 km from the end where they spread out across the topsets. Some "over bank" flow occurs along the length of the distributary as evidenced by the levees. Ebb flow has a much more diffuse pattern (Fig. 8B). During the beginning of the ebb, water drains off the topsets into the channel taking the shortest, most direct route. At lower ebb stage, flow becomes more channelized and is confined in the distributary.

Velocity in channel

A study of velocity was untertaken at the lake outlet and in the delta channel to determine the flow conditions that would likely entrain and transport channel bed material. Two different methods of current measurement were used: (1) four days of current profiles, taken at 30-minute intervals over a flood or ebb cycle, (2) readings taken at 7.5-minute intervals, with a tethered meter, one meter from bottom (19 days).

Current profiles (Hydro Products, Inc. Savonius Rotor with a direct readout for current speed (model ≠460A) and direction (model ≠465A)) were made from a boat anchored at the lake outlet. Each profile included measurements at 7 depths (d) (10 cm from bottom, one meter from bottom, 0.2d, 0.4d, (mean), 0.6d, 0.8d, and surface). The measurements (both magnitude and direction) at each depth were based on readings averaged over a two-minute period, thus each profile spans 15 to 20

Table 1. Range of lake stage levels (1973) in meters

Month	Maximum	Mean	Minimum
March	1.04	.73	.67
June	.67	.45	.27
Sept.	.82	.64	.42
Dec.	1.16	1.04	.67

minutes. A digital counter integrating electrical pulses over a 10-second period was used to average velocity fluctuations caused by micro- and macroturbulence (Matthes, 1947). Mean velocity and velocity at 10 cm from bottom for one complete flood cycle (June 24, 1975) are shown in Fig. 9. Peak mean velocity (47 cm/sec) occurs early in the flood cycle. In contrast, ebb examples revealed that peak velocity occurs late in the cycle. This time-velocity asymmetry was found to be characteristic of velocity curves from Pitt River as well (Ashley, 1977).

Critical shear stress necessary for sediment entrainment at the lake outlet and in the channel near the outlet was determined from Shields' diagram as modified from Briggs and Middleton (1965). A friction velocity, V_*, of 1.47 cm/sec is necessary to move sediment (mean grain size = 0.25 mm) at the outlet while V_* = 1.54 cm/sec is required to move material (mean grain size = 0.32 mm) in the southern delta channel.

The log velocity law (Prandtl-Von Karman equation) (Inman, 1963) was used to calculate the basal shear stress from the lake profile data (June, July, and August). Results showed that a critical shear (friction) velocity 1.47 cm/sec was seldom reached in the outlet during this time period. Time series measurements were taken to determine if this were true for other seasons and for the lake channel as well.

The continuously recorded velocity measurements were made by a positively buoyant meter (General Oceanics, Inc. Film Recording current meter (model ≠2010)) anchored to the channel bottom but free to sway with changing currents. The meter was located in the middle of the delta channel approximately 1 km from the lake outlet and recorded on film instantaneous readings of magnitude and direction of flow (one meter off bottom) at 7.5-minute intervals. The meter was placed in channel on several occasions, but only one record (April 15 - May 4, 1976) was readable. Portions of this record are shown in Figure 10; Table III summarized the proportion of total time devoted to ebb (60 %) and flood flow (40 %). It is important to note that although total time of ebb flow is longer than flood, velocities are significantly lower. For instance, about 1 % of ebb time flow is greater than 40 cm/sec in contrast to 13 % of time under flood flow.

Analysis of the 19 days of data found that peak velocity and average velocity were higher on flood flows than on ebb (Table IV). It is inferred from this that mean velocity (0.4d measured from bed) is also higher on the flood. In both river and lake data the mean velocity of a profile was found to equal or exceed the

Table 2. Estimated peak discharges for Pitt system

Season	FLOOD m^3 sec^{-1}	EBB m^3 sec^{-1}
Winter	2400	2080
Freshet	1800	950

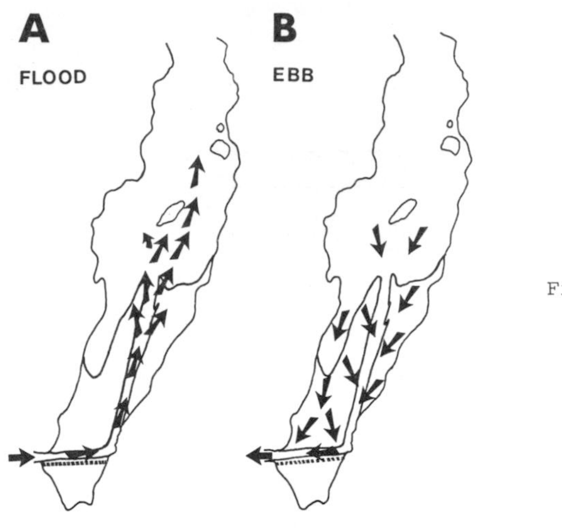

Figure 8. Tidal flow pattern. - A. Flood: flow is channelized until 2 km from end where it spreads over topsets in over-bank flow. Flow pattern into lake appears to be that of a simple jet oriented to the northeast, i.e., east of Goose Island. - B. Ebb: flow drains off topsets by shortest route to delta channel then along channel to outlet.

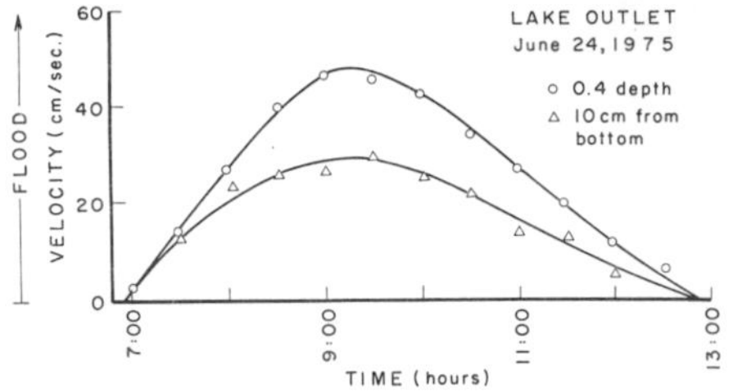

Figure 9. Profile data from lake outlet, June 24, 1975. Mean velocity (0.4 depth) reaches maximum of 47 cm/sec. Time-velocity is asymmetric, i.e., peak is reached early, then decreases gradually.

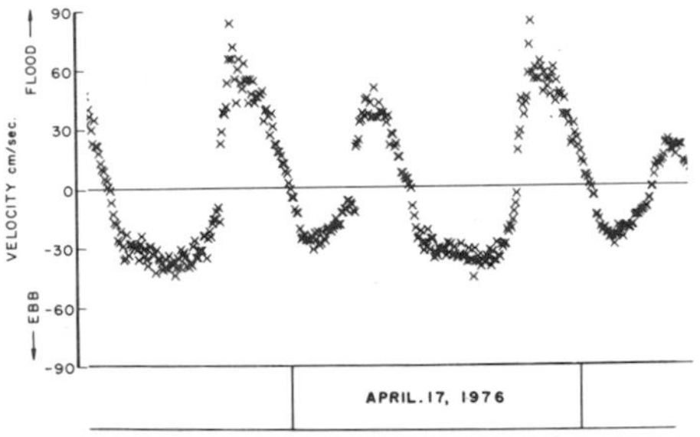

Figure 10. Computer plot of "continuously" recorded velocity data from southern delta channel. Each X represents an instantaneous velocity (magnitutde and direction) measurement at 7.5-minute intervals. Measurements are one meter from bottom. Note flood velocities are higher than ebb. April 16-18, 1976.

velocity at 1 m. The use of the log velocity law on Pitt River velocity profile data (Ashley, 1977) demonstrated that mean velocity of 32 cm/sec was necessary to obtain a critical velocity ($V_* = 1.77$) at base of flow. It follows that a slightly lower mean velocity (approximately 30 cm/sec) would be necessary to create the $V_* = 1.54$ needed to entrain sediment in the delta channel. It can be seen in Table IV that more flood time (20.8 %) is above 30 cm/sec compared to total ebb time (18.1 %). However, the proportion changes drastically at velocity of 20 cm/sec; flood 26 %, ebb 39.5 %. Thus, the data show that flood-oriented currents flow at a higher velocity (in paritcular, above critical velocity), but for a shorter time, than corresponding ebb-oriented flows. This explains the "upstream" transport of sand into the lake and why the sandwaves in the channel are flood-oriented.

In conclusion, a complex interaction of the tidal prism and varying discharge of Fraser River and Pitt basin results in a flood-dominated system. Highest peak discharges and related basal shear stresses occur during flood (winter) flows. Thus, net sediment transport would occur during the winter. The greater efficiency for entraining and moving sediment under flood flow supports the conclusions based on the morphology of the delta, that it is presently active and being constructed under flood conditions by Fraser derived sediments.

SEDIMENTS

Stratigraphy of delta and lake bottom

Sediments of Pitt tidal delta can be grouped into three general environments: topset, foreset, and bottomset-lake bottom. The topset beds consist of fine sand to coarse silt and are horizontally laminated. The foresets consist of silt and clay layers, some of which are rhythmically layered, whereas the bottomset-lake bottom beds are laminated clays. Of the 160

Table 3. Summary of velocity measurements (at one meter above bed) in lake channel April 15 - May 4, 1976: the proportion of time devoted to flood and ebb at 10 cm/sec intervals of velocity

Velocity (cm/sec)	EBB Data Pts.	EBB Hrs.	EBB Cum Hrs.	EBB % Total	FLOOD Data Pts.	FLOOD Hrs.	FLOOD Cum Hrs.	FLOOD % Total
80	0	0	0	0	3	0.75	0.75	0.08
70	0	0	0	0	14	3.50	4.25	0.47
60	0	0	0	0	29	7.25	11.50	1.30
50	2	0.50	0.50	0.05	166	41.50	53.00	5.90
40	45	11.25	11.75	1.30	244	61.00	114.00	12.75
30	602	150.50	162.25	18.10	291	72.75	186.75	20.80
20	764	191.00	353.25	39.50	183	45.75	232.50	26.00
10	555	138.75	492.00	55.00	273	68.25	300.75	33.50
0	175	43.75	535.75	60.00	230	57.50	358.25	40.00
Total	2143	535.75	535.75	60.00	1433	358.25	358.25	40.00

samples in the study area, 60 were grab samples on delta topsets and in the delta channel and 100 were cores (3.5 cm in diameter) taken from delta foresets and bottomset-lake bottom (Figure 11). Cores ranged from 24 cm to 53 cm in length.

Topset beds consist of a monotonous section of laminated silts and sands. No crossbeds and only a few graded beds were noted. On the other hand, the foreset beds were found to contain a variety of bedding structures. The nature of the layering ranges from well developed rhythmic silt and clay layers to stringy and discontinuous clay laminations (Fig. 12). In the rhythmites, silt layers are thicker than clay, but the absolute thickness of the individual layers varies with distance from the delta. Silt layers range in thickness from 1.3 cm at the distributary mouth to 0.16 cm, 2 km north of the mouth. Clay layers seldom are thicker than 0.16 cm near the delta and thin to an average thickness of 0.08 cm in the lake bottom sediments beyond Goose Island. All cores showed evidence that gas (methane?) had escaped after sampling. Houbolt and Jonker (1968) found gas "pockets" were ubiquitous in Lake Geneva sediments. Cores, cut open soon after sampling, have continuous layering with scattered gas vesicles (Fig. 12). Cores which were stored allowing gas to escape slowly, have streaky, uneven and discontinuous layers with no vesicles.

The best-developed rhythmic layering was found in cores to the west of the main delta lobe. The thickness of the layers was found to decrease abruptly at about 10 cm (or approximately 30 laminations) from the top of the sediment section (Fig. 12). Couplets below the 10 cm level are about 1 cm thick whereas those above are only 0.2 cm thick. The decrease (80 %) in thickness occurs gradually over 4 - 6 couplets and suggests a sharp decrease in sediment reaching the site. This can be interpreted as a shift in locus of sedimentation from the west to east side of Goose Island. Another interpretation is that sedimentation has decreased over all the lake, but without rhythmic layering the change is not evident. Since

Table 4. Summary of velocity measurements (one meter off bottom in lake channel) (April 15 - 30, 1976); velocity in cm/sec

Date (April, 1976)	Max flood vel.	Ave. flood vel.	Max ebb vel.	Ave. ebb vel.	Date (April, 1976)	Max flood vel.	Ave. flood vel.	Max ebb vel.	Ave. ebb vel.
15	55.5	35.0			23	15.0	10.0	31.0	22.0
						45.0	38.0	32.0	22.0
16	41.0	27.0	30.5	20	24	52.0	37.5	32.0	26.5
	71.0	39.0	40.0	30		24.0	14.0	28.0	22.5
17	42.0	31.5	27.0	20	25	51.0	32.0		
	67.0	40.0	38.0	31		34.0	21.0	38.0	27.0
18	22.0	13.0	27.0	18	26	47.5	31.0	29.0	22.5
	68.0	40.5	41.0	30		51.0	30.0	38.0	28.5
19	16.0	10.0	19.0	17	27	42.5	29.5	29.0	20.5
	68.0	40.0	45.0	28		59.0	33.0	38.5	29.5
20			23.0	16	28	38.0	26.0	26.0	20.0
	15.0	10.0	42.0	28		54.0	32.0	38.0	29.0
21			28.5	22	29	33.0	23.5	27.5	19.0
	61.0	42.5	36.0	28		55.0	35.0	39.0	29.0
22			31.0	23	30	30.0	22.0	26.5	20.0
	61.0	38.0	33.0	23		59.0	37.0	36.5	28.0

the configuration of the delta has been constant since at least 1860 (Richards, 1860), the latter explanation is favored.

The rhythmic layering was previously noted by Johnston (1922) in his work on Pitt Lake. His interpretation was that the alternating laminations were tidal in origin. It appears more likely that they are annual layers (varves). However, the mechanism involved is directly opposite to the generally accepted model for varve formation: coarse sediment is deposited during spring run-off and fine sediment accumulates during winter. In Pitt Lake the coarse layer (silt and fine sand) is brought in as bedload and suspended load during winter (November - March) when discharges of Fraser and Pitt systems are low and thus tidal effect is great. The fine layer (mainly clay) is deposited during the rest of the year. Presumably clay (9ϕ) is supplied to the lake mainly during spring run-off (May - July) and continues to settle during summer and fall. The volume of sediment represented in an average couplet for the entire foreset-bottomset-lake bottom area shown in Figure 3 was calculated to be $150 \pm 20 \times 10^3$ tonnes. Calculations were based on average thickness of layers near tops of cores and their approximate areal distribution.

^{137}Cesium dating

An unexpected "spin-off" of atmospheric nuclear testing (1952 - 1972) is the subsequent use of radioactive isotopes released during the tests (such as ^{137}Cs) for dating of recent sediments (Pennington et al., 1973; Robbins and Edgington, 1975; Ritchie et al., 1975). ^{137}Cesium was created during the nuclear testing and disseminated throughout the world by air currents and rainfall. In fresh water, Cs is preferentially adsorbed, or "fixed", onto the micaceous component of the sediment (Francis and Brinkley, 1976), presumably bonded in grain boundary sites. The strength of those bonds is sufficient so that further movement is limited under normal conditions in the natural environment (Davis, 1963; Tamura, 1964). Thus, the variation in ^{137}Cs content present in the stratigraphic column can be compared with

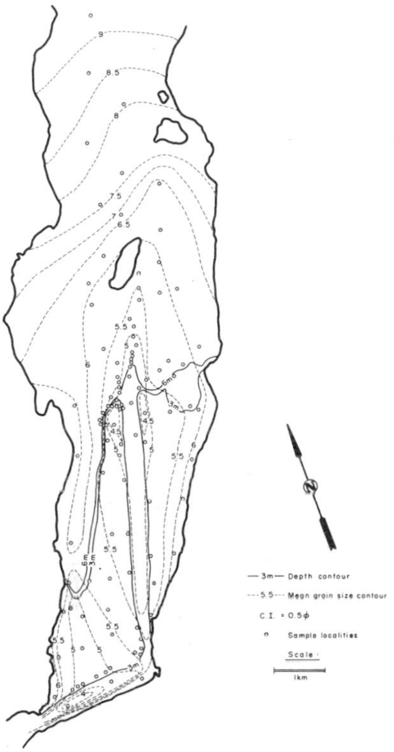

Figure 11. Mean grain size distribution map. C.I. is 0.5 ϕ. Grain size distribution reflects the flood flow pattern (Fig. 8A). The pattern is a good example of sedimentation by diffusion from a simple jet with little, if any, reworking by waves or tidal currents.

the local ^{137}Cs activity record (usually measured in rainfall or in milk) to determine sedimentation rate.

In order to determine the present annual sedimentation rate on the Pitt tidal delta, 11 large diameter (6.3 cm) cores were taken on the delta topsets and foresets (Fig. 3; Fig. 13). A Kullenberg gravity corer (weighing 130 kg) was dropped from an anchored raft. Core lengths ranged from 15 cm to 85 cm and three cores with undisturbed bedding were chosen for dating (samples ≠3, ≠8, and ≠11). Mean grain size variation of the cores is limited, ranging from coarse silt to very fine sand (50 - 70μ). As Cs is associated with micas or illites and as these minerals most often occur in the finer fractions, total ^{137}Cs content would be expected to vary with grain size. Thus the homogeneous grain size of the Pitt samples made them particularly appropriate for dating by the ^{137}Cs dating technique.

An exceptionally good record (Fig. 14) was found in core ≠3 collected off the mouth of the distributary channel in 50 m of water (Fig. 3). Average rate of sediment accumulation at this site from 1954 to 1972 was 1.8 cm/yr (no correction was made for compaction). The fluctuation in ^{137}Cs content is consistent with that measured in Vancouver (milk) (G. Griffiths, pers. comm.). Lake stratigraphy from Lake Windermere, England (Pennington, et al., 1973) and fallout recorded in Tallahatchie River watershed (Ritchie, 1973) showed a similar record. All show a minor peak in 1959 and a major one in 1963 with the ^{137}Cs concentration dropping off to low levels thereafter. Core ≠11 showed high levels of ^{137}Cs equal to core ≠3 on the delta front, but in a compressed record indicating a slow sedimentation rate of 2 mm/yr. The source of this cesium is probably slope wash from the nearby shore. However, core ≠ 8 (Fig. 3) contained no excessive ^{137}Cs, indicating no deposition occurred during the last 25 years at that site on the delta.

The above records are considered indisputable evidence for recent deposition and slow, but regular sedimentation on the delta. In addition, the difference in records from the three sites gives further information on the pattern of present day deposition which is examined in the following section.

Grain size analysis

A study of the grain size distribution of Pitt Lake sediments was carried out in order to gain insight into the nature of sedimentary processes active on the delta and lake bottom. Sediment samples were collected with a Dietz-LaFond grab sampler and a Phleger corer. A total of 190 samples were analyzed by one or more of the following analytical methods.

(1) Rapid Sediment Analyzer (R.S.A.) - settling tube with automatic recording of weight accumulated versus time (Woods Hole Settling Tube, Univ. of R.I.).

(2) Sieving - standard sieving procedure (Folk, 1968) using sieves with 0.5 ϕ intervals.

(3) Quantimet 720 - image analyzing computer (Perrie and Peach, 1973), Brock University.

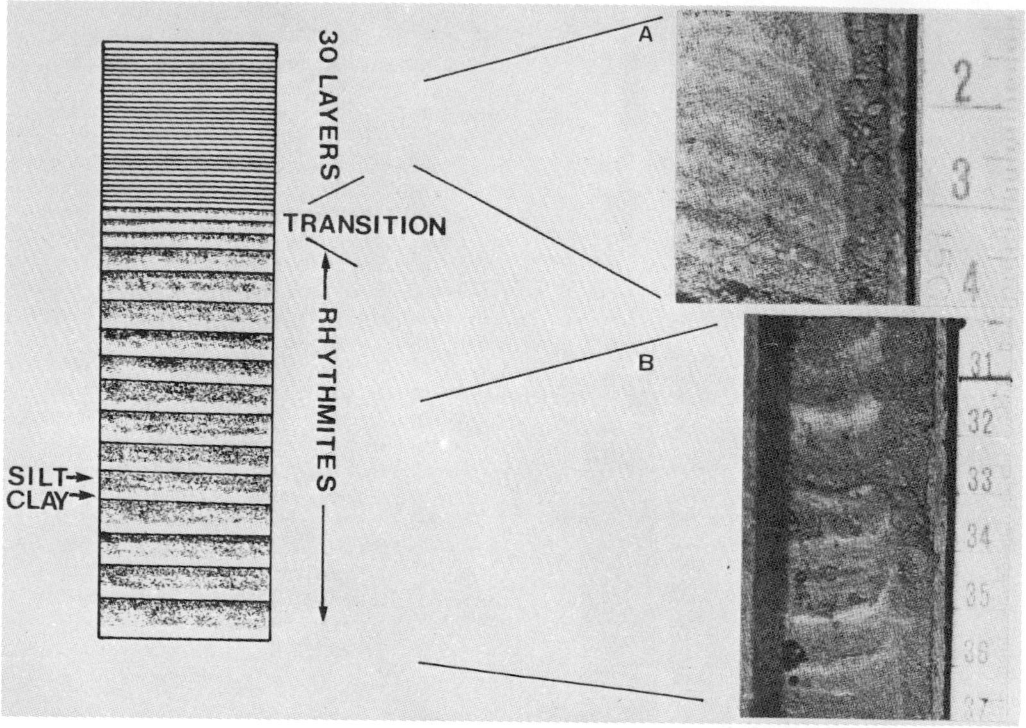

Figure 12. Diagrammatic sketch of stratigraphy showing change from regular rhythmites (B) through a few transitional couplets into thinly bedded sediments (30 laminations) of a 3.5 cm diam. core (A). Top of stratigraphic section showing a sharp decrease in sedimentation rate (A). Rhythmites (B) are interpreted as varves deposited by the following mechanism: silt deposited during winter when tidal effect is great and clay during the freshet when tidal effect is minimal. Note large vesicles formed during escape of gas (methane?).

(4) Sedigraph 5000 (Micrometrics, Inc.) - particle size analyzer which measures the concentration of particles remaining suspended as a function of settling time using a finely collimated beam of x-rays (Olivier, et al., 1970/71).

(5) Hydrometer - standard method for grain size analysis of soils (A.S.T.M. D422-63).

Weight percent for each 0.5ϕ class was used to compute statistical parameters (using method of moments) and cumulative probability plots.

SEDIMENTOLOGICAL PROCESSES

Grain size distribution

Figure 11 depicts with size contours (C.I. = 0.5 ϕ) the aerial distribution of mean grain size of each grab sample and sample from the top of each core. Mean grain size is fine (5.5 ϕ) in the deep areas of the channel at the outlet and right-angle bend, with coarser (2 ϕ - 4 ϕ) material in between. The rest of the channel distributary out to the delta tongue is coarse silt with a mean grain size of 5 ϕ (.03 mm).

In general, mean grain size contours follow the channel, diverging near the end, mimicking the flood flow pattern (Fig. 8A). The one exception to this generality the triangle of coarser sediment in the middle of the southern topsets, is probably a result of removal of some fines during ebb flow. Ebb drainage channels on the southern margin of this area (Fig. 3; Fig. 4A) indicate that ebb-oriented flow has a pronounced effect on this portion of the delta. The coarsest sediment, with a mean grain size of 0.044 mm (4.5 ϕ) to 0.075 mm (3.7 ϕ), is found at the northern end of the delta topsets immediately adjacent to the channel.

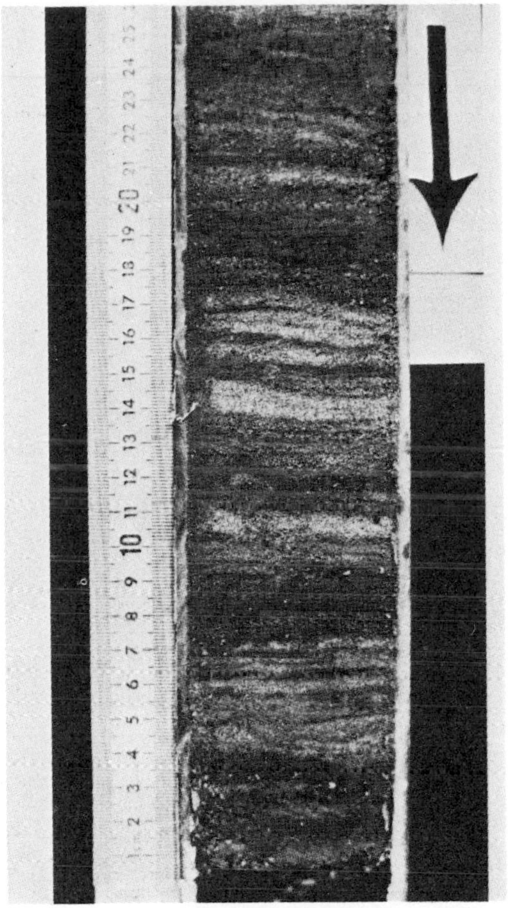

Figure 13. Photo of core ≠3 (Fig. 3) which was dated by ^{137}Cs. Note that although the stratification is not rhythmic, it is undisturbed. Core diameter is 6.3 cm.

As this material is slightly coarser than that in the channel from which it originated, winnowing is suspected.

Subsamples were taken from top, the middle, and bottom of 15 cores from the foresets and bottomsets to determine any vertical change in grain size within the couple hundred years of sedimentation present. Cores within 0.5 km of the end of the distributary channel show increasing grain size from bottom to top of core. Cores 0.5 km to 7 km (farthest removed sample) away from the channel show a slight upward decrease in grain size (generally less than 0.5 ϕ). The implication from these changes in grain size stratigraphically is that the delta is prograding, but at a lesser rate than in the past.

On the delta foresets and bottomsets, mean grain size becomes finer away from the end of the distributary channel and the contour pattern indicates sedimentation occurs mainly to the east side of Goose Island. The distribution pattern depicted in Figure 11 is a reasonably good example of sedimentation by diffusion from a simple jet with little if any reworking by waves and tidal currents.

The application of the theory of submerged free jets to delta formation has been strongly influenced by the work of Bates (1953). He believed that the diffusion pattern depends on the relative densities of the two fluids (moving and stationary bodies of water). However, one of the major limitations of Bates' theory is the assumption that there are no boundary effects. Clearly, the basic assumptions for any theoretical model of sediment diffusion by jet flow (delta formation) should directly reflect the complexities of the natural environment and not an ideal situation.

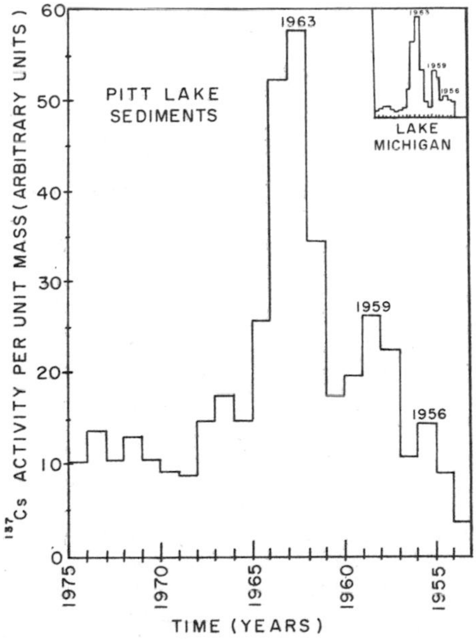

Figure 14. Plot of ^{137}Cs dating results from core ≠3 from Pitt Lake delta foresets. Cesium concentration per unit mass plotted against time (depth) results in a graph similar to the estimated annual flux of ^{137}Cs to surface of Lake Michigan (Robbins and Edgington, 1975) shown in the inset.

A recent laboratory and computer model study by Ramsayer (1974) incorporated the effect of lateral and basal shear on jet flow under steady uniform flow. He explains morphological features such as levees and distributary bars found at a delta distributary mouth by a model which predicts the distribution of effective bed shear stress. Time dependent runs using fine sand produced a sediment distribution pattern strikingly similar to that of the Pitt (Fig. 11). Thus, the Pitt appears to fit the general model of deposition from a jet. However, one feature of the model that is missing in the Pitt, is the distributary mouth bar. Jopling (1960), suggests that if the frontal slope over which the flow expands is less than 10°, the flow will not separate from the boundary at all (assuming the flow is homopycnal) and thus deposition in the form of a bar would not occur. This seems to explain the lack of bar development on the Pitt delta as the foreset slope is less than 4° and the system possesses homopycnal flow for most of the year.

In conclusion, the areal distribution of mean grain size substantiates the interpretation that flood flow is the dominant current in moving sediment across the delta. Sediment is then dispersed by jet flow oriented in a north-east direction into the lake.

Sediment dispersal and accumulation

The observations of delta morphology, the presence of bedforms in the channel, and the distribution (pattern) of grain size enable the processes involved in sediment dispersal to be delineated.

The single distributary channel, flanked by levees, which leads to a fan-shaped delta foreset indicates that the channel is the avenue for sediment movement. The fact that the channel is deeply incised along the southern margin and that the carbon-dated material (4.645 ± 95 B.P.) from the southern topsets was buried by only 60 cm of sediment suggests that the flow is channelized in this reach and little sediment is brought up onto the delta surface. Velocity data and orientation of large-scale bedforms reflect the dominance of the flood current moving through the channel.

North of the right-angle bend the distributary shallows and channel banks become less steep. Small flood-oriented sand waves (10 m in spacing) are found in the channel for the first 2 - 3 km where flow is more-or-less confined. However, in the last 3 km, only ripples are found on the channel bottom and channel bank slopes are low enough for flow to spill and spread over the topsets. Mean grain size distribution reflects this pattern of flow. Once sediment is on the delta surface it is winnowed and moved by wave-driven currents in addition to the tidal currents. Large waves and swells were observed to occur frequently in the lake, particularly during the winter months, shifting sediment in the form of ripples. Orientation of these ripples, which cover the outer topsets, reflects the direction of currents at the time. Wind-generated currents were observed to incorporate and carry sediment into the lake as a plume.

The mechanism of sediment transport beyond the delta i.e. across lake bottom was not monitored directly. However, as medium silt is found as far out as 5 km, some type of density flow mechanism is probable for at least part of the year.

Stratigraphic data presented previously suggests that a total volume of ($150 \pm 20 \times 10^{3}$ tonnes) of sediment is accumulating annually in the southern half of Pitt Lake. ^{137}Cs dating has confirmed the existence of a steady sediment flux since at least 1954. 50 % of the annual sediment accumulation is coarser than 5 ϕ and thus probably moves in the form of bedload and periodic suspension. This volume (75×10^{3} tonnes) is on the same order of magnitude as the annual sediment flux calculated for the Pitt River (South) by Ashley (1977). It is suspected that the other 50 % consists of: (1) material that is continually reentrained at points along the river and delta channels by flood flows and eventually arrives at the lake; and (2) fine silt and clay that is washed in from the Pitt watershed. ^{137}Cs dating substantiates, first, the interpreation of the foreset rhythmites as annual couplets (varves) and, second, that the sedimentation on delta topsets is slow.

The possibility of a sharp decrease in sedimentation rate approximately 30 years ago is suggested by the stratigraphy (Fig. 12). Most cores taken away from the active delta channel show a fining upward in grain size indicating a slight waning in delta growth rate. Although definitive evidence is lacking, this apparent decrease in sedimentation rate is intriguingly coincident with the initiation of large-

scale dredging in lower Fraser River. This dredging probably has had the effect of increasing the cross-sectional area of Fraser estuary and thus decreasing the magnitude (and thus competency) of tidally induced currents in Pitt River.

CONCLUSIONS

The Pitt tidal delta is presently building into the lower end of Pitt lake. Its unusual position can be readily explained in terms of tidal dynamics. Unequal velocities of tidal currents (flood is greater) have caused landward transport of sediment up Pitt River (South) from the Fraser and into the lake. A complex interaction of Fraser discharge, Pitt basin drainage, and the tidal prism creates unequal tidal flow on a seasonal basis (winter has strongest flows).

Both the geomorphology and grain size distribution reflect the dominant flood flow pattern. Basically flow is channelized as it enters the lake and remains in the channel until its bank slopes are shallow enough to allow overflow onto delta topsets. Annually 150 ± 20 x 10^3 tonnes of sediment (1 % of Fraser's total load; Mathews et al., 1970) are being deposited as varved couplets. The greatest thickness is on delta foresets in the vicinity of the main channel, with the best developed stratigraphy occurring adjacent to the delta lobe. The mechanism of varve sedimentation is unique and is directly opposite to the generally accepted model for varve formation. In Pitt Lake the coarser silty layer forms during the winter when discharges of the Fraser and Pitt systems are low and thus tidal effect great. The clay layer accumulates during summer when tidal flow is inhibited by high river discharges.

The present-day delta has been constructed during the last 6000 years at an average rate of 1.28 $m \cdot yr^{-1}$; however, delta growth would be expected to have decreased exponentially. ^{137}Cs dating and varve chronology suggest that the present growth rate is on the order of centimeters per year. A slight decrease in mean grain size upsection in cores (representing a couple hundred years sedimentation) also suggests a waning delta growth. Varve stratigraphy has revealed a sharp decrease in sedimentation rate 30 years before present. This decrease may be a response to large-scale dredging in the lower Fraser estuary.

Despite bidirectional flow and daily and seasonal fluctuations in lake stage, the delta has a simple constructional morphology. The morphology, the sand waves in channel, and sediment dispersal pattern are dominated by flood conditions. Weaker ebb currents make only minor modifications despite the average discharge of 1700 $m^3 \cdot sec^{-1}$ which drains daily from the lake.

ACKNOWLEDGEMENTS

Much appreciation is extended to W.H. Mathews, M. Church, I.J. Duncan, and W.C. Barnes who reviewed earlier versions of this manuscript.

I am grateful to L.E. Moritz (TRIUMF, U.B.C.) for ^{137}Cs analysis and to M. Church for help in developing several computer programs. Appreciation is expressed for the services and research facilities provided by the Department of Geological Sciences at the University of British Columbia. National Research Council of Canada Grant (A-1107) (Prof. W.H. Mathews provided support for both field work and preparation of the manuscript.

REFERENCES CITED

Ashley, G.M., 1977, Sedimentology of a freshwater tidal system, Pitt River-Pitt Lake, B.C.: (unpublished Ph.D. thesis), Univ. of B.C., Vancouver, B.C., Canada.

A.S.T.M., 1963, Standard method for grain size analysis of soils: A.S.T.M. D422-63.

Bates, C.C., 1953, Rational theory of delta formation: Am. Assoc. Pet. Geologists, v.37, no. 9, p. 2119-2162.

Briggs, L.I. and Middleton, G.V., 1965, Hydromechanical principles of sediment structure formation: S.E.P.M., Spec., Pub. 12, p. 5-16.

Davis, J.J., 1963, Cesium and its relationship to potassium in ecology: in Schultz, V., and Klements, A.W., Jr., (eds.), Radioecology Reinhold New York, p. 539-556.

Dyck, W., Fyles, J.G., and Blake, W.Jr., 1965, Geol. Sur. Canada Radiocarbon Dates IV: Radiocarbon, v. 7, p. 24-46.

Folk, R.F., 1968, Petrology of Sedimentary Rocks: Univ. of Texas, Hemphills, Austin, Texas.

Francis, C.W., and Brinkley, F.S., 1976, Preferential adsorption of ^{137}Cs to micaceous minerals in contaminated fresh water sediment: Nature, v. 260, p. 511-513.

Houbolt, J.J.H.C. and Jonker, J.B.M., 1968, Recent sediments in the eastern part of the Lake of Geneve (Lac Leman): Geol. Mijnbouw, v. 47, p. 131-148.

Inman, D.L., 1963, in Shephard, F.P. (ed.), Submarine Geology, Harper and Row, New York, N.Y., 557 p.

Johnston, W.A., 1922, The character of the stratification of the sediments in the recent delta of Fraser River, British Columbia, Canada: Jour. Geology, v. 30, no. 2, p. 115-129.

Jopling, A.V., 1960, An experimental study on the mechanics of bedding: (Ph.D. thesis) Harvard University, Cambridge, Massachusetts.

Mathews, R., 1973, A palynological study of postglacial vegetation changes in the University Research Forest, southwestern British Columbia: Canadian Jour. of Botany, v. 51, no. 11, p. 2085-2103.

Mathews, W.H., Fyles, J.G., and Nasmith, H.W., 1970, Postglacial crustal movements in southwestern British Columbia and adjacent Washington State: Canadian Jour. of Earth Sci., v. 7, no. 2, p. 690-702.

Matthes, G.H., 1947, Macroturbulence in natural stream flow: Am. Geophy. Union Trans., v. 28, p. 255-262.

Olivier, J.P., Hicken, G.K., and Orr, C. Jr., 1970/71, Automatic particle size analysis in subsieve range: Pow. Technology, v. 4, 257-263.

Pennington, W., Cambray, R.S., Fisher, E.M., 1973, Observations on lake sediments using fallout ^{137}Cs as a tracer: Nature, v. 242, p. 324-326.

Perrie, L.A. and Peach, P.A., 1973, Gelatin coated microscope slides in sedimentary size analysis: Jour. Sed. Petrology, v. 43, p. 1174-1175.

Ramsayer, G.R., 1975, Experimental and theoretical study of deltaic sedimentation: (Ph.D. thesis), the Univ. of Rochester, Rochester, New York.

Reineck, H.E. and Singh, I.B., 1975, Depositional Sedimentary Environments: Springer-Verlag, New York, 438 p.

Richards, G.H., 1860, Fraser River and Burrard Inlet: British Admiralty Chart, 1922.

Ritchie, J.C., McHenry, J.R., and Gill, A.C., 1973, Dating recent reservoir sediments: Limnology and Oceanography, v. 18, p. 254-263.

Ritchie, J.C., Hawks, P.H., and McHenry, J.R., 1975, Deposition rates in valleys determined using fallout Cesium-137: Geol. Soc. America Bull., v. 86, p. 1128-1130.

Robbins, J.A. and Edgington, D.N., 1975, Determination of recent sedimentation rates in Lake Michigan using Pb-210 and Cs-137: Geochim. Cosmo. Chem. Acta, p. 285-304.

Ryder, J.M. and Church, M., 1972, Paraglacial sedimentation: consideration of fluvial processes conditioned by glaciation: Geol. Soc. America Bull., v. 83, p. 3059-3072.

Scruton, P.C., 1960, Delta building and the deltaic sequence: in Shepard, F.P., Phleger, F.B., Andel, T.H. (eds.), Recent Sediments, northwest Gulf of Mexico, Am. Assoc. Petroleum Geologist, Tulsa, Oklahoma.

Tamura, T., 1964, Selective sorption reaction of cesium with mineral soil: Nuclear Safety, v. 5, p. 262-268.

Water Survey of Canada, 1966, Vancouver, British Columbia: unpublished stage and discharge records.

Deglaciation and glaciolacustrine sedimentation conditions, Okanagan Valley, British Columbia, Canada

J. SHAW & J. ARCHER
University of Alberta, Edmonton, Alta., Canada

1 INTRODUCTION

The Okanagan valley is one of a large number of structural trenches in the Canadian Cordillera. The pattern of deglaciation in these trenches includes the important element of ice-lobe stagnation caused by isolation from sources of ice accumulation (Flint 1935, Nasmith 1962, Fulton 1965). Isolation of lobes was commonly a result of abrupt changes in valley trends. Several authors have stressed the importance of stagnant ice to the distribution and tectonic structures of the glacigenic valley fills (Flint 1935, Nasmith 1962, Fulton 1965). In the first systematic study of the glaciolacustrine sediments in the North Thompson valley, Fulton (1965) described sedimentary structures, including cross-lamination, and reported an upward thinning of varve thickness. We studied the glacial sediments of the Okanagan valley in the vicinity of Skaha Lake and the southern portion of Okanagan Lake (Shaw 1975, 1977, Shaw and Archer 1978). Considerable rotation and faulting of the lacustrine sediments confirm the finding of earlier authors that deposition occurred over stagnant ice. Shaw (1977) concluded that the deep lake basins with marginal terraces of lacustrine sediment resulted from the melting of lobes of stagnant ice several hundreds of meters thick in the axial zone of the valley. However, the upward decrease in thickness and in grain size of the varved units corresponds to the classical model of glaciolacustrine sedimentation beyond a retreating ice margin (Shaw 1977). A further complication arises due to the presence of relatively coarse sediment deposited in winter by high energy turbidity currents (Shaw 1977, Shaw and Archer 1978). We will present new evidence and a conceptual framework to explain both the apparently incompatible observations of stagnation and frontal retreat and the unexpected occurrence of coarse sediment within the winter component of glaciolacustrine varves.

2 PHYSIOGRAPHY

The area of study occupies a deep valley trending north to south (Fig. 1). The width of the valley is about 6 km and adjacent ridges stand approximately 1000 m above the valley floor. Large ribbon lakes occupy basins that are up to 240 m deep. The Okanagan trench makes two right angle bends in the reach between Squally Point and Kelowna which is immediately north of the study area (Fig. 1). These abrupt changes in valley trend are thought to have created such resistance to glacier flow from the north that a major lobe became isolated in the study area. A prominent bench along the valley sides in the study area marks the distribution of glaciolacustrine sediments. The terraces are best developed in the area of predicted stagnation (Fig. 1).

3 THE ENVIRONMENT DURING DEGLACIATION

Nasmith (1962, Plate II) presented an interpretive sketch of the surficial features south of Skaha Lake. The valley floor is occupied by a broad area of kettled outwash which was subsequently dissected by the Okanagan River. Fulton (1965) described lacustrine silts in the Thompson valley which overlie interbeds of gravel and till. Disturbance occurs throughout this sequence and he stressed the importance of gravitational sliding to the formation of disturbed structures. The above observations illustrate that supraglacial deposition on stagnant ice was an important process during the deglaciation of at least two valleys of central British Columbia.

3.1 The Harrop Section

Most of the sections examined in the cliffs bordering Skaha Lake and the southern part of Okanagan Lake expose only lacustrine sediments. However, the section at the Harrop site (Fig. 1) shows not only lacustrine sediments at the surface but also a complex lower sequence. Seven recorded sections over a horizontal distance of some 200 m show a high degree of lateral variability in this lower complex (Fig. 2). Three types of diamictite are found. One is compact with a sandy matrix, is blue grey in colour, and contains rounded clasts. This diamictite occurs at the base of those sequences in which it occurs. A second type of diamictite is friable with a sandy silt matrix and contains small-scale stratification. It is found above the first type in section 4, and is noted at section 6 in which the compact diamicton is not seen. A third type of diamictite is competent, is light buff in colour, and contains rip-up clasts of silt, faulted sands and rounded rock clasts. The individual bodies of this diamictite form lobes intercalated with stratified gravels and sands (Fig. 2, sections 5 and 6). The third diamictite is found higher in the sequence than the other two.

Stratified gravels, sands and silts are extensive and often contain erosional surfaces. In some cases, the gravels are cross-stratified in relatively thick foreset beds and, in others, are horizontally stratified. Both clean, openwork gravels and matrix-rich gravels are found. Sands and silts show horizontal stratification, cross-lamination, and parallel lamination. Faulting is common within the stratified units at a variety of scales.

The uppermost deposits along the full length of the Harrop section are primarily lacustrine silts similar to those exposed elsewhere within the prominent lacustrine bench in the study area.

The Harrop section shows an expected sequence of sedimentation given the decay of a stagnant ice lobe in an overdeepened valley. The lower diamictites are interpreted as tills released by basal melt out, or possibly, in the case of the compact massive diamicton, by lodgement. The compact nature and basal stratigraphic position of the blue till are as expected for lodgement or melt out from the basal, poorly attenuated, debris-rich ice facies (Shaw in press). The overlying stratified diamictite is interpreted as a basal melt-out till from the foliated ice of the highly attenuated facies. The overlying complex of stratified sediments and lobate diamictites illustrates deposition in a supraglacial complex. The diamictites may be flow tills released directly from ice, or may be slumped supraglacial stratified sediments. In either case, their intercalation with stratified gravels illustrates slumping into a fluvial complex. The wide range of grain size and the presence of both cross-stratified gravel foresets and horizontally bedded gravels represent highly variable depositional conditions with streams and local ponds on a wasting ice surface. Extensive faulting and warping confirm deposition of the diamictite/stratified complex over buried ice.

The overlying lacustrine sediments illustrate transgression of a water body over the subaerial, supraglacial surface. The sequence of events - stagnation, downwasting, and lacustrine transgression - allows some generalisations to be made on the environmental conditions during deglaciation.

3.2 Discussion

The role of supraglacial debris in the formation of till depositional sequences and glacigenic landforms has been given great emphasis in models of glacial deposition. However, these models are severely limited when deglaciation occurs under active permafrost. Once the supraglacial debris cover is equal in thickness to the active layer, any further aggradation of debris must be of material transported from an area of surface ablation. This aggradation will cause the development of syngenetic permafrost and no further surface melting of glacier ice occurs (Fig. 3). In an extensive stagnant ice lobe of the type under discussion, surface melting is rapidly halted by the development of a complete debris cover. However, basal melting will either continue or, if the glacier is cold based, will commence with climatic amelioration, and till will be mainly deposited by a process of undermelting (Shaw in press). Undermelting causes lowering of the supraglacial surface and transgression will occur in areas conducive to lake formation. Shaw (1975) described such a transgression interpreted from sediments of Glacial Lake Edmonton, Alberta. The sequence at the Harrop site with basal melt-out tills, a supraglacial subaerial complex, and overlying lacustrine sediments corresponds exactly to that predicted for deglaciation in an enclosed basin under permafrost conditions. A complication is added with transgression by a freshwater body since permafrost must degrade beneath relatively deep bodies of fresh water (Johnston and Brown 1964). Transgression therefore causes melting at the upper surface in addition to ongoing undermelt. At the valley sides, where the

ice is relatively thin, complete melting may occur in a few tens of years. Disturbance of the oldest sediments during melt out, with overlying undisturbed sediments, explains the numerous unconformities noted in the lacustrine sediments (Flint 1935, Fulton 1965, Shaw 1977).

The presence of a stagnant ice lobe in the study area, isolated by abrupt changes in valley trend upglacier, raises the issue of the prevailing conditions above the obstruction. Although there is no direct evidence, continuity alone suggests that a zone of shearing occurred between active and stagnant ice. A similar conclusion was drawn for conditions during deglaciation in New England (Koteff 1974, Figs. 1-6). The resultant upward component of flow in the active ice leads to a complex zone in which the balance between glacier flow and ablation is in part interdependent with supraglacial accumulation and slumping. Where supraglacial debris causes aggradation of permafrost, ablation is halted, but continued upward flow in the underlying ice leads to uplift of the shearing zone (Hooke 1970). Slumping from this raised block may cause renewed surface ablation and lowering but the uplift of the shearing zone causes reverse glacier surface slopes (Hooke 1970, Fig. 6) and will contribute to the development of marginal stagnation and the successive development of upglacier shearing zones. The shearing zones when active act as sediment sources and the successively developed shear zones produce the effect of a retreating ice margin. However, a further factor is of extreme importance. If the shear zone should remain active and major supraglacial streams are developed, then constant stream reworking of the supraglacial sediments and downwasting of the ice downglacier of the shear zone will lead to the formation of a glacial lake with a proximal delta and distal lake bed deposits all of which overlie stagnant ice (Fig. 4). The Okanagan system was an outlet for a major system of interior valleys (Fulton 1969), therefore major supraglacial drainage is to be expected. The extreme changes of flow direction required for ice to pass through the bends in the Kelowna-Squally point reach are conducive to early stagnation of the downglacier lobe and maintenance of an active shear zone immediately above the reach. Consequently, the hypothetical conditions are highly plausible and were partially described on the basis of sedimentological evidence alone (Shaw 1977, Fig. 14, stage b).

4 THE LACUSTRINE DEPOSITS

The glaciolacustrine deposits of the Okanagan valley in the vicinity of Penticton were described by Shaw (1975, 1977). The sequences interpreted as lake bed deposits show varves decreasing in thickness with height. The summer sub-units of these varves are typically silts which are commonly graded and may include cross-laminated coarse silts and fine sands deposited by turbidity underflows. These deposits are similar to glaciolacustrine sediments deposited during summer in modern and ancient environments. Convolute bedding in the summer deposits was attributed to slumping on the lake bed caused by differential compaction. The winter units are quite unlike those predicted by the classical model of quiescent conditions in glacial lakes during winter. Shaw (1977) and Shaw and Archer (1978) described winter units containing grainflow deposits, antidune bedding, and cross-lamination in sands and silts. Large rip-up clasts of silt and clay are also found in these units. A section recorded recently northeast of Penticton illustrates the complexity of the winter and late-summer units.

4.1 The Northeast Penticton section

The succession at Northeast Penticton (Fig. 1) appears initially to be lacking any systematic pattern (Fig. 5). However, if four major textural and structural units are considered, the sequence lends itself to interpretation in terms of the environmental conditions outlined. Stratified silts form persistent units throughout the sequence. These are of the order 10-20 cm in thickness and are generally fairly uniform in grain size but may include distinct laminae of coarse silt or fine sand. Rather diffusely graded microunits are common in these beds. The stratified silts are used for reference and are numbered (Fig. 5). The silts are commonly overlain by a thin, relatively indistinct bed of silty clay (see units 1, 3, 4, 5, 6). The second major stratification type, convolute lamination, commonly with included clay clasts, is often found above the thin, silty clay bed (see the sequences above units 3, 4 and 5). Continuous unctious clay beds, generally of the order one centimetre thick, are found overlying either the convolute laminations or the stratified silts. However, the clays are intimately associated with sands and, in some cases, massive sandy silts. The sands are often cross-laminated and commonly lie on erosion surfaces. Internal grading and rip-up silt clasts are found in the sands which become more pronounced

towards the top of the sequence. A repetitive succession of stratified silts, convolute lamination, and interbedded clay and sand or silt beds can be recognised.

The succession at Northeast Penticton corresponds in some measure to those at Summerland and East Okanagan Lake (Shaw 1977) and the section at Kickininee (Shaw and Archer 1978). The stratified silts are interpreted as summer deposits produced by sedimentation from meltwater overflows, interflows and underflows. The systematic appearance of convolute lamination above the first clay deposition at the end of the summer melt season has not been previously reported. Consolidation of the summer silts involves an upward discharge of pore water. Deposition of clays above the silts inhibits water removal. Under-consolidation with high pore water pressures will favour subaqueous mass-movement, even on relatively gentle slopes (Morgenstern 1967). The convolute lamination is, therefore, attributed to subaqueous slumping triggered by high pore water pressures.

The unctious clay laminae are interpreted as winter layers deposited from suspension. As such, they correspond to the classical concept of the glaciolacustrine varve. However, the associated sands, which were also deposited in the winter season, pose special problems. First, the sands are not found in adjacent summer layers and must have been introduced from outside the immediate area of deposition. Second, the current lamination, erosional surfaces, and rip-up clasts illustrate deposition from underflows which were much more powerful than any summer events. Finally, the internal grading illustrates that several underflow events occurred per season. Any explanation of these sands must cover their origin in winter, their source, the triggering mechanism which initiated the underflows, and the occurrence of several events per winter season.

The obvious source of the winter sands lies in the deltaic deposits formed at the margin of the active ice (Fig. 4). A complicated hydraulic situation is expected in this zone and the precise form and flow characteristics of the glacial drainage system cannot be given. However, some general principles can be suggested which are sufficient to explain the initiation of winter turbidity underflows. The observations at the Harrop section led to the conclusion that subglacial melting was occurring in the Okanagan trunk glacier. Some connectivity is expected between the lake (Fig. 4) and the subglacial drainage system (Röthlisberger 1971). Lake levels in glaciated areas tend to fall during winter when the major component of inflow, surface meltwater, is substantially reduced (Gilbert 1975). The englacial and subglacial drainage system cannot be expected to respond as quickly as lake level changes, and the subsequent pressure imbalance will produce strong flow from the glacier system, through the deltaic sediments, and into the lake. Upward seepage through sands and silts produces a quick condition with resultant failure on even low slopes (Andresen and Bjerrum 1967). Failure of sands in this fashion is reported to occur by regressive sand slides. The proposed hydraulic and sedimentological conditions explain the initiation of underflows during winter. Regressive sand slides explain the occurrence of a number of events per season. As the winter season is one of generally low lacustrine suspended sediment concentrations, the density contrast between the ambient lake water and a winter turbidity underflow is maximised. This may partly account for the high flow powers necessary to produce the erosion and traction transport illustrated by the winter sands.

5 CONCLUSIONS

The conditions proposed for deglaciation in the Okanagan valley appear not to have modern analogues. The pattern of deglaciation is, therefore, a matter of interpretation. Stagnation of ice lobes with a zone of shearing at the junction of active and stagnant ice is proposed. Accumulation of debris on the surface of stagnant lobes under active permafrost conditions prevented surface melting, and bottom melting played a major role in the release of till from ice and the lowering of the supraglacial surface. Downmelting of the stagnant portion of the glacier and the constant supply of surface material in the zone of shearing led to the formation of a glacial lake over stagnant ice, but with many of the attributes of a proglacial lake. Undermelting and lowered lake levels led to seepage and the release of slides in deltaic sands and silts during winter. These slides were responsible for the coarse beds found within the winter component of glaciolacustrine varves.

6 ACKNOWLEDGEMENTS

Grants from the National Research Council are gratefully acknowledged. We wish to thank Fran Litschko for typing the manuscript and Inge Wilson for drafting the figures.

7 REFERENCES

Andresen, A. & L. Bjerrum 1967, Slides in subaqueous slopes in loose sand and silts. In A.F.Richards (ed.), Marine Geotechnique, p. 221-239. Chicago, Univ.Illinois Press.

Flint, R.F. 1935, "White Silt" deposits in the Okanagan Valley, British Columbia, Royal Soc.Canada Trans. 29:107-114.

Fulton, R.J. 1965, Silt deposition in late-glacial lakes of southern British Columbia, Am.Journ.Sci. 263:553-570.

Fulton, R.J. 1969, Glacial lake history, southern interior plateau, British Columbia, Geol.Surv.Canada Paper 69-37.

Gilbert, R. 1975, Sedimentation in Lillooet Lake, British Columbia, Can.Journ.Earth Sci. 12:1697-1711.

Hooke, R. LeB. 1970, Morphology of the ice-sheet margin near Thule, Greenland, Journ.Glaciol. 9:303-324.

Johnston, G.H. & R.J.E. Brown 1964, Some observations on permafrost distribution of a lake in the Mackenzie Delta, N.W.T., Canada, Arctic 17:163-175.

Koteff, C. 1974, The morphologic sequence concept and deglaciation of Southern New England. In D.R.Coates (ed.), Glacial Geomorphology, p. 121-144. Binghamton, State Univ.New York.

Morgenstern, N.R. 1967, Submarine slumping and the initiation of turbidity currents. In A.F.Richards (ed.), Marine Geotechnique, p. 189-220. Chicago, Univ.Illinois Press.

Nasmith, H. 1962, Late glacial history and surficial deposits of the Okanagan Valley, British Columbia, Brit.Columbia Dept.Mines and Petr.Res.Bull. 46.

Röthlisberger, H. 1972, Water pressure in intra- and subglacial channels, Journ. Glaciol. 10:177-203.

Shaw, J. 1975, Sedimentary successions in Pleistocene ice-marginal lakes, Soc.Econ. Paleontol.Mineral.Spec.Publ. 23:281-303.

Shaw, J. 1977, Sedimentation in an alpine lake during deglaciation, Okanagan Valley, British Columbia, Canada, Geogr.Ann. 59A: 221-240.

Shaw, J. & J. Archer 1978, Winter turbidity deposits in Late Pleistocene glaciolacustrine varves, Okanagan Valley, British Columbia, Canada, Boreas, 7:123-130.

Shaw J. in press, Deglaciation in perma-frozen regions, Boreas.

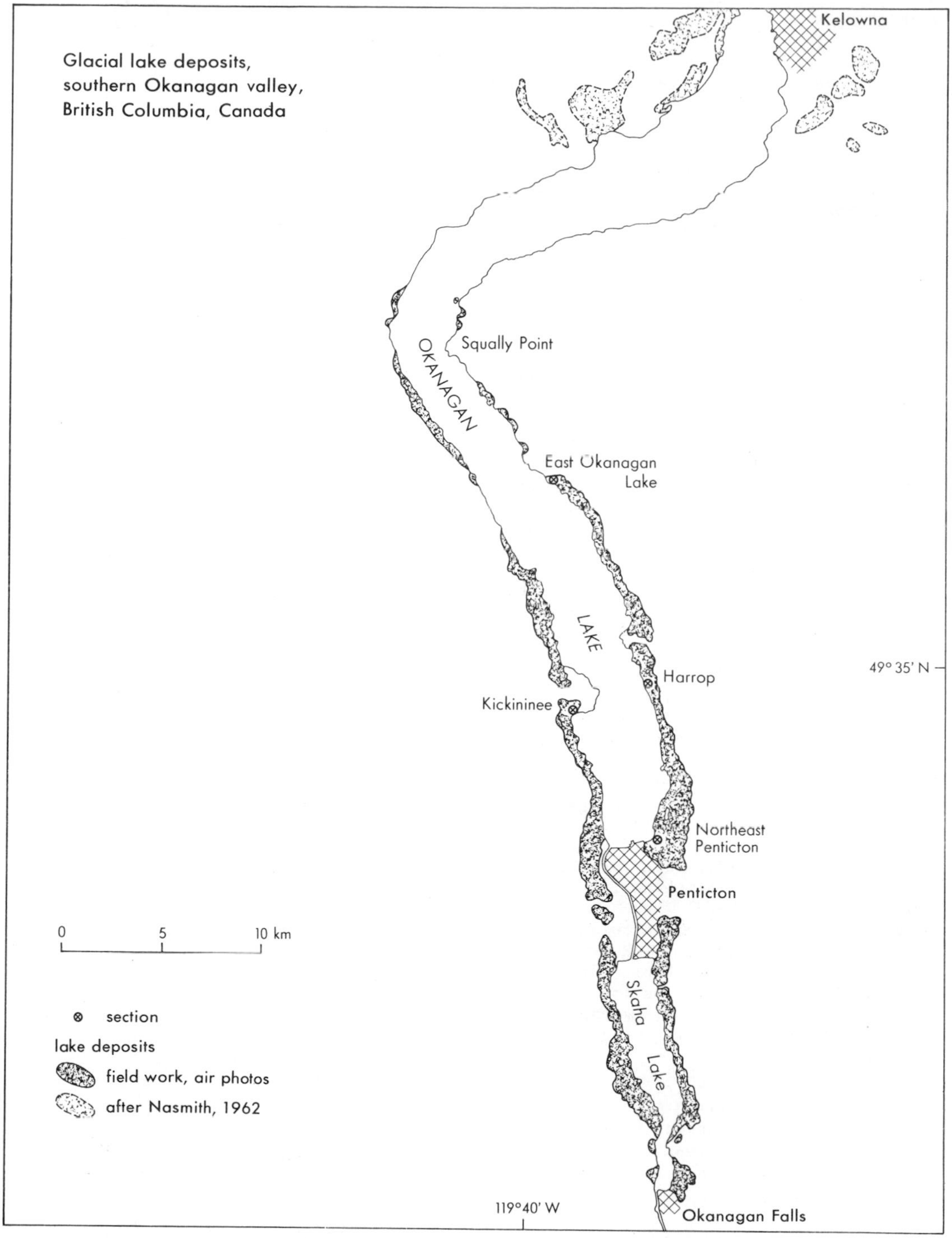

Figure 1 Glacial lake deposits in the Okanagan valley, Kelowna to Okanagan Falls

Harrop section

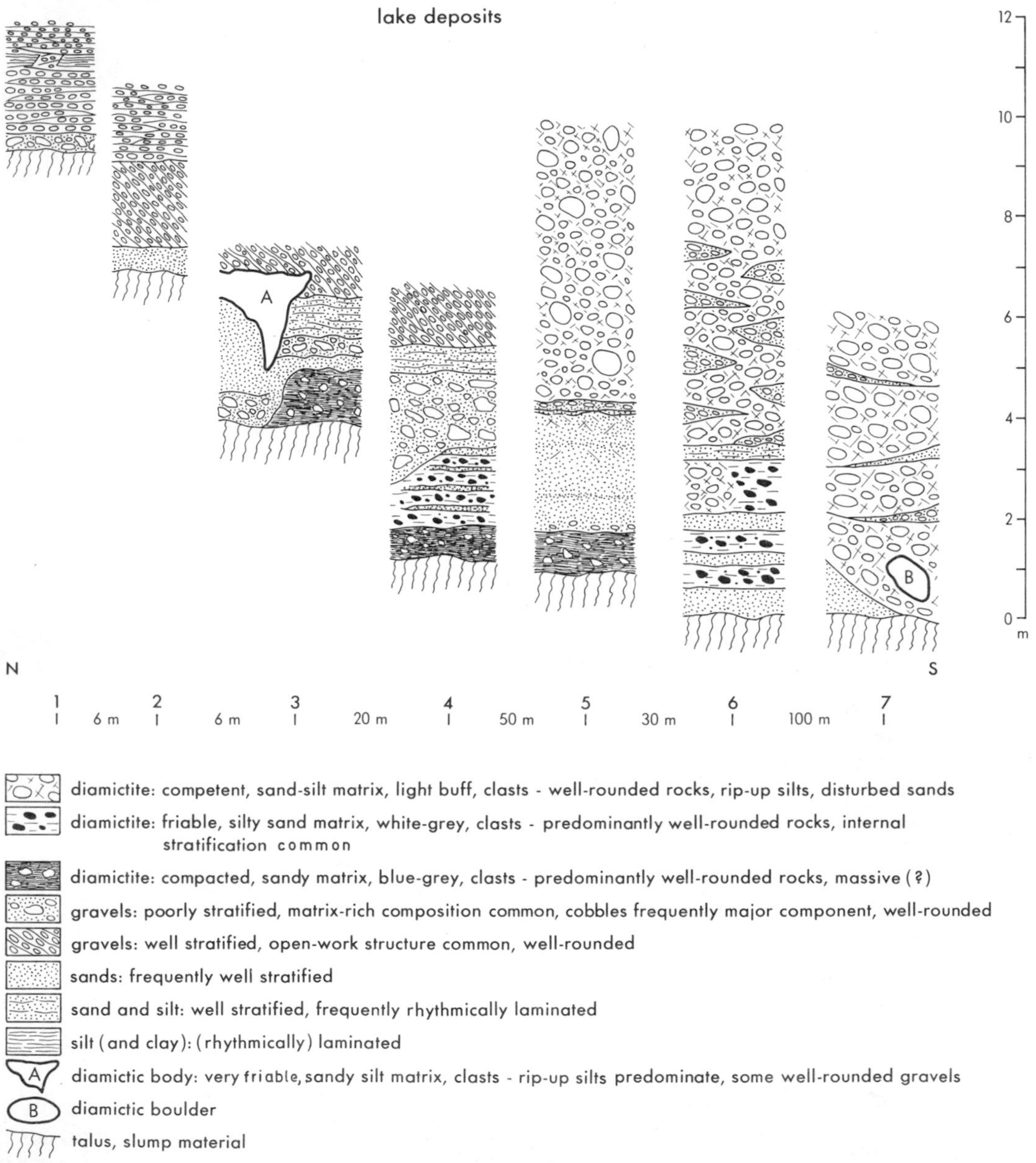

Figure 2 The Harrop section, map reference 130946 Summerland (B.C., Canada) 82E/12E CTS

Deglaciation under permafrost conditions

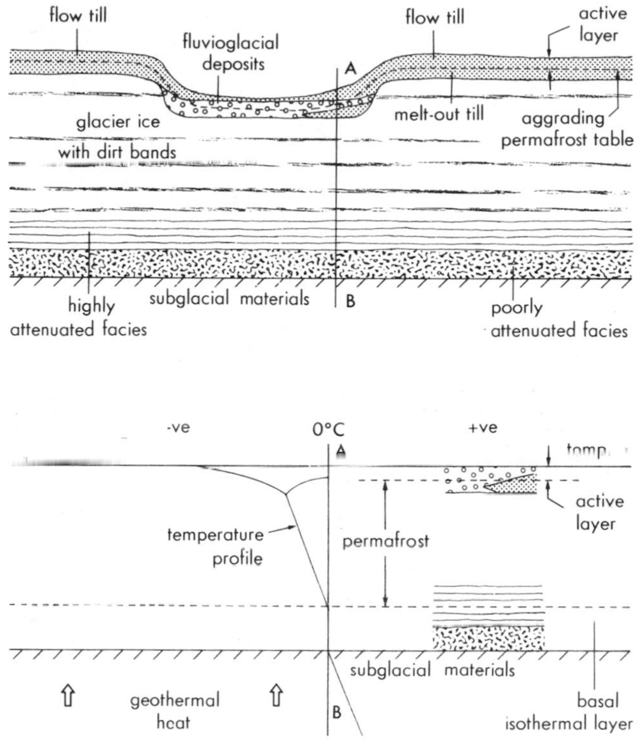

Figure 3 Deglaciation under permafrost conditions during climatic amelioration. Note the development of syngenetic permafrost in the supraglacial sediments and an approximately isothermal basal layer at pressure melting temperatures

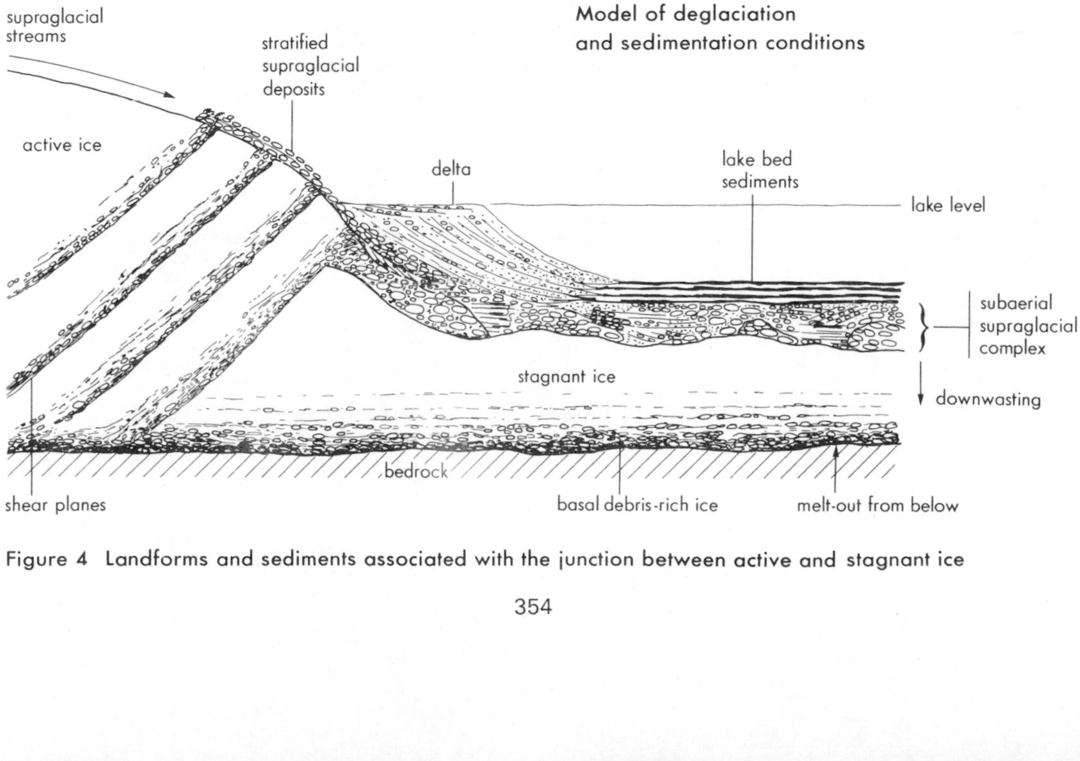

Figure 4 Landforms and sediments associated with the junction between active and stagnant ice

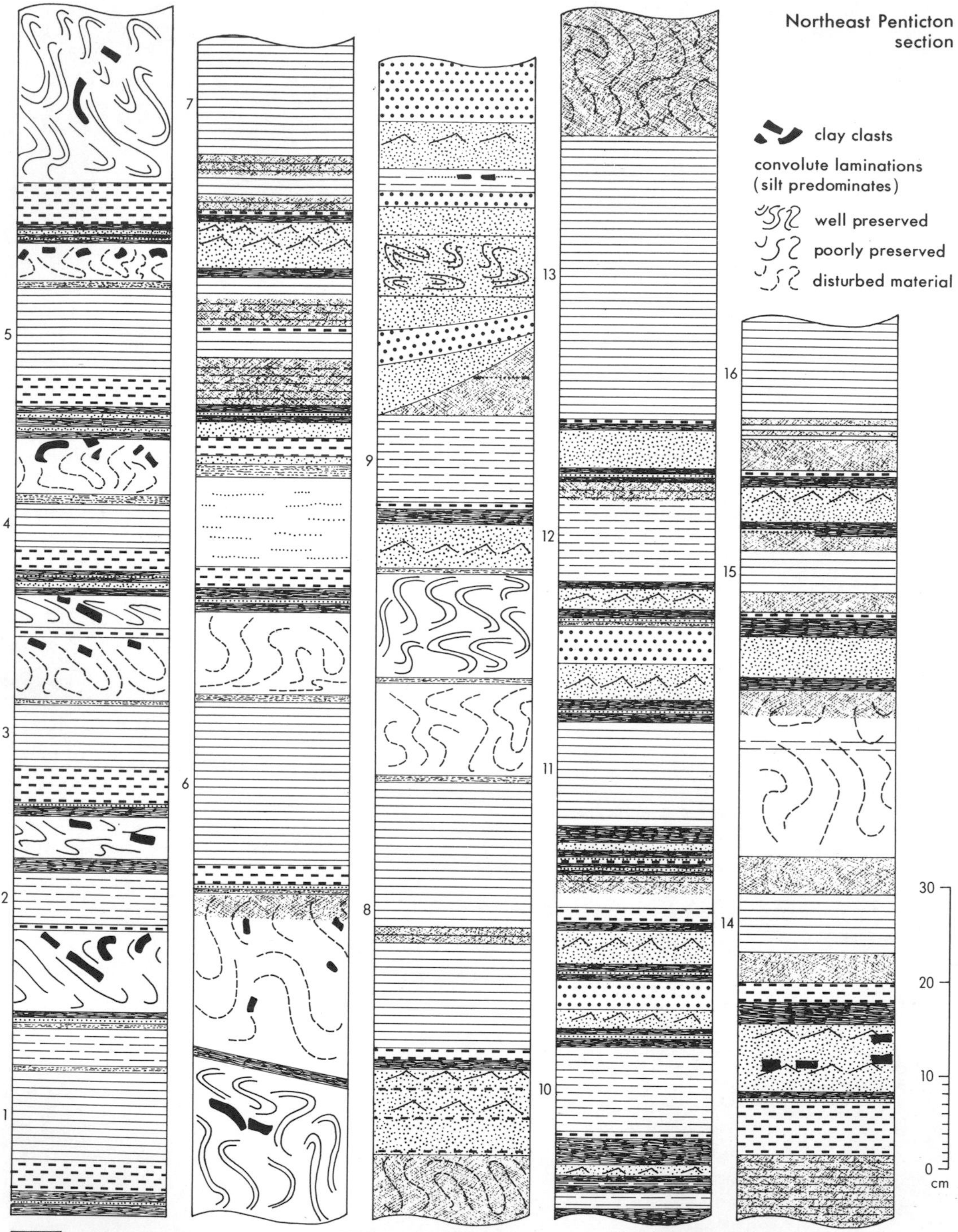

Figure 5 The northeast Penticton section - graphic log. Map reference Summerland, 134869, 82E/12E CTS

Deltaic sedimentation in glacial Lake Blessington, County Wicklow, Ireland

J.M. COHEN
Trinity College Dublin, Dublin, Ireland

1 INTRODUCTION

This paper is an outline of research presently being carried out on processes of glaciolacustrine sedimentation. Sedimentary features of a particularly large and complex delta are described and criteria for possible depositional mechanisms are presented.

1.1 Study Area

Glacial Lake Blessington was formed during the later stages of the Midlandian (Weichselian) glaciation and occupied large river basins on the western side of the Wicklow mountains. The topographically low area which the lake occupied was exposed as the midland ice from the central lowland plain to the west and the local ice cap from the Wicklow mountains withdrew. The lake existed during the last three maximal phases of the last glaciation in Ireland, (fig.1). The corresponding positions of the associated ice masses is shown by the limits of deposition of limestone rich till from the midland ice and granite rich till from the local ice of the Wicklow mountains. Synge (1977) showed that fluctuations in lake level were associated with these ice movements which determined the distribution of ice marginal deltas at various levels around the lake boundaries.

During its maximum extent, the lake was approximately 24 kms. long and 5 kms. wide, being markedly elongate in shape and orientated roughly north to south. The lake was bound on its western side by a long ridge composed of Ordovician slate. Three breaches in this ridge provided meltwater entrances into the lake from the western boundary with the midland ice, (Farrington, 1957). Corresponding ice marginal deltas associated with these meltwater entrances are the Moanaspick delta to the north, the Blessington delta in the centre and the Pollaphuca delta to the south.

The main area of study in this investigation is the Blessington delta. This delta is about 5 kms. wide and is the largest of glacial Lake Blessington. Excavations for sand and gravel have exposed about ½ sq.km. of the internal structure of the delta. From the proximity of the ice shown by Synge (1977) and the presence of flow tills, the delta is clearly ice marginal. As it will be shown later it was formed in a high energy environment.

The lakeward facing slope seen in unexposed portions of the delta corresponds roughly with foreset dips seen in cross sectional exposures. Ice contact slopes can be seen on the western side of the delta. Subaerial or subglacial fluvial erosion has resulted in the incision of large channels in the delta and has also resulted in rare preservation of topset sediments. The delta surface is draped by a till unit which closely follows the underlying morphology.

2 DESCRIPTION OF THE SEDIMENTS

The delta has been subdivided primarily into bottomset and foreset sediments. Foreset sediments consist mainly of beds of gravel with consistent dip angles. Bottomset sediments show lower dip angles and a predominant content of current bedded sand. Only those sections are described which represent the principle significant features of the delta.

2.1 Foreset Sediments

Foresets are widely distributed around the exposed portion of the delta. The textural range passes from coarse gravel (-6.0 ϕ)

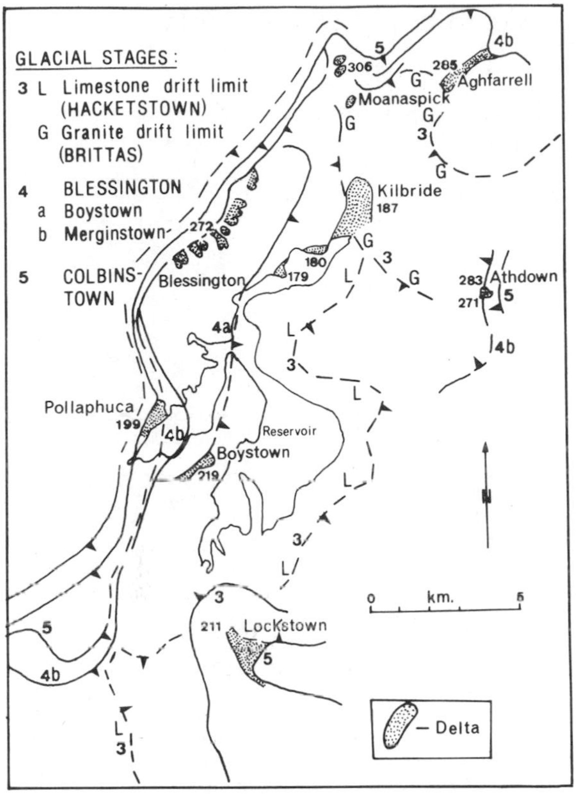

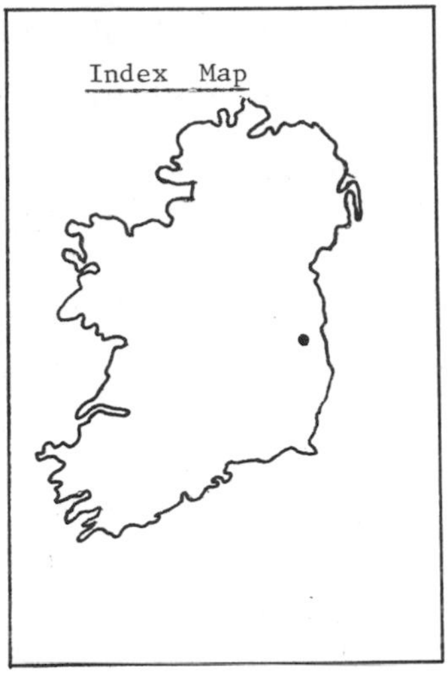

Figure 1. Map showing the location of deltas of glacial Lake Blessington associated with respective ice marginal positions, (redrawn from Synge, 1977). The numbers refer to heights in metres of delta surfaces. Triangular symbols indicate lake side of ice margins.

to coarse sand (0.5 Ø). Occasional large boulders up to 75 cms. (b axis) in diameter are present within the coarser beds. Bed thicknesses range from 0.25 to 1.50 metres and generally have subparallel dip angles ranging from 15 to 35 degrees. Coarse poorly sorted beds are structureless in contrast to finer grained beds which are frequently parallel laminated. The process of avalanche is thought to have been particularly active in the deposition of foreset beds.

The maximum single exposure thickness of foresets observed is 26 metres and is capped by 2 metres of massively laminated topset gravels. The top of this exposure lies at a height of 272 metres O.D. and is thought to closely correspond to the lake level which existed during the deposition of the major portion of the delta.

2.2 Delta progradation

Delta progradation is demonstrated by the occurrence of coarsening upward sequences in the bottomsets. Interdigitation of foreset gravel units become more frequent towards the top of bottomset sequences. Contacts between foresets and bottomsets are clearly transitional with lateral decreases in dip angles as foresets grade into bottomsets.

2.3 Vertical changes of dip angles

Abrupt change of dips between foreset and bottomset sediments are explained by shifting directions of meltwater input over the delta. This alteration can be seen in figure 2 where a succession of foresets (A) is overlain by 4 metres of bottomset current bedded sand (B), succeeded by 4 metres of foresets (C). A shift in the meltwater input which deposited beds A to a more distant area, created a lower energy depositional environment at the original site and deposited beds B. A return of high energy deposition represented by foresets C, was caused by either a return of the original input or by the generation of a new one.

Figure 2. The effect of changing directions of meltwater input is shown by the change of dip between foresets A and C and bottomsets B. Foreset current directions from right to left.

2.4 Processes of deposition on foreset slopes

The main process for the deposition of the coarse grained steeply dipping foreset beds is thought to be avalanche. Allen (1970a) showed that with larger sediment supply rates, occasional avalanche turns into continuous sliding. Where sediment supply is small, sediment builds up at the top of the foreset slope until failure occurs, producing intermittent slumps. In the studied delta sediment supply was obviously large, as shown by the great thickness and coarse calibre of the foreset sediments and continuous sliding was probably quite common. It also seems quite possible that variations in the sediment supply caused by fluctuations in meltwater discharge and and shifting directions of input could have resulted in accumulations of sediment which caused intermittent slumping.

The activity of bottomflowing turbidity currents is suggested to explain the formation of current bedded sediment in deep water. Cosets of dune cross bedding (fig.3) found within foreset sequences occur at elevations which indicate lake depths up to 34 metres. The cosets are separated by regressive climbing ripples. The occurrence of regressive climbing ripples in context with the foreset environment conforms closely with the situation predicted by Jopling (1960). He related the origin of this structure to flow separation that occurs when a current expands over a sharp discontinuity such as a delta front. Reverse circulation is directed towards and up the foreset slope

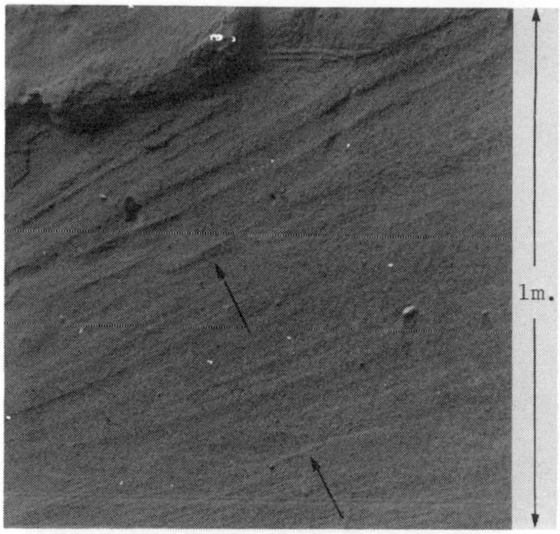

Figure 3. Cosets of dune cross bedding separated by regressive 'A' type climbing ripples, (arrows).

producing an upslope angle of climb. Jopling suggested a flow velocity not less than 0.6 metres per second for the formation of this structure. This at least suggests a minimum velocity for the flow of turbidity currents down the foreset slope.

It does not seem possible to isolate the process of turbidity current flow from avalanche on the foreset slopes. During avalanche sediment would be thrown temporarily into suspension. This would result in an interchange of particles between the avalanche and the turbidity current flow.

A succession of fine grained rhythmites within a foreset sequence contain features which suggest the activity of discrete types of mass flow under low energy flow conditions. The rhythmites (fig.4) consist of a 31.7cms. thick succession containing 19 couplets of clay and silt or fine sand, occasionally separated by sand partings and diamicton subunits. Other features present include intraformational crumplings. The presence of such fine grained sediment within a predominently high energy environment indicates that a period of sudden reduction in meltwater input over this part of the delta occurred. This can be explained by a change in the direction of meltwater input.

Current lineations can be seen on the bedding surfaces of silt and fine sand laminae. Both subparallel longitudinal lineations and transverse markings have been identified, (see fig.5). The longitudinal lineations consist of broad flat low profile ridges and are irregularly spaced. The ridges are between 0.5 to 1.2 cms. in width and are raised rarely more than 1 mm. above the surrounding troughs. The longitudinal lineations consist of broad flat low profile ridges and are irregularly spaced. The ridges are between 0.5 to 1.2 cms. in width and are raised rarely more than 1 mm. above the surrounding troughs. The longitudinal lineations are subparallel to the extent that the orientations vary up to 10 degrees on a single bedding plane. Lineation orientations from successive bedding planes are orientated roughly in the same direction with a directional deviation of about 15 degrees. The mean orientation corresponds with the direction of maximum dip of the surrounding foresets. The lineation ridges are composed of slightly coarser sediment than the adjacent troughs. This indicates that the sediment comprising the ridges is current sorted and the feature may therefore be referred to as current lineations.

The transverse markings (fig.5a) are sharp crested and closely spaced ridges orientated at right angles to the longitudinal lineations. Ridge crests often have profile heights up to 3 mms. and are generally spaced between 3 to 9 mms. apart. The ridges are generally symmetrical although asymmetric profiles have also been found with lee sides steeper than stoss sides. Longitudinal lineations are often superimposed upon transverse markings. The preservation of transverse markings in these cases, indicates that lower current velocities were responsible for the formation of the superimposed lineations.

Current lineations similar to those described are quite common in ancient turbidites, eg. Kuenen (1957), Dzulynski and Sanders (1962), Dzulynski and Smith (1963), Dzulynski and Walton (1965), Pettijohn and Potter (1964), Rich (1950). Erosional markings similar to the transverse markings described above have also been produced experimentally by Allen (1969) and Mantz (1978). Allen (1969) produced transverse markings at mean flow velocities of 138 cms./sec. and with decreasing flow strength the subparallel lineations were formed. This supports the deduction made earlier that transverse markings were produced at higher flow velocities. Unfortunately, it is not possible to apply experimental current velocities used to produce these structures directly to natural phenomena. The variables involved are those which define the Reynolds Number of flow, which can not be determined from field evidence.

Sand partings are quite common within the silt laminae of the rhythmite couplets. Coarse granular sand partings (see fig.5c) are rarely thicker than one grain thickness, whereas medium sand partings may be up to 3 mms. in thickness. From a bedding plane view, the partings can be seen to be

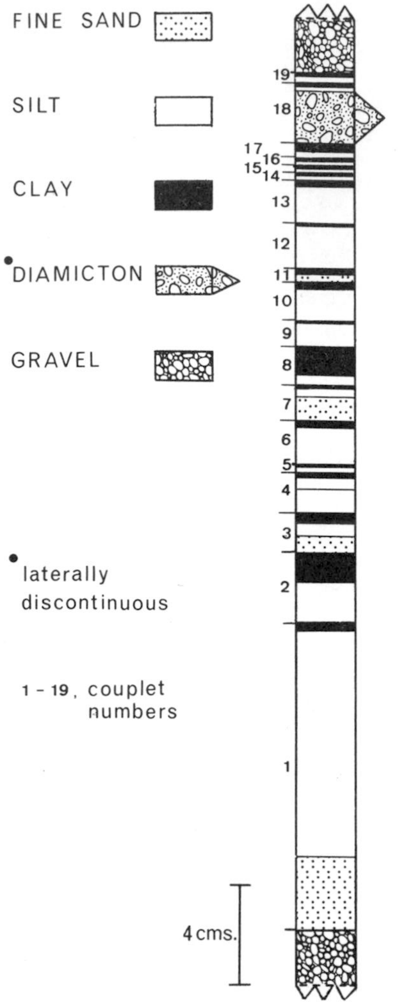

Figure 4. A vertical sequence of rhythmites deposited on a foreset slope.

extremely well sorted with a distinctive unimodal grain size distribution.

In contrast to the sand partings the diamicton subunits are distinctively poorly sorted consisting of fine pebble clasts supported in a clay matrix. The diamicton beds are laterally discontinuous with a lenticular appearance in sections cut transverse to the current flow directions. The thickness of the diamictons range from 0.5 to 3.0 cms.

A distinction can be made between the unimodal sand partings and the polymodal diamicton subunits within the concept of subaqueous mass transport mechanisms. According to Bagnold (1954, 1956) two

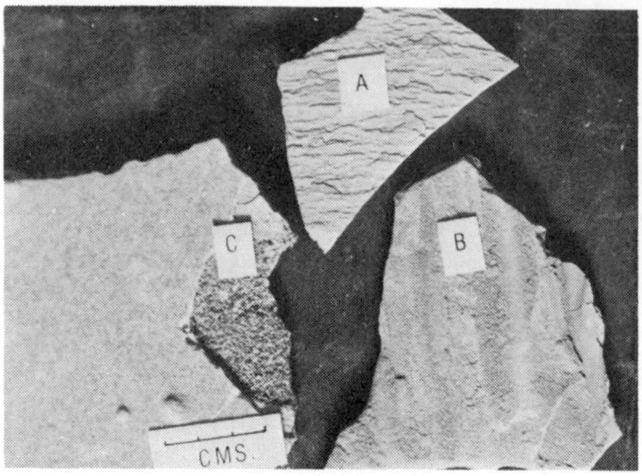

Figure 5. A. transverse markings, B. longitudinal parallel lineations, C. granular sand parting.

different types of behaviour are likely to be characteristic of mass transport on unstable slope environments. The first type, which conforms closely with the unimodal property of the sand partings, is characterised by flows of low viscosities in which dispersive pressures are generated from actual impacts between the grains. The second type of flow, which resulted in the deposition of the diamicton subunits, has a high viscosity due to the addition of mud to the flow. This results in the generation of dispersive pressures as a result of ordered shear velocity changes as dispersed grains approach one another. A further dilatent effect is introduced by the non-Newtonian behaviour of fine grained suspensions. Bagnold (1954) classified the two types of flow as grain flow with varying degrees of viscosity, Carter (1975) suggested the limitation of the use of the term grain flow to lower viscosity flows and applied the term slurry flow to higher viscosity flows. The latter terminology appears to be a more suitable classification to describe the processes of deposition active in this case. The formation of the sand partings is therefore attributed to the process of grain flow and the diamicton units to the process of slurry flow. As Carter emphasised however, both grain flow and slurry flow together with avalanche and turbidity current flow belong to the general process of subaqueous mass transport and a continuum between these four transport mechanisms must be envisaged.

Intraformational crumplings are also present in this rhythmite sequence, (fig.6). The sediments appear to be literally crumpled to produce isoclinal and recumbent folds similar to the roll up structures found by Gustavson et al. (1975). The deformation incorporates only a limited number of

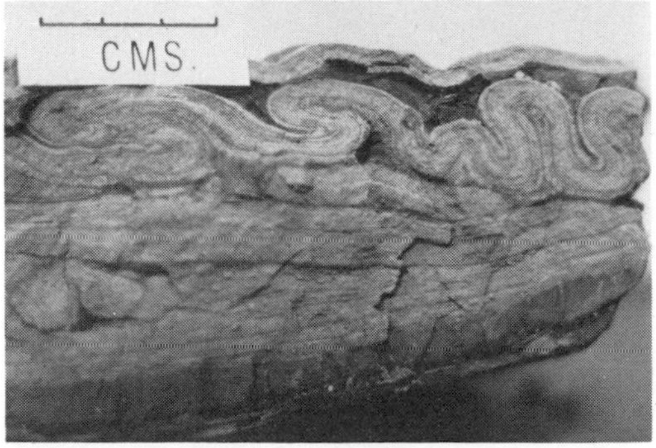

Figure 6. Intraformational crumplings in rhythmite sediments

laminae and does not extend throughout the complete succession. This indicates that the structure is syndepositional. The axes of the folds run almost transverse to the maximum forset dip direction. This indicates that the feature was probably produced by a movement of the cohesive sediment down the foreset slope resulting in compression and overturning of the strata. This form of deformation is closely similar to structures observed in the Silurian turbidites in Aberysthwyth, Wales. These have been described by Rich (1950) who suggested that intraformational crumplings may be used as a criteria for the recognition of slope depositional environments. Rich also used the presence of flow markings similar to those described earlier, together with the repeated alteration of uniform relatively thin bedding of silt and clay and the general absence of ripple marks as features consistent with a slope depositional environment.

It may therefore be concluded that the apparently anomalous position of the rhythmite succession within the foresets never the less contains features consistent with a foreset environment of deposition.

3 BOTTOMSET SEDIMENTS

These sediments contain a range of primary sedimentary structures including, 'A' and 'B' type climbing ripples, drape laminations, parallel laminated units, structureless beds, dune cross bedding, graded beds and flame structures. The texture of the sediments range from fine to coarse sand. Coarser sediments are present in transitional areas where foreset gravels interdigitate with finer grained bottomset sediments. A coarser fraction is also added by the presence of dropstones in the bottomsets.

Generally the bottomset beds dip at angles less than 5 degrees, however, steeper dip angles up to 15 degrees have been found in very localised areas. These exceptionally steep angles are associated with local undulations of thick vertical sequences of bottomset beds. The undulations usually exist as isolated anticlines with 'wavelengths' up to 35 metres and amplitudes up to 2 metres. In some cases the dip created by these undulations can be found truncated by bottomset beds with opposing dip directions. This suggests that the feature is depositional and implies that currents had to flow over an undulating lake floor to continue deposition. It seems possible that this irregular bed geometry was formed by the deposition of lobes of sediment from local meltwater inputs.

3.1 Objectives and techniques

The ultimate object in this study of bottomset sediments is to be able to formulate a lateral facies subdivision of bottomset sediments based upon the proximity of depositional sites to foreset locations. Grain size analysis has been extensively carried out on bottomset sediments in order to assist in the facies subdivision.

Moment statistics have been used to compute grain size parameters from grain size distributions. Grain size analysis has been carried out at 0.5 ∅ intervals using the sieve and hydrometer techniques. At present cluster analysis is being applied as a possible alternative to the use of grain size parameters in this research, in order to identify groups of texturally similar sediments. This technique has the advantage of incorporating complete grain size distributions and is unaffected by the modality of the distribution. The results of this part of the investigation are not yet avaliable.

3.2 Faulting

Faults are widely distribued in the bottomset sediments and consist of normal high angle displacements. Fault dipsrange between 50 degrees to nearly vertical, with displacements up to 40 cms. The displacements are generally linear, although occasional convex upwards faults have been recognised. The absence of reverse faults serve to distinguish the origin of the disturbance from that caused by the melting of buried ice, (McDonald and Shilts, 1975). In two separate sections, major faults have been recognised with consistent lakeward dip directions. This indicates that the faulting may have been caused by gravitational movements

towards the unsupported side of the delta. In some cases syndepositional faulting is quite clearly shown by undisturbed sediment overlying fault complexes. In general the faults continue to the top of the exposures suggesting that postdepositional faulting was active.

3.3 Process of deposition of bottomset sediments

Bottomset sediments are present at heights which indicate water depths up to 38 metres and are laterally extensive for distances up to 350 metres from the nearest foresets. The only mechanism which could be envisaged to account for this large distribution of sediment is that of bottom flowing turbidity currents. Additional evidence supporting the activity of these currents is the presence of truncation planes and erosional scours in bottomset sediments. Erosional planes are thought to be an effect of directional changes of input in bottomset areas of deposition.

Overturned flame structures and graded beds in bottomsets also suggest the activity of turbidity currents. In one case, a series of flame structures situated at the base of a graded beds are consistently overturned in a direction corresponding with the direction of current flow. Similar structures have been described by Sanders (1965) as; "pointed and curved wisps of mud drawn up from the bottom". Shrock (1948) attributed their origin to the interaction between a turbulent suspension and a cohesive but hydroplastic bottom.

3.4 Vertical sequence of bedforms in bottomset sediments

A sequence of bottomsets is shown in figure 7 which exemplifies the range of bedforms commonly found in the sediments. The vertical sequence consists of a two part section. The lower portion is situated 40 metres to the northwest of the upper portion and is more proximal to the laterally equivalent foresets. These foresets are situated 170 metres to the north of the lower portion. The sequence is not continuous as an overlap could not be identified between the upper and lower portions, although the elevations overlap. This sort of problem has been experienced quite often in the field work and is mainly due to the undulating nature of the bottomsets, mentioned earlier. The sections are cut roughly parallel to the general flow directions.

Climbing ripples are well exemplified in this sequence and are also most informative. As shown by Allen (1971) and Walker (1969), the angle of climb bears a relationship to the relative contribution of sediment falling out of suspension to that carried as bedload. According to Allen (1971), an increase in the ratio of sediment falling out of suspension to the rate of deposition as bedload would cause an increase in the angle of climb. The 'A' type climbing ripples are characterised by a lack of preservation of stoss side laminae with angles of climb between 5 to 20 degrees and 'B' type show complete preservation of stoss side laminae with angles of climb between 20 degrees to nearly vertical, (Jopling and Walker, 1968). Deposition totally from suspension is represented by drape laminations, (Southard et al., 1972). The clay drapes are similar to those described by Gustavson et al.,(1975). The clay thicknesses range from 1 to 3 mms. and extend laterally at constant thicknesses.

The sequence shows vertical changes in bedform and mean grain size which immediately indicates changes in flow power. Periods of no flow are represented by the drape laminations. Exceptionally high flow velocities are represented by coarser units of parallel laminations and dune cross bedding.

The general sequence represents separate and irregular flow events during which there were fluctuations in current velocity.

In contrast to this sequence, glaciolacustrine deltaic sequences described by Gustavson et al. (1975), demonstrates complete single flow events, represented by vertical transitions between successive drape laminations consisting of 'B' to 'A' to 'B' type climbing ripples. In that case, the lower 'B' type ripples represents the commencement of flow to a maximum velocity represented by the 'A' type ripples. The subsequent reduction in flow power towards the end of the flow event is represented by the upper 'B' type climbing ripples which show an increase in angle of climb towards the top of the unit.

In this case however, single decellerating flow events cannot be recognised. Vertical increases in angle of climb within single units of climbing ripples are rare in these bottomsets. This may be attributed to the occurrence of successive accelerations in flow velocity before the current decelerated sufficiently to cause an increase in angles of climb. Pulsations in flow may be explained by short term changes in meltwater discharge. Allen (1971) showed that short term fluctuations in meltwater discharge can control vertical changes in climbing ripple types and may be visualised to occur in a matter of hours or tens of hours. Rapid rates of sedimentation during the formation of climbing ripples are evident from the

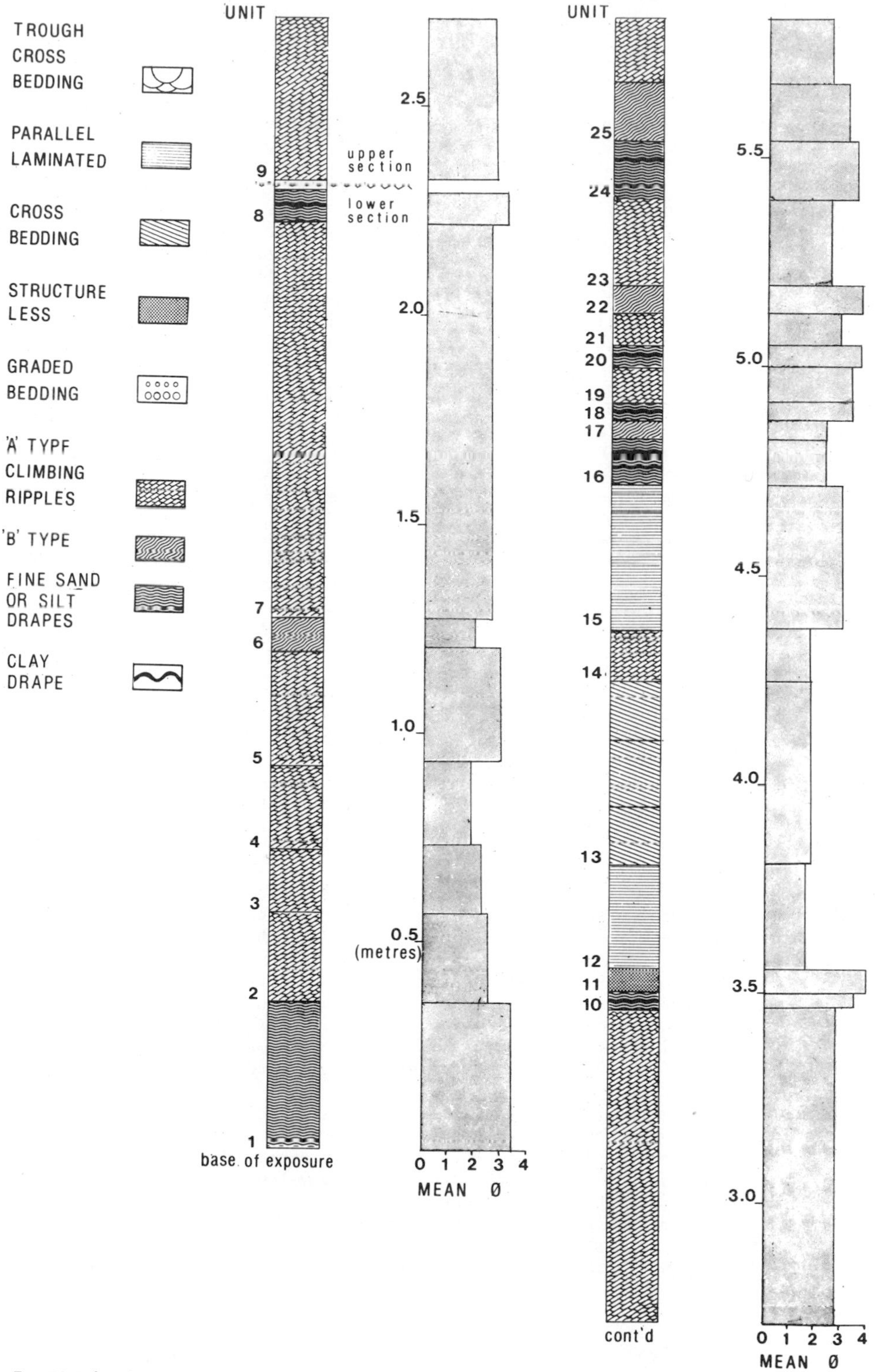

Figure 7. Vertical sequence of bottomset sediments

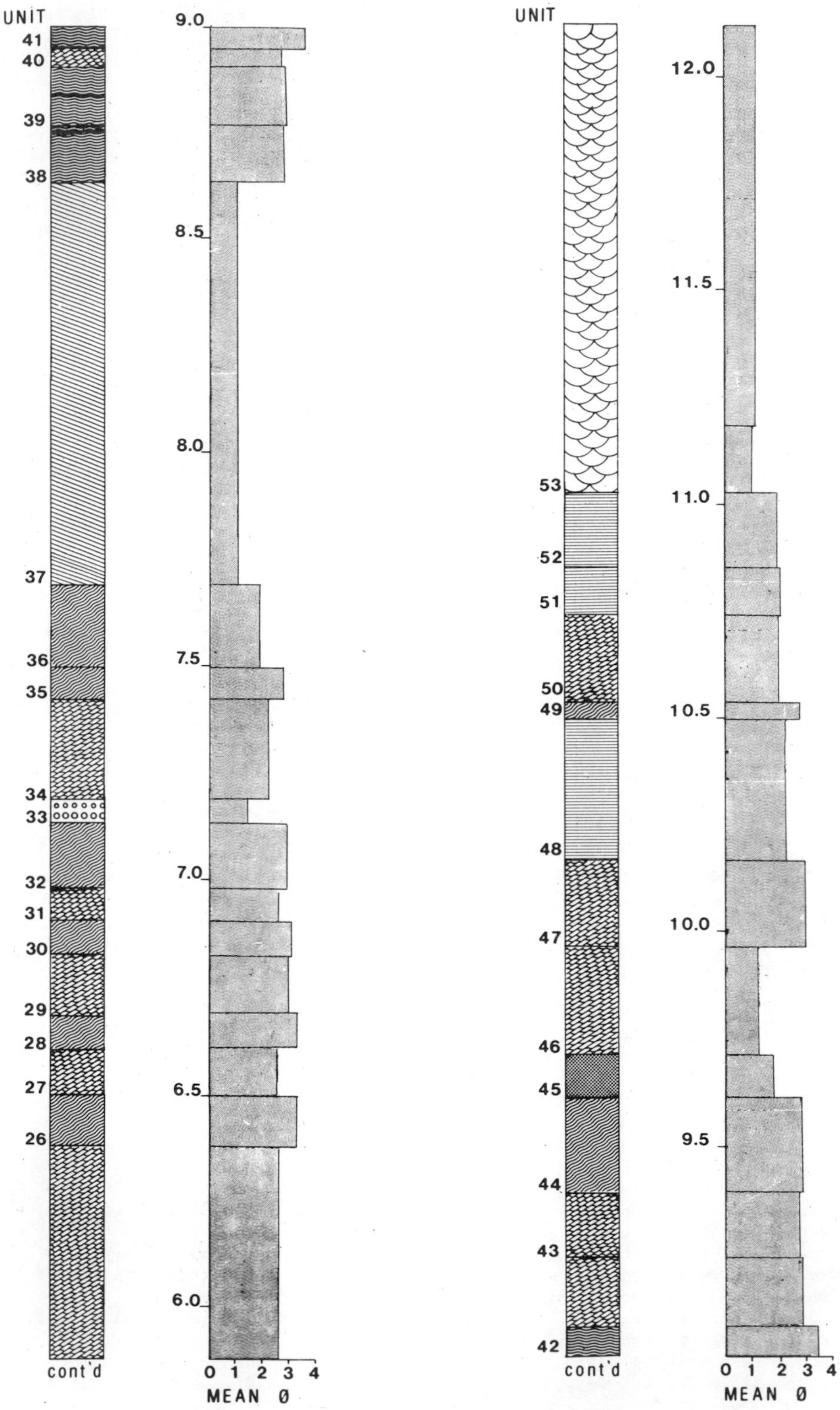
UNIT
41
40
39
38
37
36
35
34
33
32
31
30
29
28
27
26
cont'd
9.0
8.5
8.0
7.5
7.0
6.5
6.0
0 1 2 3 4
MEAN Ø
UNIT
53
52
51
50
49
48
47
46
45
44
43
42
cont'd
12.0
11.5
11.0
10.5
10.0
9.5
0 1 2 3 4
MEAN Ø

experimental results of Southard et al. (1972) which indicate that thicknesses of climbing ripples in the order of 15 cms. could be deposited in about 2 hours.

It therefore seems probable that great thicknesses of bottomset sediments, such as the sequence illustrated in figure 7, represents an extremely rapid rate of sedimentation. It is not clear whether changes in the direction of meltwater input or seasonal variations in discharge controlled the actual turbidity current flow event. Subsequently, it is not clear as the duration of no flow represented by the drape laminations.

3.5 The occurrence of dune cross bedding in bottomset sediments.

This feature deserves special consideration in the context of bottomset sediments deposited by turbidity currents. As far as the author is aware, dune cross bedding in glaciolacustrine bottomset sediments have only been recorded on one occasion in the past, (Huddart, 1970). In the Blessington delta, dune cross bedding is quite common in the bottomset sediments and is represented by cosets of cross bedding as well as by single units. In all cases where exposure permits, a general dune profile can be seen associated with the cross bedding. In the sequence illustrated in figure 7, a composite unit consisting of three cosets of cross bedding (unit 13) can be identified. Foresets of the cross bedding are tabular with a concentration of coarser grains towards the base. The unit can be traced laterally for 32 metres and shows a lenticular profile with a maximum thickness of 1.10 metres at the centre of the dune. Figure 3 shows cosets of dune cross bedding similar to unit 13 except for the presence of the regressive climbing ripples.

According to Allen (1970b) dunes are virtually absent from turbidite sequences. In normal fluvial deposits, declining flow power produces sequences of sedimentary structures passing from plane beds to dune cross bedding to climbing ripples. In turbidite sequences, the dune interval is virtually always omitted. Allen (1970b) attributed this absence to the unavailability of suitable grain sizes for transport and suggested a lower size limit of 2 Ø necessary for the formation of dunes in turbidites. In all cases where dune cross bedding has been found in this delta, the mean grain size is equal to and coarser than 2 Ø. This supports Allen's arguement and also substantiates the evidence that turbidity currents were responsible for the deposition of delta bottomsets. It must also be considered that the supply of suitable grain sizes for transport, controls to an extent the bedform type in this glaciolacustrine environment.

4 SUMMARY

The delta represents a high energy depositional environment in which sediment distribution was controlled to a large extent by shifting directions of meltwater inputs. This has resulted in sudden vertical alterations between foreset and bottomset sediments.

Superimposed upon this major control was the effect of fluctuations in meltwater discharge which resulted in minor variations in bedform and texture in vertical sequences.

Foresets were deposited mainly by avalanche of coarse grained sediments. Observations suggest the existance of certain types of mass transport mechanisms including, grain flow, slurry flow, avalanche and turbidity current flow in the foreset environment. It does not seem possible that only these discrete types of mass transport mechanisms operated. Depending upon the relative concentration of fines in suspension, intermediate flow types must have operated.

5 RESEARCH IN PROGRESS

Experimental investigations are being carried out to evaluate the relative contributions of certain variables in controlling the competency of turbidity currents which are evidently responsible for the deposition of the delta bottomset beds.

The experiments are being carried out in a small scale model of a glacial lake, in which continuously flowing low density turbidity currents are controlled. The principle selected variables include: concentration of suspended sediment, mean grain size and sorting of the sediment. Secondary variables include: temperature, discharge and flume width.

By repeating experiments under varying conditions of flume width and discharge, the scale of the experiment can be altered. This is being used to test if a constant relationship between the principle selected variables and the currents' competency exists regardless of scale and may therefore be applied to nature.

The design of the model basically consists of a longitudinal flume tank which opens into an expansion tank at the distal end. The complete length of the model is 5 metres.

The centre of the longitudinal channel is split along its length by a third wall. Flows are restricted to one side of the divide. By varying the position of the third wall the flow width can be controlled. The purpose of the expansion tank is to dissipate flow remaining after deposition along the length of the flume and hence avoid rebound effects. The weight of sediment deposited along the base of the flume is measured at equal intervals and compared for various runs. This is taken as a measure of distal thinning which is a function of the currents' competency. A continuous turbidity current is generated by pumping continuously mixed sediment suspensions into the tank at the proximal end. Two mixing tanks operate alternately in order to generate a continuous supply of sediment suspensions.

6 ACKNOWLEDGEMENTS

I wish to thank Dr.D.Huddart for his continuous advise during this research project and for critically reading the manuscript.

7 REFERENCES

Allen, J.R.L. 1969, Erosional current marks of weakly cohesive mud beds, Jour.Sed.Pet., 39(2):607-623.

Allen, J.R.L. 1970a, The avalanching of granular solids on dunes and similar slopes, J.Geol., 78:326-351.

Allen, J.R.L. 1970b, The sequence of sedimentary structures in turbidites, with special reference to dunes, Scott.J.Geol., 6(2):146-161.

Allen, J.R.L. 1971, A theoretical and experimental study of climbing-ripple cross-lamination, with a field application to the Uppsala esker, Geografiska Annaler, 53:157:187.

Bagnold, R.A. 1954, Experiments on a gravity free dispersion of large solid spheres in a Newtonian fluid under shear, Proc.Royal Soc.Lond., A225:49-63.

Bagnold, R.A. 1956, The flow of cohesionless grains in fluids, Phil.Trans,Royal Soc.Lond., A249:235-297.

Dzulynski, S. & J.E.Sanders, 1962, Current marks on firm cohesive mud bottoms, Trans. Connecticut Acad. Arts Sciences, 42:57-96.

Dzulynski, S. & A.J.Smith, 1963, Convolute lamination, its origin, preservation and directional significance, Jour.Sed.Pet., 33(3):616-627.

Dzulynski. S. & E.K.Walton, 1965, Sedimentary features of flysch and greywackes. Amsterdam, Elsevier.

Farrington, A. 1957, Glacial Lake Blessington, Irish Geography, 3:216-222.

Gustavson, T.C., G.M.Ashley, & J.C.Boothroyd 1975, Depositional sequences in glaciolacustrine deltas. In A.V.Jopling and B.C. McDonald (eds.), Glaciofluvial and glaciolacustrine sedimentation, p.264-280, S.E.P.M., Spec. Publ., No.23.

Huddart, D. 1970, Aspects of glacial sedimentation in the Cumberland lowland. Unpubl. Ph.D. thesis, Reading.

Jopling, A.V. 1960, Origin of regressive ripples explained in terms of fluid mechanic processes, U.S.Geol.Survey Prof.Paper, 424.D:15-17.

Jopling, A.V. & R.G.Walker, 1968, Morphology and origin of ripple-drift cross-lamination, with examples from the Pleistocene of Massachusetts, Jour.Sed.Pet., 38:971-984.

Kuenen, Ph.H. 1957, Sole marks of graded greywacke beds, Jour.Geol., 65:231-256.

Mantz, P.A. 1978, Bedforms produced by fine cohesionless, granular and flakey sediments under subcritical water flows, Sedimentology, 25:83-103.

McDonald, B.C. & W.W.Shilts 1975, Interpretation of faults in glaciofluvial sediments. In A.V.Jopling & B.C.McDonald (eds.), Glaciofluvial and glaciolacustrine sedimentation, p.264-280, S.E.P.M., Spec. Publ., No.23.

Pettijohn, F.J. & P.E.Potter 1964, Atlas and glossary of primary sedimentary structures. Berlin, Springer-Verlag.

Rich, J.L. 1950, Flow markings, groovings and intra-stratal crumplings of slope deposits, with illustrations from the Silurian rocks of Wales, Bull.Am.Assoc. Pet.Geol., 34(4):717-741.

Sanders, J.E. 1965, Primary sedimentary structures formed by turbidity currents and related resedimentation mechanisms. In G.V.Middleton (ed.), Primary sedimentary structures and their hydrodynamic interpretation, p.192-219, S.E.P.M., Spec. Publ., No.12.

Shrock, R.R. 1948, Sequences in layered rocks, New York, McGraw-Hill Book Co.

Synge, F.M. 1977, Blessington. In D.Huddart (ed.), Guidebook for excursion A14, South East Ireland, X INQUA Congress, p.28-30.

Southard, J.B., G.M.Ashley & J.C.Boothroyd 1972, Flume simulation of ripple drift sequences, Geol.Soc.Am., Abs. with programs, 4(7):672.

Walker, R.G. 1969, Geometrical analysis of ripple-drift cross lamination, Can.J.Earth Sciences, 6:383-391.

Paleogeography of Lake Biwa, Japan – and deep drilling site investigations in ancient lakes

SHOJI HORIE
Kyoto University, Shinga-ken, Japan

1 GEOSCIENTIFIC FEATURES OF THE LAKE BIWA BASIN

Lake Biwa which is located in the isthmus of the Japanese Archipelago has complex geohistory. One most interesting fact is the bending of the general strike of the Paleozoic strata which shows an "S" shaped feature along the outline of the Paleo-Biwa Basin. This feature is regarded as having been created in the Pre-Mesozoic Era; it suggests that an embryo of Lake Biwa begun its geological structure far back from the Tertiary Period. Geophysically speaking, that isthmus was noted as a twisted point of the Japanese Archipelago caused by tension (Fig. 1 after Omori, 1910). Accordingly, a tectonic lake structure of ancient origin such as Biwa might have had its origin (geologic structure) before the Cenozoic Era. The following facts also emphasize its importance and peculiarity. In the result of the Bougeanomaly measurement throughout the Japanese Islands, one of the most remarkable negative anomaly was found in the Lake Biwa Basin (Tsuboi et al., 1954). On the strength of that finding, the writers tried another measurement by using an underwater gravi-meter in Lake Biwa itself. It demonstrates the existence of -55 mgal in the northernmost part of the lake which coincides geomorphologically with the deepest part of the present bathymetry. This fact suggests the existence of very thick lacustrine sediments, probably as a result of the continuous subsidence and the deposition of lacustrine material since the Tertiary Period.

Another interesting fact is the deep-seated earthquake zone. It is the most distinguished earthquake area in the Japanese Archipelago. It crosses the Honshu Island just at the area of the above-mentioned isthmus. Great earthquakes that have ever occurred in the Japanese Islands including the Lake Biwa Basin have their epicenters mostly in this zone. It must be noted that such striking features of gravity anomaly and seismic activity are both closely connected with Lake Biwa's ancient origin antedating the Mesozoic Era.

2 CLIMATIC CHANGES AND ANCIENT LAKE SEDIMENTS

Ancient lake sediments have a significance in their long historic record of the natural environmental succession, particularly of the change of climate throughout the globe. During recent years, ocean sedimentary cores have been studied by many workers (Emiliani, 1978), not only for the reconstruction of past climates but for various purposes. Similarly, lacustrine sedimentary study is quite useful for the investigation of the past climate.

Speaking in general on the glacial geologic study, points worthy of notice are the minor classification of Glacial/Interglacial ages and Stadial/Interstadial and the age determination of cirque glacier deposits, though many difficulties do exist. For the former, terrestrial evidence such as glacial-till stratigraphy is inadequate because of the lack of a complete continuous record of glaciation and deglaciation. Fluvioglacial deposits are also in the same situation. On the other hand, a lake in an inland area is sensitive to the change of climate and its sediments keep a continuous record deposited in the quiet environment.

Regarding the latter, organic material which is useful for radiocarbon dating is usually lacking in the till forming moraines by a cirque glacier. However, if we can find evidence that correlates morainic areas with lake districts located outside the glacial region, the age determination is made possible by the dating of lake sed-

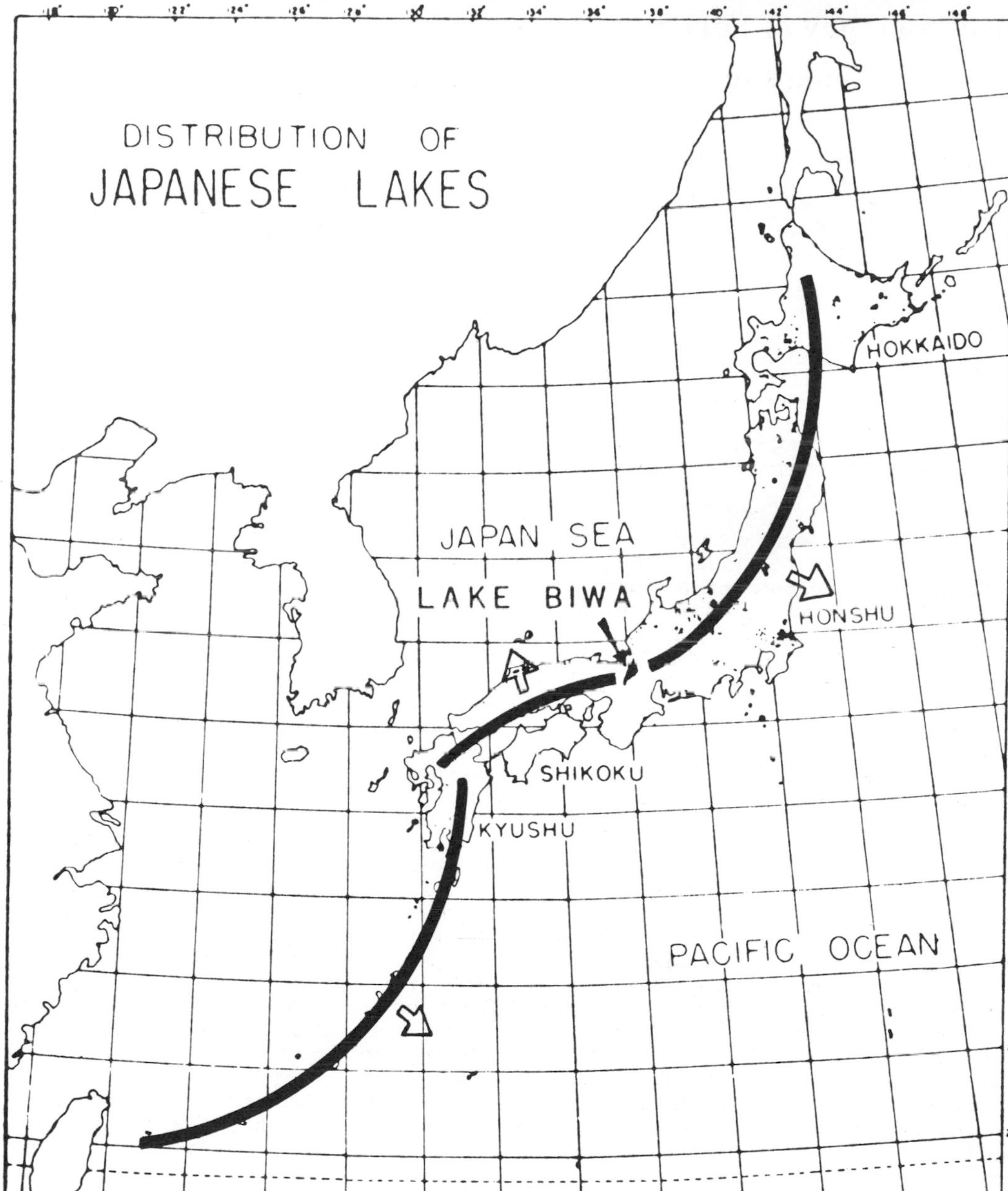

Fig.1 Location of Lake Biwa in connection with the geo-scientific feature of the Japanese Archipelago.

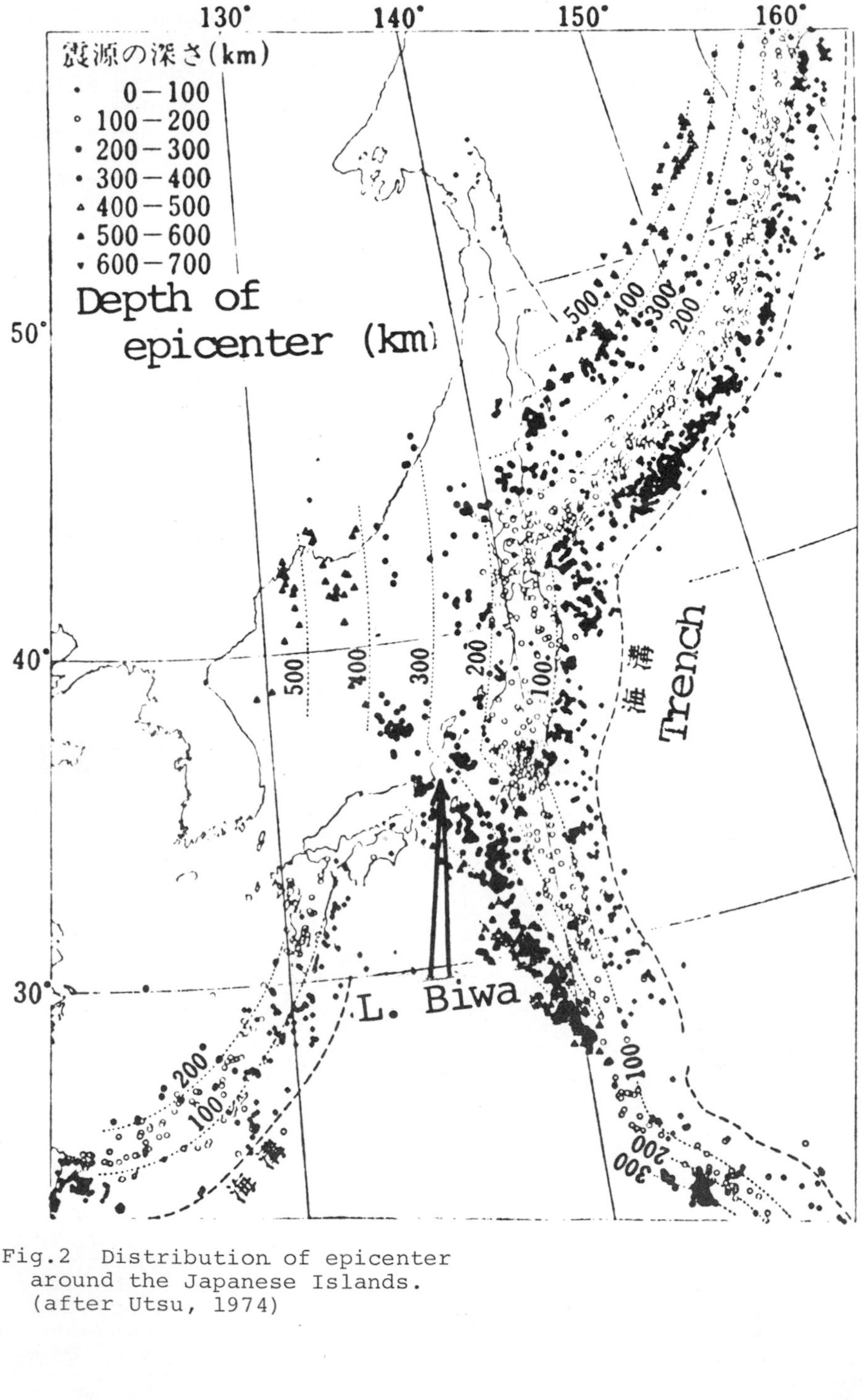

Fig.2 Distribution of epicenter around the Japanese Islands. (after Utsu, 1974)

iments.

In order to cover these problems, the writer tried his paleolimnological work on ancient lake sediments such as those of Lake Biwa.

One of the most useful analyses is pollen-chronology. In their recent work (Fuji & Horie, 1977), the writers discussed the climatic transition exhibited in the 200-meter core covering last 500,000 years. It is comparable with the work of Emiliani on the deep ocean deposits. However, as regards the question of glaciation, the glacial advance is controlled not only by the temperature indicated by pollen composition but by the amount of precipitation. Therefore, Yamamoto (1976, 1977) analysed the grain size variation in the same core and clarified the variability of the size of composing grains.* Although granulometric variation depends also on an uplift around the lake basin, in some example like the one at the core depth of 80 meters, microfossil remains coincide with the change of the size of grain, thereby indicating the usefulness of a grain size study. On the basis of this fact, two pieces of evidence on fossil pollen grains and granulometric composition were combined by the writer together with the writer's glacial geologic record; then he reached to a conclusion that both temperature and increased rainfall are beneficial to the formation of glacial moraines accompanied with the glacial advance. Furthermore, by tentatively assigning a low temperature/increased rainfall period to each moraine, the writer has been led to the inference on the age (Horie, 1978) of each glaciation. In the writer's field work, the age determination was occasionally based on the tephra on a tiny cirque moraine above though such evidence is obscure. Furthermore, a comparison of the grade of river dissection and weathering, vegetation development with the present-day nivation moraine situated just below the perennial snow on the top of the cirque floor, also indicates the age of each moraine. However, the absolute age determination of these moraines is now possible through the stratigraphy of Lake Biwa, if an exact correlation is carried out. At the same time, inter-morainic correlations are expected to be advanced in the near future.

The above-stated pollen-granulometry data are supported by diatom analysis. Mori and Horie (1975) worked on a quantitative study on diatoms in 98 samples, each 5 cm long, which were taken at intervals of ca. 25cm from the upper 3.3-30 m part of the core.

This part is included in the A diatom zone, so that it is not unusual that the major species existing in large quantity are M. solida, S. carconensis and var. pusilla. However, it was recognized that both species varied in quantity and in quality at some horizons. M. solida varied especially remarkably at four points, while S. carconensis at three points. The abundance diagram of each species is given in Fig. 3, in which an arrow marks, such as Ms1 or Sc1, indicates a variation point.

In samples of 30 m-Ms4 (= Sc3) point, the abundance of S. carconensis was roughly the same as in the present bottom sediments of Lake Biwa. But valves of this species silicify more strongly than those of recent times. M. solida was roughly the same in abundance in samples of 30-27 m in core depth and in the present bottom sediments, but in samples of 26.7 m-Ms4 point, the species scarcely occurred. It may be that, in the period of 30 m-Ms4 point, the lake was under the climate that is the same as or a little cooler than the present day.

In samples of Ms4-Ms2 point, M. solida occurred with about two times (in the period of Ms4-Ms3 point) and three times (in the period of Ms3-Ms2 point) as much abundance as in the present bottom sediments. The most conspicuous feature is that nearly all M. solida showed a fine cell wall form, differing from the normal one in the following points. The form shows slightly large valves and a thin mantle, of which pore and pore row arrangement are regular, like M. islandica subspec. helvetica. Since a chain colony composed of both forms has been found, it is true that the fine cell wall form represents one state of polymorphous M. solida. Such deformation of M. solida can be assumed to be an adaptation to a fall of water temperature. S. carconensis showed no quantitative change at Sc3 point, but the silicification of valves became weaker. These facts suggest that the climatic condition of the lake alternated from cool in the Ms4-Ms3 period to warm in the Ms3-Ms2 period.

At about the last stage in the period of Ms3-Ms2 point, that is, at Sc2 point, S. carconensis proliferated abruptly and this condition continued to Sc1 point. The abundance in the period of Sc2-Sc1 point was about three times as much as in the present bottom sediments. On the other hand in the period of Ms2-Ms1 point, the fine cell wall form of M. solida decreased rapidly, and the normal form increased. These facts suggest that the water temperature of the lake rose slightly and this rise would have led to stronger circulation of the lake water, so that the trophogenic layer had enough nu-

*For estimating past rainfalls, buried forests in a closed lake such as Lake Yogo (Horie et al., 1975) or recurrence surfaces are useful, as the writer will discuss later.

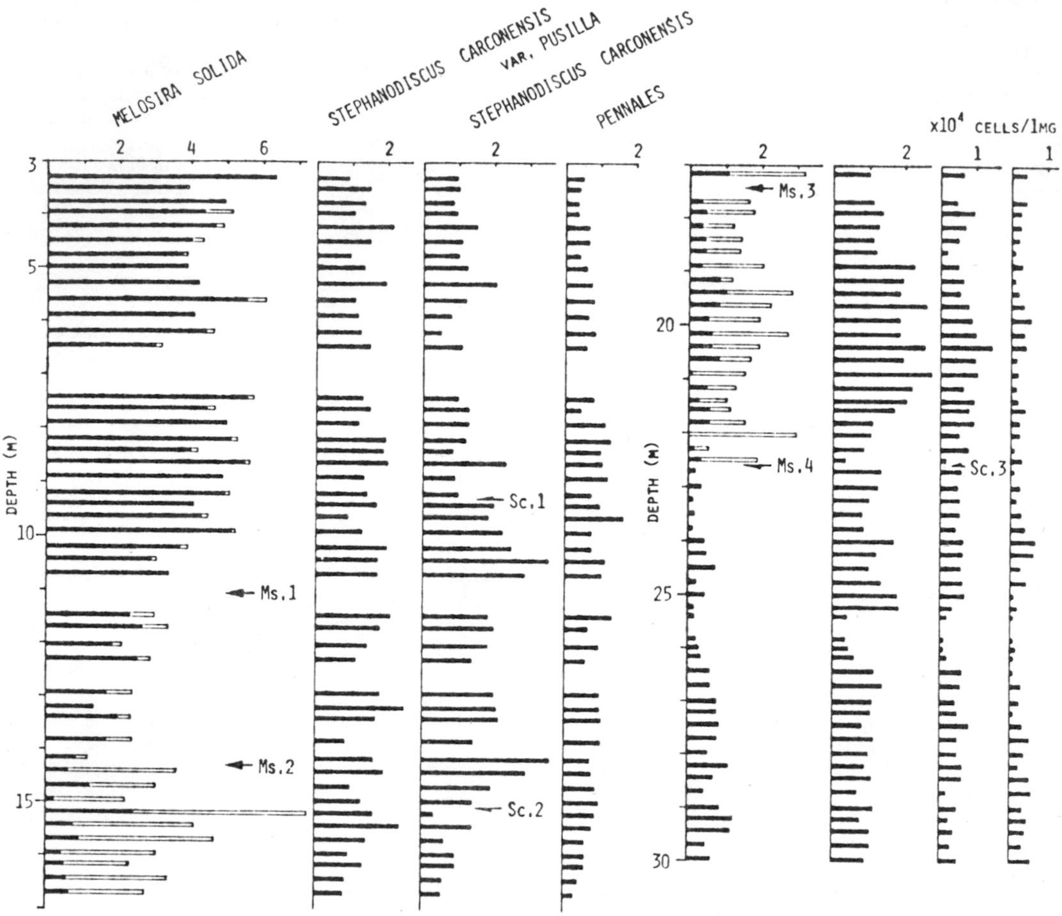

Fig.3 Diatoms diagram on the upper 30 m part of the 197.2m core. Blanked line in M. solida diagram: the fine cell wall form; arrow marks: the variation point of each species. (Mori & Horie 1975)

trients to progagate S. carconensis.

In the period of Msl-3.3 m, the normal form of M. solida occurred with about four times as much abundance as in the present bottom sediments and the fine cell wall form became very scarce, while S. carconensis decreased at Scl point. Consequently, it can be concluded that the climate of Lake Biwa was warm in the period of Scl-3.3 m. And we can refer the period after Scl point to the post-glacial time. This feature is a newly found fact; thus the microfossil composition together with grain size data is extremely useful for reconstructing the Pleistocene climatic succession.

The writer has already tried such work in both Lake Biwa and Japanese high mountains (Horie, 1965, 1976) and proposed the Hotaka Glacial in place of the Würm Glacial (Horie, 1978). This nomination is the first work of Glacial Geology based on paleolimnological features of ancient lake deposits.

3 CONCLUDING REMARKS

The faster sedimentation rate of Lake sediments, unlike that of ocean deposits, is uniquely valuable, because we can reconstruct chronology in minute details. In addition, a long historical record is kept in a long core in a lake; thus, ancient lake sediments not only in Biwa but in Baikal, Caspi, Tanganyika and other lakes of this type are valuable.

Although Paleolimnology is thus in progress as a new scope of science, it is quite important to have an exact correlation between small cirque moraines and the horizon of ancient lake sediments. Our solution to this question is a key for the study of climatic changes during the Pleistocene Epoch.

4 REFERENCES

Emiliani, C. 1978, The cause of the ice ages, Earth and Planetary Sci.Lett. 37: 349-352.

Fuji, N. & S.Horie 1977, Palynological study of a 200-meter core sample from Lake Biwa, Central Japan. I. Palaeoclimate during the last 600,000 years, Proc.Japan Acad. 53:139-142.

Horie, S. 1965, Late Pleistocene glacial fluctuations and changes of sea level in the Japanese Islands and their tentative correlation with oscillations in North America and Europe, VIth Internat.Quaternary Congr. (Warsaw, 1961), Rept. p. 175-184.*

Horie, S., S.Kanari & K.Nakao 1975, Buried forest in Lake Yogo-ko and its significance for the study of past bioenvironments, Proc. Japan Acad. 51:669-674.

Horie, S. 1976, Lake Biwa sediment stratigraphy and the glacial evidences on Japanese high mountains, Proc. Japan Acad. 52:203-206.

Horie, S. 1978, Nomination on the Late Glacial Age in Japan. Oriental morainic features and Lake Biwa sedimentary sequence, Proc. Japan Acad. 54:213-216.

Mori, S. & S.Horie 1975, Diatoms in a 197.2 meters core sample from Lake Biwa-ko, Proc. Japan Acad. 51:675-679.

Omori, F. 1910, Great earthquake in Japan, Shinsai Yobō Chōsa-kai Hōkoku 68:164-165.

Tsuboi, C., A.Jitsukawa & H.Tazima 1954, Gravity survey along the lines of precise levels throughout Japan by means of a Worden Gravimeter (V), Bull.Earthq.Res. Inst.Tokyo Univ.Supplement 4:198.

Utsu, T. 1974, Distribution of epicenters around the Japanese Islands, Kagaku 44: 739-746.

Yamamoto, A. 1976, Paleoprecipitational change estimated from the grain size variations in the 200 m-long core from Lake Biwa. In S.Horie (ed.), Paleolimnology of Lake Biwa and the Japanese Pleistocene, 4:179-203.

Yamamoto, A. 1977, The structure of density and grain size variations seen near the 130 m-layer in the 200 m-long core sample from Lake Biwa. In S.Horie (ed.), Paleolimnology of Lake Biwa and the Japanese Pleistocene, 5:125-136.

*There are some misprints in p. 181.

Contributions related to the field excursion

Introductory remarks

Ch. SCHLÜCHTER
ETH-Hönggerberg, Zurich, Switzerland

The early days of "Quaternary Research" in Switzerland have been characterized by the geognosts' endeavour to explain why and how large boulders of granite from the Central Massifs of the Alps occur in association with the youngest deposits in the Alpine Foreland. This 17th and 18th century-geology of the erratic boulders (in German they are called "erratische Blöcke, Findlinge, Geisberger, Teufelssteine") has been also an important part of research far into the 20th century. Original geo-fantastic explications (using devil-monitored processes) have gradually been replaced by demonstrations based on observations (Guler von Weineck 1616, Lang 1708, Gruner 1773, Kuhn 1787, von Buch 1811, Hugi 1825, 1843, Schimper 1837, Desor 1840, Venetz 1830, 1861, de Charpentier 1834, 1835, 1841, Agassiz 1840, Bachmann 1870).

The break-through of the idea that the transporting medium of the erratic boulders was glacier ice only and has extended far outside the Alps proper has activated substantial research during the second half of the last century on the acting agent itself: the glaciers of the Alps. This activity has culminated in the publication of Heims "Gletscherkunde (1885)" and in the "Gletscherkonferenz" at Gletsch in 1899 (Böhm 1901).

Activities in Quaternary Geology in this century have more and more been dominated by geotechnically oriented investigations. But, in fact, these close interrelationships between geotechnical applications and the evolution of knowledge on glacial geology dates back to Scheuchzer (1725) - when he described a profile of Pleistocene deposits for the first time after having visited the construction site for the deterioration of the River Kander (Berner Oberland). - The predominantly applied character of glacial geology during the last four decades is also reflected by the extensive use of Pleistocene deposits as construction material (Fig. 9). These large-scale excavations and the hydrogeological investigations then have contributed considerably to our knowledge of the Pleistocene deposits and stratigraphy, mainly in the Alpine Foreland (von Moos 1953, Jäckli 1962, Schindler 1968, Schlüchter 1976). Another marker-date in Quaternary Research in Switzerland is the publication of the first volume of Hantke's "Eiszeitalter" (1978).

The beginning of a new phase of research has been initiated by a field excursion of the International Geological Correlation Programme (Project 73/1/24 "Glaciations in the Northern Hemisphere") in 1976 (Frenzel 1978). This excursion has been a reintroduction of scientific contacts between Switzerland and the highly advanced research processes abroad. What the IGCP excursion has been for the further development of Quaternary Research aiming at chronostratigraphy, the excursion "Moraines and Varves" is for the further lines of research in the field of glacial and glaciofluvial sedimentology. It has not marked an occasion to present our "full-scale theories" on every subject (which are not completed at all), but we have received a tremendous stimulus to re-begin with and re-think of our research in the field of glacial geology.

The field excursion "Moraines and Varves" has traversed the Alpine Forland, then come across the perialpine lakes, visited high alpine areas and recent glaciers (cf. Fig. 3). The main topics under discussion have been:

- sedimentation in perialpine lakes
- morainic deposits in inneralpine valleys
- glacial erosion
- recent glaciers
- glacial and glaciofluvial deposits in the Alpine Foreland

Some of the problem spots visited during the excursion are illustrated as follows (cf. also "Guidebook to the field excursion", Schlüchter 1978):

Fig. 1 : View from Findelen towards the Matterhorn near Zermatt: Glacial erosion is marked by clearly developed striations on the bedrock knobs. The loose material on the bedrock is washed debris of sub-, in- and supraglacial transport.

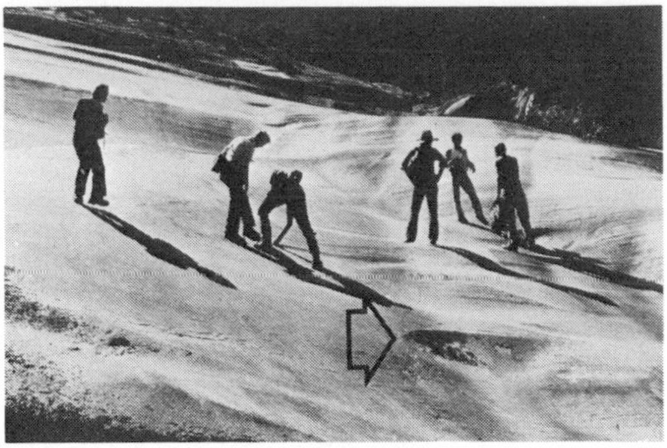

Fig. 2 : "Helle Platten" at Handegg on the northern side of Grimselpass: a fossil glacierbed is preserved and not covered by vegetation. Granitic material has been eroded by the glacier flow and typical glacier sole relief has been carved with crescentic gauges (arrow). Glacier movement from left to right.

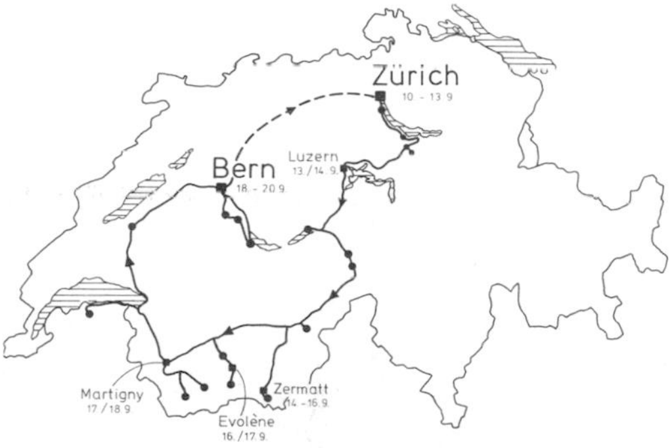

Fig. 3 : Index map of Switzerland: the route of the field excursion "Moraines and Varves" in 1978.

Fig. 4 : The Pyramids at Euseigne (Valais at the confluence of two glacier lobes enormous amounts of till has been accumulated - the same in lateral positions downvalley (arrows). The sediment shows in- and subglacial transport characteristics but is structurally not of the ground-moraine type. It is interpreted here as subglacial meltout till in sublateral "levee position".

Fig. 5 : The Glacier de Tsidjiore Nouve: Dr. F. Röthlisberger looks at the lense of ice pushed on top of the huge lateral moraine. The glacier is obviously reactivated here and "overflows" the subfossil lateral moraine (cf. papers by F. Röthlisberger & W. Schneebeli and by K. Gripp, this volume). The internal structure ...

Fig. 6 : ... can be observed in a crevasse perpendicular to the main lateral moraine (which is to the right). The pressed-up structure of the scale is clearly visible.

Fig. 7 : The Glacier de Tsidjiore Nouve: view downglacier from the same place where Fig. 6 has been taken: the surface of the pressed scale exhibits small longitudinal ridges reflecting the debris-rich bands. - The surface of the "non-pressed" ice is to the right (G) and the pressed-up character of the scale becomes evident ("laterale Press-Schuppen", cf. article by K. Gripp, this volume).

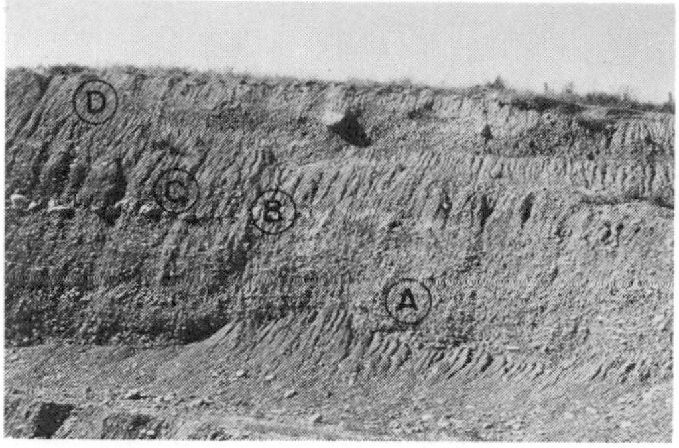

Fig. 8 : The gravel pit at Jaberg (Alpine Foreland): sediment accumulation as an example of the complex relationships of glacial (morainic) and glaciofluvial (gravel) accumulations. - Two accumulation cycles are present: the lower gravel unit (A) is covered by a relict unit of basal lodgement till (Grundmoräne, B) which is replaced to the left by a boulder-horizon as erosional relict. The upper gravel unit (C) is completely covered by an extensive unit of basal lodgement till (D). These depositional cycles with glaciofluvial gravel continuing upwards into basal lodgement till are characteristic of accumulations during glacial advance phases ("Vorstoss Schotter mit zugehöriger Grundmoräne", cf. Schlüchter 1976, 1978). - Both cycles belong to the Last Glaciation (the alpine "Würm").

Fig. 9 : Gravel pit at Thalgut (Alpine Foreland): Glaciofluvial/glacial accumulation cycle. Note the well developed coarsening upward sequence of the gravel and the gradual transition to basal lodgement till. The lodgement till is divided into two "phases" by a thin layer of basal melt-out debris. - The coarsening upwards of the gravel is one of the significant properties of a glacio fluvial gravel body which is genetically and sedimentologically defined as "Vorstoss-Schotter".

Fig. 10 : Gravel pit at Thalgut: close-up of the section shown in Fig. 9. - Both the coarsening upward of the gravel and the transition from dominantly fluvial to dominantly glacial (basal lodgement till, B) can be observed (transition = T).

Fig. 11 : Example of a moraine as a geomorphic feature: lateral moraine ridge ("Seiten- oder Ufermoräne") on Längenberg, south of Berne (Alpine Foreland). The glacier depositing the moraine was to the right of the vallum.

REFERENCES

Agassiz, L., A. Guyot & E. Desor 1840, Etudes sur les glaciers, 346 p. avec Atlas de 32 p. par J. Bettanier, Neuchâtel 1840 et Paris 1847. - cf.: Carozzi, A.V. (Ed. & Transl.) 1967, Louis Agassiz, Studies on Glaciers, New York (Hafner Publishing Company).

Bachmann, I. 1870, Die Kander im Berner Oberland. Ein ehemaliges Gletscher- und Flussgebiet, Bern (Dalp).

Böhm, A. 1901, Geschichte der Moränenkunde, K.K. Geogr. Gesellsch. in Wien, Abhandlungen, vol. 3/no. 4, 334 p.

von Buch, L. 1811, Ueber die Ursachen der Verbreitung grosser Alpengeschiebe, Abhandlungen der physikalischen Klasse der Königlich-Preussischen Akademie der Wissenschaften in Berlin aus den Jahren 1804-1811, p. 161-186, Berlin 1815.

de Charpentier, J. 1834, Annonce d'un des principaux résultats des recherches de Mr. Venetz, ingénieur des Ponts et Chaussées du Canton du Valais, sur l'état actuel et passé des glaciers du Vallais, Verhandlungen der allg. Schweiz Ges. für die ges. Natw. in ihrer 19. Jahresversammlung zu Luzern, p. 23-24, Luzern 1835.

de Charpentier, J. 1834, Sur la cause probable du transport des blocs erratique de la Suisse, Annales des mines, 3e série, t. 8, 219-236, Paris 1835.

de Charpentier, J. 1841, Essai sur les glaciers et sur le terrain erratique du bassin du Rhône, Lausanne (Marc Ducloux).

Desor, E. 1840, Sur les blocs erratiques, Bibliothèque universelle de Genève, nouvelle (3è) série, t. 30, 397-401, Genève.

Frenzel, B. (Hrsg.) 1978, Führer zur Exkursionstagung des IGCP Projektes 73/1/24 "Quaternary Glaciations in the Northern Hemisphere" vom 5. bis 13. Sept. 1976 in den Südvogesen, im nördlichen Alpenvorland und in Tirol, Bonn-Bad Godesberg (DFG).

Gruner, G.S. 1773, Die Naturgeschichte Helvetiens in der alten Welt, Bern (Abraham Wagner).

Guler von Weineck, J. 1616, Raetia, Zürich.

Hantke, R. 1978, Eiszeitalter. Die jüngste Erdgeschichte der Schweiz und ihrer Nachbargebiete, Band 1, 468 p., Thun (Ott Verlag AG).

Heim, A. 1885, Handbuch der Gletscherkunde, Bibliothek geographischer Handbücher, XVI + 560 p., Stuttgart (F. Ratzel).

Hugi, Fr. J. 1825, Ueber die Granitblöcke am Jura, Zweyter Jahresbericht der naturhist. Kantonal-Gesellsch. in Solothurn, Solothurn.

Hugi, Fr. J. 1843, Die Gletscher und die erratischen Blöcke, XVI + 256 p., Solothurn.

Jäckli, H. 1962, Moränen als Baugrund und Baustoff, Strasse und Verkehr, Nr. 9/ 1962, Solothurn (Vogt-Schild AG).

Kuhn, B. Fr. 1787, Versuch über den Mechanismus der Gletscher, Magazin für die Naturkunde Helvetiens, Bd. 1, 119-136, Zürich (Hrsg. von A. Höpfner).

Lang, K.N. 1708, Historia Lapidum figuratorum Helvetiae eiusque viciniae, in qua enarrantur omnia eorum genera ... - Venedig.

von Moos, A. 1953, Der Baugrund der Schweiz, Schweiz. Bauzeitung 1953, Nr. 50, 3-12, Zürich (J. Frey AG).

Penck, A. & Brückner, E., 1901-1909, Die Alpen im Eiszeitalter, Leipzig (Tauchnitz).

Scheuchzer, J.J. 1723, Herbarium Diluvianum Collectum. Lugduni Batavorum, Sumptibus Petri Vander Aa. Bibliopolae, Civitatis atque Academiae Typographi.

Schimper, W.Ph. 1837, Ueber die Eiszeit. Brief an Prof. Agassiz, Actes de la Société helvétique des sciences naturelles, 22e session, Neuchâtel 1837, 38-51, Neuchâtel.

Schindler, C. 1968, Zur Quartärgeologie zwischen dem untersten Zürichsee und Baden, Eclogae geol. Helv., vol. 61/2, 395-433, Basel (Birkhäuser).

Schlüchter, Ch. 1976, Geologische Untersuchungen im Quartär des Aaretals südlich von Bern (Stratigraphie, Sedimentologie, Paläontologie), Beiträge zur Geol. Karte der Schweiz, N.F. 148. Lfg., Bern (Stämpfli).

Schlüchter, Ch. 1978, Guidebook for the Excursion, INQUA-Commission on Genesis and Lithology of Quaternary Deposits, Symposium 1978, ETH-Zürich, Switzerland, 112p., Zürich.

Venetz, I. 1830, Sur l'ancienne extension des glaciers, et sur leur retraite dans leurs limites actuelles, Actes de la Société helvétique des sciences naturelles, 15è réunion à l'Hospice du Grand Saint-Bernard 1829, Lausanne 1830.

Venetz, I. 1861, Memoire sur l'extension des anciens glaciers, N. Denkschr. allg. Schweiz. Ges. Natw., 18.

Synoptic history of the Quaternary of Switzerland

RENÉ HANTKE
ETH-Zentrum, Zurich, Switzerland

During the maximum extent of glaciation more than 95% of the area of Switzerland was covered, whereas the Würm glaciers occupied about 85% only. During the warm period separating the Riss and the Würm glaciations, the ice covered only 2% of this area, less than the 3% cover of today. Only the history of the Würm glaciation, the retreat of these glaciers and the last 10'000 years of postglacial time are known in detail. From a small part of the Swiss Plain and from the northwest of Switzerland, some information about the last interglacial and the Riss glaciation is available. Records of older glaciations are scarce and ambiguous; not even their number is known.

According to PENCK and BRUECKNER, gravel deposition during cold phases is followed by erosion during interglacials. This model, which led to the theory of the four glaciations, has been applied too stringently in Switzerland. Criticism regarding this model has arisen during the last few decades. Recently, it has even been suggested - from a very local point of view, of course - to abandon "once and forever the scheme of the four pseudo-glaciations".

Traces of an oldest, Late Pliocene, cold period can be observed in the northwest of Switzerland, in the Ajoie region and in the Delémont basin. Deposition of Vosges Gravels so far from their source area was only possible under climatic conditions which prevented the growth of trees in the southern Vosges. The highest mountains in that area may even have been covered by ice caps. Catastrophic summer meltwaters could have transported coarse debris more than 60 km to the southeast. A following warmer period is documented by a corresponding flora (with Liquidambar and Podogonium, an extinct genus), succeeding the Vosges Gravels and still showing Pliocene characteristics. Another cold period occurred at the Pliocene/Pleistocene boundary. South of Basle, on the Blauen mountain, at the eastern margin of the Laufen basin and in the Canton of Baselland, the "Wanderblock"-Formation is quite a puzzling deposit. It consists of boulders, most commonly silicified Buntsandstein (Lower Triassic) and crystalline rocks from the Southern Black Forest, in particular from the Kandern fault region. The size of these boulders - the largest ones are more than 1 m in length - cannot be explained by fluvial transport. There is no trace of a corresponding drainage basin. The smaller, well-rounded pebbles seem to be reworked material from Buntsandstein conglomerates. Further, the brown argillaceous matrix of the "Wanderblock"-Formation cannot be the result of fluvial deposition. However, all these facts are easily explained if we assume the "Wanderblock" formation to be a deeply weathered till and the large blocks erratic boulders. In the southeastern part of the Canton of Baselland,

Black Forest pebbles occur together with Alpine - but perhaps younger - material. Therefore, it is possible that a contemporaneous Alpine glaciation had already covered large parts of the Swiss Plain; probably in similar proportions as during later glaciations.

This "Wanderblock Glaciation" must have taken place before the more than 800 m of vertical movement occurred on the Kandern fault in the southern Black Forest. At the same time the Vosges gravels were overthrusted by the outermost fold of the Jura mountains, the chain of Montagne-Bürgerwald, in the southern Sundgau region. Between Waldshut and Basle, joints parallel to the Kandern-Fault led to the formation of the Hochrhein valley and caused a drastic change in the drainage network. The Aar, which collected the Swiss Alpine rivers, could not follow its old course towards the Danube. It turned at Waldshut and followed the joints, excavating a new valley towards Basle, then continued through the Sundgau to reach the Doubs and finally the Rhône. During the following cold period the Sundgau gravels, with Helvetic pebbles, resisted erosion and alteration - siliceous limestones, quartzites, gneisses, red granites of the Rigi conglomerates, amphibolites and radiolarites - have been deposited in the Sundgau west of Basle.

The association of heavy minerals in the Lower Rhine region shows that during the next cold period the drainage network changed again: The waters from the Swiss Alps flowed past Basle through the Rhine Graben. During this glaciation the Higher Deckenschotter was deposited at the margins of glaciers which covered a pronounced relief. Formerly they were believed to be remnants of a single continuous outwash plain.

In the Hochrhein valley the succeeding interglacial is proven by a fossil flora containing Tsuga pollen. This shows that the climatic conditions allowed the growth of temperate and humid forests. During the following glaciation, the Alpine glaciers again managed to advance into the Hochrhein valley. The outwash plains of the Lower Deckenschotter, formed in a similar way as the Higher Deckenschotter, again attest a marked relief. The glaciers caused pronounced overdeepening. Depressions were filled during the retreat of the glaciers. These sediments, deeply weathered during the subsequent interglacial, are covered by the till of the glacier advance. This till, containing large boulders, can be followed up to the terminal moraines of Möhlin and of Liestal, 20 km east, and 12 km southeast of Basle respectively, which are believed to be of Riss age.

Characteristic erratics allow the various glacier systems: Rhône, Rhône/Aar, Aar/Reuss, Linth/Rhine and Rhine, of the Riss glaciation to be distinguished.

Let's turn to the Rhône glacier. In the Lake of Geneva basin it divided into two branches. The main branch, flowing via Geneva and cutting through the Southern Jura chains, reached down to Lyons. The other branch covered the main part of the Swiss Plain. Different lobes flowed over the Jura mountains, filling the valleys there and reaching Salins and Ornans, 18 km southeast of Besançon. The upper limit of erratic boulders indicates that as it left the Alps, the Rhône glacier surface was at 1550 m at the Prealpine and at the Jura border at 1450 m. In the foreland of the Alps and in the Jura mountains there were only a few nunataks. However, most of these, reaching above the regional snowline, contributed local ice to the main glaciers.

In the ice-free regions, plants and animals which were adapted to the cold could survive. After the retreat of the ice these species could spread and from the southwest, the south and the southeast of the Alps surviving refugee species could return.

The first stages of retreat of the Riss glaciers appear near Laufenburg and near Koblenz in the Hochrhein valley. At this stage the Rhine glacier separated from the Helvetic ice. A more important stage exists near Turgi and a last

one near Baden. The forests reconquered the land in a rather short time. This is proven by several interglacial palynological profiles.

The history of the Würm glaciation could formerly only be deciphered starting from its maximum. Now, evidence of the history of advance has been discovered in the Alpine foreland. Lacustrine deposits and lignites believed to belong to the Riss/Würm interglacial, proved to be, by radiocarbon and palynological methods, of Würm age. They demonstrate the formation of peat during several warmer stages interrupting the glacier advances of the Würm glaciation. However, it is still too early to establish a continuous forest evolution for the whole region.

For example, we know that in the Alps the Rhône and Rhine glaciers did not fill their valleys at first: the accumulation zones of the future tributaries were closer to the Thalweg than those of the main glaciers. The main glaciers could only push back the tributaries or even dam them at a later date.

Several attempts have been made to synchronize the retreat stages of the different Würm glacier systems and of their many lobes. Thanks to detailed investigations in every valley, we are now beginning to understand more of this mechanism.

In the western Swiss Plain, for example, the Rhône glacier reached 200 m higher up than formerly believed. This corresponds to the ice level at the Jura border, documented there by the highest lateral moraines. This fact has not only effects on the history of the retreat of the Rhône glacier and its tributaries, but also on the history of their advances.

Regarding other glacier systems of Switzerland, too, we must change our concept of the retreat of the Würm ice. Numerous recent palynological and radiocarbon investigations provide new information allowing a synchronisation of the final phases of the Würm glaciation and the delineation of the climatic history of the Late Würm and the Holocene.

Cooler	Warmer climate predominant
Würm advances with lignites	
	Riss/Würm - Interglacial
Retreat Stages	
Outermost moraines and gravels of the High Terrace	
	Deep weathering
Gravels in the Valley Bottom	
Lower Deckenschotter	
	Flora with Tsuga
Upper Deckenschotter	
	Drainage changed to the North
Sundgau gravels	
	Drainage changed to the West
Wanderblock-Formation	
	Flora with Liquidambar and Podogonium
Vosges gravels	
	Vosges sands with Populus

Table 1 : Schematic summary of the Pleistocene chronology of Switzerland reaching back to the Upper Pliocene

Genesis of lateral moraine complexes, demonstrated by fossil soils and trunks; indicators of postglacial climatic fluctuations

FRIEDRICH ROETHLISBERGER
Eidgenössische Anstalt für das forstliche Versuchswesen, Birmensdorf ZH, Switzerland

WALTER SCHNEEBELI
Universität Zürich, Zurich, Switzerland

1 INTRODUCTION

Climate is a vital condition of our lives. Climatic fluctuations can be of catastrophic impact. Thus today we witness, for instance, the gradual expansion southward of the Sahara desert; a critical rainfall shortage in the Sahel region is drying out wells and killing people and animals. Only ruins scattered in the desert remain of the Roman settlements from the time when, 2000 years ago, vast regions of Northern Africa constituted the major corn belt of the Roman Empire.

Sea level, too, has risen globally by dozens of meters since the glacial maximum of the last Ice Age about 18,000 years ago. If the remaining polar ice masses were to melt, sea level would rise even further and submerge fertile lowlands everywhere, along with numerous coastal settlements. Within the last few years, a series of sensational reports on climatic changes in the future have appeared, drawing unsound generalizations from individual case studies. The attention of the public was caught with apocalyptic visions of an imminent new ice age. The question whether at present we live in an interstade with increasingly warm climate, or whether a new climatic deterioration is soon to follow, has not been conclusively answered by any scientific study so far. For the fundamental question of the causes of climatic fluctuations remains an unsolved problem of science up to this day. Many theories and hypotheses have been brought forward. Some scientists attribute climatic fluctuations to cosmic causes such as sun spot activity, while others trace them to processes on and in the earth itself, such as volcanism.

Yet any prediction presupposes two achievements:

- a detailed reconstruction of the climatic changes of the past,
- and an insight into the causes and principle laws of climatic change.

The latter presupposes the former.

Past climatic change can be traced in recent nature in many respects. Among the indicators are the ice masses of Greenland and Antarctic, shore terraces stemming from past fluctuations of sea level, marine fossils such as corals and foraminifera, pollen deposits in marshes, glacier moraines, and tree rings. In the 1976 Autumn/Winter issue of the periodical of the Swiss Alpine Club, two dissertations by Walter Schneebeli and Friedrich Roethlisberger were published which feature new approaches in the scientific reconstruction of the history of climatic change and glaciation fluctuations in the postglacial period.

The geographical unit of investigation is the area extending from the Val de Bagnes to Zermatt, both in the Swiss Canton of Valais. In addition to well-known traditional research methods, the authors have been focusing on the following procedures:

- identification of periods of favorable climate by way of excavation and subsequent dating of fossil soils buried beneath moraines;
- x-ray-dendroclimatological analysis of timber and entire tree-trunks preserved by glaciers and moraines.
- historical and topographic-archeological study of mule trails on glacierized mountain passes.

All of these approaches were used to identify not only maxima, but also minima of glaciation. While advances of glaciers are evidenced by existing moraines, glacial mi-

FIG. 1: AERA OF INVESTIGATION

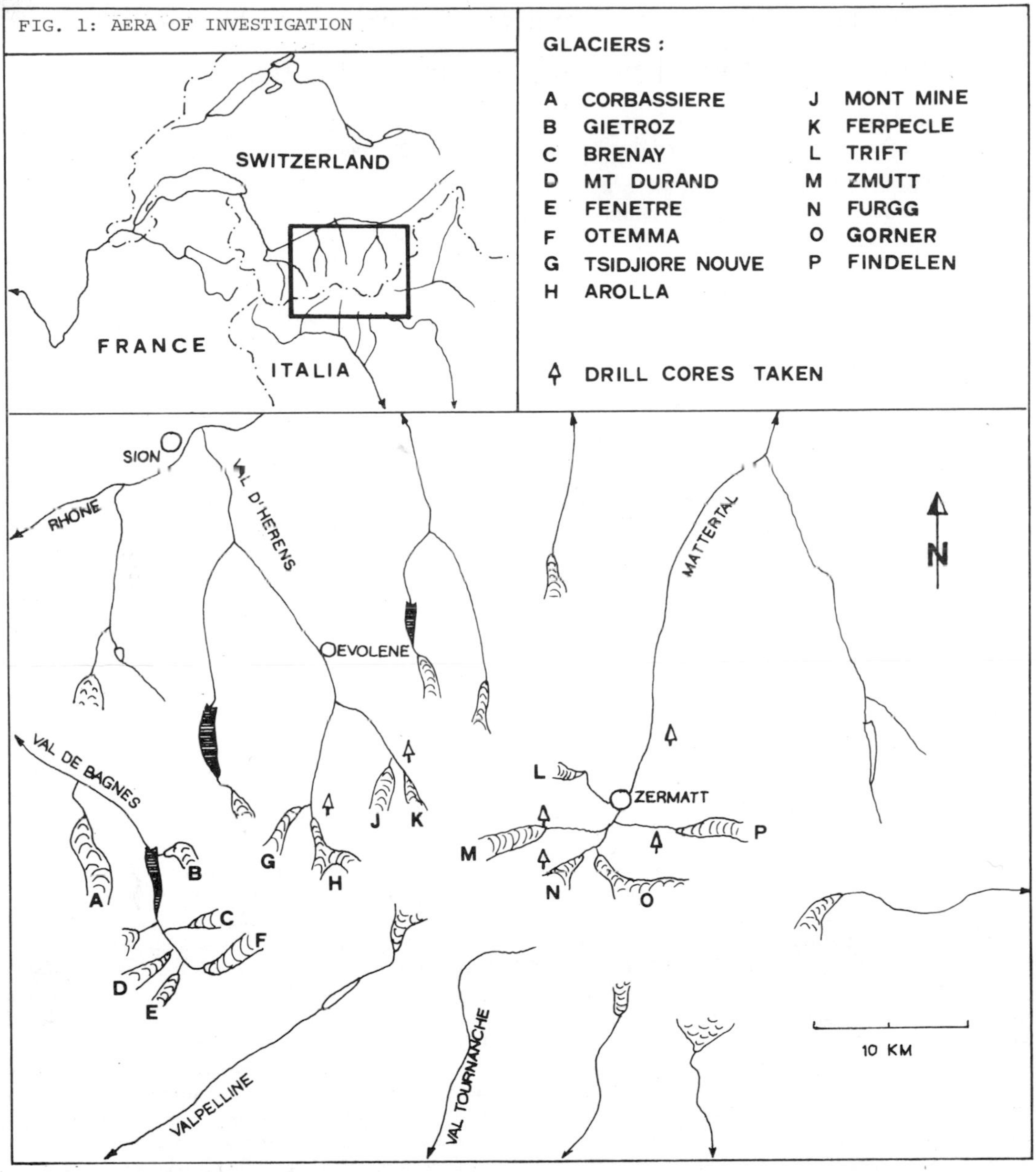

nima after stages of retreat have to be determined on the basis of other indicators. After shrinkage, the deglaciated area is covered by new vegetation or, later, taken into possession by man. Forests grow, and may be cleared and turned into alpine pastures. In addition, mountain trails may be cut and spring water wells build. These products of nature and civilization are overridden and plowed under by the glacier in the course of its following readvance. During the subsequent glacier shrinkage they can be recovered. From the location and distribution of these objects, it will be possible to infer the point of farthest possible retreat of the glacier at the time of growth or installation of these objects. It is in this way that glacial minima are identified and defined.

The present article is a summary of a part of the studies: Schneebeli/Roethlisberger (76) and present historical documents for glacier movements and genesis of moraines, shown by fossil soils and tree-trunks.

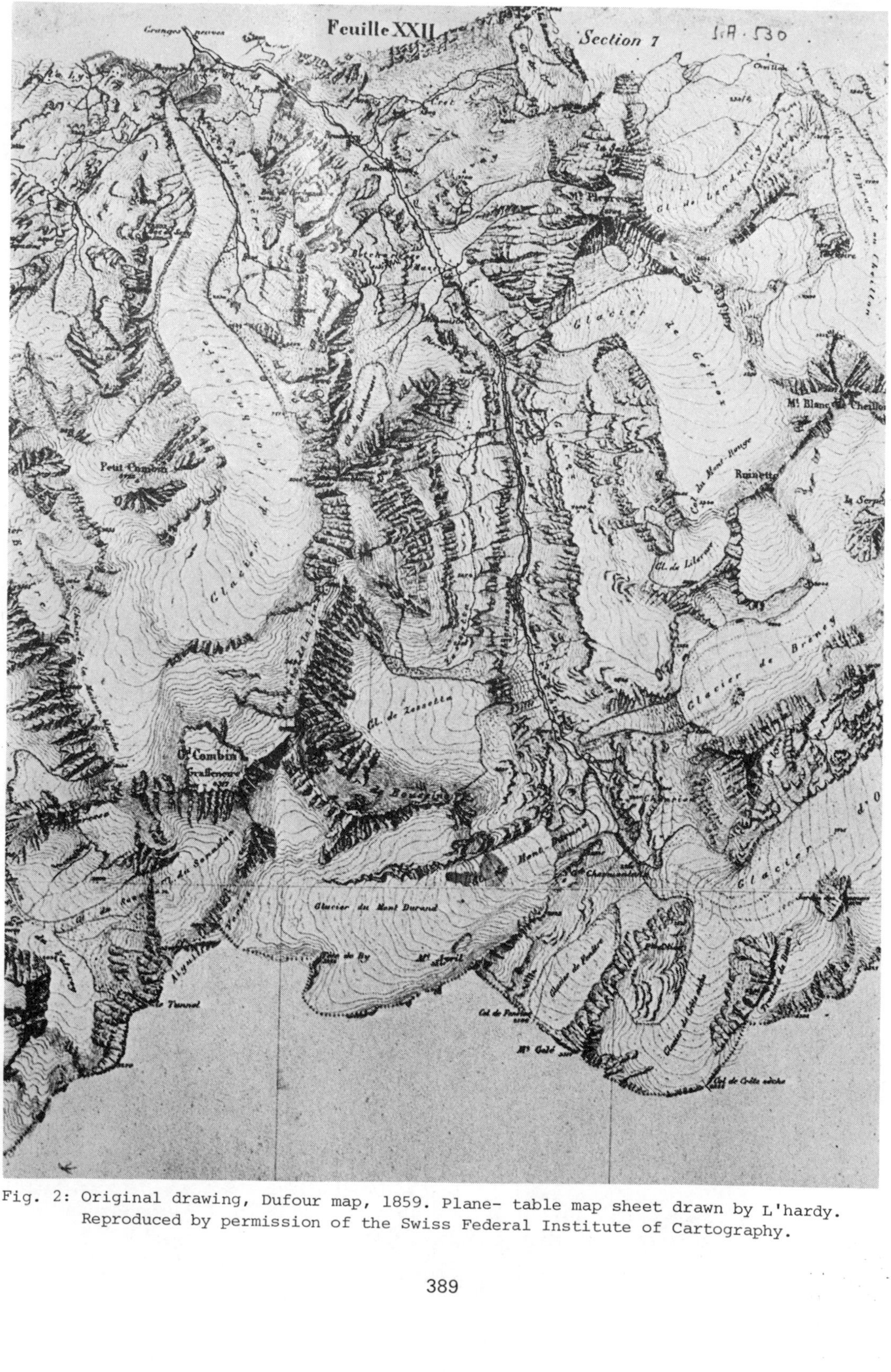

Fig. 2: Original drawing, Dufour map, 1859. Plane- table map sheet drawn by L'hardy. Reproduced by permission of the Swiss Federal Institute of Cartography.

Fig. 3
Deglaciation in the upper Val de Bagnes, 1859–1965.

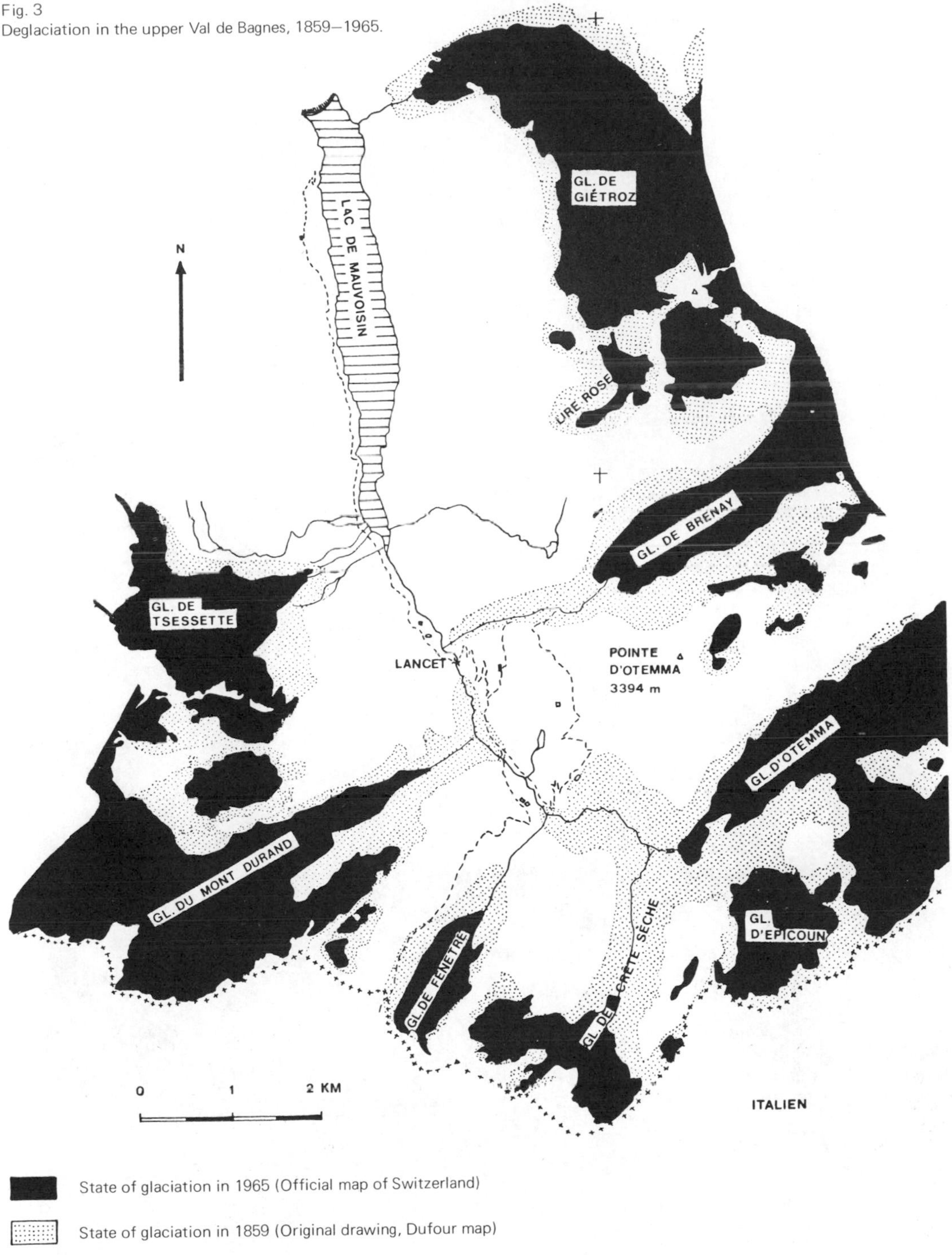

Upper Val de Bagnes

Water-color by H. C. Escher von der Linth

1820

Gl. d'Otemma

Gl. de Fenêtre

Gl. du Mont Durand

Gl. de la Tsessette

1974

Gl. de Crête Sèche

Fig. 4
Panoramic picture of the upper Val de Bagnes. A water-color from 1820 by H. C. Escher is contrasted to a modern photograph taken by W. Schneebeli in 1974.

Figure 5:
Glacier de Tsidjiore Nouve to the right, Glacier d'Arolla in the background left (August 1, 1835). In the foreground, the village of Arolla. The white arrow marks the endmoraine at 1817. In the creek bed of the Glacier de Tsidjiore Nouve, larch timber, roots, and entire tree-trunks were found. Dating of one of the trunks yielded the C^{14}-age of 2940 $\pm$ 150 years before present. The larches had been overridden by the glacier which in a quick advance reached the extent, at least, of 1890. Among the fossil larch forests of Europe located at high altitude and still in proximity of the margins of the present glacier, the forest covered by the moraines of the Glaciers de Tsidjiore Nouve and Arolla is one of the oldest found so far (8400 $\pm$ 200: Ly - 749).
(Buehlmann, Drawing Book 10, Number 256. Reproduced by permission of the Graphische Sammlung of the Federal Institute of Technology, Zurich).

2 GLACIAL MAXIMA OF MODERN TIMES

Historical documents give evidence of a series of rapid glacier advances in the Alps, due to a general climatic deterioration beginning towards the end of the 16th century. A period of cool climate, characterized by several glacial maxima around 1600, 1640, 1820, and 1850, extended through the mittle of the last century. This period has become known in geological research by the term "Little Ice Age". The introduction of the exaggerating term "Little Ice Age" has to be attributed to the fact that during this period, many Alpine glaciers were extending farther than at any time since the end of the last Ice Age, approx. 10,000 years ago. Further research has shown, however, that within the area of the Canton of Valais alone, the glaciers have matched or surpassed the extent of glaciation of the "Little Ice Age" almost a dozen times since the end of the last Ice Age.

The most recent climax among the glacial maxima in modern times has fortunately been recorded for posterity in the first official

Figure 6:
Glacier de Ferpècle and Glacier du Mont Miné in the uppermost part of the Val d'Hérens. In the background left, the white peak of Dent Blanche. Today, the two glacier tongues seen confluent here are separate and drastically retracted (cf. National Map of Switzerland, Sheet 283). A big larch trunk with about 300 year rings found in the till 500 m in front of the actual glacier tongue,gave an age of 9010 $\pm$ 100 BP (B-3194). (Buehlmann, Drawing Book 10, Number 253 (July 31, 1835). Reproduced by permission of the Graphische Sammlung of the Federal Institute of Technology, Zurich).

map of Switzerland, the so-called Dufour map. Figure 2 shows the original planetable map sheet drawn by the cartographer L'hardy in 1859. His contour-line map, originally drawn at a scale of 1:50,000 and edited in two color hatching. Figure 3 shows the respective states of glaciation of the upper Val de Bagnes in 1859 and in 1965. Depending on altitude and exposure, the various glaciers did shrink in very uneven proportions. Thus the front of the Crête Sèche has retreated more than three kilometers, whereas the tongue of the Glacier de Giétroz, originating at higher altitude, has retreated only 300 meters.

Even more impressively than on the map, deglaciation of the upper Val de Bagnes is illustrated by the two panoramic pictures of Figure 4.

During the interval between the glacial maxima of 1820 and 1850, glaciers were generally retreating only slightly. From the year 1835, there exist very beautiful and interesting drawings by J.R. Buehlmann. The artist was frequently travelling the Valais Alps, in search of engraving and lithograph subjects.

This kind of historical documents and illustrations can help to reconstruct the glacier movements over the last few hundred years as f.ex. Zumbuehl (1976) showed in his excellent study about the "Unterer Grindelwaldgletscher".

Figure 7:
Left lateral moraines of the Glacier de Tsidjiore Nouve near Arolla. Age of the deposits, from left to right: older than 8000 BP, 2500 BP, 1500 BP, and 900 BP. In the valley background is the terminal moraine of 1817.

3 DATING OF GLACIAL MAXIMA BY MEANS OF FOSSIL SOILS (fAh)*

A general period of unfavorable climate in the Valais in modern times is documented by historical records. Beyond the moraines deposited in modern times, however, there are not infrequently additional, older moraine walls which indicate that the glaciers repeatedly reached the extend of glaciation of 1850 in earlier times, too. These older moraines are clearly distinguished from more recent ones by their blurred shape, by their vegetation and by a state of soil formation that is advanced considerably beyond the moraines of the 19th century.

What methods of moraine dating are feasible? The moraines corresponding to the last distinct advance, around 1850, are in many places already covered by new vegetation to a considerable extent. On the bedrock, new soils have developed. It can be safely assumed that the same process occured in earlier periods of favorable climate, too, and that therefore during glacier advances, such soils were regularly covered by new deposits and consequently fossilized. Thus, moraine soils can be dated by means of the C^{14}-method, provided the soil contains organic material in sufficient amounts. In an Alpine context, the critical reservations advanced by Geyh against accuracy and relevancy of C^{14}-dating of soils (Geyh 1970) are only marginally pertinent. In most cases, the soil formation processes under consideration here are shortterm, involve only little production of humic acid, and hence entail little or no contamination of fossil soils with foreign, more recent material.

* The word soil is used in this context for all stades of soil development. Often you find only a small A-horizon (fAh) without a B-horizon.

3.1 Development of fossil soils underneath accretions and superpositions of moraines.

In Fig. 8, processes of soil fossilization are schematically demonstrated:
- The valley is covered with soil and vegetation (a).
- During a period of climatic deterioration, a glacier advances (b). The soil is overlain with moraine and till material, and fossilizes.
- The subsequent stage of ice shrinkage is illustrated in Fig. 8(c). The moraine deposition is progressively being covered with vegetation. Depending on the duration of the "interstade", an organic soil of variable thickness develops.
- In Figure 8(d), a second period of climatic deterioration is shown. The old moraine and the soil that has formed on its surface is overlapped by a new moraine. The two marginal sections of the diagram illustrate the two principal possibilities:
  . As shown on the left, the old moraine may be completely overwhelmed by new moraine material.
  . In the case illustrated on the right, the new moraine buries only part of the old moraine

One single glacier frequently provides

Schematic illustration of soil fossilization

Fig. 8

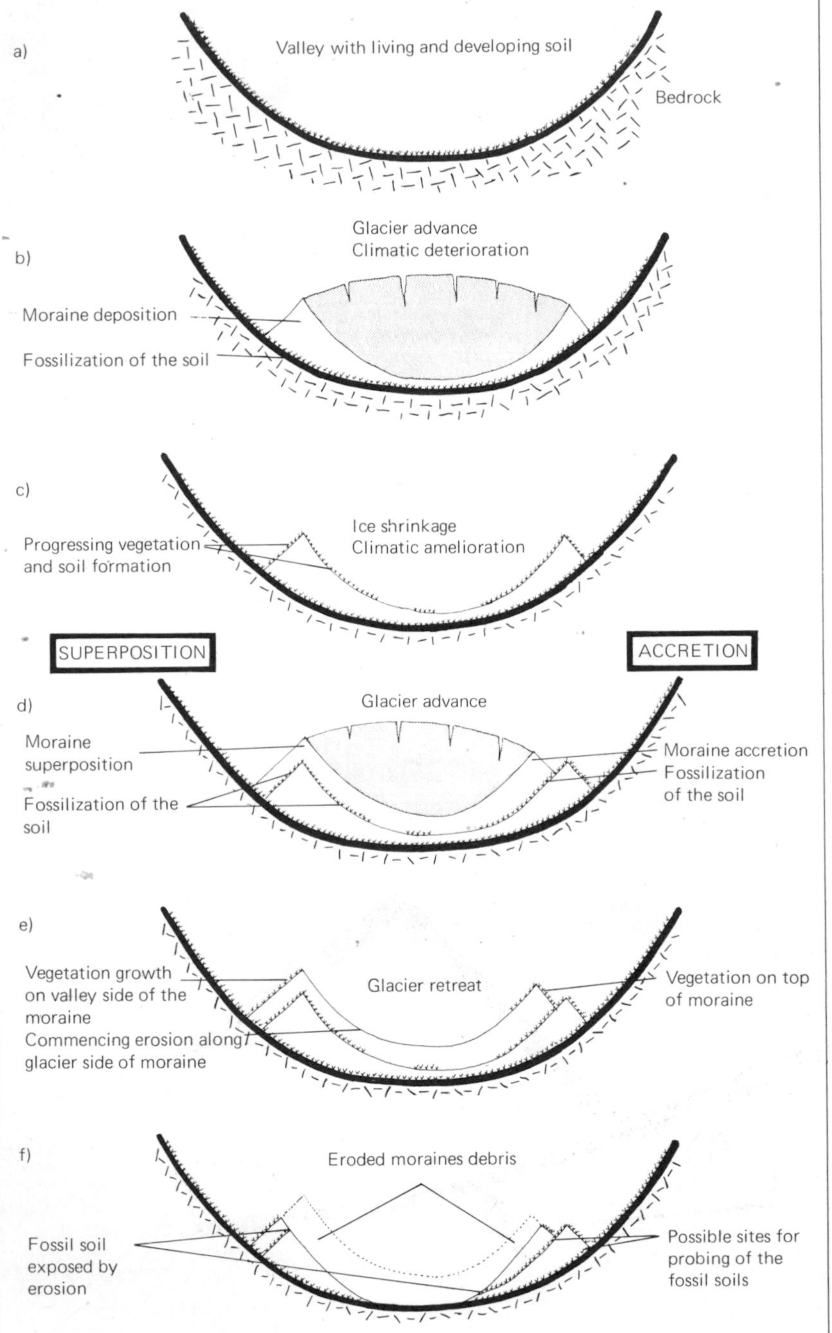
a)
Valley with living and developing soil
Bedrock
b)
Glacier advance
Climatic deterioration
Moraine deposition
Fossilization of the soil
c)
Ice shrinkage
Climatic amelioration
Progressing vegetation and soil formation
SUPERPOSITION
ACCRETION
d)
Glacier advance
Moraine superposition
Fossilization of the soil
Moraine accretion
Fossilization of the soil
e)
Glacier retreat
Vegetation growth on valley side of the moraine
Commencing erosion along glacier side of moraine
Vegetation on top of moraine
f)
Eroded moraines debris
Fossil soil exposed by erosion
Possible sites for probing of the fossil soils

examples of both superpositions (overlapping) and accretions (non-overlapping). Depending on exposure, altitude, geological properties of the deposits, and climatic conditions, the process of vegetation renewal gains momentum more or less rapidly. Development of vegetation is decisive for the production of humus and for the suitability of the C^{14}-method. Figures 8(e) and 8(f) refer to a new "interstade". As a consequence of climatic amelioration, vegetation and soil may develop on the second, more recent moraine, as it had been the case during the first "interstade" (Fig. 8(c)). However, it is equally likely that the moraine walls are partially or completely worn down. In this case, the previously buried soils are exposed along the glacier side of the moraine. If the new moraine does not completely overlap the old, the fossil soil can be laid open along the fissure between the bedrock and the old moraine. Digging in the cleft between the two moraines will be equally successful. The resulting C^{14}-age of this soil will point to the "interstade".

3.2 Searching and examples for fossil soils.

Within the perimeter of the glacier front, the soils rarely cover large areas in a completely continuous way. Like islands, rock spurs and large boulders from early ablation phases of the moraine intersperse and protrude from vegetation. Glacier milk and torrents swollen by thunder-showers are apt to destroy the thin layer of soil.

Fig. 9

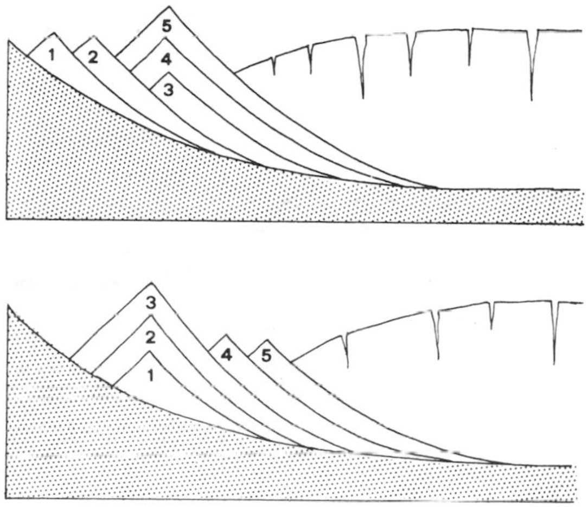

Possibilities of superpositions and accretions of moraines

Fig. 10

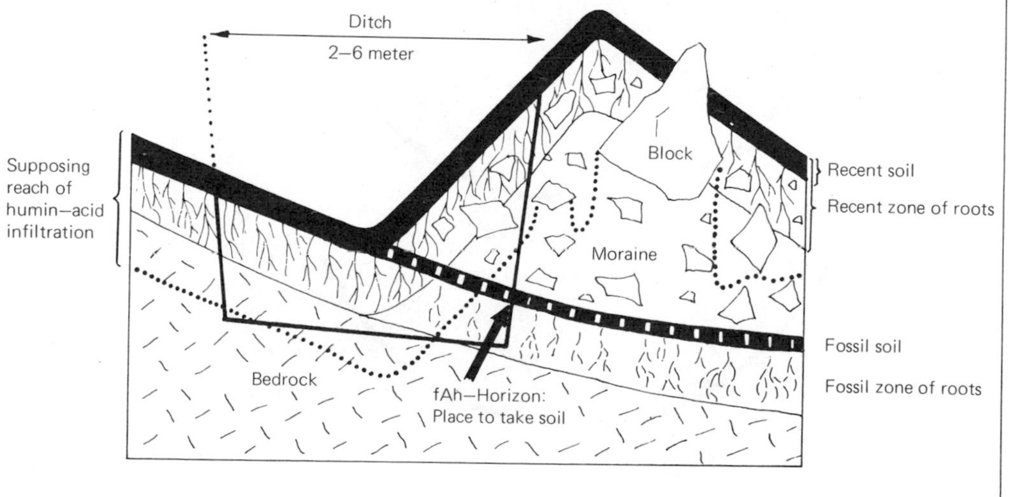

Figure 11:
Fossil soil on an eroding slope at the right lateral moraine of the Findelen Glacier, 2660 m above sea level.

Fig. 12

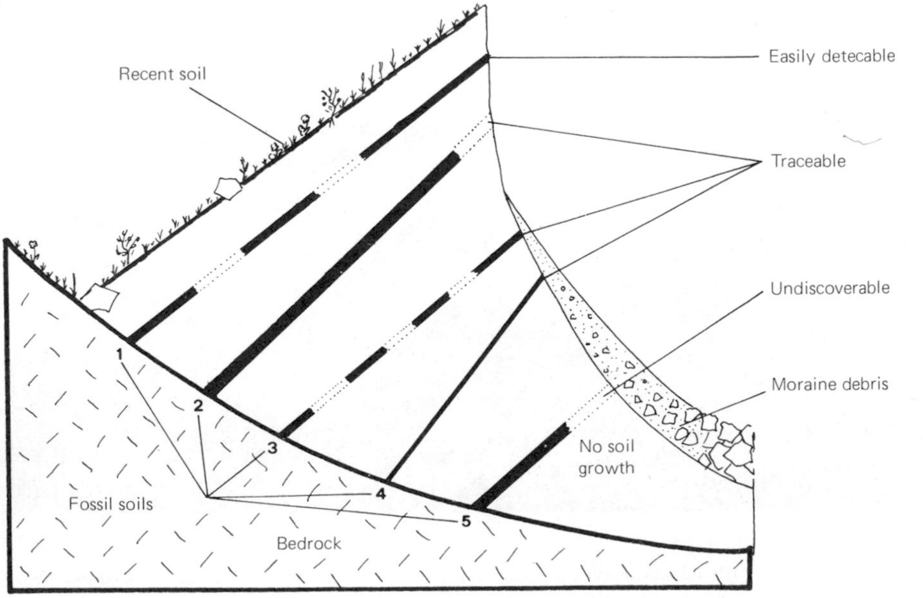

Figures 13 and 14:
Findelen Glacier, right side. Several moraine walls leave from the main ridge. They are dated by fossil soils between 2700 and 900 BP.

Figures 15 and 16:
Findelen Glacier. The two illustrations show different intensity of erosion. The main wall was built up during the advances of 2500, 1500 and 900 BP. The moraine walls of the Little Ice Age (Fig. 15), situated to the glacier side of the moraine, lie here below the ice level of earlier times. In a next glacier advance with superposition, different soil developments will be fossilized; so all fossil soils have to be interpreted, as nature is teaching us.

During intermittent minor advances of the glacier, the soil is stripped away wherever it is directly exposed to the pressure of the ice masses. After the glacier has deposited moraine material, it is difficult to predict whether an undamaged fossil soil can be excavated. Digging in the moraine is extremely cumbersome. Often large boulders have to be blasted. Stubborn optimism and dogged perseverance in the face of failures is asked for if the climatologist is to find his "gold", the fossil soils.

Figure 10 shows schematically at which points underneath an accretion samples are taken.

Less laborious, but more severely exposed to the danger of falling rock is the sampling on the steep slopes at the glacier side of the moraines, which are invariably in the process of erosion. Locating the fossil soil horizons often presents great difficulties; although they are of brown color, they are in most cases covered with a layer of greyish-white moraine silt. Thus on the glacier side of the moraine, almost always the entire face has to be worked with the pickaxe from bottom to top.

Figure 12 shows six moraine generations interbedded with five fossil soils. Fossil soil (1) lies open on the surface. The profile of soil (2) does not show in the cross-section taken here, but could be found farther up- or downstream. Fossil soils (3) and (4), partly eroded long ago, are concealed by moraine debris apposited to the steep glacier side of the moraine. Soil (3) is covered only by the uppermost "layer" of till which is of moderate thickness and can be removed. Soils (4) and (5), on the other hand, are only theoretically traceable.

The "Findelen Profile":
An interesting example of moraine*superposition found so far has been located at a point along the left lateral moraine of the Findelen Glacier above Zermatt (illustrated in Figs. 13 to 19). The planetable map sheet inserted here, from the Dufour map shows glaciation as it stood in 1859. The arrow points to the site where 7 layers of fossil soil have been found. Figure 17 shows a general cross-section.

Figure 19 shows the profile face of the moraine. In the background is the glacier which has considerably shrunk. The arrow points to a human figure near the lowest one of the eight layers of fossil soil discovered. The figure adjacent to the picture gives age and height, measured from the moraine ridge down, of the respective soils.

The relatively moderate age of the lowest fossil soil - only 2565 ± 195 years BP - is surprising. If there is evidence for nine glacier advances during this time, all of them nearly matching the glacial maximum of 1850 - how many may have occured during the first 7500 years since the end of the Ice Age, which unfortunately are not as equally well documented. The questions arise where the fossil soils corresponding to those earlier glacier advances could be located; whether they are presently undiscovered because of their being buried deeper below moraine debris (cf. Figure 12); or whether those older moraines had been shoved away altogether by the expanding glacier or eroded by water during "stadials".

Superpositions and accretions can be found all over in the Alps, especially in the western part. Some instructive examples are presented in the illustrations of the Glacier de Ferpècle and Mont-Miné (Figs. 22, 23, 24, 25, 26) as in the Photos of the Glacier de Tsidjiore Nouve (Figs. 22, 28, 29), where on the left lateral side recent ice is pushed over the moraine ridge, already described by Whalley (1973).

* Remark:
Regarding the classification of moraines in alpine/valley glacier environment the following has to be remembered: the term moraine (Moräne) is usually a synonym of the term lateral moraine (Seiten-/Randmoräne) and therefore it has more a geomorphological than lithological meaning.

Fig. 17

Findelen: Schematic cross section

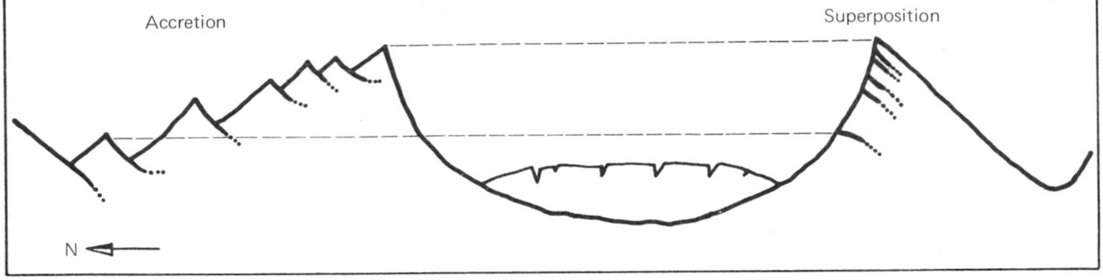

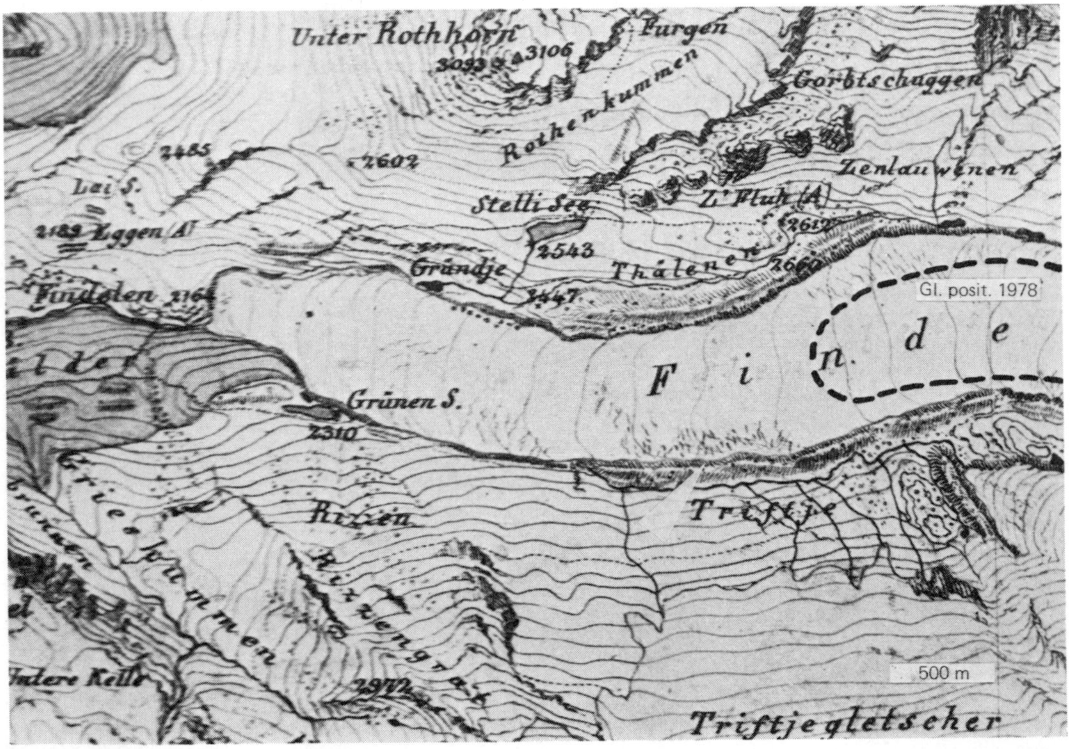

Figure 18:
Original survey sheet of the Dufour map (Number 535, Zermatt section) drawn in 1859 by Bétemps. The map scale is 1:50,000. Grey and brown colors indicate uneven density of vegetation along the left lateral moraine. In 1859, the glacier did not reach the ridge of the left lateral moraine. Hence moraines shown in grey color derive from glacier advances in modern times (1850-1870), whereas moraines marked by brown color (for greater density of vegetation) must be older. This is the case for Findelen profile (1), indicated by the arrow.
(Published by permission of the Swiss Federal Institute of Cartography).

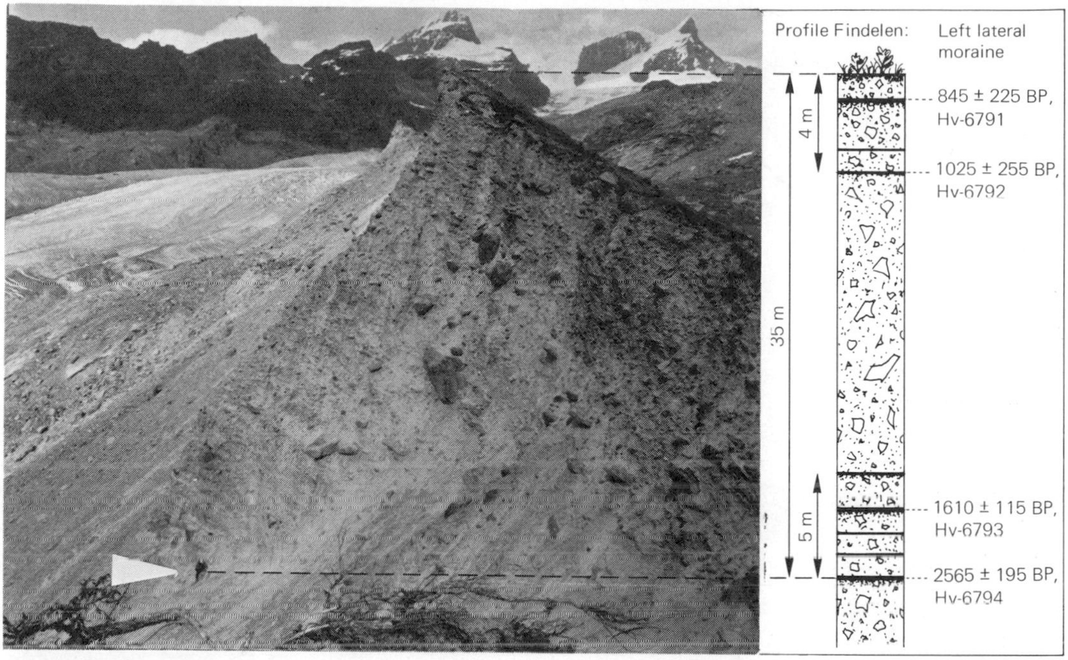

Figure 19

Figure 20:
Findelen Glacier: Fossil soil in the right lateral moraine

Figures 22 and 23:
Glacier de Ferpècle and Mont Miné 1877 (Fig. 22) and 1970 (Fig. 23). The arrows point to trunks and fossil soils as shown in Figs. 24, 21, 25 and 26.

Figure 24:
Ferpècle - Mont Miné Glacier: Fossil soil in moraine 900 $\pm$ 75 BP (Hv - 6800).

Figure 21:
Mont Miné Glacier: Larch trunk in the right lateral moraine between two fossil soils, 40 m below the moraine edge.
Age: 1555 $\pm$ 60 BP (Hv-6799)

Figure 25:
Glacier de Ferpècle (in the background): The upper fossil soils (arrows) of a sequence of four soil horizons. The lowermost lies about 50 m under the moraine edge with a C^{14}-age of 2480 ± 70 BP (Hv-6820).

Figure 26:
Glacier du Mont Miné (down): Sampling of a fossil soil.

Figure 27:
Glacier de Tsidjiore Nouve (Arolla). Arrow 1 locates Figs. 28 and 29, arrow 2 Fig. 32, arrow 3 Figs. 33 and 34.

Figure 28:
Glacier de Tsidjiore Nouve: Left lateral moraine. A: Old wall; B: The old wall is partially superposed by a new moraine; C: Glacier ice completly covered with superglacial till.

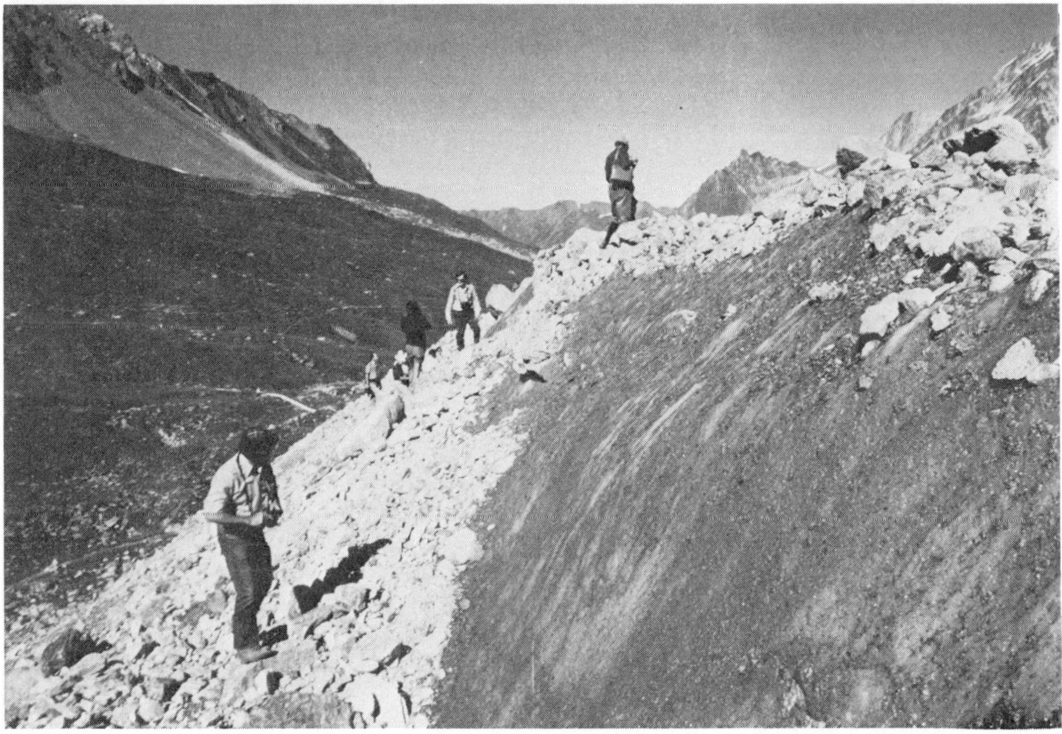

Figure 29:
The ice of Fig. 28, C, overlaps further up the old main moraine wall.

3.3 Solifluction lobes as approach for moraine datings

If we look at figure 32 we see cross over the Glacier de Tsidjiore Nouve with the big right lateral moraine against us covered by solifluction lobes, probably developed with glacier advances. This solifluction lobes are used also for climatological chronologies (Furrer, 1971, 1972, 1975) by dating the different overflowed fossil soils, illustrated in figures 30/31.

This solifluction lobes are often to find at the outside of moraine walls as for exampel on the Findelen Glacier shown in figures 35/36. To dig, to sample and to date the overflowed fossil soil gives a minimum age of the moraine. In figures 32 and 33 a situation on the left lateral moraine of the Glacier de Tsidjiore Nouve shows superposition of moraines in combination with a solifluction lobe, due to glacier advance.

All this illustrations of chapter 3 shall show the dynamic genesis of moraines and the possibility to date them by their different fossil soils.

Fig. 30

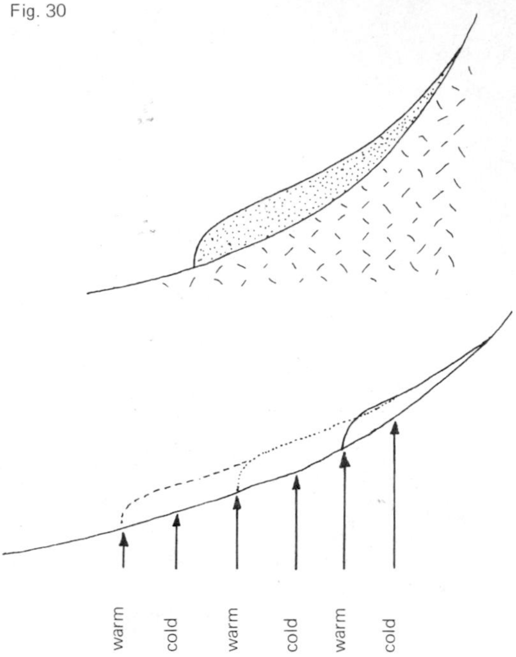

Fig. 31

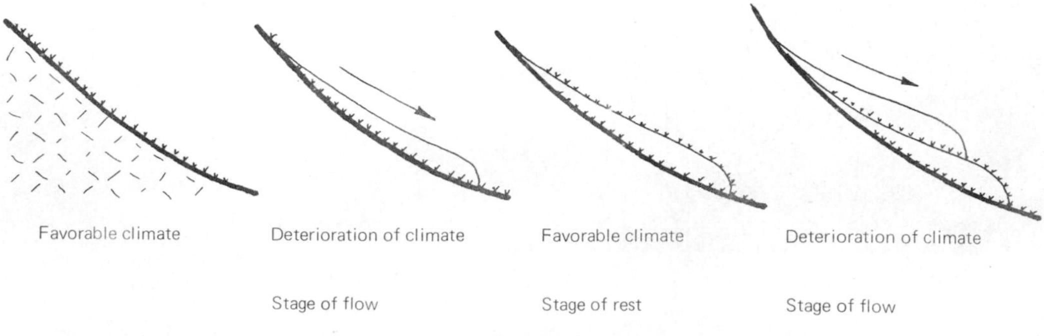

Figure 32:
View across the Glacier de Tsidjiore Nouve. Solifluction lobes probably due to the overriding of the lateral moraine ridge by the glacier and fed by morainic debris.

Figures 33 and 34:
Tsidjiore Nouve, left lateral moraine:
Combination between superposition of moraines and solifluction lobes, schematic illustrated in fig. 34.

Fig. 34
Tsidjiore Nouve 1: Cross section
Superposition and solifluction lobe

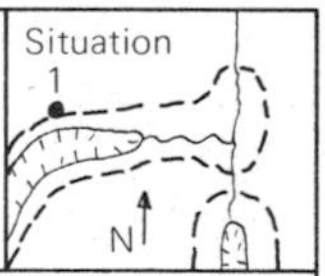

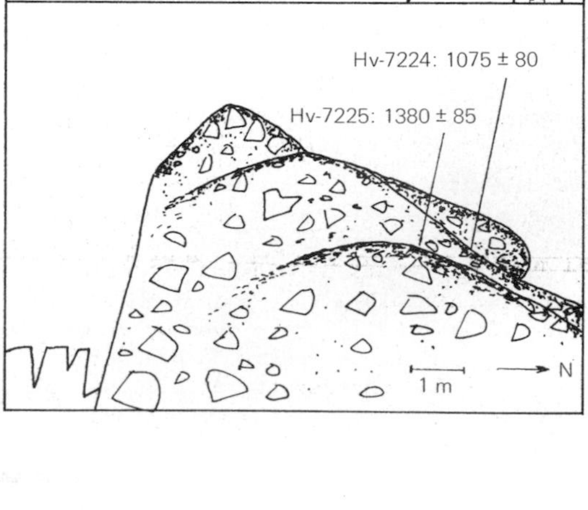

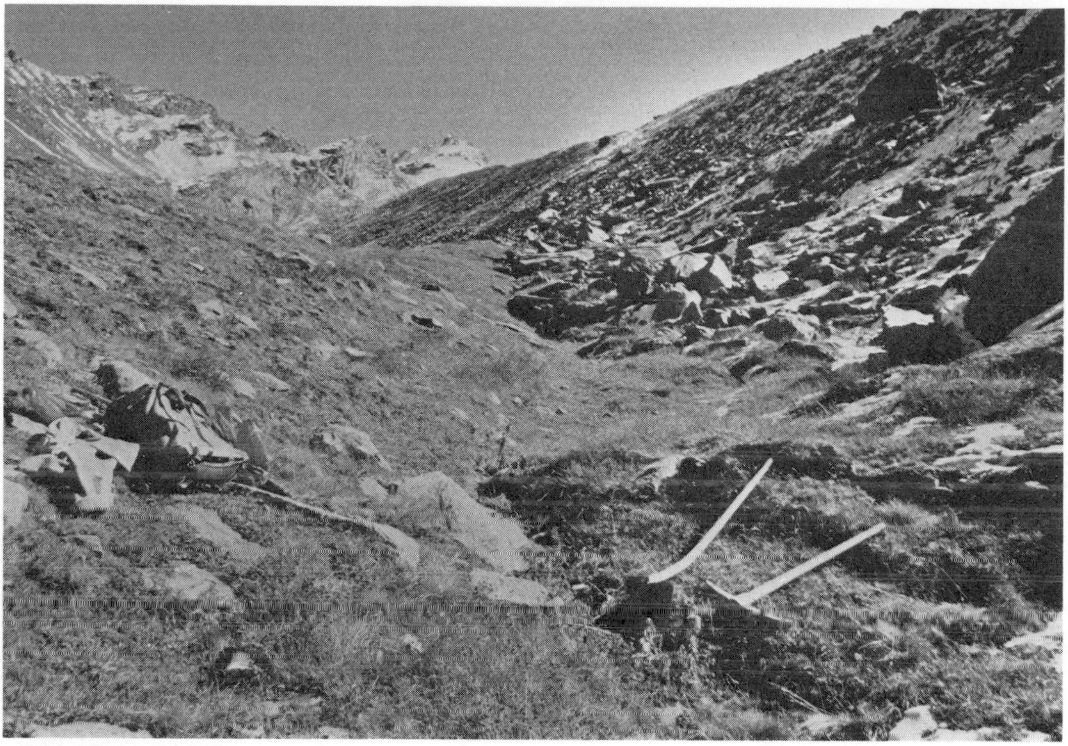

Figures 35 and 36:
Findelen Glacier; Solifluction lobes at the outside of the right lateral moraine, schematically illustrated in fig. 36.

Fig. 36

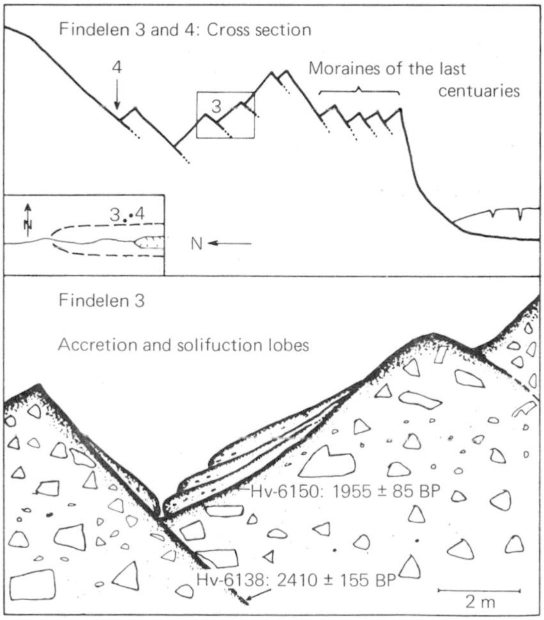

3.4 The research value of fossil soils.

H.-N. Müller (1975) gave a carefully study of a sequence of four superposed fossil soils from the inside slope of the left lateral moraine of the Rossboden Glacier (Simplon, Valais). In the analysis of the 1,2 m deep profile, he proves by determination of organic C- and by Ca/Mg-content, by ph-values and the density of pollen, that the brown horizons found in the eroded moraine

slopes are in fact soils of old moraine surfaces (fAh), fossilized by glacier advances. Fossil soils contain radioactive carbon-14 which stems from plants and burrowing animals once populating the living soil.

With decomposition of these, C - 14 isotope concentration begins to decrease. The half - life period is 5568 years. If the remaining amount of radioactive C - 14 contained in a fossil soil sample is measured, the absolute age of the sample can be determined, on the basis of the general rate of decay of C - 14 and of the standard isotopic abundance of a sample of known age.

In figure 37, the process of soil formation and of its subsequent overlapping with till is sketched by way of a vertical time chart.

If carbon-14 age as measured could be reliably equated with one of the three theoretical definitions of soil age (A_1, $\bar{A}$, A_2), fossil soils could be regarded as ideal yardsticks for dating and perfect indicators of climatic fluctuations. However, this is not at all the case. Soil formation presupposes the decay and consequent transformation into humus of organic material. Hence the decrease of isotope concentration in dead matter has to be marginally reckoned with already in the initial phase of soil formation. During the entire process of soil formation, additional dead substance is continuously deposited and adds to and affects the rate of disappearance. To complicate matters, the dead material does not remain stratified in neat layers. Burrowing animals carry more recent material downwards, and vice versa. Moreover, the soil is perturbated by drain water and frost (cryoturbation). Soil profile samples therefore contain indetermineate amounts of organic substance of unequal age.

The weighted means of the age groups involved yield the so-called average true residence time ($\bar{a}$), This age ($\bar{a}$) is not identical to the so-called mean radiometric residence time ($\bar{a}_r$), i.e., the sample age as determined by radiocarbon dating, because decay of radioactive C^{14} is not a linear but an exponential process.

The radiometric age of a soil sample always corresponds to a phase of favorable climate. Due to the short duration of the soil formation process in between advances of the

Fig. 37

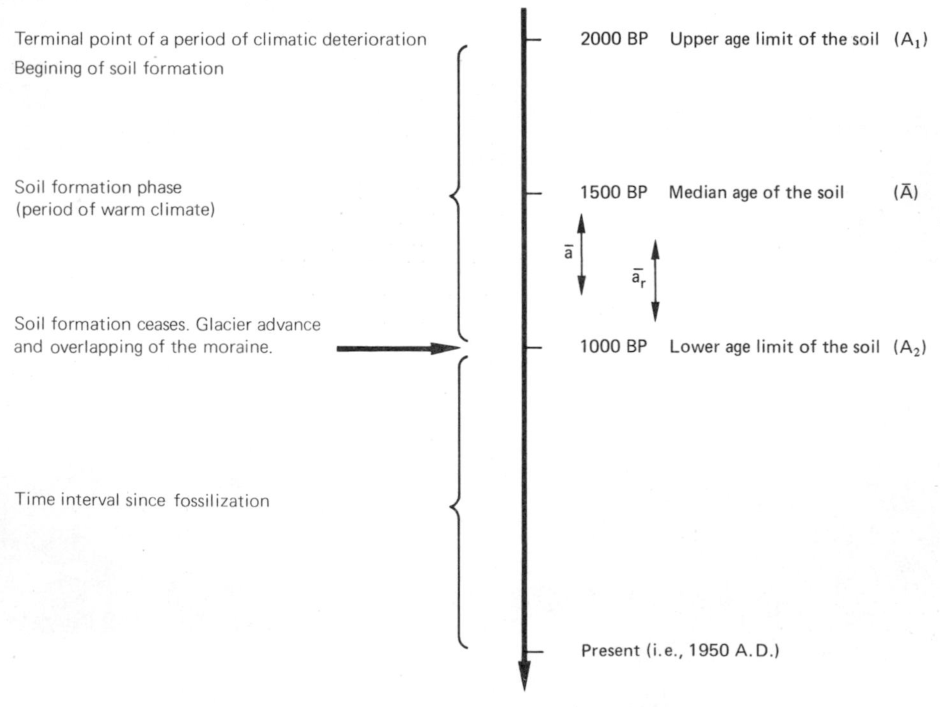

Figure 38:
Overridden larch trunk by the Glacier de Tsidjiore Nouve (Ly-750:2940 ± 150 BP).

glacier, the apparent carbon-14 age must be close to the terminal point of the soil formation process, i.e. the starting-point of a climatic deterioration and consequent a renewed glacial advance.

4 WOOD SAMPLES AND TREE TRUNKS FROM GLACIERS AND MORAINES

During the postglacial period, advancing glaciers have frequently knocked over and overridden trees. During later ice shrinkage phases, such trees occasionally reappear (Roethlisberger 1973, 150-239). Already in 1760, such a case is reported from the Lower Grindelwald Glacier by Gruner:

"In the center part of the glacier, all the way up on the flanks of Viescherhorn and Eiger, a good number of larch trunks (Pinus larix) protrude from the ice. Since the wood of this species is known to harden in wet conditions, those trees may have been lying there for centuries." (Gruner, Vol. III, 1760, Ch. V, in: Dollfuss-Ausset, 1864, Vol. I, 43). Since that time, reports of tree and wood findings from glaciers and moraines have multiplied. Even at present (1976), the Mont Miné and Zmutt Glaciers keep releasing large pieces of wood at their front and through the tunnel. Unfortunately many specimens thus retrieved have been burned or otherwise lost. Only the development of the carbon-14 dating method by Libby (1952) has opened up the possibility of thorough scientific analysis of wood samples from glaciers. Thus in 1961, H. Roethlisberger and Oeschger succeeded in dating a glacier advance in 1200 A.D., by means of radiocarbon analysis of tree stumps knocked over while growing. Many further wood specimens found on other glaciers have since been dated. In the course of the discussion of these results, two conflicting interpretations have been brought forward. While some scientists believe in a reliable causal connection between the dates determined and historic advances of the glaciers, others point to the possibility that the wood analyzed may have been transported on or near the glacier tongue by avalanches, landslides or heavy storms. This group consequently denies any direct connection between carbon-14 age of wood specimens and historic glacier advances.

The following considerations may guide us in the evaluation of wood samples and tree-trunks found on glaciers:

- Forestation of the glacier perimeter:
  If the area directly in front of the glacier snout is presently wooded, the same is likely to have been the case during a glacier shrinkage in earlier times. Hence it can be presumed that in the course of advances of the glacier under consideration, trees were regularly overridden.
- Characteristic tree-ring pattern:
  If the tree-ring pattern of typical cross-sections is nearly concentric, it can be assumed that the tree was standing in a flat area, as it is characteristic of the glacier perimeter. If, on the other hand, the medulla is dislodged towards the periphery of the cross-section, it is more likely that the tree originally was growing on a slope. This is to take as a general rule only (75 %) and does not mean to exclude the possibility of the occurrence of "eccentric" tree-ring patterns in flat surroundings (caused by the influence of boulders, wind pressure and other growth impediments) or of concentric profiles of trees from slopes.
- State of conservation of the wood:
  Lying openly exposed on the glacier or in front of the glacier snout, wood will generally rot, in Alpine climatic conditions, within 30-60 years, as is suggested by the experience gathered by the authors within the area of investigation indicated above. Trees released by the Findelen, Ferpècle and Arolla Glaciers between 1930 and 1950 and still remaining on the till surface, are today in a state of near-complete decay.
- Proximity of avalanche trains which seem to be able, at present, to transport living trees onto the glacier tongue or its surrounding.
  It is possible for such avalanche-transported trunks to be buried underneath the advancing glacier almost immediately, as has been observed in 1971/72 in the case of the Glacier de Saleina and Glacier du Trient. The state of preservation allows an informed judgement as to how long the wood might have been lying near the glacier tongue before being overridden by the glacier.

In connection with "avalanche wood" and extent of glaciation, the following further considerations are to be taken into account. Due to the laws of movement of the glacier (as evidenced by surface moraine formation), wood deposited on the tongue of the glacier by avalanches or storms will in most cases stay on the surface of the glacier, where

it will begin to rot. In the summer 1973, the author discovered bones, horseshoes and leather gear of a mule on the Theodul Glacier (near the trail leading up to the Theodul Pass). Teeth, one by one frozen off the jaws, and all remaining bones were widely scattered over an area measuring 90 by 150 feet (The animal must have been lost in a crevasse). No muscle or tender tissue was left. The findings would seem to suggest that objects remaining on the surface of the ice are subject to full-rate decomposition. The state of conservation of recovered wood specimens is thus an important indicator providing clues as to former states of glaciation. Since only objects completely swallowed by crevasses may end up in the ground moraine, it can be safely concluded that fossil wood which does not show signs of rot cannot have been deposited on top of the glacier tongue but must stem from the area immediately in front of the glacier snout; it must either have been deposited on the perimeter or originate from trees once growing there. Thus if well-preserved wood is released by the glacier front, circumstances of the finding and C^{14}-age point to a state of glaciation which in all likelihood was smaller than at the time of the finding. In sum, it must be repeated that every finding must be evaluated individually.

To recapitulate, C^{14}-age and site of tree-trunks permit in many cases direct conclusions concerning the position of the glacier front at an earlier time. This in turn allows a judgment on the climatic conditions of the time. Among the numerous wood specimens found within the area of our investigation, 40 have been dated so far and evaluated with regard to the identification of historic states of glaciation.

The scientific evaluation and utilization of fossil wood is far from being restricted to radiocarbon dating. Leonardo da Vinci (1452-1519) already noticed a direct correlation between tree-ring width and precipitation. The Swedish botanist Carl von Linné (1707-1778), studying oaks near their northern distribution limits, stated a correlation between tree-ring width and average summer temperatures. The Swiss forest ranger Kasthofer (1822, 319/20) conducted an interesting study when he compared samples of fossil Swiss stone pine (Pinus cembra) that had once been overridden by the Unteraar Glacier with living specimens. Kasthofer remarks that the people from the Grimsel Pass lodge used to stock up on their short supply of wood with roots, tree stumps and entire trunks taken or even dug up from the glacier. In addition, he relates the following observations:

"In Summer 1820, the author found two tree stumps of four foot diameter each, buried below between 2 and 4 feet of rocks, silt, and podsol. No signs of rot were detectable. The reddish wood was damaged on the surface only, and there was still - after a thousand years, perhaps - the characteristic smell of Swiss stone pine. A cross-section showed clear-cut, remarkably wide tree-rings which bore lively testimony to the tree's vitality in times long past. Many times the author has investigated tree stumps of stone pine recently cut in the forests of the Grindelwald, Gadmen and Gasteren Valleys, at an elevation of 1500 and 2000 feet below that of the glacier; but never, not even on particularly fertile soil, did he find evidence of a rate of growth comparable to the one attained by those buried fossil trees from ancient times".

These observations are remarkable indeed. Kasthofer was one of the first to compare tree-ring width, site and climatic conditions of living and fossil trees, finding as a result that the trees recovered on glaciers must stem from an earlier period of favorable climate, as is evidenced by their wide tree rings. In this, he was close to the recognition of how tree-ring series of characteristic pattern can be utilized for dating.

In between dendrochronology and dendroclimatology became an important appliance for many scientists, as more as now by x-ray densitometry additional informations beyond tree-ring width are taken out (density minima and maxima of early and late wood, width and percentage of early and late wood). Especially the density maxima gives information about temperature (localities in humid high altitude) and precipitation (dry localities).

All the trunks found in moraines, tills and marshes are analysed by x-ray densitometry at the Swiss Federal Institute of Forestry research. So in reality, it is not so important, whether we can find out in fact if timbers are in context with a glacier advance or not, but important is that they are well conserved and can be analysed and interpreted by x-ray dendroclimatology.

5 HIMALAYA AND ANDES

Already Kick (1956) mentioned superpositions and accretions of moraines in the Karakorum. He describes the Chogo Lungma Glacier with his big lateral moraines, built up by several advances. The fresh last superposition covers a much older moraine nucleus.*)

In Summer 1977 the author worked in the Cordillera Blanca (Huaraz, Peru), where also superposed fossil soils and trunks in moraines and tills were found. Eight C^{14}-dates were done. A paper about this work is in preparation. This result shows, that the genesis of moraines of alpine glaciers with superposition and accretion shown by fossil soils and trunks can also be found in other glaciated areas of the world.

6 CONCLUSION

Various examples and illustrations from the Wallis (Switzerland) demonstrate the genesis of moraines of the alpine glacier typ. The big lateral moraines, normally dated to the time of the Little Ice Age, are in reality much older, as Kick showed already in 1956 in the description of the Chogo Lungma Glacier (Karakorum, Pakistan) and Röthlisberger (in prep.) also verified in a study of the moraines of the Cordillera Blanca Glaciers (Peru). Very clear stratifications by fossil soils in the inside slopes of lateral moraines prove, that these walls were built up at least by the advances of the last 3500 years (C^{14}-cates of fossil soils), but probably already during the last 9000 years (C^{14}-dates of fossil trunks 500 m in front of the actual glacier front). - Fossil soils and fossil trunks permit to construct glacier extensions of the postglacial period. The analysis of fossil trunks by x-ray densitometry and the resulting climatological interpretations will give us in a year by year scale temperatur variations of the last 10,000 years.

BIBLIOGRAPHY

Beeler, F. 1977, Geomorphologische Untersuchungen am Spät- und Postglazial im Schweizerischen Nationalpark und im Berninapassgebiet (Südrätische Alpen). Diss. Uni Zürich, in: "Ergebnisse der wissenschaftl. Untersuchungen im Schweizerischen Nationalpark", Bd.XV, 77, 276p.

Bezinge, A. 1976, Troncs fossiles morainiques et climat de la période Holocène en Europe. Bulletin de la Murithienne, 93, 93-111.

*) Kick writes (Zeitschrift für Gletscherkunde, III, 3, p. 339-340): "Dieser über 30 km lange Zug einer auffallend frischen Ufermoräne zu beiden Seiten und auch an den Seitengletschern bezeugt einen allgemeinen Eisrückgang im ganzen Chogo Lungma-Becken, der seit mehreren Jahrzehnten ziemlich kontinuierlich vor sich gegangen sein muss. Stellenweise sind wohl einzelne Rückzugsstufen vorhanden, aber im ganzen so unbedeutend, dass kein längerer allgemeiner Stillstand oder Vorstoss fürs ganze Gebiet seit dem letzten, der grossen Ufermoräne entsprechenden Hochstand, erkennbar ist. Dies gilt auch unter Berücksichtigung der Vorsicht gebietenden Eigenart dieses Gletschers, dass nämlich verschiedene Gletscherstände ihren Moränenschutt über die gleiche Ufermoräne geschüttet haben. Diese letztere Tatsache wurde ungefähr bei km 20 deutlich. Dort streckt sich etwa aus der Mitte des Aussenhanges der Grossen Ufermoräne quer zu ihr eine "Endmoränenzunge" heraus (Bild 11). Ihre Oberfläche ist mit kniehohem Gesträuch bewachsen, was zusammen mit dem viel fortgeschritteneren Verwitterungsgrade ein wesentlich höheres Alter bezeugt, als das der frischen Ufermoräne selbst. Die Erklärung kann nur sein, dass über den Ufermoränenwall einmal ein Eislappen hinüberhing, der seine "Endmoräne" zurückliess, später aber durch neues Anwachsen des Hauptgletschers die Ufermoräne von neuem überschüttet und höher gebaut wurde. Unter dem heutigen frischen Schuttmantel verbirgt sich demnach in der Grossen Ufermoräne ein wesentlich älterer Kern. An einer anderen Stelle weiter oben war durch Abrutschung an der Gletscherseite der Grossen Ufermoräne die gleiche Zusammensetzung freigelegt. Die Stelle mit der "Eislappen-Endmoräne" liegt an einer mehrere 100 m langen Einbuchtung des Talgehänges, die viel Platz für ein Ausbiegen des Hauptgletschers geboten hätte. Dieser aber nützte die Möglichkeit nicht aus, sondern zog mit seinem Rand geradeaus weiter. Es sieht so aus, als hätte sich der Gletscher mit der Grossen Ufermoräne seinen eigenen Grenzwall hochgebaut, dem er sich immer wieder anpasste. Verschiedene Hochstände haben hier ihre seitliche Begrenzung an dieser gleichen Ufermoräne eingehalten, während ausserhalb und darunter ein breites "Ufertal" eisfrei blieb."

Bieler, P.L. 1976, Etude paléoclimatique de la fin de la période quaternaire dans le bassin lémanique. Archives acad. des sciences, vol. 29, fasc. 1, 5-53.

Bircher, W., Lageänderungen von Gletscherzungen im Saastal. Diplomarbeit, Geogr. Institut Universität Zürich, 86p.

Bocquet,G. 1969, Morphologie et Glaciers dans le haut Val de Bagnes. Dipl.Arb.Uni. de Grenoble. Faculté des lettres et sciences humaines. Inst. de Géographie Alpine.

Burga, C.A. 1975, Spätglaziale Gletscherstände im Schams. Eine glazialmorphologisch-pollenanalytische Untersuchung am Lai da Vons (Graubünden). Diplomarbeit, Geogr. Inst. Zürich, 72 p.

Burri, M. 1974, Histoire et préhistoire glaciaire des vallées des Drances, Valais. Eclogae Geologicae Helvetiae, Vol.67/1

Fries, M. 1977, Ehemalige Gletscherstände im Val Maighels. Diplomarbeit,Geogr. Inst.Uni.Zürich.

Furrer, G., Bachmann, F., Fitze, P. 1971, Erdströme als Formelement von Solifluktionsdecken im Raume Munt Chavagl/Munt Buffalora, Schweiz.Nat.-Park. Ergebnis der wissenschaftl. Untersuchungen im Schweizer Nat.-Park, Band XI.

Furrer, G. 1972, Bewegungsmessungen an Solifluktionsdecken. Zs.f.Geom.Suppl., Bd.13.

Furrer, G., Leuzinger, H., Ammann, K. 1975, Klimaschwankungen während des alpinen Postglazials im Spiegel fossiler Böden. Vierteljahresschrift der naturf.Ges. in Zürich, Jg.120, Heft 1, 15-31.

Furrer, G., Gamper-Schollenberger, B., Suter, J. (in press.), Zur Geschichte unserer Gletscher in der Nacheiszeit. Geogr.Inst.Uni.Zürich.(Z.Glk)

Gasenzer, H. 1964, Morphologische Untersuchungen in einem Gletschervorfeld (Bifertengletscher). Diplomarbeit,Geogr. Inst.Uni.Zürich.

Geyh, M.A. 1970, Möglichkeiten und Grenzen der Radiokohlenstoff-Altersbestimmung von Böden: Methodische Probleme. Mitt.Dtsch. Bodenkundl.Gesellschaft 10.

Geyh, M.A. 1971, Die Anwendung der 14C-Methode. Clausthaler Tektonische Hefte 11, Verl.Ellen Pilger, 118 p.

Geyh, M.A., Benzler J.H. and Roeschmann, G. 1972, Problems of Dating Pleistocene and Holocene Soils by Radiometric Methods. Intl. Soc. Soil Sci. and Israel Universities Press.

* Haas, Ph. 1978, Untersuchung zur Gletschergeschichte im Val d'Anniviers. Diplomarbeit, Geogr.Inst.Uni.Zürich, 103p.

Hagen, T. 1944, Der Gletscherausbruch von Ferpècle. Die Alpen, 269-274.

Hagen, T. 1948, Geologie des Mont Dolin und des Nordrandes der Dent Blanche-Decke zwischen Mont Blanc de Cheilon und Ferpècle (Wallis). In: Beiträge zur Geologischen Karte der Schweiz, Kümmerly & Frey, Bern, 64p.

Holzhuser, H. 1978, Zur Geschichte des Fieschergletschers. Dipl.Arbeit,Geogr. Inst.Uni.Zürich, 120p.

Kasthofer, K. 1822, Bemerkungen auf einer Alpen-Reise über den Susten, Gotthard, Bernardin, und über die Oberalp, Furka und Grimsel ... Nebst Betrachtungen über die Veränderungen in dem Klima des Bernischen Hochgebirgs. Aarau, 354p. (Klima, 273-349).

Kick, W. 1956, Der Chogo Lungma Gletscher (Superposition and accretions of moraines). Zeitschr.f.Gletscherkunde III, 3, 1956, p. 338-340.

King, L. 1974, Studien zur postglazialen Gletscher- und Vegetationsgeschichte des Sustenpassgebietes. Basler Beiträge zur Geographie, Heft 18, Helbing und Lichtenhahn, Basel, 123p.

Kinzl, H. 1932, Die grössten nacheiszeitlichen Gletschervorstösse in den Schweizer Alpen und in der Mont Blanc-Gruppe. Zeitschrift für Gletscherkunde, XX, Band, Leipzig, 269-397p.

Lamprecht, A. 1978, Die Beziehung zwischen Holzdichtewerten von Fichten aus subalpinen Lagen des Tirols und Witterungsdaten aus Chroniken im Zeitraum von 1370-1800. Diplomarbeit, Geogr.Inst.Uni.Zürich und Eidg.Anstalt für das forstliche Versuchswesen, 77p.

Lehner, K. 1963, Zermatter Sagen und Legenden. Mengis, Visp, 192p.

Lüthi, A. 1978, Zermatt und die Hochalpenpässe. Eine geländearchäologische Untersuchung. Tscherrig AG, 134p.

Maisch, M. 1977, Glazialmorph.Untersuch.im Val Maighels.Dipl.Arb.Geogr.Inst.Uni.Zürich

Maisch, M. 1978, Gletschergeschichtliche Vorgänge im Raum Sertig. Davoser Zeitung 15.Aug., Nr. 188.

Messerli, B., Zumbühl, H.J., Ammann, K., Kienholz, H., Oeschger, H., Pfister, Ch., Zurbuchen, M. 1976, Die Schwankungen des Unteren Grindelwaldgletschers seit dem Mittelalter. Ein interdisziplinärer Beitrag zur Klimageschichte. Zeitschrift für Gletscherkunde und Glazialgeologie, Bd.XI, Heft 1, 1975, 3-110p.

* Gruner, G.S. 1760, Die Eisgebirge des Schweizerlandes. 3 vol., Bern, total 680p.

Müller, H.-N. 1975, Untersuchungen ehemaliger Gletscherstände im Rossbodengebiet, Simplon VS. Dipl.-Arbeit, Geogr.Institut Uni.Zürich, 15-17p.

Müller, H.-N. 1975, Fossile Böden (fAh) in Moränen (Gäli Egga, Rossbodengebiet Simplon VS). Bulletin de la Murithienne, 92, Sion, 21-31p.

Patzelt, G. 1973, Die postglazialen Gletscher- und Klimaschwankungen in der Venedigergruppe (Hohe Tauern, Ostalpen); mit sechs Pollendiagrammen von S.Bortenschlager. Zeitschr.f.Geomorphologie, Suppl.Bd. 16, 25-72p.

Patzelt, G. 1977, Der zeitliche Ablauf und das Ausmass postglazialer Klimaschwankungen in den Alpen. **Erdwiss. Forschung**, 13, 249-259p.

Pfister, Ch. 1976, Die Schwankungen des Untern Grindelwaldgletschers im Vergleich mit historischen Witterungsbeobachtungen und Messungen. Zeitschr.f.Gletscherkunde XI/1, 1975, 74-90p.

Portmann, J.P. 1977, Variations glaciaires, historiques et préhistoriques dans les Alpes suisses. Les Alpes; 4, 145-172p.

Renner, F. 1977, Ehemalige Gletscherstände im Witenwasseren- und Muttental, Urseren. Diplomarbeit Geogr.Inst.Uni.Zürich , 64p.

Röthlisberger, F. 1973, Blüemlisalpsagen und Gletscherpässe im Raume Zermatt-Ferpècle-Arolla. Ein Beitrag zu Klima-Schwankungen im Postglazial mit einem Anhang über Holzfunde aus Gletschern. Diplom.Geograph.Inst.Uni.Zürich, 300p.

Röthlisberger, F. 1974, Etude des variations climatiques d'après l'histoire des cols glaciaires. Le col d'Hérens (Valais, Suisse). Bolletino comitato glaciologico Italiano, 22, 9-34.

Röthlisberger, F. 1974, Datations des moraines d'après des horizonts de sol fossile: Glacier de Findelen. In: Les variations climatiques postglaciaires dans les Alpes (Schneebeli, Leuzinger, Müller, Röthlisberger). Congr.Int.Glac.Soc.Courmayeur, Uni.Zürich, 12-14p.

Röthlisberger, F. 1976, Gletscher- und Klimaschwankungen im Raum Zermatt, Ferpècle und Arolla. In: 8000 Jahre Walliser Gletschergeschichte, ein Beitrag zur Erforschung des Klimaverlaufs in der Nacheiszeit. Die Alpen, 3/4, 61-152p.

Röthlisberger H., Oeschger, H. 1961, Datierung eines ehemaligen Standes des Aletschgletschers durch Radioaktivitätsmessung an Holzproben und Bemerkungen zu Holzfunden an weiteren Gletschern. Zeitsch. Gletscherk. Glazialgeol., 4(3). 191-205p.

Schneebeli, W. 1974, Morphologische und geschichtliche Untersuchungen der neuzeitlichen Gletscherschwankungen im hinteren Val de Bagnes. Diplomarbeit Uni.Zürich,164p.

Schneebeli, W. 1976, Untersuchungen von Gletscherschwankungen im Val de Bagnes. In: 8000 Jahre Walliser Gletschergeschichte, ein Beitrag zur Erforschung des Klimaverlaufs in der Nacheiszeit. Die Alpen 3/4, 5-57 p.

Schollenberger, B. 1976, Moränenwälle im Hintern Rosegtal. Diplomarbeit,Geogr. Inst.Uni.Zürich, 45p.

Suter, J., Suter, U. 1976, Glazialmorphologische Untersuchungen in der Val Bever. Diplomarbeit,Geogr.Inst.Uni.Zürich, 110p.

Schweingruber, F.H., Schär, E., Bräker, O.U. 1978, Jahrringe aus sieben Jahrhunderten. Saaner Jahrbuch, 7, 1977, in press.

Schweingruber, F.H., Fritts, H.C.,Bräker, O.U., Drew, L., Schär, E., The x-ray technique as applied to dendroclimatology. Tree ring bulletin. (In press.)

Schweingruber, F.H., Bräker, O.U., Schär,E., X-ray densitometric results for subalpine conifers and their relationship to climate. British arch.rep. (in press.)

Schweingruber, F.H., Bräker, O.U., Schär, E. Dendroclimatic studies on conifers from central Europe and England. Boreas. in prep. 30 p. (in press.)

Vögele, A.E. 1976, Untersuchung postglazialer Gletscherstände im Dischmatal (Davos, Graubünden). Diplomarbeit, Geogr.Inst. Uni.Zürich

Vuagneux, R. 1976, Untersuchung Spät- und postglazialer Gletscherstände im Raume Flüelapass. Diplomarbeit, Geogr.Inst. Uni.Zürich, 57p.

Wegmüller, S. 1977, Pollenanalytische Untersuchungen zur spät- und postglazialen Vegetationsgeschichte der französischen Alpen (Dauphiné), Bern, Haupt, 185p.

Whalley, W.B. 1973, An exposure of ice on the distal side of a lateral moraine. Journal of Glaciology, Vol.12, No.65, p.327-329.

Winistorfer, J. 1977, Paléographie des stades glaciaires des vallées de la rive gauche du Rhône entre Viège et Aproz (Valais). Bull. de la Murith. 94, 3-65.

Zoller, H. 1977, Alter und Ausmass postglazialer Klimaschwankungen in den Schweizer Alpen. Erdwiss. Forschung,13, 271-281p.

Zumbühl, H.J. 1976, Die Schwankungen des Untern Grindelwaldgletschers in den historischen Bild- und Schriftquellen des 12. bis 19. Jahrhunderts. Zeitschrift für Gletscherkunde XI/1, 1975, 12-50 und 95-100.

Penecontemporaneous deformation structures in a Pleistocene periglacial delta of western Swiss Plâteau

A. PARRIAUX
Federal Institute of Technology, Lausanne, Switzerland

ABSTRACT

A sandy and silty formation on the top of a periglacial lacustrine delta presents a series of deformation of the "convolute bed" and "ball and pillow structure" types. The author situates the paleogeographical and sedimentological frame. Then, he describes in detail the deformations. Some geotechnical parameters of concerned soils are given. At last, the relationships between the sedimentation and folding are examined.

1. INTRODUCTION

An outcrop of sand and silt showing spectacular deformations has been discovered on the shelf of a Pleistocene periglacial delta at Granges-Marnand in the Broye valley. The scarcity as well as the beauty of distortion figures make this case worth being described in detail.

In the western part of Swiss Plateau near lemanic depression, the Broye basin diverts the water to the north through the lakes of Morat and Neuchâtel. The rocky substratum is entirely molassic. The north and south parts are cut in shaly and sandy rocks whereas the center part only erodes sand- and siltstones. Consequently the nature of quaternary deposits is very sandy and silty.

During quaternary glaciations, the Broye valley was located in the inner part of the country occupied by the Rhône glacier which divided into two tongues. The first one flew to Lyon and the other one northeastwards in the direction of Soleure. This last one occupied two different glacial valleys which were separated by the Jorat-Vully molassic Plateau :

- in the northwest, the valleys of Venoge, of Orbe-Thielle, of Lake of Neuchâtel and the Aar valley
- in the southwest, the valleys of Broye then of Aar (Jäckli 1970).

The Broye valley was fed with ice by two transfluence cols in the edge of the Leman, then it guided the ice flow creating important troughs separated by steps.

During the last retreat, periglacial lakes have surrounded the melting tongue depositing deltaïc terraces actually perched on slopes.

During the fusion, the lacustrine level has lowered. The lakes always occupied vaster areas in front of the glacier. That is during one of the last glacio-lacustrine periods that the delta of Granges-Marnand has taken form.

2. LOCAL PALEOGEOGRAPHY

The periglacial lake in which the Granges delta has been sedimented, extended over the sides and in front of the Broye glacier. In the north, it rested probably against the tongue of the Neuchâtel Lake glacier which was flowing into the low Broye valley by transfluence over Estavayer depression (Parriaux 1978).

The water level was determined by morphology and sediments of the delta at 505 meters.

At this retirement period, an ice tongue still filled the bottom of the main valley (Fig. 1).

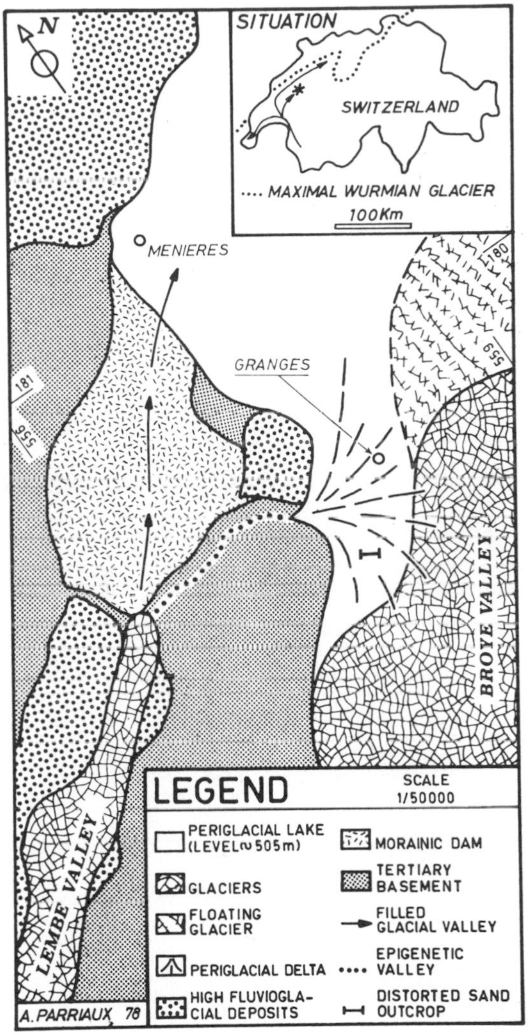

Figure 1 : local paleogeography of Lembe delta

Westwards, the secondary glacial valley of Lembe was dammed by a morainic body which impeded the natural water flow northeaster. This meltwater which was accumulated behind the dam, created another periglacial lake with a level about 560 meters. In the beginning, its water flew out over the morainic dam until such time as it made its way towards east to the 505 m lake. An epigenetic course was rapidly eroded through the thin molassic ridge. The flow carried a great solid discharge which occured from the frehly deposited material of Lembe valley. Little by little an important delta has grown at the outlet of the new Lembe river.

3. LITHOSTRATIGRAPHY OF DELTAIC SEDIMENTS

The base of deltaïc deposits is unknown. Its contact with the molassic ground decreases rapidly towards the axis of the plain. In the studied area, its cote is probably located several tens of meters under the inferior level indicated in figure 2.

The visible part begins with gravely horizon of foreset beds (a-b) not much inclined downstream. Gradually the dip of beds increases (b-c) for adjoining up to about 30°, always eastwards. Above it, a series with horizontal sedimentation (c-s) of topset beds deposited which begins with a very continuous gravely level (c-d). Above it, the grain size reduces very quickly and passes to fine sand levels and to ripple laminated silt (f-s) in which deformations are observed. The sedimentology of this sandy unity is thoroughly described in the figure 2. We meet spectacular ripple lamination of very different types where a prominent part is played by the climbing ripple structures.

Finally, a gravely complex (s-t) eroded the sandy series by cutting well marked channels. The sedimentation is very coarse. Many cross bedding can be observed.

We note that the main gravely member was touched by tilt movements which created a normal fault system. The reason of these deformations must be searched in the fusion of dead ice blocks at the base of sediments.

4. DESCRIPTION OF CONTORTED BEDS

The penecontemporaneous deformations which affected the central part of the sandy member of the deltaïc series can be seen on an outcrop of about hundred meters, on the periphery of a quarry. The dip of the layer is practically horizontal (picture 1).

Up-stream, the structures disappear and it does not subsist but small deformations in lower levels surmounted of ripple laminated sands. Down-stream, on the other hand, the structures seem continuing under the weathering soils which gradually cover the series in its whole.

Like the majority of penecontemporaneous deformation, those of the Lembe delta only concern a particular horizon in entirely disturbing its first sedimentological struc-

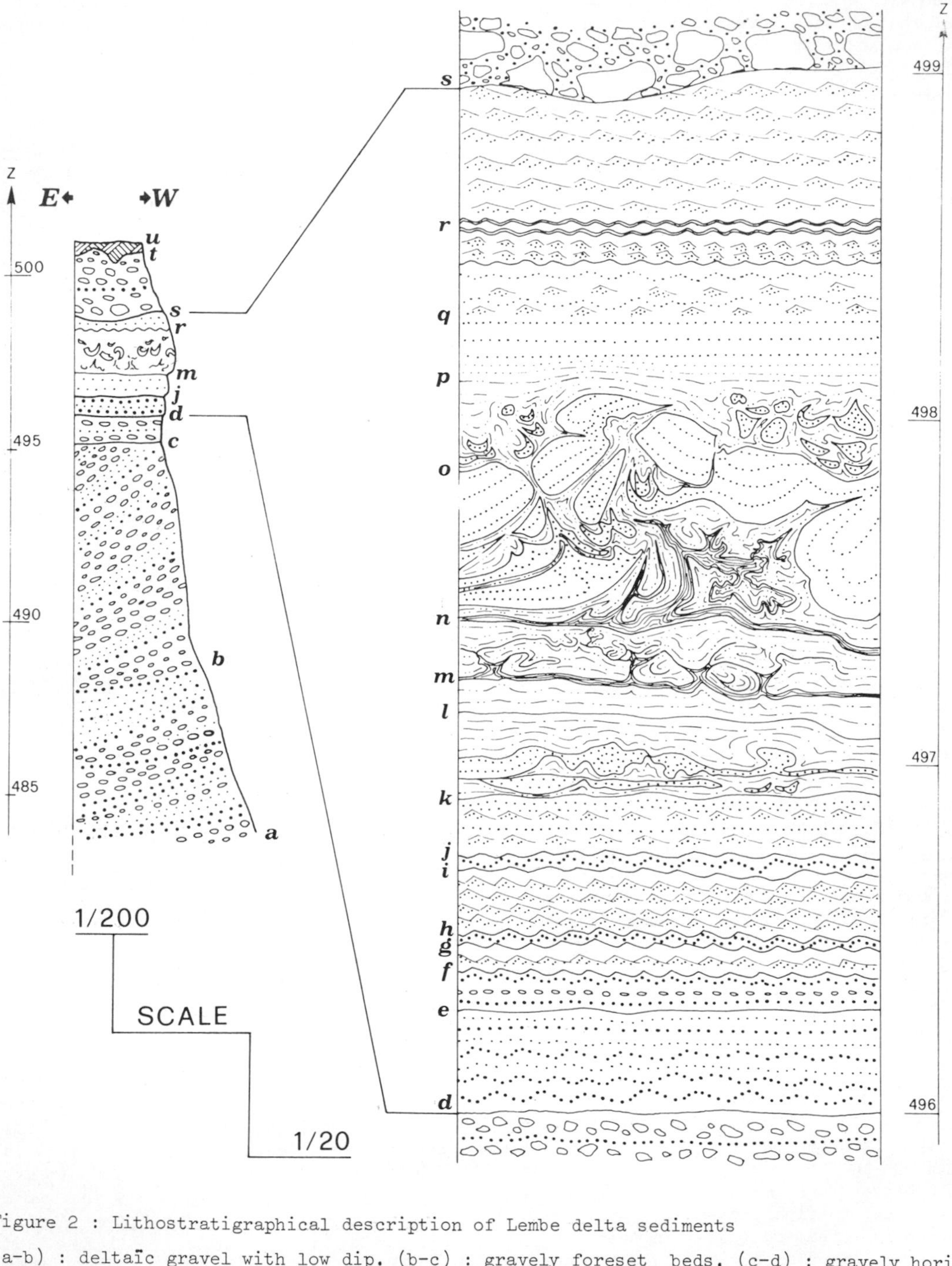

Figure 2 : Lithostratigraphical description of Lembe delta sediments

(a-b) : deltaïc gravel with low dip. (b-c) : gravely foreset beds. (c-d) : gravely horizontal topset. (d-e) : bedded middle sand with few ripples. (e-f) : coarse sand with fine gravel. (f-g) : fine sand with ripple, paleocurrent → E. (g-h) : ripple laminated middle sand → E. (h-i) : fine sand with ripple → E. (i-j) : cross bedded middle sand → E. (j-k) : fine and middle sand with ripple → E. (k-l) : small convoluted fine sand and sandy silt, convolution disappear upwards. (l-m) : very fine sand a little deformed. (m-n) : silty

sand with small but very well formed convolutions. (n-o) : main convoluted bed with big deformation structures in fine sand, silt and sandy silt. (o-p) : Small scale convoluted f. sand and silt. (p-q) : undeformed horizontal bedded fine sand. (q-r) : very fine sand with ripple, paleoc.→ E at the base,→ W at the top. (r-s) : fine sand with great wavelength ripple, paleoc.→ E. (s-t) : coarse channel gravel, discordant. (t-u) : top soil.

---

tures, while the edges and the levels which surround it are practically not affected (picture 1, 9 A-L)

The convolutions gradually appear upwards (fig. 2). The level (k-l) only shows a light flow of a silt which induces some waves in a sandy horizon. Then the deformations temporarily vanish up to a bed (m-n) with very well formed but small contortions. Here, a silty level has been injected in fine and long tongues inside a silty sand, dividing it into syncline-folded cells.

Above, we penetrate inside the principal distorted level (n-p). This layer was composed in the origin of a silty base surmounted of two main fine sand layers with plane beds lamination separated by more silty levels (n-o). On the top, small lenses of fine sand were stratified in a sandy and silty body (o-p).

This original sedimentation is entirely modified by a practically continuous series of vertical displacements of material. On the one hand, the sandy horizon penetrated into the silty mud breaking and folding in very variable shaped-synclines.

On the other hand, the silty level of the base concentrated in different points and migrated upwards through sandy series by extremely complex anticlines.

According to the zones these deformations can be bound either to "convolute beds" or to "ball and pillow structure" from recent definitions (Reineck and Singh 1973). An extract of the most interesting fragments of the outcrop is presented on the pictures 9 A-L.

4.1 Sandy synclines

The deformations which affect the sandy levels are of very different intensities. In a few zones, the bed collapsed in bulk without important folding. We only notice cuts due to the injection of a silty dyke with curvature of sand laminae (pict. 9 I).

When deformation increases, sandy cells become isolated and folded in open often assymetric synclines with a sudden interruption of layers on the top. The axes of the folds are approximatively vertical, sometimes tilted. The extension of the structure in the axial direction is limited (pict. 2).

In a still more advanced step of the deformation appear sandy nuclei with an entirely spherical central part, retired within itself (picture 3).

In such cases, we sometimes observe very closely packed isoclinal folds in the core of the structure (picture 4).

Generally, the material which occupies the center of the synclines is mainly sandy whereas the edge is silty. Notwithstanding we notice an exception : the picture 5 shows a thin sandy level which is folded in ball and separates its silty core from the outer fine grained body.

4.2 Silty anticlines

The vertical ascent of the silty material was performed by anticlinal foldings in diapiric fold type with very complicated shapes (fig. 2, picture 6). Their morphology can be designed as it follows :
In the lower part, the mud firstly migrates horizontaly from the center of the synclines towards the discontinuities of the sandy ceiling where it heaps up in concave flanks anticlines. In the center of them, the beds are intensively compressed and minutely folded. On its top, the flanks draw nearer to one another and go through the overlying formation like plastic tongues which mold the syncline sides and cut the sand beds. This injection can keep on very high in the series where to its vertical mouvement comes and conjugates horizontal displacement. These steps (picture 7) probably represent the mud arrival at the contact with current water. They often correspond to the silty recovering which cuts the upper part of the sandy synclines.

After appearing progressively, the deformations decrease in the same manner upwards where they still affect a level with small sand beds and create a collection of cells tilted and folded in all directions (picture 8).

5. SOME MECHANICAL PROPERTIES OF DISTORTED SOILS

Some geotechnical parameters have been determined on probes of the upper sand and of

the lower silt (n-o). These datas are gathered on table 1 and on figure 3.

Table 1 : geotechnical datas

	P A R A M E T E R S	UPPER FINE SAND	LOWER SANDY SILT
GRAIN SIZE DISTR. (see fig. 3)	Median diameter (mm)	0.11	0.04
	Effective diameter D_{10}	0.045	0.008
	Uniformity coefficient	3.2	6.2
	Curvature coefficient	0.77	1.56
	Skewness coefficient	1.11	0.87
SHEAR BOX (sat. soil)	Internal friction coef.	31.º	29.º
	Cohesion (Kg/cm^2)	0.06	0.10
	Liquid limit %	21.8	19.5

Note : Both samples are issued from the level (n-o),(fig. 2). The lower one is a mixture taken in the central part of an injection body.

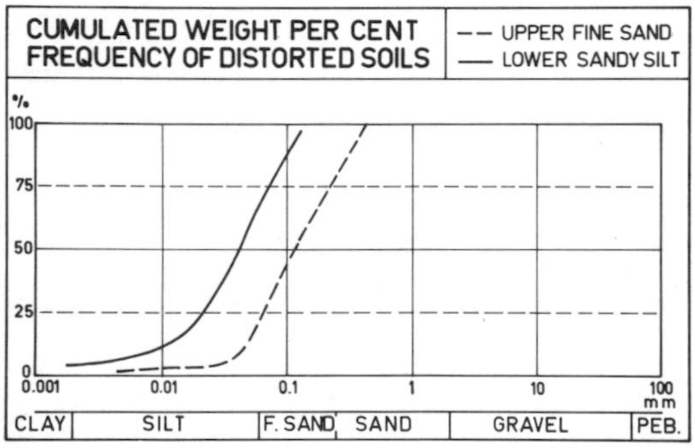

Figure 3 : grain size distribution of both deformed beds (determined by the Laboratory of soil mechanics, Federal Inst. of Techn. Lausanne)

We notice that the silt is poor in clay but it becomes very fluent at saturated state. Otherwise the soils do not show any exceptional property. The obtained values are currently measured on similar sediments without such deformations occuring.

6. GENESIS OF THE CONVOLUTIONS

Similar cases have been discovered in several places in the world. They concern sediments of different age and of very variable depositional environment. Their description constitutes a large documentation from which an enough complete bibliography is gathered in the recent book of Reineck and Singh (1973).

Particularly, we can mention the descriptions of natural cases made by Emery (1950), Einsele (1963), Wunderlich (1967), Coleman (1969) etc... Several reconstitution tests in laboratory have led to interesting results. Among them, the first experiences made by Kuenen (1958) have allowed the artificial confection of ball and pillow structures with the help of vibration of the test container. Then, Mc Kee (1962) and Mc Kee + Goldberg (1969) have tried to determine the influence of several parameters which promote the formation of convolute beds.

From these different works, we can come to the conclusion that the genesis of convolute bed deformations is mainly due to the phenomenon of density inversion. Indeed, the deposition of a sand bed over a layer of silt or clay supersaturated with water provokes the sinking of the upper level by its own weight by pushing the mud and its imbibition water upwards in order to restore the gravific stability. This phenomenon is more especially accentuated as the load is unequally disposed when the upper sediment is wavy.

There is no doubt that this mechanism is the one which acted in the case of Lembe delta. Let us examine in detail what would have been the relationship between sedimentation, erosion and folding.

Firstly, inside the synclinal structures, there is no angular discrepancy between the sand laminae. Consequently the deformation occured after the depositing of the main part of the sand level as mentioned by Emery (1950). During and after the folding phase, the erosion would probably have truncated the upper part of certain synclines (picture 2). It would also have momentaneously stopped the vertical progression of silt tongues and allowed their migration at the contact with the water (picture 7).

Then, a new slice would have deposited, followed by a second phase of deformation creating a series of sandy synclines wrapped in the lower silt by its flowing reactivation. This way, step by step, these sediments supersaturated with water would have deformed. Such a model involves a rapid formation in order that the mobility of the lower fluid layer remains sufficient to allow its migration very high in the series.

7. CONCLUSION

Thanks to the discovery of the penecontemporaneous deformation of the periglacial Lembe delta, we have been able to describe in detail very well shaped structures situated in their sedimentological frame.

These deformations affect sandy and silty terrains extremely poor in clay. However, the latter is particularly fluent in the saturated state.

According to the proposed model, the contorted sediments would have formed slice by slice the deformation immediately occuring after the sedimentation of each step.

BIBLIOGRAPHY

Coleman, J.M. 1969, Brahmaputra River : Channel processes and sedimentation, Sediment. Geol. 3, 129-239.

Einsele, G. 1963, "Convolute bedding" und ähnliche Sedimentstrukturen im rheinischen Oberdevon und anderen Ablagerungen, Neues Jahrb. Geol. Paleontol. Abhandl. 116, 162-198.

Emery, K.O. 1950, Contorted strata at Newport Beach, California, J. Sediment. Petrol. 20, 111-115.

Jäckli, H. 1970, La Suisse durant la dernière période glaciaire, Atlas de la Suisse, feuille 6, 1/550'000.

Kuenen, Ph.H. 1958, Experiments in Geology, Trans. Geol. Soc. Glasgow, 23, 1-28.

McKee, E.D., Reynolds, M.A., Baker, C.H. 1962, Laboratory studies on deformation in unconsolidated sediment, U.S. Geol. Surv. Profess. Papers 450D, D 151-D 155.

McKee, E.D., Goldberg, M. 1969, Experiments on formation of contorted structures in mud, Geol. Soc. Am. Bull. 80, 231-244.

Parriaux, A. 1978, Quelques aspects de l'érosion et des dépôts quaternaires du bassin de la Broye, Ecl. Geol. Helvet. 71,1.

Reineck, H., Singh, I. 1973, Depositional sedimentary Environment, Springer-Verlag, Berlin.

Wunderlich, F. 1967, Die Entstehung von " convolute bedding an Platenrändern", Senkenberg. Lethae 48, 345-349.

LEGEND OF PICTURES

PICTURE 1 : Disposition of the convoluted strata in the upper deltaïc sediments.

PICTURE 2 : Gently folded syncline in a plane bed laminated sand with truncation of its upper part.

PICTURE 3 : Closely folded fine sand cell with a circular appearance of its gravely center.

PICTURE 4 : Strongly folded core of a synclinal structure in fine sands.

PICTURE 5 : Ball of unusual composition with the same silty material inside and outside of its folded sandy edge (horizontal lines result from the brushing of the outcrop).

PICTURE 6 : Complex convolute bedding in a big anticlinal zone.

PICTURE 7 : Silty anticlinal tongue with horizontal migration at its top.

PICTURE 8 : Small scale convolute bedding at the upper part of the main deformed layer.

PICTURE 9 : Extracts of the picture combination of the deformed horizon (a complete stereographic cover is deposited in the Laboratory of Geology, FIT, Lausanne).

1
2
5CM
3
5CM
4
2CM

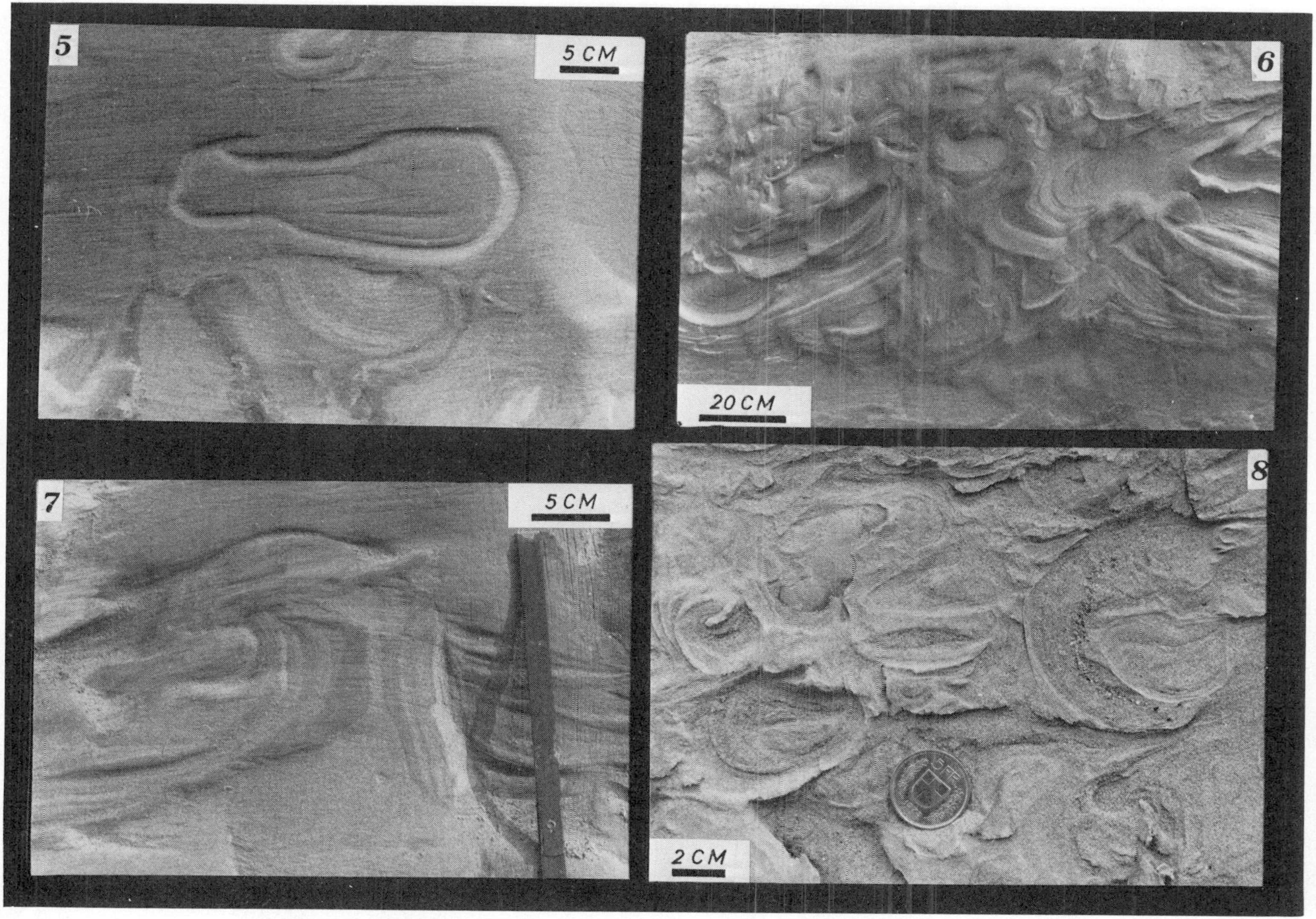
5
5 CM
6
20 CM
7
5 CM
8
2 CM

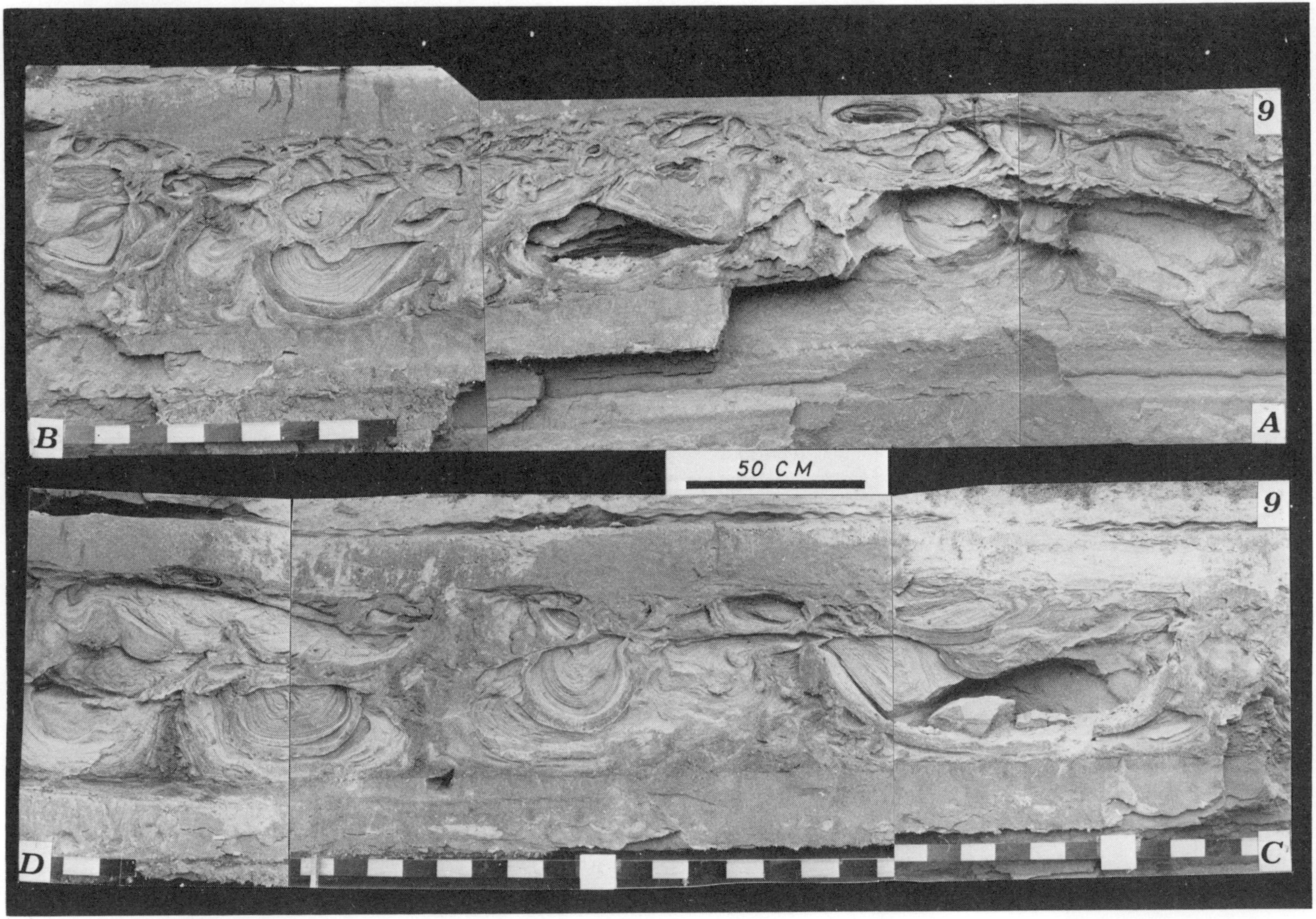

A
B
C
D
9
9
50 CM

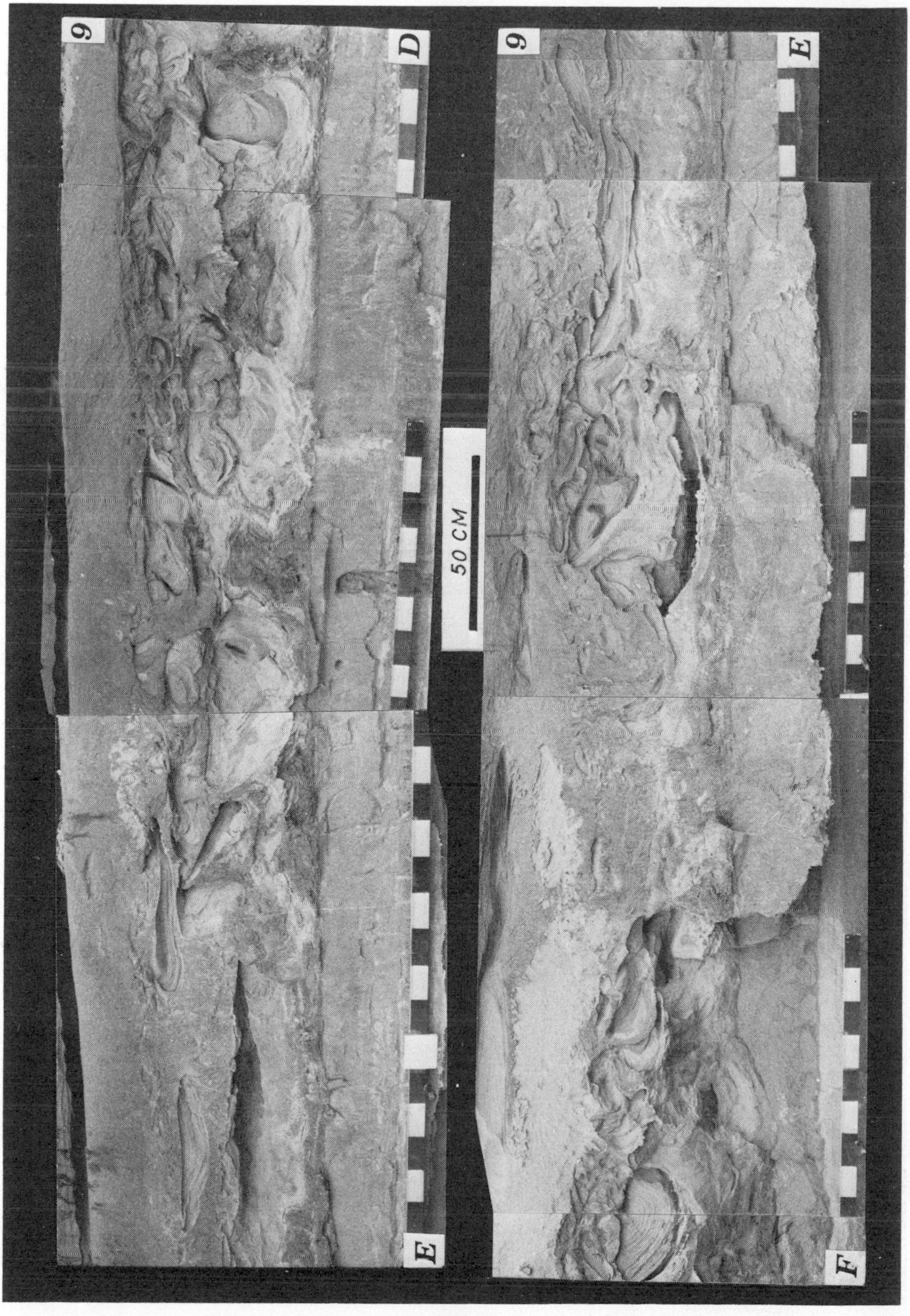
9
D
9
E
50 CM
E
F

9
G
F
50 CM
9
I
H

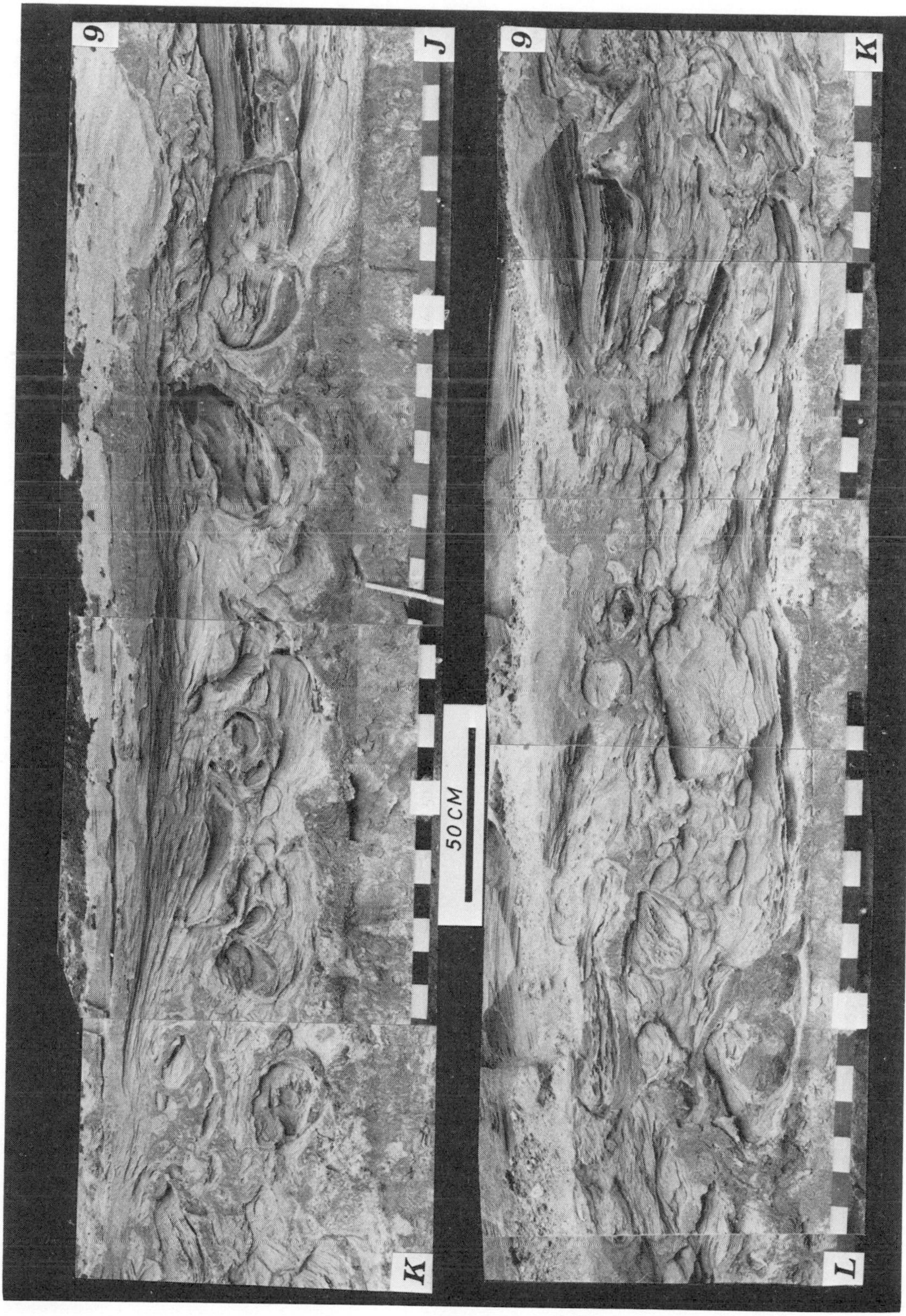
50 CM
K
L
J
9
K
K
9

Intrastratal contortions in a glacio-lacustrine sediment sequence in the eastern Swiss Plain

Ch.SCHLÜCHTER
ETH-Hönggerberg, Zurich, Switzerland

U.KNECHT
University of Zurich, Zurich, Switzerland

ZUSAMMENFASSUNG

Die meisten Beispiele von Wickelstrukturen, die in der Literatur beschrieben worden sind, stammen aus marinen "flyschartigen" Sedimenten verschiedenen Alters. Das Vorhandensein von Wickelstrukturen im letzteiszeitlichen, glazilimnischen Delta von Ebnet bei Diessenhofen (Kt. Zürich) ist wegen der Grösse der eingewickelten Sedimentpakete, der Vollkommenheit der Strukturen und der relativ grobkörnigen Zusammensetzung der betroffenen Schichten erwähnenswert. Von Bedeutung ist das Querprofil in Figur 6, das den strukturellen Uebergang vom longitudinalen "Sandrücken" des Deltarinnenrandes in die "verwickelte Linse" und weiter in die ungestörten strömungsgeschichteten Lagen zeigt. Im Vergleich zu den Resultaten von Coleman & Gagliano (1965) handelt es sich bei den Ebnet-Sanden um typische Ablagerungen der Deltafront, im Besonderen des "Subaqueous Levee" - Milieus. - Die beschriebenen Wickelstrukturen sind syn- oder metasedimentär entstanden durch das Ueberfordern der Tragfähigkeit des weichen, lockeren Sandes infolge von plötzlicher Auflagerung der Sedimentfracht eines dichten "(high-density-) underflow" (Dichteinversion) während Hochflut und/oder durch das Auftreten von Verflüssigungserscheinungen. Wir betrachten das in Figur 6 dargestellte Beispiel als Beweis für die Richtigkeit der von Artyushkov (1963) aufgestellten Bedingung, dass nämlich Instabilität im Ablagerungsgeschehen eintritt, wenn $\Delta\rho \cdot \Delta h > \tau_b$ ist.

Im Ebnet-Delta treten Wickelstrukturen in verschiedenen Horizonten mehrmals in unterschiedlicher Intensität auf. - Ihr Auftreten steht nicht mit klimatischen Erscheinungen (wie Permafrost-bedingter Kryoturbation) in ursächlichem Zusammenhang.

1 INTRODUCTION

During fieldwork by the second author, large lenses of contorted sediment have been recognized in the gravel pits of Ebnet and Totenmaa, near Diessenhofen, Canton of Zurich. The discovery of these sedimentary complications has led to a controversial discussion as to the nature of the contortions. They have been interpreted as being climatically induced (cryoturbations), mostly in a desperate need for climatostratigraphic information in the sediment-sequences of the area for a more precise reconstruction of Upper Pleistocene events.

With regard to the speculative character of the discussion on the "true nature" of the intrastratal contortions we consider it being necessary to present a short description of the site and an interpretation based on sedimentological comparisons. Furthermore, we give it as an interesting addendum to the foregoing article by A. Parriaux because both, the paleogeography and the morphodynamic situation is essentially the same at Granges-Marnand and at Diessenhofen.

2 PALEOGEOGRAPHY AND GEOLOGICAL SETTING

The location of the study area and of the main gravel pit at Ebnet is given in Figure 1 and 2. The position of the maximum extension of the Rhine Glacier during the Last Glaciation ("Würm") is schematically indicated in Figure 2.

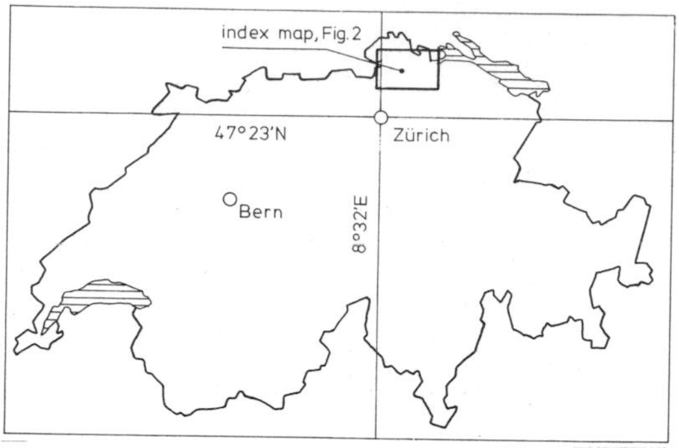

Figure 1: The location of the study area in Eastern Switzerland

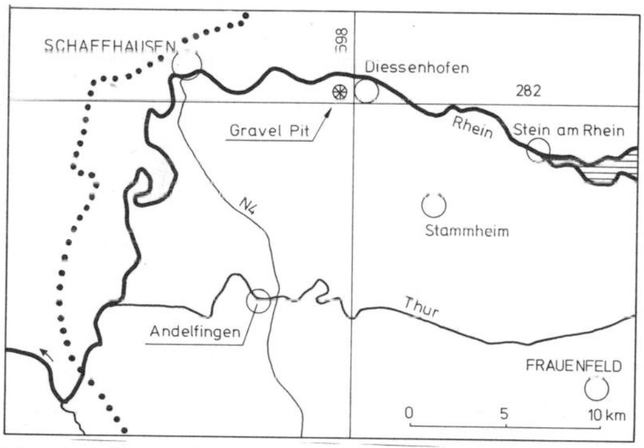

Figure 2: Index map of the study area. The maximum extension of the Rhine Glacier during the Last Glaciation ("Würm") is indicated by the dotted line.

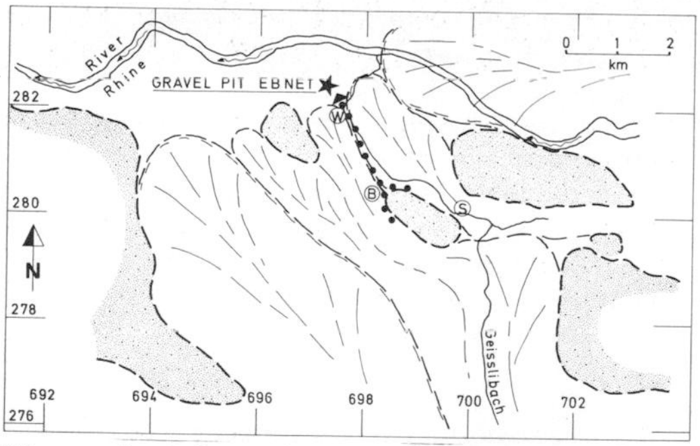

Figure 3: Schematic reconstruction of the lobe-type desintegration of the Rhine Glacier during early stages of retreat. - Schaded areas: Molasse bedrock forming hills; dotted line: meltwater drainage system feeding the glaciolacustrine delta at Ebnet; W: Willisdorf; B: Basadingen; S: Schlattingen.

Bedrock morphology is characterized in the area by east-west oriented erosional forms: bedrock hills with (partly) overdeepened depressions in between (Fig. 3). Bedrock consists predominantly of sand- and silt-stone with conglomerates of the Upper Fresh-water Molasse (Miocene).

The depositional history of the Pleistocene sediments in the area is complicated because several lobes of the Rhine Glacier developed individuel erosional and depositional activities. As a result complicated lateral and vertical facies - interrelationships occur (cf. Fig. 3). Deposits older than the Last Glaciation are scarce or totally absent in the area of the gravel pit. These are restricted to the cover of the Molasse-hills or to deeper parts of the valleys.

The retreat of the ice from its maximum extension (as indicated in Fig. 2) to the next well developed stadial has been complicated glacio-dynamically by the existence of individual ice lobes of the same glacier (Fig. 3 and Hofmann & Hantke 1964): at a given time during back- and downwasting of the icemass the front of the glacier extending westwards became irregular and the individual lobes were formed. Meltwater was preferably collected at the contact of the glacier lobes or along bedrock ridges. The direction of flow of the Geisslibach (Fig. 3) before its deviation to the northeast seems to follow such a drainage pattern.

3 THE PROGLACIAL DELTA AT EBNET

The lobe-type desintegration of the icefront led to the formation of an (extensive?) lake in front of the receeding glacier. One of the most effective meltwater collectors draining the area of Willisdorf - Basadingen - Schlattingen has accumulated the proglacial delta at Ebnet. Quarry work has revealed the existence of two clearly seperated depositional facies within the delta: a lower unit of fore-set deposits and an upper unit of braded river ("sandur") accumulations (Fig. 4 & 5). - Drilling in the area of the gravel pit indicates that the fore-set beds continue vertically below the base of the gravel pit and are underlain, probably, by morainic sediments. - The age of the deposits is uncertain, but the whole sequence in the pit is best explained to have been formed within a tardi-glacial chronosystem of the Last Glaciation ("Würm").

4 THE INTRASTRATAL CONTORTIONS

4.1 Occurrence and description

The lower sediment unit exposed in the Ebnet gravel pit exhibits well developed fore-set beds (dip of 10° - 15° to the NNW). At some places top-set beds are preserved. The delta itself is composed of a complex set of bars of cross-bedded sand. This slice-like build-up of the whole deltacomplex is due to slight shifting overflows from channels and over the accumulating area(s). The individual bars exhibit large scale cross-bedding (Fig. 6). In between two bars, areas of current stratified sand occur with well developed sets of climbing ripples being the most significant sedimentary structure (Fig. 7 & 10). Within fine grained beds "intrastratal contortions" occur repeatedly in different lateral and vertical positions. One "unit of contortions" can be traced laterally for several meters but may extend for more than twenty meters. An example of the "larger units" is given in Figure 6.

In general two principal forms of contortions can be distinguished:

1. "folds" with well developed anticlinal structures, the synclinal part being broad the anticlinal part being narrow. The anticlines are often squeezed in the middle part and extended at the top (primitive form of mushroom-shape), or are clearly peaked (Fig. 8, 9, 10).

2. "balls" (convolution balls) as pieces of broken-off beds and sunken in the underlying strata, sometimes completely embedded (Fig. 6, 8, 9 & 10).

Transitional forms between 1 and 2 and between contorted and undisturbed sediment are shown in Fig. 6, 7 and 11. - The material of embedding is in all observed cases cohesionless fine to medium sand, whereas the embedded "broken-up beds" are either gravelly sand or sandy gravel (Fig. 8, 9, 10, 15).

The grain size distributions of the sediment involved are given in Figure 11 and the sample localities are indicated in Figure 12 & 13. Basically, there are two grain-size associations present: 1) fine to medium size sand (unimodal distribution) and 2) sandy gravel (bi- to polymodal distribution). Of interest, and most probably of genetic significance, is the mixture of the two grain-size associations, producing gravelly sand.

In Figure 11 the measured grain-sizes are compared to the range of grain-size distribution of material sensitive to liquefaction (after diagram in Pettijohn, Potter & Siever 1973 : 372). This comparison indicates that the sand in the Ebnet gravel pit has to be classified as highly sensitive to liquefaction. The sediment parameters of the samples (parameters after Folk & Ward 1957) are given in Table I.

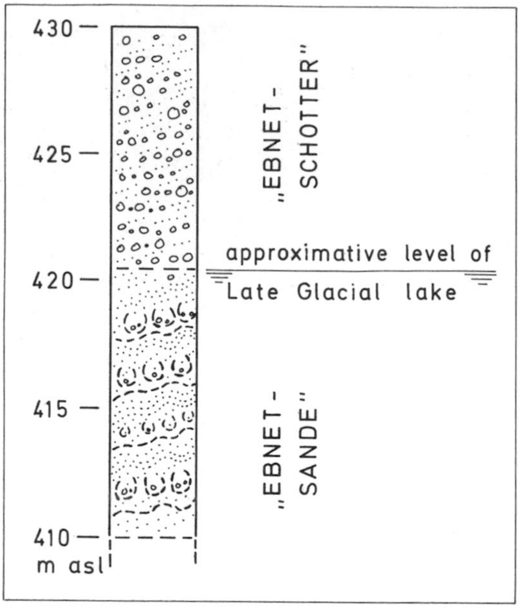

Figure 4: Schematic reconstruction of the lithostratigraphy exposed in the Ebnet gravel pit.

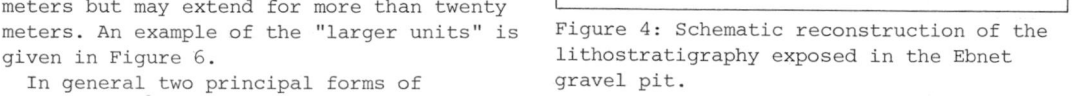

Figure 5: Transition from "Ebnet Sande" to "Ebnet Schotter" or the changing depositional environment from subaqueous (bottom) to subaerial (top). Main flow direction is towards observer. Arrow indicates approximate position of Late Glacial lake level.

Figure 6: Example of facies transect from longitudinal bar (B) with large scale cross-bedding to the lense of convoluted sediment (C) and to the undisturbed current bedded levee deposits (L). The height of the contortions at its maximum is approx. 2,5 m. - Details see Figure 7, 8 and 9. - Direction of overflows from the channel to the levee proper is parallel to the pit face shown here (left to right).

The following characteristics are common to all well developed convolutions:

1. they are associated in up-current direction to large-scale cross-bedding (cf. Fig. 6).

2. they exhibit significant differences in their grain-size composition: lower and convoluted material = finer; upper and broken-up strata = coarser (cf. Fig. 8, 9, 10, 13, 14).

3. some of the broken-off balls exhibit a distinct graded-bedding (cf. Fig. 8, 9, 10).

Compared to the data given by Dzulynski & Walton (1965 : 179 - 181) the convolutions at Ebnet must be classified as "jumbo-sized" with regard to the thickness of the convoluted bed (up to 2,5 m) and to the grain sizes involved (cf. Fig. 11).

Figure 6 gives a well developed and characteristic example of the "facies associations": the large-scale crossbedding to the left (up-current) is followed by a lense of contorted sediment which is replaced to the right (down-current) by evenly undisturbed current-bedded sand with well developed climbing-ripples. The contortions are bound at the base by a layer of silty sand (in Fig. 6, 8, & 9 visible as a dark layer). - The transition towards the overlaying deposits from the contorted layer is shown in Figure 11, 13, 14, 15. A pronounced decrease in grain size composition can be noticed and a transition from cross- to current-bedded strata is obvious. Then follows another set of convolutions. Interesting initial stages of forming convolutions are shown in Figure 11 and 14 where embryonal "folds" start to develop.

The paleogeographic setting as well as the undisturbed primary sediment structures indicate that the sediments under discussion have been deposited in a subaequeous delta with a (considerable) net accumulation rate of predominantly sand. Towards the top of the delta the amount of gravel increases and then becomes dominant (Fig. 4, 5, 14) resulting in subaerial accumulations as braided river (or sandur) deposit. The transition into subaerial deposits on top of the subaequeous sand indicates that the water depth during the accumulation of the sand was minimal (less than 10 m, cf. Fig. 4).

4.2 Definition

The intrastratal contortions described here can be traced laterally where they are in transition with non-contorted sediment. They appear perpendicular to the main current direction as vertically and laterally more or less extended "lenses" of contorted strata (cf. Fig. 6). Their internal structures ("folds & balls") justify a definition as convolute laminations (Pettijohn, Potter & Siever 1973). Sanders' (1965) restriction, however, "of differential deformation between successive laminae - the lower strata being more deformed than the upper one" holds true for the described structures (Fig. 8, 9, & 10). There is an increasing order in sedimentation pattern upwards within the contorted sediment lens (Fig. 10, 13, 15).

Figure 7: Transition from contorted sediment to undisturbed current bedded strata.

The broken-off balls are classified as "convolution balls", referring to Ten Haaf (1956).

4.3 Interpretation

The genetic interpretations of convolute laminations are almost as many as known occurrences. Most studies deal with convolutions in marine flysch-like sediments of different age (Davies 1965, Pettijohn, Potter & Siever 1973). - The deposits of the Ebnet Delta belong to the glacio-lacustrine delta front environment. Meltwater carrying a high amount of sand and gravel has been transported in (several?) main channels and occasionally dispersed as overflow to the subaqueous levee. The sand deposits with large scale cross-bedding may be considered as transition from the channel

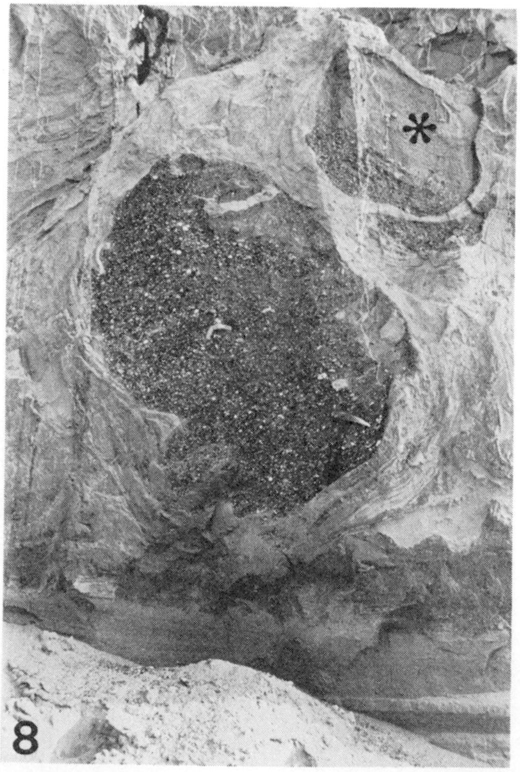

Figure 8 and 9: Details of the "jumbo-sized" convolutions of Figure 6. Note the texture (sand) and the structure of the liquefied material and how far upwards the embedding layers can be traced. Some of the "broken-off" balls are graded bedded (*). The contorted bed is limited at the base by a layer of silty sand (dark in the photograph). The cohesion of this layer has prevented the convolutions to develop further down.

to the subaqueous levee, and the "even current bedding" with pronounced current-ripple lamination belongs to the subaqueous levee proper (cf. Fig. 7, 10, 13, 15). In this interpretation, the large-scale cross-bedded unit (Fig. 6 to the left) acts as a bar, flanking the channel. At average current conditions sand in suspension is transported from the channel as overflow-load to the levee and deposited (as "fallout") due to the decreasing current velocity. At occasional "heavy flooding" coarser material also overflows the transitory longitudinal bar and is deposited there on top of the extremely loosely packed sand causing density inversion with ρ of upper layer (= cumulative bulk density of suspended material of density underflow) $\gg$ ρ of lower layer (= of loosely packed sand). This instantly increased overburden causes the structure of the lower sand to collapse. For a short time, the bearing capacity of the lower sand unit is hold only by the interstitial water pressure until the structure collapses and the water begins to migrate upwards. - As the underlying sand is in a more or less liquified state parts of the heavier (gravelly) sediment can penetrate as involution balls into the liquefied sand downwards. This continues until a new equilibrium is reached and the density gradient is normal again. The "excess" interstitial water has drained upwards and/or the reduction by volume due to liquefaction of the lower bed is replaced by sunken involution balls.

A second factor - besides the load casting - seems to have contributed to the formation of the convolutions: the channel boundary is given by a longitudinal bar

Figure 10: Well developed involution ball of sandy gravel (G) embedded in sand; to the left almost undisturbed cross bedded gravelly sand (open arrow). The contorted bed is followed by undisturbed current bedded strata. Behind Dr. Fitze's head the next set of involutions is visible (black arrow).

which seperates the channel proper from the subaqueous levee. This bar creates a difference in elevation to the levee. Thus, the observation put forward by Artyushkov (1963) becomes valid: if $\Delta\rho \cdot \Delta h > \tau_b$, then involution will occur. - $\Delta\rho$ is the difference between the cumulative bulk densities of the two layers, Δh is the difference in elevation along interface (between the channel-boundary bar and the levee) and τ_b is the shear between the two layers. Figure 6 is a perfect illustration of the interrelationship of $\Delta\rho$, Δh and τ_b. The difference in height (Δh) as well as $\Delta\rho$ is obviously large so that the considerable convolutions can occur (Pettijohn, Potter & Siever 1973).

The observations at Ebnet correspond to the results presented by Coleman & Gagliano (1965) and their description of the subaqueous levee environment with a dominance of current produced primary structures with abundant convolute laminations.

The input of clastic material to the Ebnet delta-front occurred as density underflows fed and controlled by the melting processes at the ice margin. Periods of regular currents caused deposition of undisturbed sand with current-bedding (cf. Fig. 10, 13, 15). During periods of increased meltwater pulses the delta channel was overflown by high-density underflows which dumped their coarse load after having passed the highest point of the longitudinal (channel -) bar. Locally, erosion along the longitudinal bar can

Table I: Sediment characteristics of the deposits in the Ebnet gravel pit (parameters after Folk & Ward, 1957)

Sample No.	M_Z	σ_I	K_G	S_{KI}
37751	0,57	2,43	0,97	-0,46
37752	3,20	0,74	1,11	0,32
37753	1,43	1,81	1,02	-0,56
37754	0,77	2,12	1,02	-0,38
37755	2,80	0,70	0,94	-0,02
37756	3,40	1,06	1,30	0,26
37759	-2,90	2,60	0,90	0,43

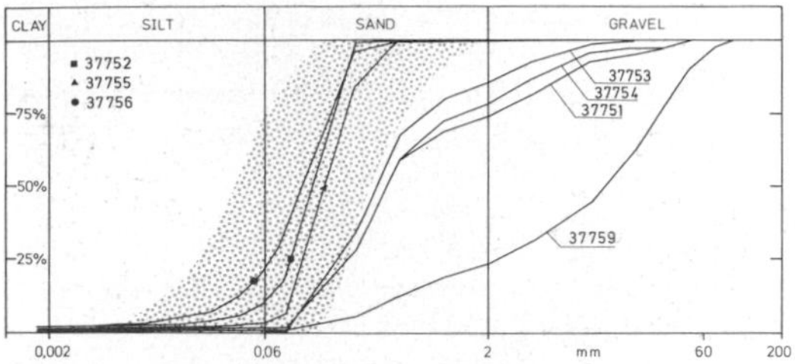

Figure 11: Grain size distributions of the sediments involved in the convoluted strata at Ebnet. The numbers refer to the laboratory invoices of the Institute of Foundation Engineering. For the sample locations: see Figure 12 and 13. - The shaded area comprises sediments sensitive to liquefaction (after diagram in Pettijohn, Potter and Siever 1973 : 372).

take place during this process and thus the "mixed" gravelly sand (cf. Fig. 11, samples No. 37751, 37753, 37754) is produced. - The idea of the density underflow mechanism is supported by the presence of graded bedding in many involution balls (cf. Fig. 8, 9, 10).

4.4 Syn- or postdepositional origin of convolutions

If the collapsing of the original bed was caused by overloading (load casting) or by the action of the shock wave of a density underflow inducing instant liquefaction of the underlying sand cannot be distinguished. The properties of the material and the resulting convolutions allow both processes to have taken place and facilitated each other. If considering the structures shown in Figure 10 then the convolution has been created almost syndepositional - and hence is due to loading pressures (Dzulynski & Smith 1963). Considering the initial fold structure (Fig. 12, 14) and Figure 7 then the contortion is more likely to be meta- to postdepositional - and hence due to quick-sand movement and upward migration of interstitial water (Selley et al. 1963) or to intrastratal laminar flow (Williams 1960). - Fluid drag as an involution producing process seems to be of minor importance at Ebnet (cf. Saunderson 1975). An example where fluid drag may have served as initiation of parts of the convolutions is given in Figure 13 with a small embryonal roll-up structure (arrow).

If syn- or metadepositional origin only has to be considered for the formation of the described structures cannot be answered definitely yet. But the complications in sediment development are obviously due to an increase of hydrostatic pressure and

Figure 12: Sample locations of No. 37753, 37754, 37755 and 37759, cf. Fig. 11 for grain size distribution curves.

Figure 13: Sample locations of No. 37751 and 37752, cf. Fig. 11 for grain size distribution curves. - Note the increasing order in sedimentation pattern up to the next contorted bed.

Figure 14: "Initial fold" associated to the edge of lense of convoluted bed. Darker laminae are silty sand.

Figure 15: Contorted bed of 1,5 m thickness with incorporated gravel as elongated involution balls. Note the undisturbed underlying and overlying current bedded sand.

a resulting change of packing (and a subsequent upwards movement of partly liquefied material) in the "lower sand layer" as it is evidenced by the shown structural "modifications" (cf. Terzaghi 1956). The sediment structures at Ebnet are thus explained as characteristic features of the subaqueous levee environment (Coleman & Gagliano 1965) and their genesis has not been influenced by climatic causes such as permafrost.

5 CONCLUSIONS

Most cases of convolute laminations as described in the literature are found in marine flysch-like deposits of different age. The occurrence of convolute lamination in the proglacial lacustrine delta at Ebnet is noteworthy because of the "jumbo-sized" convolutions and the coarser grain size involved. The relationship between contorted and undisturbed strata is significant as a facies association in the delta front: the subaqueous levee sediment association. An example is given with a transect from the longitudinal bar along the channel to the adjoining contorted lense and then to the current-bedded undisturbed levee sediments in Figure 6. The convolutions are caused by syn- and/or metadepositional load casting (of soft sand/stressed bearing capacity) by sudden density underflows (density inversions) and most probably by liquefaction effects. We consider the example given in Figure 6 as a valid demonstration of Artyushkov's (1963) assumption that depositional instability will occur if $\Delta\rho \cdot \Delta h > \tau_b$. - No climatic significance is given by the occurrence of convolute laminations.

6 ACKNOWLEDGEMENTS

This paper is part of a study under contract No. 295 J of the Institute of Foundation Engineering and Soil Mechanics at the Federal Institute of Technology. The permission by the Director, Prof. H.J. Lang to carry out the investigations and to publish the results is gratefully acknowledged. We are also indebted to Prof. Dr. G. Furrer, Zurich, for encouraging the publication. - Dr. V.P. Lautridou, Caën, has provided us with information on the existing literature and Dr. M. Sturm, Zurich, has acted as a critical but constructive reviewer: many thanks to both of them.

7 REFERENCES

Artyushkov, Ye.V. 1963, O vozmozhnosti vozniknoveniya i obshchikh zakonomernostyakh razvitiya knvektivnoy neustoychivosti v osadochnykh porodakh. Doklady Akad. Nauk SSR 153: 162 - 165 (1963) (Possibility of convective instability in sedimentary rocks and the general laws of its development. Earth Sci. Sec., p. 26 - 28).

Coleman, J.M. & Sh.M. Gagliano 1965, Sedimentary structures: Mississippi River Deltaic Plain. - In: Primary sedimentary structures and their hydrodynamic interpretation. - SEPM Spec. Publ. 12 : 133 - 148.

Davies, H.G. 1965, Convolute lamination and other structures from the Lower Coal Measures of Yorkshire. - Sedimentology 5 (1965) : 305 - 325, Amsterdam (Elsevier).

Dzulynski, S. & A. Smith 1963, Convolute lamination, its origins, preservation, and directional significance. - J. Sediment. Petrol., 33 : 616 - 627.

Dzulynski, S. & E.K. Walton 1965, Sedimentary features of flysch and greywackes. - Developments in Sedimentology No. 7, 300 p. - Amsterdam (Elsevier).

Folk, R.L. & W.C. Ward 1957, Brazos River bar: a study in the significance of grain size parameters. - J. Sediment. Petrol, 27 : 3 - 26.

Hofmann, F. & R. Hantke 1964, Erläuterungen zum Atlasblatt 38 (Diessenhofen) des Geol. Atlas der Schweiz 1:25'000. - Bern (Kümmerly & Frey).

Pettijohn, F.J., P.E. Potter & R. Siever 1973, Sand and Sandstone XVI + 618 p. - New York, Heidelberg, Berlin (Springer).

Sanders, J.E. 1965, Primary sedimentary structures formed by turbidity currents and related sedimentation mechanisms. - In: Primary sedimentary structures and their hydrodynamic interpretation. - SEPM Spec. Publ. 12 : 192 - 219.

Saunderson, H.C. 1975, Sedimentology of the Brampton Esker and its associated deposits: an empirical test of theory. - In: Jopling, A.V. & B.C. McDonald (Editors) 1975, Glaciofluvial and glaciolacustrine sedimentation, SEPM Spec. Publ. 23 : 155 - 176, Tulsa, Oklahoma, USA.

Selley, R.C., D.J. Sherman, J. Sutton & J. Watson 1963, Some underwater disturbances in the Torridonian of Skye and Raasay. - Geol. Mag. 100 : 224 - 244.

Ten Haaf, E. 1956, The significance of convolute lamination. - Geol. Mijnbouw, 18 : 188 - 194.

Terzaghi, K. 1956, Varieties of submarine slope failures. 8th Texas Conf. on soil mechanics and foundation Eng. Proc., Univ. Texas Bur. Eng. Research Spec. Publ. 29, 41 p.

Williams, E. 1960, Intra-stratal flow and convolute folding. - Geol. Mag., 97 : 208 - 214.